AF587324

Infosys Science Foundation Series

Infosys Science Foundation Series in Mathematical Sciences

The *Infosys Science Foundation Series in Mathematical Sciences* is a sub-series of The *Infosys Science Foundation Series.* This sub-series focuses on high quality content in the domain of mathematical sciences and various disciplines of mathematics, statistics, bio-mathematics, financial mathematics, applied mathematics, operations research, applies statistics and computer science. All content published in the sub-series are written, edited, or vetted by the laureates or jury members of the Infosys Prize. With the Series, Springer and the Infosys Science Foundation hope to provide readers with monographs, handbooks, professional books and textbooks of the highest academic quality on current topics in relevant disciplines. Literature in this sub-series will appeal to a wide audience of researchers, students, educators, and professionals across mathematics, applied mathematics, statistics and computer science disciplines.

More information about this series at http://www.springer.com/series/13817

Muhammad Akram

Fuzzy Lie Algebras

Muhammad Akram
Department of Mathematics
University of the Punjab
Lahore, Pakistan

ISSN 2363-6149 ISSN 2363-6157 (electronic)
Infosys Science Foundation Series
ISSN 2364-4036 ISSN 2364-4044 (electronic)
Infosys Science Foundation Series in Mathematical Sciences
ISBN 978-981-13-3220-3 ISBN 978-981-13-3221-0 (eBook)
https://doi.org/10.1007/978-981-13-3221-0

Library of Congress Control Number: 2018961200

This Springer imprint is published by the registered company Springer Nature Singapore Pte Ltd.
The registered company address is: 152 Beach Road, #21-01/04 Gateway East, Singapore 189721, Singapore

I dedicate to my dear parents!

Foreword

Lie algebras (also termed as infinitesimal groups) appeared in mathematics at the end of the nineteenth century through the works of Sophus Lie and Wilhelm Killing in connection with the study of Lie groups. They also occurred in implicit form somewhat earlier in mechanics. The term "Lie algebra" itself was introduced by H. Weyl in 1934. Since the algebra and topology of a Lie group are closely entwined, Lie algebras which are regarded as tangent spaces at the identity element of the associated Lie group are used to study the structure of Lie groups. Thus, the employment of Lie algebra rids of the topological complexity. Hence, the role of Lie algebras increased in proportion to the place taken by Lie groups in geometry and also in classical and quantum mechanics. The apparatus of Lie algebras is not only a powerful tool in the theory of finite groups but also a source of elegant problems in linear algebra. The notion of Lie superalgebras was introduced by Kac in 1977 as a generalization of the theory of Lie algebras. In 1985, Filippov introduced the concept of n-Lie algebras ($n \geq 2$). The definition when $n = 2$ agrees with the usual definition of a Lie algebra.

Fuzzy set theory was proposed in 1965 by Lofti A. Zadeh from the University of California, Berkeley. Fuzzy set theory has been developed by many scholars in various directions. Azriel Rosenfeld discussed fuzzy subgroups in 1971, and his paper led to a new area in fuzzy mathematics. Since then many mathematicians have been involved in extending the concepts and results of abstract algebra to the broader framework of the fuzzy setting.

This book introduces readers to fundamental theories such as fuzzy Lie subalgebras, fuzzy Lie ideals, anti-fuzzy Lie ideals, fuzzy Lie superalgebras, and hesitant fuzzy Lie ideals over a field. The concepts of nilpotency of intuitionistic fuzzy Lie ideals, intuitionistic fuzzy Killing form, m-polar fuzzy Lie algebras, $(\in, \in \vee q)$-fuzzy Lie ideals, and rough fuzzy Lie algebras are also presented. Another goal of this book is to present fuzzy ideals and Pythagorean fuzzy ideals of n-Lie algebras.

Therefore, this book presents a valuable contribution for students and researchers in fuzzy mathematics, especially for those interested in fuzzy algebraic structures. The author is a reputed researcher in the fields of fuzzy algebras, fuzzy graphs, and fuzzy decision-making systems. I believe that he will be appreciated by both the experts and those who aim to apply the large collection of ideas on classical and quantum mechanics, quantum field theory, computer vision, and mobile robot control that he supplies in this book.

Yazd, Iran

Bijan Davvaz
Yazd University

Preface

A fuzzy set, as a superset of a crisp set, owes its origin to the path-breaking work of Zadeh since 1965. It was introduced to deal with the possibility of partial degrees of truth between "absolutely true" and "absolutely false." Zadeh's remarkable idea has found many applications in several fields, including decision making, networking, computer science, discrete mathematics, economics, management, biomedical sciences, engineering, differential equations, pattern recognition, automation, and robotics. In pure mathematics, Rosenfeld used the concept of a fuzzy subset of a set in order to introduce the notion of a fuzzy subgroup of a group in 1971. Rosenfeld's paper spearheaded the development of fuzzy abstract algebra, which reached many other algebraic structures (rings, semirings, hemirings, nearrings, vector spaces, etc.).

Many generalizations of Zadeh's fuzzy sets followed this fundamental concept, including, but not limited to, type-1 fuzzy sets, interval-valued fuzzy sets, intuitionistic fuzzy set, interval-valued intuitionistic fuzzy sets, vague sets, hesitant fuzzy sets, and bipolar fuzzy sets. The work presented here intends to apply these concepts to Lie algebras and Lie superalgebras. This monograph deals with certain structures of fuzzy Lie algebras, fuzzy Lie superalgebras, and fuzzy n-Lie algebras. It is based on a number of papers by the author, which have been published in various scientific journals. This book may be useful for researchers in mathematics and for social scientists. I believe that this book will also be helpful to students and researchers who are interested in learning fuzzy Lie algebraic structures.

In agreement with this purpose, the contents of this book are as follows.

In Chap. 1, I present a concise review of fuzzy set theory, Lie algebras, and Lie superalgebras. Then, I describe certain specialized concepts, including fuzzy Lie subalgebras, fuzzy Lie ideals, anti-fuzzy Lie ideals, fuzzy Lie superalgebras, and hesitant fuzzy Lie ideals over a field.

In Chap. 2, I present certain concepts, such as intuitionistic fuzzy Lie subalgebras, Lie homomorphisms, intuitionistic fuzzy Lie ideals, special types of intuitionistic fuzzy Lie ideals, intuitionistic (S, T)-fuzzy Lie ideals, nilpotency of intuitionistic (S, T)-fuzzy Lie ideals, and intuitionistic (S, T)-fuzzy Killing form.

In Chap. 3, I present notions of certain concepts, such as interval-valued fuzzy Lie ideals, quotient Lie algebras via interval-valued fuzzy Lie ideals, interval-valued intuitionistic fuzzy Lie ideals, and interval-valued fuzzy Lie superalgebras.

In Chap. 4, I introduce more concepts, such as (α, β)-fuzzy Lie subalgebras, implication-based fuzzy Lie subalgebras, $(\alpha, \beta)^*$-fuzzy Lie subalgebras, interval-valued $(\in, \in \vee q)$-fuzzy Lie ideals, and (γ, δ)-intuitionistic fuzzy Lie algebras.

In Chap. 5, I present the concept of fuzzy Lie ideals of a Lie algebra over a fuzzy field. I describe $(\in, \in \vee q_m)$-fuzzy Lie subalgebras over a fuzzy field. I discuss vague Lie subalgebras over a vague field. I also present anti-fuzzy Lie sub-superalgebras over an anti-fuzzy field.

In Chap. 6, I present properties of bipolar fuzzy Lie ideals, bipolar fuzzy Lie sub-superalgebras, bipolar fuzzy bracket product, solvable bipolar fuzzy Lie ideals, and nilpotent bipolar fuzzy Lie ideals.

In Chap. 7, I deal with properties of $m-$polar fuzzy Lie subalgebras and $m-$polar fuzzy Lie ideals.

In Chap. 8, I deal with concepts like soft intersection Lie algebras, fuzzy soft Lie algebras, $(\in_\alpha, \in_\alpha \vee q_\beta)$-fuzzy soft Lie subalgebras, bipolar fuzzy soft Lie algebras, and $(\in, \in \vee q)$-bipolar fuzzy soft Lie algebras. In Chap. 9, I deal with rough fuzzy Lie subalgebras, rough fuzzy Lie ideals, fuzzy rough Lie subalgebras, and rough intuitionistic fuzzy Lie subalgebras.

Lastly, in Chap. 10, I put forward some properties of fuzzy subalgebras and ideals of n-Lie algebras. I also present the notion of Pythagorean fuzzy ideals of n-Lie algebras.

Acknowledgements I am grateful to the administration of the University of the Punjab, who provided the facilities which were required for the successful completion of this monograph. I thank the researchers worldwide whose contributions are referenced in this book, especially Lotfi A. Zadeh, K. T. Atanassov, John Mordeson, D. S. Malik, Wieslaw Dudek, Bijan Davvaz, and Y. B. Jun. I also would like to acknowledge the assistance of my students Miss Hafsa Masood and Mr. Ghous Ali.

Lahore, Pakistan Muhammad Akram

Contents

About the Author

Dr. Muhammad Akram is Professor at the Department of Mathematics, University of the Punjab, Pakistan. He earned his PhD in fuzzy mathematics from the Government College University, Pakistan. His research interests include numerical solutions of parabolic PDEs, fuzzy graphs, fuzzy algebras, and new trends in fuzzy set theory. He has published five monographs and over 265 research articles in international peer-reviewed journals. He has been an editorial board member of 10 international academic journals and a reviewer/referee for 115 international journals, including Mathematical Reviews (USA) and Zentralblatt MATH (Germany). Seven students have successfully completed their Ph.D. research work under his supervision. Currently, he is supervising six Ph.D. students.

List of Figures

List of Tables

Chapter 1
Fuzzy Lie Structures

In this chapter, we present a concise review of fuzzy set theory, Lie algebras and Lie superalgebras. Then, we describe properties of certain concepts, including fuzzy Lie subalgebras, fuzzy Lie ideals, anti-fuzzy Lie ideals, fuzzy Lie superalgebras and hesitant fuzzy Lie ideals over a field. This chapter is due to [4, 41, 49, 89, 133].

1.1 Introduction

1.1.1 Fuzzy Sets

In 300 BC, a Greek philosopher and scientist, Aristotle, formulated the *law of excluded middle* which states that "an element in a set is either asserted or denied". A *Cantor set* (crisp set) is described using a membership function that assigns a value in binary terms to each element of the universe X. Most of our conventional tools for explicit reasoning, modeling, and computing are precise, deterministic, and crisp in character, that is, yes-or-no characterization instead of more-or-less type. For illustration, in the set theory, an element is either a member or a nonmember; in optimization, a result is either feasible or not. This phenomenon signifies that the system is unequivocal, and it involves no obscurity. But, on the other hand, real situations are generally uncertain or ambiguous in various ways. In the present world, there exists information that is incomplete, unclear, uncertain, inexact, equivocal, or ambiguous in nature. In the frameworks of crisp set theory, it is exceptionally hard to answer a few questions since they do not have absolutely true answers. Humans, in any case, can give satisfactory answers, which are presumably true. Nature has given us a mind that permits a specific uncertainty in the depictions of our environment. This level ought to be figured utilizing imprecision, uncertainty, and the vulnerability of facts that were applied. Expert systems should also be able to cope with unreliable and incomplete information and with different expert opinions. There are situations in which the significance of a word or image cannot be captured to an exact degree by

M. Akram, *Fuzzy Lie Algebras*, Infosys Science Foundation Series,
https://doi.org/10.1007/978-981-13-3221-0_1

a crisp set. The term *frosty*, for instance, means the range of cold temperatures, which may change in various circumstances, certainly, lacks a sharp boundary. Similarly, the set of individuals one may call ones *good friends* is not as pointedly characterized as would be required if traditional sets were to be utilized. The representation of this sort of system is based on the idea called *fuzziness*. Fuzziness exists when a piece of information contains vagueness, uncertainty, or ambiguities. The word *fuzzy* actually means "unable to think clearly". For instance, the words like, creditworthy clients, tall women, intelligent men, and yellow roses, are fuzzy in nature.

A *fuzzy set*, as a superset of a crisp set, owes its origin to the work of Zadeh [139] in 1965 that has been introduced to deal with the notion of *partial truth* between "absolute true" and "absolute false". Zadeh's remarkable idea has found many applications in several fields, including chemical industry, telecommunication, decision making, networking, computer science, discrete mathematics, economy management, biomedical engineering, groups, rings, semirings, hemirings, nearrings, vector spaces, differential equations, pattern recognition, automation, and robotics. In 1971, Rosenfeld [115] used the concept of a fuzzy subset of a set to introduce the notion of a fuzzy subgroup of a group. Rosenfeld's paper spearheaded the development of fuzzy abstract algebra.

Definition 1.1 A *fuzzy set* μ in a universe X is a mapping $\mu : X \to [0, 1] \subset \mathbb{R}$.

In the theory of fuzzy sets, the membership degrees of elements range over the interval $[0, 1]$. The membership degree expresses the degree of belongingness of elements to a fuzzy set. The membership degree 1 denotes that an element completely belongs to its corresponding fuzzy set, and the membership degree 0 denotes that an element does not belong to the fuzzy set. The membership degrees in the interval $(0, 1)$ denote the partial membership to the fuzzy set.

Example 1.1 Let X be a set of network servers $X = \{S_1, S_2, S_3, S_4, S_5\}$. Every server provides a different signal strength and frequency from other network servers which cannot be persuaded in binary terms. These strategies are uncertain in nature. The degree of signal strength and frequency of each network server is shown in Table 1.1. Corresponding to signal strength and frequency, two fuzzy set μ and ν on X can be defined.

$$\mu = \{(S_1, 0.8), (S_2, 0.7), (S_3, 0.9), (S_4, 0.5), (S_5, 0.65)\},$$
$$\nu = \{(S_1, 0.5), (S_2, 0.7), (S_3, 0.4), (S_4, 0.6), (S_5, 0.8)\}.$$

Table 1.1 Fuzzy sets on network servers

Server	Signal strength	Frequency
S_1	0.8	0.5
S_2	0.7	0.7
S_3	0.9	0.4
S_4	0.5	0.6
S_5	0.65	0.8

Definition 1.2 The *fuzzy empty set* and the *fuzzy whole set* in a set X are denoted by 0_X and 1_X and defined as $0_X(x) = 0$ and $1_X(x) = 1$ for all $x \in X$, respectively.

Fuzzy set operations are operations on fuzzy sets. These operations are generalization of crisp set operations.

Definition 1.3 Let μ and ν be two fuzzy sets in the universe of discourse X, then the operations over fuzzy sets are defined as follows:

- $\mu \subseteq \nu \Rightarrow \mu(x) \leq \nu(x)$,
- $\mu = \nu \Leftrightarrow \mu(x) = \nu(x)$,
- $\mu \cap \nu(x) = \min\{\mu(x), \nu(x)\} = \mu(x) \wedge \nu(x)$,
- $\mu \cup \nu(x) = \max\{\mu(x), \nu(x)\} = \mu(x) \vee \nu(x)$,
- $\mu \cdot \nu(x) = \mu(x) \cdot \nu(x)$,
- $\mu + \nu(x) = \mu(x) + \nu(x) - \mu(x) \cdot \nu(x)$,
- $\mu^c(x) = 1 - \mu(x)$

for all $x \in X$.

Remark 1.1 There is a difference between theory of crisp sets and the theory of fuzzy sets.

1. The *law of excluded middle* is not valid in fuzzy set theory, that is,

$$\mu \cup \mu^c \neq 1_X .$$

2. *Law of contradiction* is not valid in fuzzy set theory, that is,

$$\mu \cap \mu^c \neq 0_X .$$

Definition 1.4 Let $\{\mu_i\}_{i \in I}$ be a collection of fuzzy subsets of X. Then, we define the fuzzy subsets $\bigcap_{i \in I} \mu_i$ and $\bigcup_{i \in I} \mu_i$ by:

$$\left(\bigcap_{i \in I} \mu_i\right)(x) = \inf_{i \in I}\{\mu_i(x)\} \text{ for all } x \in X,$$
$$\left(\bigcup_{i \in I} \mu_i\right)(x) = \sup_{i \in I}\{\mu_i(x)\} \text{ for all } x \in X.$$

Definition 1.5 Let μ be a fuzzy set on a universe of discourse X. For any $t \in [0, 1]$, a *t-cut set*(t-level set) of μ is defined as

$$U(\mu, t) = \{x \in X \mid \mu(x) \geq t\}.$$

The set

$$U^+(\mu, t) = \{x \in X \mid \mu(x) > t\}$$

is called the *strong t-cut set* of μ. Note that $U(\mu, t)$ and $U^+(\mu, t)$ are crisp sets.

Definition 1.6 A *fuzzy relation* on a nonempty set X is a fuzzy set $\lambda : X \times X \rightarrow [0, 1]$. Let μ be a fuzzy set on X and λ fuzzy relation in X. We call λ is a fuzzy relation on μ if

$$\lambda(x, y) \leq \min\{\mu(x), \mu(y)\} \quad \text{for all } x, y \in X.$$

Definition 1.7 Let λ be a fuzzy relation on a universe of discourse X. For any $t \in [0, 1]$, a *t-cut set*(t-level set) of λ is defined as

$$U(\lambda, t) = \{(x, y) \in X \times X \mid \lambda(x, y) \geq t\}.$$

Note that $U(\lambda, t)$ is a crisp set.

Definition 1.8 Let λ be a fuzzy relation on a universe X. Then, λ is called a *fuzzy equivalence relation* on X if it satisfies the following conditions:

(a) λ is fuzzy reflexive, i.e., $\lambda(x, x) = 0$ for each $x \in X$,
(b) λ is fuzzy symmetric, i.e., $\lambda(x, y)$=$\lambda(y, x)$ for any $x, y \in X$,
(c) λ is fuzzy transitive, i.e., $\lambda(x, z) \geq \bigvee_y (\lambda(x, y) \bigwedge \lambda(y, z))$.

Definition 1.9 Let $\lambda_1(X \rightarrow Y)$ and $\lambda_2(Y \rightarrow Z)$ be two fuzzy relations. The *max–min composition* $\lambda_1 \circ \lambda_2(X \rightarrow Z)$ is the fuzzy relation defined by the membership function

$$\lambda_1 \circ \lambda_2(x, z) = \bigvee_y (\lambda_2(x, y) \wedge \lambda_1(y, z))$$

for all $(x, z) \in X \times Z$ and for all $y \in Y$. Note that $\lambda_1 \circ \lambda_2 \neq \lambda_2 \circ \lambda_1$.

Proposition 1.1 *Let λ, λ_1 and λ_2 be fuzzy relations on a nonempty set X.*

1. *If λ is symmetric, then so is λ^{-1},*
2. *λ is symmetric if and only if $\lambda^{-1} = \lambda$,*
3. *If λ_1 and λ_2 are symmetric relations on X, then $\lambda_1 \cup \lambda_2$, $\lambda_1 \cap \lambda_2$, λ_1^c, $\lambda_1 + \lambda_2$, $\lambda_1 \cdot \lambda_2$ are symmetric,*
4. *If λ_1 and λ_2 are reflexive relations on X, then $\lambda_1 \cup \lambda_2$, $\lambda_1 \cap \lambda_2$, λ_1^c, $\lambda_1 + \lambda_2$, $\lambda_1 \cdot \lambda_2$ are reflexive,*
5. *If λ_1 and λ_2 are transitive relations on X, then $\lambda_1 \cup \lambda_2$, $\lambda_1 \cap \lambda_2$, λ_1^c, $\lambda_1 + \lambda_2$, $\lambda_1 \cdot \lambda_2$ are transitive,*
6. *If λ_1 and λ_2 are reflexive(symmetric, transitive) relations, then their composition $\lambda_1 \circ \lambda_2$ may not be reflexive(symmetric, transitive).*

Proposition 1.2 *Let λ be a fuzzy relation on a nonempty set X.*

1. *λ is reflexive if and only if every t-cut set of λ is reflexive,*
2. *λ is symmetric if and only if every t-cut set of λ is symmetric,*
3 *λ is transitive if and only if every t-cut set of λ is transitive.*

Definition 1.10 Let f be a mapping defined on a set X. If μ is a fuzzy set in X, then the fuzzy set ν in $f(X)$ defined by

$$\nu(y) = \sup_{x \in f^{-1}(y)} \mu(x)$$

for all $y \in f(X)$ is called the *image* of μ under f. If ν is a fuzzy set in $f(X)$, then the fuzzy set $\mu = \nu \circ f$ in X, i.e., the fuzzy set defined by $\mu(x) = \nu(f(x))$ for all $x \in X$, is called the *pre-image* of ν under f.

Definition 1.11 Let f be a mapping from a set X into set Y.

1. Let μ be a fuzzy set in Y. The inverse image of μ, denoted by $f^{-1}(\mu)$, is the fuzzy set in X defined by $f^{-1}(\mu)(x) = \mu(f(x)) \quad$ for all $x \in X$.,
2. Let μ be a fuzzy set in X. The image of μ is denoted by $f(\mu)$ and is the fuzzy set in Y such that

$$f_{\sup}(\mu)(y) = \begin{cases} \sup_{x \in f^{-1}(y)} \mu(x), & \text{if} f^{-1}(y) \neq \emptyset, \\ 0, & \text{otherwise} . \end{cases}$$

Proposition 1.3 *Let μ and μ_i be fuzzy sets, and let ν be a fuzzy set in Y. Let $f : X \to Y$ be a function. Then,*

1. $f(f^{-1}(\mu)) = \mu$, *if f is surjective,*
2. $f(\emptyset_X) = \emptyset_Y$,
3. $f(1_X) = 1_Y$, *if f is surjective,*
4. $f^{-1}(1_Y) = 1_X$,
5. $f^{-1}(\emptyset_Y) = \emptyset_X$,
6. $f(\cup \mu_i) = \cup f(\mu_i)$.

Definition 1.12 Let G be a group. A fuzzy subset μ of G is called a *fuzzy subgroup* of G if the following conditions are satisfied:

(1) $\mu(xy) \geq \min\{\mu(x), \mu(y)\}$ for all $x, y \in G$,
(2) $\mu(x^{-1}) \geq \mu(x)$ for all $x \in G$.

Note that by (1) and (2), we obtain $\mu(e) \geq \mu(x)$ and $\mu(x^{-1}) = \mu(x)$ for all $x \in G$.

1.1.2 Lie Algebras

A Lie algebra is one of the basic notions of mathematics. It is so-named in honor of Sophus Lie (1842–1899), a Norwegian mathematician who pioneered the study of this branch. A Lie algebra has been used by electrical engineers, mainly in the mobile robot control [46].

A *Lie algebra*(pronounced "lee") is a vector space L over a field $\mathbb{F}$ (equal to $\mathbb{R}$ or $\mathbb{C}$) on which $L \times L \to L$ denoted by $(x, y) \to [x, y]$ is defined satisfying the following axioms:

(L1) $[x, y]$ is bilinear,
(L2) $[x, y] = -[y, x]$ for all $x, y \in L$,
(L3) $[[x, y], z] + [[y, z], x] + [[z, x], y] = 0$ for all $x, y, z \in L$ (Jacobi identity).

Example 1.2 (1). The vector space $\mathfrak{R}^3$ with the bracket defined by $[x, y] = x \times y$, where $\times$ is the cross-product, is a classical example of a Lie algebra. If $\{e_1, e_2, e_3\}$ is the standard basis, then the bracket operation is completely determined by the relations: $[e_1, e_2] = e_3$, $[e_2, e_3] = e_1$, $[e_3, e_1] = e_2$. The relation $[e_2, e_1] = -e_3$ follows from the above by (L2),

(2). Any vector space with defined an associative multiplication of vectors is a Lie algebra with respect to the commutator $[x, y] = xy - yx$.

We note that the operation $[x, y]$ in a Lie algebra is not associative; i.e., it is not true in general that $[[x, y], z] = [x, [y, z]]$. A subspace H of L closed under $[\cdot, \cdot]$ will be called a *Lie subalgebra*. A subspace J of L is called a *Lie ideal* of a Lie algebra L if $[x, y] \in J$ for all $x \in J$, $y \in L$, i.e., $[J, L] \subseteq J$. Let L_1 and L_2 be Lie algebras over a field $\mathbb{F}$. A linear transformation $f : L_1 \to L_2$ is called a *Lie homomorphism* if $f([x, y]) = [f(x), f(y)]$ for all $x, y \in L_1$. A *quotient Lie algebra* is a Lie algebra that is the quotient of a Lie algebra L and one of its Lie ideal J, denoted L/J. Let J be a Lie idea of L, then for all $x, y \in L$ and $\alpha \in \mathbb{F}$, $x + J, y + J \in L/J$, we define

(i) $(x + J) + (y + J) = (x + y) + J$,
(ii) $[x + J, y + J] = [x, y] + J$,
(iii) $\alpha(x + J) = \alpha x + J$.

For any $x \in L$, we define the function $adx : L \to L$ putting $adx(y) = [x, y]$. It is clear that this function is a linear homomorphism with respect to y. The set $H(L)$ of all linear homomorphisms from L into itself is made into a Lie algebra by defining a commutator on it by $[f, g] = f \circ g - g \circ f$. The function $ad : L \to H(L)$ defined by $ad(x) = adx$ is a Lie homomorphism which is called the *adjoint representation* of L. The adjoint representation $adx : L \to L$ is extended to $\overline{adx} : J^L \to J^L$ by putting

$$\overline{adx}(\gamma)(y) = \sup\{\gamma(a) \mid [x, a] = y\}$$

for all $\gamma \in J^L$ and $y \in L$.

A Lie algebra L is said to satisfy the *descending chain condition* for Lie ideals if for any sequence of Lie ideals $J_1, J_2, \cdots, J_i, \cdots$ of L such that

$$J_1 \supseteq J_2 \supseteq J_3 \supseteq \cdots \supseteq J_i \cdots,$$

there exists an element $n \in \mathbb{N}$ such that $J_m = J_n$ for each $m \in \mathbb{N}$, $m \leq n$. $\mathbb{N} = \{1, 2, \ldots\}$ always denotes the set of natural numbers.

L is called *Artinian* if it satisfies the descending chain condition on its Lie ideals. Similarly, L is called *Noetherian* if it satisfies the ascending chain condition on its Lie ideals. A Lie ideal C of a Lie algebra L is said to be a *fully invariant* Lie ideal if $f(C) \subseteq C$ for all $f \in \mathrm{End}(L)$, where $\mathrm{End}(L)$ is the set of all endomorphisms of L. A Lie ideal C of Lie algebra L is said to be *characteristic* if $f(C) = C$, for all $f \in Aut(L)$, where $\mathrm{Aut}(L)$ is the set of all automorphisms of a Lie algebra L. The mapping $K : L \times L \to \mathbb{F}$ defined by $K(x, y) = Tr(adx \circ ady)$, where Tr is the *trace* of a linear homomorphism, is a symmetric bilinear form which is called the *Killing form*. It is not difficult to see that this form satisfies the identity $K([x, y], z) = K(x, [y, z])$.

1.1.3 Lie Superalgebras

The notion of Lie superalgebras was introduced by Kac [85] in 1977 as a generalization of the theory of Lie algebras. A Lie superalgebra has been played an important role in the study of mathematics and physics.

Definition 1.13 Suppose that V is a vector space and $V_{\bar{0}}$, $V_{\bar{1}}$ are its (vectors) subspaces. Let $V = V_{\bar{0}} \oplus V_{\bar{1}}$ be the direct sum of the subspaces. Then, V (with this decomposition) is called a *Z_2-graded vector space* if each element v of a Z_2-graded vector space has a unique expression of the form $v = v_{\bar{0}} + v_{\bar{1}}$ $(v_{\bar{0}} \in V_{\bar{0}}, v_{\bar{1}} \in V_{\bar{1}})$. The subspaces $V_{\bar{0}}$ and $V_{\bar{1}}$ are called the *even part* and *odd part* of V, respectively. In particular, if v is an element of either $V_{\bar{0}}$ or $V_{\bar{1}}$, v is said to be *homogeneous*.

Definition 1.14 A Z_2-graded vector space $\mathscr{L} = \mathscr{L}_{\bar{0}} \oplus \mathscr{L}_{\bar{1}}$ with a Lie bracket

$$[\ ,\ ] : \mathscr{L} \times \mathscr{L} \xrightarrow{\text{bilinear}} \mathscr{L}$$

is called a *Lie superalgebra*, if it satisfies the following conditions:

(1) $[\mathscr{L}_{\bar{i}}, \mathscr{L}_{\bar{j}}] \subseteq \mathscr{L}_{\overline{i+j}}$ for $i, j \in Z_2$,
(2) $[x, y] = -(-1)^{|x||y|}[y, x]$ (antisymmetry),
(3) $[x, [y, z]] = [[x, y], z] + (-1)^{|x||y|}[[x, z], y]$ (Jacobi identity),

where x, y are homogeneous elements and $|x|$ is 0 or 1 according to whether x is in $\mathscr{L}_{\bar{0}}$ or $\mathscr{L}_{\bar{1}}$. The subspaces $\mathscr{L}_{\bar{0}}$ and $\mathscr{L}_{\bar{1}}$ are called the even and odd parts of $\mathscr{L}$, respectively. Therefore, a Lie algebra is a Lie superalgebra with trivial odd part.

Definition 1.15 If $\varphi : \mathscr{L} \to \mathscr{L}'$ is a linear map between Lie superalgebras $\mathscr{L} = \mathscr{L}_{\bar{0}} \oplus \mathscr{L}_{\bar{1}}$ and $\mathscr{L}' = \mathscr{L}'_{\bar{0}} \oplus \mathscr{L}'_{\bar{1}}$ such that

(4) $\varphi(\mathscr{L}_i) \subseteq \mathscr{L}'_i$ $(i \in Z_2)$ (preserving the grading),
(5) $\varphi([x, y]) = [\varphi(x), \varphi(y)]$ (preserving the Lie bracket).

Then φ is called a *homomorphism* of Lie superalgebras.

1.2 Fuzzy Lie Ideals

Definition 1.16 Let V be a vector space over a field $\mathbb{F}$. A fuzzy subset μ of V is called a *fuzzy subspace* of V if the following conditions are satisfied:

(1) $\mu(x+y) \geq \min\{\mu(x), \mu(y)\}$ for all $x, y \in V$,
(2) $\mu(\alpha x) \geq \mu(x)$ for all $x \in V$, $\alpha \in \mathbb{F}$,

Note that by (2), we obtain $\mu(-x) \geq \mu(x)$ and $\mu(0) \geq \mu(x)$ for all $x \in V$.

Lemma 1.1 *If μ is a fuzzy subspace of a vector space V, then*

(1) $\mu(x) = \mu(-x)$,
(2) $\mu(x-y) = \mu(0) \Longrightarrow \mu(x) = \mu(y)$,
(3) $\mu(x) < \mu(y) \Longrightarrow \mu(x-y) = \mu(x) = \mu(y-x)$

for all $x, y \in V$.

Theorem 1.1 *For a fuzzy subset μ of a vector space V, the following statements are equivalent.*

(1) *μ is a fuzzy subspace of V,*
(2) *Each nonempty $U(\mu, t)$ is a subspace of V.*

This theorem firstly proved in [86] is a consequence of the transfer principle for fuzzy sets described in [90].

Definition 1.17 A fuzzy set μ, i.e., a map $\mu : L \rightarrow [0, 1]$, is called a *fuzzy Lie subalgebra* of L over a field $\mathbb{F}$ if it is a fuzzy subspace of L such that

(3) $\mu([x, y]) \geq \min\{\mu(x), \mu(y)\}$

hold for all $x, y \in L$ and $\alpha \in \mathbb{F}$.

Example 1.3 The real vector space $\Re^3$ with $[x, y] = x \times y$, where $x, y \in \Re^3$, is a real Lie algebra. Define a fuzzy set μ on $\Re^3$ by

$$\mu(x) = \begin{cases} 0.9, & \text{if } x = (0, 0, 0), \\ 0.6, & \text{if } x = (c, 0, 0), \quad c \neq 0, \\ 0.2, & \text{otherwise.} \end{cases}$$

By direct calculations, it is easy to see μ is a fuzzy Lie algebra.

Definition 1.18 A fuzzy set $\mu : L \rightarrow [0, 1]$ is called a *fuzzy Lie ideal* of L if

(1) $\mu(x+y) \geq \min\{\mu(x), \mu(y)\}$,
(2) $\mu(\alpha x) \geq \mu(x)$,
(3) $\mu([x, y]) \geq \mu(x)$

hold for all $x, y \in L$ and $\alpha \in \mathbb{F}$.

Proposition 1.4 *Every fuzzy Lie ideal is a fuzzy Lie subalgebra.*

The converse of Proposition 1.4 is not true, in general. The fuzzy set μ defined in Example 1.3 is a fuzzy Lie subalgebra, but it is not a fuzzy Lie ideal.
The following Lemma is obvious.

Lemma 1.2 *Let μ be a fuzzy Lie ideal of L, then*

(1) $\mu(0) \geq \mu(x)$,
(2) $\mu([x, y])) \geq \max\{\mu(x), \mu(y)\}$,
(3) $\mu([x, y]) = \mu(-[y, x]) = \mu([y, x])$,
(4) $\mu(x - y) = \mu(0) \Rightarrow \mu(x) = \mu(y)$,
(5) $\mu(x - y) = \mu(x) = \mu(y - x)$, *if* $\mu(x) < \mu(y)$ *for all* $x, y \in L$.

Theorem 1.2 *Let μ and ν be two fuzzy Lie ideal of L. Then, $\mu \bigcap \nu : L \to [0, 1]$ and $\mu + \nu : L \to [0, 1]$ are fuzzy Lie ideals of L.*

Proof Let $x, y \in L$ and $\alpha \in \mathbb{F}$. Then

$$\begin{aligned} \left(\mu \bigcap \nu\right)(x + y) &= \inf\{\mu(x + y), \nu(x + y)\} \\ &\geq \inf\{\min\{\mu(x), \mu(y)\}, \min[\nu(x), \nu(y)]\} \\ &= \inf\{\min\{\mu(x), \nu(x)\}, \min\{\mu(y), \nu(y)\}\} \\ &= \min\left\{\left(\mu \bigcap \nu\right)(x), \left(\mu \bigcap \nu\right)(y)\right\}, \\ \left(\mu \bigcap \nu\right)(\alpha x) &= \inf\{\mu(\alpha x), \nu(\alpha x)\} \\ &\geq \inf\{\mu(x), \nu(x)\} \\ &= \left(\mu \bigcap \nu\right)(x), \\ \left(\mu \bigcap \nu\right)([x, y]) &= \inf\{\mu([x, y]), \nu([x, y])\} \\ &\geq \inf\{\mu(x), \nu(x)\} \\ &= \left(\mu \bigcap \nu\right)(x). \end{aligned}$$

Hence, $\left(\mu \bigcap \nu\right)$ is a fuzzy Lie ideal of L. Similarly, it is easy to prove that $(\mu + \nu)$ is a fuzzy Lie ideal of L.

Theorem 1.3 *If $\{\mu_i \mid i \in I\}$ is a family of fuzzy Lie ideals of Lie algebras L, then so is $\bigcap \mu_i$.*

Proof Since each μ_i is fuzzy Lie ideal, then they all must satisfied these conditions

$$\mu_i(x + y) \geq \min\{\mu_i(x), \mu_i(y)\}, \quad \mu_i(\alpha x) \geq \mu_i(x), \quad \mu_i([x, y]) \geq \mu_i(x)$$

for all $x, y \in L$ and $\alpha \in \mathbb{F}$.

By using Definition 1.18 and infimum, $(\bigcap \mu_i)(x) = \inf_{i \in I}\{\mu_i(x)\} = \mu_k(x)$. Hence, $(\bigcap \mu_i)$ is a fuzzy Lie ideal.

Theorem 1.4 *Let μ and ν be two fuzzy Lie algebras of L. Then, the direct product $\mu \times \nu$ is a fuzzy Lie algebra of L.*

Proof Let $x = (x_1, x_2)$ and $y = (y_1, y_2)$. Then, $(\mu \times \nu)(a, b) = (\mu(a), \nu(b))$. Thus,

$$\begin{aligned}\mu \times \nu([x, y]) &= \mu \times \nu([(x_1, x_2), (y_1, y_2)])\\ &\geq \min\{\min\{\mu(x_1), \nu(x_2)\}, \min\{\mu(y_1), \nu(y_2)\}\}\\ &= \min\{(\mu \times \nu)(x_1, x_2), (\mu \times \nu)(y_1, y_2)\}\\ &= \min\{(\mu \times \nu)(x), (\mu \times \nu)(y)\}.\end{aligned}$$

Similarly, the other conditions of a fuzzy Lie algebra can be verified.

Definition 1.19 Let L be a Lie algebra. If $\mu_1, \ldots, \mu_n$ are fuzzy sets of L, then the fuzzy set

$$\left(\sum_{i=1}^{n} \mu_i\right)(x) = \bigvee\left\{\bigwedge \mu_i(x_i) \mid x = \sum_{i=1}^{n} x_i\right\} \quad \text{for all } x \in L,$$

is called the *sum* of $\mu_1, \ldots, \mu_n$.

Definition 1.20 Let L be a Lie algebra with defined multiplication of vectors. By the *product* $\mu\nu$ of two fuzzy sets μ and ν of a Lie algebra L, we mean the fuzzy set

$$(\mu\nu)(x) = \bigvee\left\{\bigwedge_{i=1}^{m}\{\mu(c_i) \wedge \nu(d_i)\} \mid x = \sum_{i=1}^{m} c_i d_i\right\} \quad \text{for all } x \in L.$$

Definition 1.21 The *sup-min product* of fuzzy sets μ and ν of L, denoted by $\ll \mu\nu \gg$ is defined as

$$\ll \mu\nu \gg (x) = \begin{cases} \left\{ \sup\limits_{x=\sum_{i=1}^{n}[x_i, y_i]} \left\{ \min\limits_{i \in \mathbb{N}}\{\min\{\mu(x_i), \nu(y_i)\}\}\right\}\right\}, & \text{where } x_i, y_i \in L;\\ 0, \quad \text{if } x \text{ is not expressible as } x = \sum\limits_{i=1}[x_i, y_i]. \end{cases}$$

Note that that $\mu\nu \subseteq \ll \mu\nu \gg$ and $\mu\nu \neq \ll \mu\nu \gg$.

Theorem 1.5 *Let ν and μ be any two fuzzy Lie ideals of L. Then, the sup-min product is also fuzzy Lie ideal of L.*

Proof Suppose $x, y \in L$. We have to show that the conditions of Definition 1.18 are satisfied.

(i) We prove that $\ll \mu\nu \gg (x+y) \geq \min\{\ll \mu\nu \gg (x), \ll \mu\nu \gg (y)\}$. Suppose that $\ll \mu\nu \gg (x+y) < \min\{\ll \mu\nu \gg (x), \ll \mu\nu \gg (y)\}$.
Then, we have

$$\ll \mu\nu \gg (x+y) < \ll \mu\nu \gg (x) \quad \text{and} \quad \ll \mu\nu \gg (x+y) < \ll \mu\nu \gg (y).$$

Choose a number $t \in [0, 1]$ such that

$$\ll \mu\nu \gg (x+y) < t < \ll \mu\nu \gg (x)$$

and

$$\ll \mu\nu \gg (x+y) < t < \ll \mu\nu \gg (y).$$

There exist $x_i, y_i, x'_j, y'_j \in L$ such that

$$x = \sum_{i=1}^{n} [x_i, y_i] \quad \text{and} \quad y = \sum_{j=1}^{m} [x'_j, y'_j]$$

and for all i, j,

$$\mu(x_i) > t, \ \ \nu(y_i) > t, \ \ \mu(x'_j) > t, \ \ \nu(y'_j) > t.$$

Since $x + y = \sum_{i=1}^{n} [x_i, y_i] + \sum_{j=1}^{m} [x'_j, y'_j]$, then we have

$$\ll \mu\nu \gg (x+y) > t.$$

But

$$\ll \mu\nu \gg (x+y) > t > \ll \mu\nu \gg (x+y), \ \text{this is a contradiction.}$$

Similarly, $\ll \mu\nu \gg (\alpha x) \geq \ll \mu\nu \gg (x)$ and $\ll \mu\nu \gg ([xy]) \geq \ll \mu\nu \gg (x)$. Hence, $\ll \mu\nu \gg$ is a fuzzy Lie ideal of L.

The transfer principle for fuzzy sets (cf. [90]) suggest the following result.

Theorem 1.6 *A fuzzy set μ of a Lie algebra L is its fuzzy Lie ideal if and only if each nonempty set $U(\mu, t) = \{x \in L \mid \mu(x) \geq t\}$ is a Lie ideal of L.*

Proof Assume that μ is a fuzzy Lie ideal of L, and let $t \in [0, 1]$ be such that $U(\mu, t) \neq \emptyset$. Let $x, y \in U(\mu, t)$. Then, $\mu(x) \geq t$ and $\mu(y) \geq t$. It follows that

$$\begin{aligned} \mu(x+y) &\geq \min\{\mu(x), \mu(y)\} \geq t, \\ \mu(\alpha x) &\geq \mu(x) \geq t, \\ \mu([x, y]) &\geq \mu(x) \geq t, \end{aligned}$$

so that $x + y \in U(\mu, t)$, $\mu(\alpha x) \in U(\mu, t)$ and $\mu([x, y]) \in U(\mu, t)$. Hence, $U(\mu, t)$ is a Lie ideal of L.

Conversely, suppose that each nonempty $U(\mu, t)$ is a Lie ideal of L. Assume that $\mu(x + y) < \min\{\mu(x), \mu(y)\}$ for some $x, y \in L$. Taking

$$t_0 := \frac{1}{2}\{\mu(x + y) + \min\{\mu(x) + \mu(y)\}\},$$

we have $\mu(x + y) < t_0 < \min\{\mu(x), \mu(y)\}$. So, $x + y \notin U(\mu, t_0)$ and $x, y \in U(\mu, t_0)$. This is a contradiction.

Hence, $\mu(x + y) \geq \min\{\mu(x), \mu(y)\}$ for all $x, y \in L$.

Similarly, we can verify other conditions.

Remark 1.2 If μ and ν are fuzzy sets of L, then $U(\mu, t) + U(\nu, t) \subseteq U(\mu + \nu, t)$ for all $t \in]0, 1]$.

Definition 1.22 Let L_1 and L_2 be two Lie algebras and f a function of L_1 into L_2. If μ is a fuzzy set in L_2, then the *pre-image* of μ under f is the fuzzy set in L_1 defined by

$$f^{-1}(\mu)(x) = \mu(f(x)) \qquad \forall x \in L_1.$$

Equivalently, if μ is a fuzzy set in $f(L_1)$, then the *pre-image* of μ under f is the fuzzy set ν in L_1 defined by

$$\nu(x) = \mu(f(x)) \qquad \forall x \in L_1.$$

Theorem 1.7 *Let $f : L_1 \to L_2$ be an epimorphism of Lie algebras. If ν is a fuzzy Lie ideal of L_2 and μ is the pre-image of ν under f, then μ is a fuzzy Lie ideal of L_1.*

Proof For any $x, y \in L_1$ and $\alpha \in \mathbb{F}$,

$$\begin{aligned}
\mu(x + y) &= \nu(f(x + y)) = \nu(f(x) + f(y)) \\
&\geq \min\{\nu(f(x)), \nu(f(y))\} = \min\{\mu(x), \mu(y)\}, \\
\mu(\alpha x) &= \nu(f(\alpha x)) = \nu(\alpha f(x)) \geq \nu(f(x)) = \mu(x), \\
\mu([x, y]) &= \nu(f([x, y])) \geq \nu(f(x)) = \mu(x).
\end{aligned}$$

Hence, μ is a fuzzy Lie ideal of L_1.

Theorem 1.8 *Let $f : L_1 \to L_2$ be an onto homomorphism of Lie algebras over same field. If μ is a fuzzy Lie ideal of L_2, then $f^{-1}(\mu)$ is a fuzzy Lie ideal of L_1.*

Proof For any $x, y \in L_1$ and $\alpha \in \mathbb{F}$,

$$\begin{aligned} f^{-1}(\mu)(x+y) &= \mu(f(x+y)) = \mu(f(x)+f(y)) \\ &\geq \min\{\mu(f(x)), \mu(f(y))\} = \min\{f^{-1}(\mu)(x), f^{-1}(\mu)(y)\}, \\ f^{-1}(\mu)(\alpha x) &= \mu(f(\alpha x)) = \mu(\alpha f(x)) \geq \mu(f(x)) = f^{-1}(\mu)(x) \\ f^{-1}(\mu)([x,y]) &= \mu(f([x,y])) \geq \mu(f(x)) = f^{-1}(\mu)(x). \end{aligned}$$

Hence, μ is a fuzzy Lie ideal of L_1.

Theorem 1.9 *Let $f : L_1 \to L_2$ be an onto homomorphism of Lie algebras over same field. If μ is a fuzzy Lie ideal of L_2, then $f^{-1}(\mu^c) = (f^{-1}(\mu))^c$.*

Proof Let μ be a fuzzy Lie ideal of L_2. Then, for $x \in L_1$,

$$f^{-1}(\mu^c)(x) = \mu^c(f(x)) = 1 - \mu(f(x)) = 1 - f^{-1}(\mu^c)(x) = (f^{-1}(\mu))^c(x).$$

That is $f^{-1}(\mu^c) = (f^{-1}(\mu))^c$.

Definition 1.23 Let L_1 and L_2 be two Lie algebras and f a function of μ is a fuzzy set in L_1, then the *image* of μ under f is the fuzzy set defined by

$$\mu^f(y) = \begin{cases} \sup\{\mu(t) | t \in L_1, f(t) = y\}, & \text{if } f^{-1}(y) \neq \emptyset; \\ 1, & \text{otherwise.} \end{cases}$$

Definition 1.24 A fuzzy set μ in L_1 has the *sup property* if for any nonempty subset $A \subseteq L_1$, there exists $a_0 \in A$ such that $\mu(a_0) = \sup\limits_{a \in A} \mu(a)$.

Theorem 1.10 *A Lie algebra homomorphism image of a fuzzy Lie ideal having the sup property is a fuzzy Lie ideal.*

Proof Let $f : L_1 \to L_2$ be a homomorphism of L_1 onto L_2 and μ be a fuzzy Lie ideal of L_1 with the sup property. Consider $f(x), f(y) \in f(L_1)$. Let $x_0, y_0 \in f^{-1}(f(x))$ be such that $\mu(x_0) = \sup\limits_{t \in f^{-1}(f(x))} \mu(t)$ and $\mu(y_0) = \sup\limits_{t \in f^{-1}(f(y))} \mu(t)$, respectively. Then

$$\begin{aligned} \nu^f((x)+(y)) &= \sup_{t \in f^{-1}(f(x)+f(y))} \mu^f(t) \\ &\geq \mu(x_0 + y_0) \geq \min\{\mu(x_0) + \mu(y_0)\} \\ &= \min\{\sup_{t \in f^{-1}(f(x))} \mu(t), \sup_{t \in f^{-1}(f(y))} \mu(t)\} \\ &= \min\{\nu(f(x)) + \nu(f(y))\}, \\ \nu^f((\alpha x)) &= \sup_{t \in f^{-1}(f(\alpha x))} \mu(t) \\ &\geq \mu(x_0) \geq \min\{\mu(x_0)\} = \nu(f(x)), \end{aligned}$$

$$\begin{aligned}\nu^f([x,y]) = \nu(f([x,y])) = & \sup_{t\in f^{-1}(f(x))} \mu^f(t)\\ \geq & \mu([x_0,y_0]) \geq \mu(x_0) = \nu^f(x).\end{aligned}$$

Consequently, ν is a fuzzy Lie ideal of L_2.

Definition 1.25 Let L_1 and L_2 be two sets and let $f : L_1 \to L_2$ be any function. A fuzzy set μ is called *f-invariant* if and only if for $x, y \in L_1$ $f(x) = f(y)$ implies $\mu(x) = \mu(y)$.

Theorem 1.11 *Let $f : L_1 \to L_2$ be a surjective Lie homomorphism. Then*

(1) if μ and ν are two fuzzy Lie ideals of L_1, then $(\mu+\nu)^f = (\mu)^f + (\nu)^f$,
(2) if μ and ν are two fuzzy Lie ideals of L_1, then $(\ll \mu\nu \gg)^f = \ll (\mu)^f (\nu)^f \gg$,
(3) if $\{\mu_i \mid i \in I\}$ is a set of f-invariant fuzzy Lie ideal of L, then

$$\Big(\bigcap_{i\in I}\mu_i\Big)^f = \bigcap_{i\in I}(\mu_i)^f.$$

Proof For $x \in L$,

$$\begin{aligned}(\mu+\nu)^f(x) = & \sup_{y=f^{-1}(x)} (\mu+\nu)(y)\\ = & \sup_{y=f^{-1}(x)} \mu(y) + \sup_{y=f^{-1}(x)} \nu(y)\\ = & \mu^f(x) + \nu^f(x).\end{aligned}$$

Similarly, we can prove other cases.

Theorem 1.12 *Let J be a Lie ideal of a Lie algebra L. If μ is a fuzzy Lie ideal of L, then the fuzzy set ν of L/J defined by*

$$\nu(a+J) = \sup_{x\in J}\mu(a+x)$$

is a fuzzy Lie ideal of the quotient Lie algebra L/J.

Proof Clearly, μ is well defined. Let $x + J, y + J \in L/J$, then

$$\begin{aligned}\nu((x+J)+(y+J)) = & \nu_A((x+y)+J) = \sup_{z\in J}\mu((x+y)+z)\\ = & \sup_{z=s+t\in J} \mu((x+y)+(s+t))\\ \geq & \sup_{s,t\in J} \min\{\mu(x+s), \mu(y+t)\}\\ = & \min\{\sup_{s\in J}\mu(x+s), \sup \mu_{t\in J}\mu(y+t)\}\\ = & \min\{\nu(x+J), \nu(y+J)\},\end{aligned}$$

$$\nu(\alpha(x+J)) = \nu(\alpha x + J) = \sup_{z\in J} \mu(\alpha x + z) \geq \sup_{z\in J} \mu(x+z) = \nu(x+J),$$
$$\nu([x+J, y+J]) = \nu([x,y]+J) = \sup_{z\in J} \mu([x,y]+z) \geq \sup_{z\in J} \mu(x+z) = \nu(x+J).$$

Hence, μ is a fuzzy Lie ideal of L/J.

Theorem 1.13 *Let J be a Lie ideal of a Lie algebra L with defined multiplication of vectors. Then, there is a one-to-one correspondence between the fuzzy ideal μ of L such that $\mu(0) = \mu(u)$ for all $u \in J$ and the set of all fuzzy Lie ideals μ of L/J.*

Proof Let μ be a fuzzy Lie ideal of L. Then, by Theorem 1.12, we can see that $\mu(a+J) = \sup_{u\in J} \mu(a+u)$ is a fuzzy Lie ideal of L/J. Since $\mu(0) = \mu(u)$ for all $u \in J$,

$$\mu(a+u) \geq \min\{\mu(a), \mu(u)\} = \min\{\mu(a), \mu(0)\}.$$

But $\mu(0) = \mu(0 \cdot a) \geq \mu(a)$. Hence $\mu(a+u) \geq \mu(a)$.
Again,

$$\mu(a) = \mu(a+u-u) \geq \min\{\mu(a+u), \mu(u)\} = \mu(a+u).$$

Therefore, $\mu(a+u) = \mu(a)$ for all $u \in J$. Thus, it follows that $\mu(a+J) - \mu(a)$. Hence, the correspondence $\mu \to \mu$ is one to one.

Suppose now μ is a fuzzy Lie ideal of L/J. Define μ on L by $\mu(a) = \mu(a+J)$. It can be shown that μ is a fuzzy Lie ideal of L. Also for any $u \in J$,

$$\mu(u) = \mu(u+J) = \mu(J)$$

shows that $\mu(u) = \mu(0)$, for all $u \in J$. This completes the proof.

Theorem 1.14 *Let L be a Lie algebra with defined multiplication of vectors and the identity element e.*

1. *If μ is a fuzzy Lie ideal of L and $t = \mu(0)$, then the fuzzy subset τ of L/J defined by $\tau(a + U(\mu,t)) = \mu(x)$, for all $x \in L$, is a fuzzy ideal of $L/U(\mu,t)$,*
2. *If J is an ideal of L and μ is a fuzzy ideal of L/J such that $\mu(x+J) = \mu(x)$ only when $x \in J$, then there exists a fuzzy ideal ν of L such that $U(\nu,t) = J$, where $t = \nu(0)$ and $\mu = \overline{\nu}$, where $\overline{\nu}$ is a quotient Lie algebra L/J.*

Proof (1). Let μ be a fuzzy Lie ideal of L. Then, by Theorem 1.6, $U(\mu,t)$ is an ideal of L. Now, τ is a well defined, because

$$\begin{aligned} x + U(\mu,t) &= y + U(\mu,t) \text{ where } x, y \in L \\ &\Longrightarrow x - y \in U(\mu,t) \Longrightarrow \mu(x-y) = \mu(0) \\ \Longrightarrow \mu(x) = \mu(y) &\Longrightarrow \tau(x + U(\mu,t)) = \tau(y + U(\mu,t)). \end{aligned}$$

Next, we show that τ is a fuzzy ideal of L/J. For any $x, y \in L$ and $\alpha \in \mathbb{F}$, we have

$$\begin{aligned}\tau((x+U(\mu,t))+(y+U(\mu,t))) &= \tau((x+y)+U(\mu,t)) = \mu(x+y) \geq \min\{\mu(x),\mu(y)\}\\ &= \min\{\tau(x+U(\mu,t)),\tau(y+U(\mu,t))\}.\\ \tau(\alpha(x+U(\mu,t))) &= \tau(\alpha x+U(\mu,t)) = \mu(\alpha x) \geq \mu(x) = \tau(x+U(\mu,t)).\\ \tau([x+U(\mu,t),(y+U(\mu,t))]) &= \tau([x+y]+U(\mu,t)) = \mu([x,y])\\ &\geq \mu(x) = \tau(x+U(\mu,t)).\end{aligned}$$

(2). Define a fuzzy subset μ of L by $\nu(x) = \mu(x+J)$ for all $x \in L$. A routine computation shows that ν is a fuzzy ideal of L. Now $U(\nu,t) = J$, because

$$x \in U(\nu,t) \Longleftrightarrow \nu(x) = t = \nu(0) \Longleftrightarrow \mu(x+J) = \mu(x) \Longleftrightarrow x \in J.$$

Finally, $\overline{\nu} = \mu$, since

$$\overline{\nu}(x+J) = \overline{\nu}(x+J) = \nu(x) = \mu(x+J).$$

This completes the proof.

Theorem 1.15 *Let $f : L_1 \twoheadrightarrow L_2$ be an epimorphism of Lie algebras. Then, μ is an f-invariant fuzzy Lie ideal of L_1 if and only if $f(\mu)$ is a fuzzy Lie ideal of L_2.*

Proof Let $x, y \in L_2$ and $\alpha \in \mathbb{F}$. Then, there exist $a, b \in L_1$ such that $f(a) = x$, $f(b) = y$, $x + y = f(a+b)$ and $\alpha x = \alpha f(a)$. Since μ is f-invariant,

$$\begin{aligned}f(\mu)(x+y) &= \mu(a+b) \geq \min\{\mu(a),\mu(b)\} = \min\{f(\mu)(x), f(\mu)(y)\},\\ f(\mu)(\alpha x) &= \mu(\alpha a) \geq \mu(a) = f(\mu)(x),\\ f(\mu)([x,y]) &= \mu([a,b]) = [\mu(a),\mu(b)] \geq \mu(a) = f(\mu)(x).\end{aligned}$$

Hence $f(\mu)$ is a fuzzy Lie ideal of L_2.

Conversely, if $f(\mu)$ is a fuzzy Lie ideal of L_2, then for any $x \in L_1$,

$$\begin{aligned}f^{-1}(f(\mu))(x) &= f(\mu)(f(x)) = \sup\{\mu(t) \mid t \in L_1, f(t) = f(x)\}\\ &= \sup\{\mu(t) \mid t \in L_1, \mu(t) = \mu(x)\} = \mu(x).\end{aligned}$$

Hence $f^{-1}(f(\mu)) = \mu$ is a fuzzy Lie ideal by Theorem 1.8.

Definition 1.26 An ideal J of Lie algebra L is said to be *characteristic* if $f(J) = J$, for all $f \in \text{Aut}(L)$, where $\text{Aut}(L)$ is the set of all automorphisms of L. Fuzzy Lie ideal μ of Lie algebra L is said to be *fuzzy characteristic* if $\mu^f(x) = \mu(x)$, for all $x \in L$ and $f \in \text{Aut}(L)$.

Lemma 1.3 *Let μ be a fuzzy Lie ideal of L. Then, for any $x \in L$, $\mu(x) = s$ if and only if $x \in U(\mu,s)$ and $x \notin U(\mu,t)$ for all $s < t$.*

Proof Suppose $\mu(x) = t$. Take $x \in U(\mu, t)$. Then, $\mu(x) = t < s$ implies $x \notin U(\mu, s)$.

Conversely, suppose $x \in U(\mu, t)$ and $x \notin U(\mu, s)$ and $\mu(x) = s \neq t$. Then, $x \in U(\mu, s)$, a contradiction.

Lemma 1.4 *Let μ be a fuzzy Lie ideal of a Lie algebra L and let $x \in L$. Then, $U(\mu, t) = U(\mu, s)$ if and only if there is no $x \in L$ and such that $s \leq \mu(x) < t$ for all $s < t$.*

Theorem 1.16 *A subset J of L is a Lie ideal if and only if the characteristic function of J is a fuzzy Lie ideal of L.*

Proof Suppose that μ is fuzzy characteristic, and let $s \in Im(\mu)$, $f \in \text{Aut}(L)$ and $x \in U(\mu, s)$. Then, $\mu^f(x) = \mu(x)$ implies $\mu(f(x)) \geq s$, whence $f(x) \in U(\mu, s)$. Thus, $f(U(\mu, s)) \subseteq U(\mu, s)$. Let $x \in U(\mu, s)$ and $y \in L$ such that $f(y) = x$. Then, $\mu(y) = \mu^f(y) = \mu(f(y)) = \mu(x) \geq s$, consequently $y \in U(\mu, s)$. So, $x = f(y) \in U(\mu, s)$. Thus, $U(\mu, s) \subseteq f(U(\mu, s))$. Hence, $f(U(\mu, s)) = U(\mu, s)$, i.e., $U(\mu, s)$ is characteristic.

Conversely, suppose that each level Lie ideal of μ is characteristic, and let $x \in L$, $f \in Aut(L)$, $\mu(x) = t$. Then, by virtue of Lemma 1.3, $x \in U(\mu, t)$ and $x \notin U(\mu, s)$, for all $s > t$. It follows from the assumption that $f(x) \in f(U(\mu, t)) = U(\mu, t)$, so that $\mu^f(x) = \mu(f(x))) \geq t$. Let $s = \mu^f(x)$ and assume that $s > t$. Then, $f(x) \in U(\mu, s) = f(U(\mu, s))$, which implies from the injectivity of f that $x \in U(\mu, s)$, a contradiction. Hence, $\mu^f(x) = \mu(f(x)) = t = \mu(x)$ showing that μ is a fuzzy characteristic.

Definition 1.27 Let $f : L_1 \to L_2$ be a homomorphism of Lie algebras which has an extension $f : J^{L_1} \to J^{L_2}$ defined by:

$$f(\mu)(y) = \sup\{\mu(x), x \in f^{-1}(y)\}.$$

for all $\mu \in J^{L_1}$, $y \in L_2$. Then, $f(\mu)$ is called the *homomorphic image* of μ.

Proposition 1.5 *Let $f : L_1 \to L_2$ be a homomorphism of Lie algebras, and let μ be a fuzzy Lie ideal of L_1. Then,*

(i) $f(\mu)$ *is a fuzzy Lie ideal of* L_2,
(ii) $f([\mu]) \supseteq [f(\mu)]$.

Definition 1.28 Let μ be a fuzzy Lie ideal in L. Define a sequence of fuzzy Lie ideals in L putting $\mu^0 = \mu$ and $\mu^n = [\mu^{n-1}, \mu^{n-1}]$ for $n > 0$. If there exists a positive integer n such that $\mu^n = 0$, then a fuzzy Lie ideal μ is called *solvable*.

Theorem 1.17 *Homomorphic image of a solvable fuzzy Lie ideal is a solvable fuzzy Lie ideal.*

Proof Let $f : L_1 \to L_2$ be a homomorphism of Lie algebras over same field. Suppose that μ is a solvable fuzzy Lie ideal in L_1. We prove by induction on n that $f(\mu^n) \supseteq (f(\mu))^n$, where n is any positive integer. First, we claim that $f([\mu, \mu]) \supseteq [f(\mu), f(\mu)]$.

Let $y \in L_2$, then

$$\begin{aligned} f(\ll \mu, \mu \gg) &= \inf\{\ll \mu, \mu \gg (x) \mid f(x) = y\} \\ &= \sup\{\sup\{\min\{\mu(a), \mu(b)\} \mid a, b \in L_1, [a, b] = x, f(x) = y\}\} \\ &= \sup\{\min\{\mu(a), \mu(b)\} \mid a, b \in L_1, [a, b] = x, f\mu(x) = y\} \\ &= \sup\{\min\{\mu(a), \mu(b)\} \mid a, b \in L_1, [f\mu(a), f\mu(b)] = x\} \\ &= \sup\{\min\{\mu(a), \mu(b)\} \mid a, b \in L_1, f\mu(a) = u, f\mu(b) = v, [u, v] = y\} \\ &\leq \sup\{\min\{\sup_{a \in f^{-1}\mu(u)} \mu(a), \sup_{b \in f^{-1}\mu(v)} \mu(b)\} \| [u, v] = y\} \\ &= \sup\{\min(f(\mu)(u), f(\mu)(v)) \mid [u, v] = y\} = \ll f(\mu), f(\mu) \gg (y). \end{aligned}$$

Thus, $f(\ll \mu, \mu \gg) \supseteq f(\ll \mu, \mu \gg) \supseteq \ll f(\mu), f(\mu) \gg = [f(\mu), f(\mu)]$.

Now for $n > 1$, we get

$$f(\mu^n) = f([\mu^{n-1}, \mu^{n-1}]) \supseteq [f(\mu^{n-1}), f(\mu^{n-1})] \supseteq [(f(\mu))^{n-1}, (f(\mu))^{n-1}] = (f(\mu))^n.$$

This completes the proof.

Definition 1.29 Let μ be a fuzzy Lie ideal in L and let $\mu^{(n)} = [\mu, \mu^{(n-1)}]$ for $n > 0$, where $\mu^{(0)} = \mu$. If there exists a positive integer n such that $\mu^{(n)} = 0$ then μ is called *nilpotent*.

Theorem 1.18 *Homomorphic image of a nilpotent fuzzy Lie ideal is a nilpotent fuzzy Lie ideal.*

Theorem 1.19 *If μ is a nilpotent fuzzy Lie ideal, then it is solvable.*

Definition 1.30 Let μ is a fuzzy ideal of Lie algebra L and $x \in L$. The fuzzy subset μ^* of L defined by

$$\mu^* = \mu(a - x) \text{ for all } a \in L$$

This is called *fuzzy coset* determined by x and μ.

Theorem 1.20 *Let μ is a fuzzy ideal of Lie algebra L. Then, L/μ, the set of all cosets of μ in L, is a Lie algebra under the following operations:*

$$\begin{aligned} &\mu_x^* + \mu_y^* = \mu_{x+y}^* && \textit{for all } x, y \in L, \\ &\alpha\mu_x^* = \mu_{\alpha x}^* && \textit{for all } \alpha \in \mathbb{F},\ x \in L, \\ &[\mu_x^* + \mu_y^*] = \mu_{[x,y]}^* && \textit{for all } x, y \in L. \end{aligned}$$

Proof The proof follows from Theorem 1.12 and Definition 1.30.

Lemma 1.5 *Let J be an ideal of L. If μ is a characteristic function of J, then μ^* is the characteristic function of $x + J$.*

Proof Let $a \in L$, then $\mu(a) = 1$ and $x + a \in x + J$. Thus, we have $\mu^*(x + a) = \mu(x + a - x) = \mu(a) = 1$. If $a \notin J$, then $\mu(a) = 0$ and $a + x \notin x + J$. Hence, $\mu^*(x + a) = \mu(x + a - x) = \mu(a)$. This implies that μ^* is a characteristic function of $x + J$.

Theorem 1.21 *If μ is a fuzzy ideal of Lie algebra L with multiplication of vectors, then the mapping $f : L \to L/\mu$ defined by $f(x) = \mu_x^*$ for all $x \in L$, is a homomorphism with the kernel $U(\mu, t)$, where $t = \mu(0)$.*

Proof That f is a homomorphism is easy to verify. Now we show that $\mu(x) = \mu(0)$ gives $\mu_x^* = \mu_0^*$. For this, let $a \in L$. Then $(a) \leq \mu(x) = \mu(0)$. If $\mu(a) < \mu(x)$, then $\mu(a - x) = \mu(a)$ by Lemma 1.5. On the other hand, from $\mu(a) = \mu(x)$ it follows that $a, x \in \{y \in L \mid \mu(y) = \mu(0)\}$. Hence, $\mu(a - x) = \mu(0) = \mu(x) = \mu(a)$. Therefore, in either case, we have shown that $\mu(a - x) = \mu(a)$ for all $a \in L$. Consequently, $\mu_x^* = \mu_0^*$. Also $\mu_x^* = \mu_0^*$ implies $\mu(x) = \mu(0)$. Now

$$\begin{aligned}\ker\ f =& \{x \in L \mid f(x) = \mu_0^*\} = \{x \in L \mid \mu_x^* = \mu_0^*\} \\ =& \{x \in L \mid \mu(x) = \mu(0)\} = U(\mu, \mu(0)).\end{aligned}$$

This completes the proof.

Theorem 1.22 *Given a homomorphism of a Lie algebra $f : L_1 \to L_2$ and μ is a fuzzy ideal of Lie algebra L_1 and ν of Lie algebra L_2 such that $F(u) \subseteq \nu$, There is a homomorphism of Lie algebra $g : L_1/\mu \to L_2/\nu$ where $g(\mu_x^*) = \nu_{f(x)}^*$ such that the diagram is commutative.*

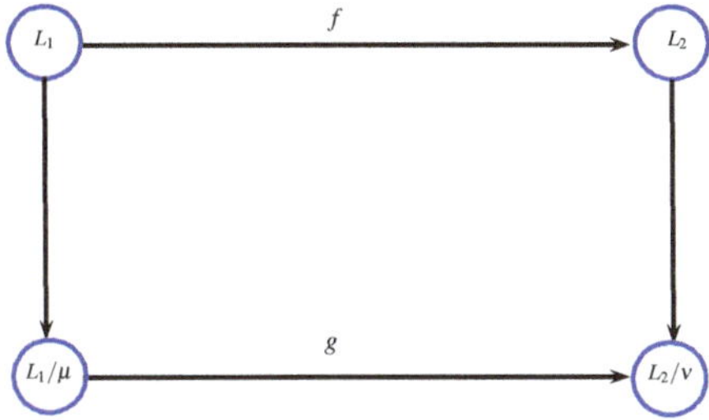

Proof Let $\mu_x^* = \mu_y^*$, then $\mu(x - y) = \mu(0)$. So

$$\begin{aligned}\nu(f(x) - f(y)) =& \nu(f(x - y)) = f^{-1}(\nu)(x - y), \\ =& \mu(x - y) = \mu(0)\end{aligned}$$

and so, $\nu(f(x) - f(y)) = \mu(0)$. Hence, $\nu(f(x)) = \nu(f(y))$ holds. Thus, g is well defined. It is easily seen that g is a homomorphism.

Theorem 1.23 (Fuzzy first isomorphism theorem) *Let $f : L_1 \to L_2$ be an epimorphism of Lie algebra and ν a fuzzy ideal of L_2. Then, $L/f^{-1}(\nu) \cong L_2/\nu$.*

Proof Define $g : L/f^{-1}(\nu) \to L_2/\nu$ by $g(f^{-1}(\nu)[x]) = \nu[f(x)]$. Suppose $f^{-1}(\nu)[x] = f^{-1}(\nu)[y]$. Then, $f^{-1}(\nu)(x-y) = f^{-1}(\nu)(0)$ and so $\nu(f(x) - f(y)) = \mu(0)$, that is, $\nu[f(x)] = \nu[f(y)]$. Hence, g is well defined. g is a homomorphism because

$$\begin{aligned} g(f^{-1}(\nu)[x] + f^{-1}(\nu)[y]) &= g(f^{-1}(\nu)[x+y]) \\ &= \nu[f(x+y)] = \nu[f(x) + f(y)] \\ &= \nu[f(x)] + \nu[f(y)] \\ &= g(f^{-1}(\nu)[x]) + g(f^{-1}(\nu)[y]) \\ g(\alpha f^{-1}(\nu)[x]) &= g(f^{-1}(\nu)[\alpha x]) \\ &= \nu[f(\alpha x)] = \nu[\alpha f(x)] \\ &= \alpha\nu[f(x)] \\ &= \alpha g(f^{-1}(\nu)[x]) \\ g([f^{-1}(\nu)[x], f^{-1}(\nu)[y]]) &= g(f^{-1}(\nu)[x, y]) \\ &= \nu[f([x, y])] = \nu[f(x), f(y)] \\ &= [\nu[f(x)], \nu[f(y)]] \\ &= [g(f^{-1}(\nu)[x]), g(f^{-1}(\nu)[y])]. \end{aligned}$$

Since f is an epimorphism, g is an epimorphism. Now, let $\nu[f(x)] = \nu[f(y)]$. Then, $\nu(f(x) - f(y)) = \nu(0)$. Therefore, $f^{-1}(\nu)(x-y) = f^{-1}(\nu)(0)$, and so g is one to one. Hence, $L_1/f^{-1}(\nu) \cong L_2/\nu$ and proof is complete.

Theorem 1.24 (Fuzzy second isomorphism theorem) *Let μ and ν be two fuzzy Lie ideals of a Lie algebra L with $\mu(0) = \nu(0)$. Then,* $\dfrac{L_\mu + L_\nu}{\nu} \cong \dfrac{L_\mu}{\mu \bigcap \nu}$.

Theorem 1.25 (Fuzzy third isomorphism theorem) *Let μ and ν be two fuzzy Lie ideals of a Lie algebra L with $\nu \subseteq \mu$ and $\mu(0) = \nu(0)$. Then,* $\dfrac{L/\nu}{L_\mu/\nu} \cong L/\mu$.

1.3 Anti-fuzzy Lie Ideals

Definition 1.31 Let L be a Lie algebra. A fuzzy subset λ of L is called an *anti-fuzzy Lie ideal* of L if the following axioms are satisfied:

(1) $\lambda(x+y) \le \max\{\lambda(x), \lambda(y)\}$,
(2) $\lambda(\alpha x) \le \lambda(x)$,
(3) $\lambda([x, y]) \le \lambda(x)$

for all $x, y \in L$ and $\alpha \in \mathbb{F}$.

Example 1.4 Let $\Re^2 = \{(x, y) : x, y \in \mathbb{R}\}$ be the set of all two-dimensional real vectors. Then, $\Re^2$ with $[x, y] = x \times y$ is a real Lie algebra. Define a fuzzy set of $\Re^2$ by

$$\lambda(x, y) = \begin{cases} 0 & \text{if } x = y = 0, \\ 1 & \text{otherwise.} \end{cases}$$

By routine computations, we can easily check that λ is an anti-fuzzy Lie ideal of $\Re^2$.

The following lemma is obvious.

Lemma 1.6 *Let λ be an anti-fuzzy Lie ideal of L, then*

(i) $\lambda(0) \leq \lambda(x) \quad \forall\, x \in L$,
(ii) $\lambda([x, y]) \leq \min\{\lambda(x), \lambda(y)\} \ \forall\, x, y \in L$,
(iii) $\lambda([x, y]) = \lambda(-[y, x]) = \lambda([y, x]) \ \forall x, y \in L$.

Theorem 1.26 *Let λ be an anti-fuzzy Lie ideal in a Lie algebra L. Then, λ is an anti-fuzzy Lie ideal of L if and only if the set $L(\lambda, t) = \{x \in L | \lambda(x) \leq t\}$, $t \in [0, 1]$, is a Lie ideal of L when it is nonempty.*

Proof Assume that λ is an anti-fuzzy Lie ideal of L, and let $t \in [0, 1]$ be such that $L(\lambda, t) \neq \emptyset$. Let $x, y \in L$ be such that $x \in L(\lambda, t)$, and $y \in L(\lambda, t)$. Then, $\lambda(x) \leq t$ and $\lambda(y) \leq t$. It follows that

$$\lambda(x + y) \leq \max\{\lambda(x), \lambda(y)\} \leq t,$$

$$\lambda(\alpha x) \leq \lambda(x) \leq t,$$

$$\lambda([x, y]) \leq \lambda(x) \leq t$$

so that $x + y \in L(\lambda, t)$, $\alpha x \in L(\lambda, t)$ and $[x, y] \in L(\lambda, t)$. Hence $L(\lambda, t)$ is a Lie ideal of L.

Conversely, suppose that $L(\lambda, t) \neq \emptyset$ is a Lie ideal of L for every $t \in [0, 1]$. Assume that $\lambda(x + y) > \max\{\lambda(x), \lambda(y)\}$ for some $x, y \in L$. Taking

$$t_0 := \frac{1}{2}\{\lambda(x + y) + \max\{\lambda(x) + \lambda(y)\}\},$$

we have $\lambda(x + y) > t_0 > \max\{\lambda(x), \lambda(y)\}$. So, $x + y \notin L(\lambda, t)$, $x \in L(\lambda, t)$ and $y \in L(\lambda, t)$. This is a contradiction. Hence, $\lambda(x + y) \leq \max\{\lambda(x), \lambda(y)\}$ for all $x, y \in L$.

Similarly, we can show that $\lambda(\alpha x) \leq \lambda(x)$ and $\lambda([x, y]) \leq \lambda(x)$. This completes the proof.

Theorem 1.27 *If λ and ρ are anti-fuzzy Lie ideals of a Lie algebra L, then the function $\lambda \vee \rho : L \to [0, 1]$ defined by*

$$(\lambda \vee \rho)(x) = \max\{\lambda(x), \rho(x)\}$$

is an anti-fuzzy Lie ideal of L.

Proof Let $x, y \in L$ and $\alpha \in \mathbb{F}$. Then,

$$\begin{aligned}(\lambda \vee \rho)(x+y) &= \max\{\lambda(x+y), \rho(x+y)\} \\ &\leq \max\{\max\{\lambda(x), \lambda(y)\}, \max\{\rho(x), \rho(y)\}\} \\ &= \max\{\max\{\lambda(x), \rho(x)\}, \max\{\lambda(y), \rho(y)\}\} \\ &= \max\{(\lambda \vee \rho)(x), (\lambda \vee \rho)(y)\},\end{aligned}$$

$$(\lambda \vee \rho)(\alpha x) = \max\{\lambda(\alpha x), \rho(\alpha x)\} \leq \max\{\lambda(x), \rho(x)\} = (\lambda \vee \rho)(x),$$

$$(\lambda \vee \rho)([x, y]) = \max\{\lambda([x, y]), \rho([x, y])\} \leq \max\{\lambda(x), \rho(x)\} = (\lambda \vee \rho)(x).$$

Hence, $(\lambda \vee \rho)$ is an anti-fuzzy Lie ideal of L.

Definition 1.32 For a family of fuzzy sets $\{\lambda_i | i \in I\}$ in a Lie algebra L, the *union* $\bigvee \lambda_i$ of $\{\lambda_i | i \in I\}$ is defined by

$$\left(\bigvee \lambda_i\right)(x) = \sup\{\lambda_i(x) | i \in I\},$$

for each $x \in L$.

Theorem 1.28 *If $\{\lambda_i | i \in I\}$ is a family of anti-fuzzy Lie ideals of Lie algebras L, then so is $\bigvee \lambda_i$.*

Proof Straightforward.

Theorem 1.29 *Let $f : L_1 \to L_2$ be an epimorphism of Lie algebras. If ν is an anti-fuzzy Lie ideal of L_2 and λ is the pre-image of ν under f, then λ is an anti-fuzzy Lie ideal of L_1.*

Proof For any $x, y \in L_1$ and $\alpha \in \mathbb{F}$,

$$\begin{aligned}\lambda(x+y) &= \nu(f(x+y)) = \nu(f(x)+f(y)) \\ &\leq \max\{\nu(f(x)), \nu(f(y))\} = \max\{\lambda(x), \lambda(y)\}, \\ \lambda(\alpha x) &= \nu(f(\alpha x)) = \nu(\alpha f(x)) \leq \nu(f(x)) = \lambda(x),\end{aligned}$$

and

$$\lambda([x, y]) = \nu(f([x, y])) \leq \nu(f(x)) = \lambda(x).$$

Hence, λ is an anti-fuzzy Lie ideal of L_1.

Definition 1.33 Let L_1 and L_2 be two Lie algebras over same field and f be a function of L_1 into L_2. If λ is a fuzzy set in L_2, then the *pre-image* of λ under f is the fuzzy set in L_1 defined by

$$f^{-1}(\lambda)(x) = \lambda(f(x)) \quad \forall x \in L_1.$$

Theorem 1.30 *Let $f : L_1 \to L_2$ be an onto homomorphism of Lie algebras over same field. If λ is an anti-fuzzy Lie ideal of L_2, then $f^{-1}(\lambda)$ is an anti-fuzzy Lie ideal of L_1.*

Proof Let $x_1, x_2 \in L_1$ and $\alpha \in \mathbb{F}$, then

$$\begin{aligned} f^{-1}(\lambda)(x_1 + x_2) &= \lambda(f(x_1) + f(x_2)) \leq \max\{\lambda(f(x_1)), \lambda(f(x_2))\}, \\ &= \max\{f^{-1}(\lambda)(x_1), f^{-1}(\lambda)(x_2)\}, \\ f^{-1}(\lambda)(\alpha x_1) &= \lambda(f(\alpha x_1)) \leq \lambda(\alpha f(x_1)) = \alpha f^{-1}(\lambda)(x_1), \\ f^{-1}(\lambda)([x, y]) &= \lambda(f([x, y])) = \lambda([f(x), f(y)]) \leq \lambda(f(x)) = f^{-1}(\lambda)(x). \end{aligned}$$

Hence, $f^{-1}(\lambda)$ is an anti-fuzzy Lie ideal of L_1.

Theorem 1.31 *Let $f : L_1 \to L_2$ be an onto homomorphism of Lie algebras over same field. If λ is an anti-fuzzy Lie ideal of L_2, then $f^{-1}(\lambda^c) = (f^{-1}(\lambda))^c$.*

Proof Let λ be an anti-fuzzy Lie ideal of L_2. Then, for $x \in L_1$,

$$f^{-1}(\lambda^c)(x) = \lambda^c(f(x)) = 1 - \lambda(f(x)) = 1 - f^{-1}(\lambda^c)(x) = (f^{-1}(\lambda))^c(x).$$

That is, $f^{-1}(\lambda^c) = (f^{-1}(\lambda))^c$.

Definition 1.34 Let λ be a fuzzy set in a Lie algebra L and f a mapping defined on L. Then, the fuzzy set λ^f in $f(L)$ defined by

$$\lambda^f(y) = \inf_{x \in f^{-1}(y)} \lambda(x)$$

for every $y \in f(L)$ is called the *image* of λ under f. A fuzzy set λ in L has the inf *property* if for any subset $A \subseteq L$, there exists $a_0 \in A$ such that $\lambda(a_0) = \inf_{a \in A} \lambda(a)$.

Theorem 1.32 *A Lie algebra homomorphism image of an anti-fuzzy Lie ideal having the* inf *property is an anti-fuzzy Lie ideal.*

Proof Let $f : L_1 \to L_2$ be an epimorphism of L_1 onto L_2 and λ be a fuzzy Lie ideal of L_1 with the inf property. Consider $f(x), f(y) \in f(L_1)$. Let $x_0, y_0 \in f^{-1}(f(x))$ be such that

$$\lambda(x_0) = \inf_{t \in f^{-1}(f(x))} \lambda(t) \quad \text{and} \quad \lambda(y_0) = \inf_{t \in f^{-1}(f(y))} \lambda(t),$$

respectively. Then,

$$\begin{aligned} \nu(f(x) + f(y)) &= \inf_{t \in f^{-1}(f(x)+f(y))} \lambda(t) \leq \lambda(x_0 + y_0) \leq \max\{\lambda(x_0) + \lambda(y_0)\}, \\ &= \max\{\inf_{t \in f^{-1}(f(x))} \lambda(t), \inf_{t \in f^{-1}(f(y))} \lambda(t)\}, \\ &= \max\{\nu(f(x)) + \nu(f(y))\}, \end{aligned}$$

$$\nu(f(\alpha x)) = \inf_{t\in f^{-1}(f(\alpha x))} \lambda(t) \leq \lambda(x_0) \leq \max\{\lambda(x_0)\} = \nu(f(x)),$$

$$\begin{aligned}\nu([f(x), f(y)]) = \nu(f([x, y])) &= \inf_{t\in f^{-1}(f([x,y]))} \lambda(t) \leq \lambda([x_0, y_0]),\\ &\leq \lambda(x_0) = \nu(f(x)).\end{aligned}$$

Consequently, ν is an anti-fuzzy Lie ideal of L_2.

Definition 1.35 Let L_1 and L_2 two be Lie algebras and f a function of λ is a fuzzy set in L_1, then the *anti-image* of λ under f is the fuzzy set defined by $f(\lambda)(y) =$

$$\begin{cases} \inf\{\lambda(t) \mid t \in L_1, f(t) = y\}, & \text{if } f^{-1}(y) \neq \emptyset, \\ 1, & \text{otherwise.} \end{cases}$$

Definition 1.36 Let L_1 and L_2 be any sets, and let $f : L_1 \to L_2$ be any function. A fuzzy set λ is called *f-invariant* if and only if for $x, y \in L_1$, $f(x) = f(y)$ implies $\lambda(x) = \lambda(y)$.

Theorem 1.33 *Let $f : L_1 \to L_2$ be an epimorphism of Lie algebras. Then, λ is an f-invariant anti-fuzzy Lie ideal of L_1 if and only if $f(\lambda)$ is an anti-fuzzy Lie ideal of L_2.*

Proof Let $x, y \in L_2$ and $\alpha \in \mathbb{F}$. Then, there exist $a, b \in L_1$ such that $f(a) = x$, $f(b) = y$, $x + y = f(a + b)$ and $\alpha x = \alpha f(a)$. Since λ is f-invariant,

$$\begin{aligned} f(\lambda)(x + y) &= \lambda(a + b) \leq \max\{\lambda(a), \lambda(b)\} = \max\{f(\lambda)(x), f(\lambda)(y)\}, \\ f(\lambda)(\alpha x) &= \lambda(\alpha a) \leq \lambda(a) = f(\lambda)(x), \\ f(\lambda)([x, y]) &= \lambda([a, b]) = [\lambda(a), \lambda(b)] \leq \lambda(a) = f(\lambda)(x). \end{aligned}$$

Hence, $f(\lambda)$ is an anti-fuzzy Lie ideal of L_2.

Conversely, if $f(\lambda)$ is an anti-fuzzy Lie ideal of L_2, then for any $x \in L_1$

$$\begin{aligned} f^{-1}(f(\lambda))(x) &= f(\lambda)(f(x)) = \inf\{\lambda(t) \mid t \in L_1, f(t) = f(x)\} \\ &= \inf\{\lambda(t) \mid t \in L_1, \lambda(t) = \lambda(x)\} = \lambda(x). \end{aligned}$$

Hence, $f^{-1}(f(\lambda)) = \lambda$ is an anti-fuzzy Lie ideal by Theorem 1.32.

Definition 1.37 An ideal J of Lie algebra L is said to be *characteristic* if $f(J) = J$, for all $f \in \text{Aut}(L)$, where $\text{Aut}(L)$ is the set of all automorphisms of L. Anti-fuzzy Lie ideal λ of Lie algebra L is said to be *anti-fuzzy characteristic* if $\lambda^f(x) = \lambda(x)$, for all $x \in L$ and $f \in\text{Aut}(L)$.

Lemma 1.7 *Let λ be an anti-fuzzy Lie ideal of a Lie algebra L, and let $x \in L$. Then, $\lambda(x) = s$ if and only if $x \in L(\lambda, s)$ and $x \notin L(\lambda, t)$, for all $s > t$.*

Proof Straightforward.

Theorem 1.34 *An anti-fuzzy Lie ideal is characteristic if and only if each its level set is a characteristic Lie ideal.*

Proof Suppose that λ is anti-fuzzy characteristic, and let $s \in Im(\lambda)$, $f \in$Aut(L) and $x \in L(\lambda, s)$. Then, $\lambda^f(x) = \lambda(x)$ implies $\lambda(f(x)) \leq s$ whence $f(x) \in L(\lambda, s)$. Thus, $f(L(\lambda, s)) \subseteq L(\lambda, s)$. Let $x \in L(\lambda, s)$ and $y \in L$ such that $f(y) = x$. Then, $\lambda(y) = \lambda^f(y) = \lambda(f(y)) = \lambda(x) \leq s$, consequently $y \in L(\lambda, s)$. So, $x = f(y) \in L(\lambda, s)$. Thus, $L(\lambda, s) \subseteq f(L(\lambda, s))$. Hence, $f(L(\lambda; s)) = L(\lambda, s)$, i.e., $L(\lambda, s)$ is characteristic.

Conversely, suppose that each level Lie ideal of λ is characteristic, and let $x \in L$, $f \in Aut(L)$, $\lambda(x) = s$. Then, by virtue of Lemma 1.7, $x \in L(\lambda, s)$ and $x \notin L(\lambda; t)$, for all $s > t$. It follows from the assumption that $f(x) \in f(L(\lambda, s)) = L(\lambda, s)$, so that $\lambda^f(x) = \lambda(f(x)) \leq s$. Let $t = \lambda^f(x)$, and assume that $s > t$. Then, $f(x) \in L(\lambda, t) = f(L(\lambda, t))$, which implies from the injectivity of f that $x \in L(\lambda, t)$, a contradiction. Hence, $\lambda^f(x) = \lambda(f(x)) = s = \lambda(x)$ showing that λ is an anti-fuzzy characteristic.

Definition 1.38 Let λ be an anti-fuzzy Lie ideal in L. Define a sequence of anti-fuzzy Lie ideals in L putting $\lambda^0 = \lambda$ and $\lambda^n = [\lambda^{n-1}, \lambda^{n-1}]$ for $n > 0$. If there exists a positive integer n such that $\lambda^n = 0$, then an anti-fuzzy Lie ideal λ is called *solvable*.

Theorem 1.35 *Homomorphic image of a solvable anti-fuzzy Lie ideal is a solvable anti-fuzzy Lie ideal.*

Proof Let $f : L_1 \to L_2$ be a homomorphism of Lie algebras. Suppose that λ is a solvable anti-fuzzy Lie ideal in L_1. We prove by induction on n that $f(\lambda^n) \supseteq [f(\lambda)]^n$, where n is any positive integer. First, we claim that $f([\lambda, \lambda]) \supseteq [f(\lambda), f(\lambda)]$. Let $y \in L_2$, then

$$\begin{aligned}
& f(\ll \lambda, \lambda \gg)(y) = \inf\{\ll \lambda, \lambda \gg (x) \mid f(x) = y\} \\
& = \inf\{\inf\{\max\{\lambda(a), \lambda(b)\} \mid a, b \in L_1, [a, b] = x, f(x) = y\}\} \\
& = \inf\{\max\{\lambda(a), \lambda(b)\} \mid a, b \in L_1, [a, b] = x, f(x) = y\} \\
& = \inf\{\max\{\lambda(a), \lambda(b)\} \mid a, b \in L_1, [f(a), f(b)] = x\} \\
& = \inf\{\max\{\lambda(a), \lambda(b)\} \mid a, b \in L_1, f(a) = u, f(b) = v, [u, v] = y\} \\
& \leq \inf\{\max\{\inf_{a \in f^{-1}(u)} \lambda(a), \inf_{b \in f^{-1}(v)} \lambda(b)\} \mid [u, v] = y\} \\
& = \inf\{\max(f(\lambda)(u), f(\lambda)(v)) \mid [u, v] = y\} = \ll f(\lambda), f(\lambda) \gg (y).
\end{aligned}$$

Now for $n > 1$, we get $f(\lambda^n) = f([\lambda^{n-1}, \lambda^{n-1}]) \supseteq [f(\lambda^{n-1}), f(\lambda^{n-1})] \supseteq [(f(\lambda))^{n-1}, (f(\lambda))^{n-1}] = (f(\lambda))^n$. This completes the proof.

Definition 1.39 Let λ be an anti-fuzzy Lie ideal in L, and let $\lambda_n = [\lambda, \lambda_{n-1}]$ for $n > 0$, where $\lambda_0 = \lambda$. If there exists a positive integer n such that $\lambda_n = 0$, then λ is called *nilpotent*.

Using the same method as in the proof of Theorem 1.35, we can prove the following two theorems.

Theorem 1.36 *Homomorphic image of a nilpotent anti-fuzzy Lie ideal is a nilpotent anti-fuzzy Lie ideal.*

Theorem 1.37 *If λ is a nilpotent anti-fuzzy Lie ideal, then it is solvable.*

Theorem 1.38 *Let J be a Lie ideal of a Lie algebra L. If λ is an anti fuzzy Lie ideal of L, then the fuzzy set $\overline{\lambda}$ of L/J defined by*

$$\overline{\lambda}(a+J) = \inf_{x \in J} \lambda(a+x)$$

is an anti-fuzzy Lie ideal of the quotient Lie algebra L/J.

Proof Clearly, $\overline{\lambda}$ is well defined. Let $x+J, y+J \in L/J$, then

$$\begin{aligned}
\overline{\lambda}(x+J)+(y+J)) = \overline{\lambda}_A((x+y)+J) &= \inf_{z \in J} \lambda((x+y)+z) \\
&= \inf_{z=s+t \in J} \lambda((x+y)+(s+t)) \\
&\leq \inf_{s,\, t \in J} \max\{\lambda(x+s), \lambda(y+t)\} \\
&= \max\{\inf_{s \in J} \lambda(x+s), \inf_{t \in J} \lambda(y+t)\} \\
&= \max\{\overline{\lambda}(x+J), \overline{\lambda}(y+J)\},
\end{aligned}$$

$$\overline{\lambda}(\alpha(x+J)) = \overline{\lambda}(\alpha x+J) = \inf_{z \in J} \lambda(\alpha x+z) \leq \inf_{z \in J} \lambda(x+z) = \overline{\lambda}(x+J),$$

$$\overline{\lambda}([x+J, y+J]) = \overline{\lambda}([x,y]+J) = \inf_{z \in J} \lambda([x,y]+z) \leq \inf_{z \in J} \lambda(x+z) = \overline{\lambda}(x+J).$$

Hence, $\overline{\lambda}$ is an anti-fuzzy Lie ideal of L/J.

1.4 Fuzzy Lie Sub-superalgebras

Definition 1.40 Let $V = V_{\overline{0}} \oplus V_{\overline{1}}$ be a $\mathbb{Z}_2$-graded vector space. Suppose that $\mu_{\overline{0}}$, $\mu_{\overline{1}}$ are fuzzy vector subspaces of $V_{\overline{0}}$, $V_{\overline{1}}$, respectively. Define

$$(\mu_{\overline{0}} \oplus \mu_{\overline{1}})(x) = \min\{\mu_{\overline{0}}(x_{\overline{0}}), \mu_{\overline{1}}(x_{\overline{1}})\},$$

where $x_{\overline{0}} \in V_{\overline{0}}$, $x_{\overline{1}} \in V_{\overline{1}}$. If μ is a fuzzy vector subspace of V and $\mu = \mu_{\overline{0}} \oplus \mu_{\overline{1}}$, then μ is called a *$\mathbb{Z}_2$-graded fuzzy subspace*.

Definition 1.41 Let μ be a fuzzy subset of Lie superalgebra $\mathscr{L}$. Then, μ is called a *fuzzy Lie sub-superalgebra* of $\mathscr{L}$, if it satisfies the following conditions:

(1) μ is a $\mathbb{Z}_2$-graded fuzzy subspace,
(2) $\mu([x,y]) \geq \min\{\mu(x), \mu(y)\}$, for any $x, y \in \mathscr{L}$,

If the condition (2) is replaced by

(3) $\mu([x, y]) \geq \max\{\mu(x), \mu(y)\}$,

then μ is called a *fuzzy Lie ideal* of $\mathscr{L}$.

Example 1.5 Let $\mathscr{L} = \mathscr{L}_0 \oplus L_1$ be a Lie superalgebra, where $\mathscr{L}_0 = \mathfrak{R}^3$, $\mathscr{L}_1 = 0$ and for any $x, y \in \mathscr{L}_0$, $[x, y] = x \times y$, where $\times$ is the cross-product, other elements of the bracket product is 0. The definition of $\mu_{\bar{0}} : \mathscr{L}_{\bar{0}} \to [0, 1]$ is

$$\mu_{\bar{0}}(x, y, z) = \begin{cases} 1.0, \ x = y = z = 0, \\ 0.6, \ x \neq 0, \ y = z = 0, \\ 0.0, \ \text{otherwise}. \end{cases}$$

Define $\mu_{\bar{1}} : \mathscr{L}_{\bar{1}} \to [0, 1]$ as $\mu_{\bar{1}}(x) = 1$. Then, extend $\mu'_{\bar{0}} : \mathscr{L} \to [0, 1]$ is

$$\mu'_{\bar{0}}(x) = \begin{cases} \mu_{\bar{0}}(x), \ x \in \mathscr{L}_{\bar{0}}, \\ 0, \qquad x \notin \mathscr{L}_{\bar{0}}, \end{cases}$$

$\mu'_{\bar{1}} : \mathscr{L} \to [0, 1]$ is

$$\mu'_{\bar{1}}(x) = \begin{cases} \mu_{\bar{1}}(x), \ x \subset \mathscr{L}_{\bar{1}}, \\ 0, \qquad x \notin \mathscr{L}_{\bar{1}}, \end{cases}$$

$\mu : \mathscr{L} \to [0, 1]$ define by $\mu(x) = \mu'_{\bar{0}}(x)$ $\mu = \mu_{\bar{0}} \oplus \mu_{\bar{1}}$
$(\mu'_{\bar{0}} \oplus \mu'_{\bar{1}})(x) = \sup_{x=a+b} \{\min(\mu'_{\bar{0}}(a), \mu'_{\bar{1}}(b))\} = \min(\mu'_{\bar{0}}(x_{\bar{0}}), \mu'_{\bar{1}}(x_{\bar{1}})) = \mu(x)$,
$\mu'_{\bar{0}} \cap \mu'_{\bar{1}} = 1_{\bar{0}}$.

Thus, μ is a fuzzy superalgebra. However, μ is not a fuzzy ideal of $\mathscr{L}$ because

$$\mu([(0, 1, 0), (1, 1, 1)]) = \mu_{\bar{0}}([(0, 1, 0), (1, 1, 1)]) = \mu_{\bar{0}}((0, -1, 1)) = 0.$$

However,

$$\max\{\mu(1, 0, 0), \mu(1, 1, 1)\} = \max\{\mu_{\bar{0}}(1, 0, 0), \mu_{\bar{0}}(1, 1, 1)\} = 0.6.$$

Let μ be the fuzzy ideal of $\mathscr{L}$. Define the fuzzy subset

$$x + \mu : \mathscr{L} \to [0, 1] \text{ by } (x + \mu)(y) = \mu(y - x),$$

which is fuzzy ideal μ.

Following proportions are trivial; hence, we omit their proofs.

Proposition 1.6 *Let μ be the fuzzy ideal of $\mathscr{L}$. Then, for any $x, y \in \mathscr{L}$, there is $x + \mu = y + \mu \Leftrightarrow \mu(x - y) = \mu(0)$.*

Proposition 1.7 *Let μ be the fuzzy ideal of $\mathscr{L}$, then $\mathscr{L}/\mu \cong \mathscr{L}/\mu_0$, where $\mu_0 = \{x \in L | \mu(u) = \mu(0)\}$ is the ideal of $\mathscr{L}$.*

Proposition 1.8 *Let μ be the fuzzy ideal of $\mathscr{L}$, then $\tau = \mathscr{L}/\mu_0$ is fuzzy ideal of $\mathscr{L}/\mu$.*

1.5 Hesitant Fuzzy Lie Ideals

An extension of fuzzy sets so-called hesitant fuzzy sets was introduced by Torra [125], to deal with hesitant situations, which were not well managed by the previous tools. The hesitant fuzzy set permits membership degree of an element to be a set of several possible values between 0 and 1. This situation is very usual in decision making, when an expert might consider different degrees of membership $\{0.67, 0.72, 0.74\}$ of the element x in the set X. The desired benefits of using the hesitant fuzzy sets are: (1) It is more convenient to express the uncertainty of information by using a set of possible values in the qualitative evaluation process; (2) the expression form of the hesitant fuzzy set is consistent with the decision makers subjective evaluation. A hesitant fuzzy set is defined in terms of a function that returns a set of membership values for each element in the domain.

Definition 1.42 Let X be a reference set, a *hesitant fuzzy set* on X is a function h that returns a subset of values in [0, 1] :

$$h : X \to H([0, 1]).$$

A hesitant fuzzy set can also be constructed from a set of fuzzy sets.

Definition 1.43 Let $M = \{\mu_1, \ldots, \mu_n\}$ be a set of n membership functions. The *hesitant fuzzy set* h_M associated with M, is defined as

$$h_M : X \to H([0, 1]),$$

$$h_M(x) = \bigcup_{\mu \in M} \{\mu(x)\}, \quad where\ x \in X.$$

It is remarkable that this definition is quite suitable to decision making, when experts have to assess a set of alternatives. In such a case, M represents the assessments of the experts for each alternative and h_M the assessments of the set of experts. Afterward, Xia and Xu [129] completed the original definition of hesitant fuzzy set by including the mathematical representation of a hesitant fuzzy set as follows:

$$E = \{\langle x, h_M(x)\rangle \mid \forall x \in X\},$$

where $h_M(x)$ is a set of some different values in [0, 1], representing the possible membership degrees of the element $x \in X$ to set E, and called $h = h_M(x)$ a hesitant fuzzy element and $H = \bigcup h_M(x)$, the set of all hesitant fuzzy elements of E. In some papers, the concepts hesitant fuzzy set and hesitant fuzzy element are used indistinctively, even though both concepts are different. A hesitant fuzzy set is a set of subsets in the interval [0, 1], one set for each element of the reference set X. A hesitant fuzzy element is one of such sets, the one for a particular $x \in X$.

Example 1.6 Let $X = \{x_1, x_2, x_3\}$ be a reference set. Also, let $h_M(x_1) = \{0.4, 0.5\}$, $h_M(x_2) = \{0.2, 0.3, 0.5\}$, $h_M(x_3) = \{0.3, 0.7\}$ denote the membership degree sets of $x_i (i = 1, 2, 3)$ to the set E, respectively. Then, E is a hesitant fuzzy set, namely

$$E = \{\langle x_1, \{0.4, 0.5\}\rangle \langle x_2, \{0.2, 0.3, 0.5\}\rangle \langle x_3, \{0.3, 0.7\}\rangle\},$$

where, $h_1 = \{0.4, 0.5\}$, $h_2 = \{0.2, 0.3, 0.5\}$, $h_3 = \{0.3, 0.7\}$ are hesitant fuzzy elements.

Definition 1.44 Let X be a reference set, then the empty hesitant set (h_o), the full hesitant set (h_1), the set to represent complete ignorance of x $(h_{[o,1]})$, and nonsense set (h_ϕ) are defined as follows:

- empty hesitant fuzzy set: $h_M(x) = \{0\}$ for all $x \in X$,
- full hesitant fuzzy set: $h_M(x) = \{1\}$ for all $x \in X$,
- complete ignorance for $x \in X$ (all is possible): $h_M(x) = [0, 1]$,
- set for a nonsense x: $h_M(x) = \phi$ for all $x \in X$.

Some operations on hesitant fuzzy set can be described as follows:

1. Lower bound: $h^-(x) = \min h(x)$,
2. Upper bound: $h^+(x) = \max h(x)$,
3. t-lower bound: $h_t^-(x) = \{\lambda \in h(x) \mid \lambda \leq t\}$,
4. t-upper bound: $h_t^+(x) = \{\lambda \in h(x) \mid \lambda \geq t\}$,
5. Complement: $h^c(x) = \{1 - \lambda \mid \lambda \in h(x)\}$,
6. Union: $(h_1 \cup h_2)(x) = \{\lambda \in h_1(x) \cup h_2(x) \mid \lambda \geq \max\{(h_1)^-(x), (h_2)^-(x)\}\}$,
7. Intersection: $(h_1 \cap h_2)(x) = \{\lambda \in h_1(x) \cap h_2(x) \mid \lambda \leq \min\{(h_1)^+(x), (h_2)^+(x)\}\}$.

Example 1.7 Let $X = \{x_1, x_2, x_3\}$ be the reference set, hesitant fuzzy elements h_1 and h_2 on X be $h_1 = \{\langle x_1, \{0.3, 0.4\}\rangle, \langle x_2, \{0.6, 0.8\}\rangle, \langle x_3, \{0.3, 0.4, 0.5, 0.7\}\rangle\}$ and $h_2 = \{\langle x_1, \{0.5, 0.6\}\rangle, \langle x_2, \{0.4, 0.5\}\rangle, \langle x_3, \{0.2, 0.3, 0.4, 0.6\}\rangle\}$, respectively. Then, we have

1. $(h_1)^-(x_1) = \min\{0.3, 0.4\} = 0.3$,
2. $(h_1)^+(x_1) = \max\{0.3, 0.4\} = 0.4$,
3. $(h_1)_{0.4}^-(x_1) = \{\lambda \in h_1(x_1) \mid \lambda \leq 0.4\} = \{0.3, 0.4\}$,
4. $(h_1)_{0.3}^+(x_1) = \{\lambda \in h_1(x_1) \mid \lambda \geq 0.3\} = \{0.3, 0.4\}$,
5. $(h_1)_{0.45}^-(x_3) = \{\lambda \in h_1(x_3) \mid \lambda \leq 0.45\} = \{0.3, 0.4\}$,
6. $(h_1)_{0.45}^+(x_3) = \{\lambda \in h_1(x_3) \mid \lambda \geq 0.45\} = \{0.5, 0.7\}$,
7. $(h_1)^c(x_2) = \bigcup_{\lambda \in h_1(x_2)}\{1 - \lambda\} = \{1 - 0.6, 1 - 0.8\} = \{0.4, 0.2\}$,
8. $(h_1 \cup h_2)(x_3) = \{\lambda \in h_1(x_3) \cup h_2(x_3) \mid \lambda \geq \max\{(h_1)^-(x_3), (h_2)^-(x_3)\}\} = \{\lambda \in h_1(x_3) \cup h_2(x_3) \mid \lambda \geq \max\{0.3, 0.2\}\} = \{0.3, 0.4, 0.5, 0.6, 0.7\}$,

9. $(h_1 \cap h_2)(x_3) = \{\lambda \in h_1(x_3) \cap h_2(x_3) \mid \lambda \leq \min\{(h_1)^+(x), (h_2)^+(x)\}\}$
$= \{\lambda \in h_1(x_3) \cap h_2(x_3) \mid \lambda \leq \min\{0.7, 0.6\}\} = \{0.2, 0.3, 0.4, 0.5, 0.6\}.$

We now define the concept of hesitant fuzzy subspace.

Definition 1.45 Let V be a vector space over a field $\mathbb{F}$. A hesitant fuzzy set h on V is called a *hesitant fuzzy subspace* of V if the following conditions are satisfied:

(i) $h(x+y) \supseteq h(x) \cap h(y)$ for all $x, y \in V$,
(ii) $h(\alpha x) \supseteq h(x)$ for all $x \in V$, $\alpha \in \mathbb{F}$.

Note that by (ii), we obtain $h(-x) \supseteq h(x)$ and $h(0) \supseteq h(x)$ for all $x \in V$.

Lemma 1.8 *If h is a hesitant fuzzy subspace of a vector space V, then*

(1) $h(x) = h(-x)$,
(2) $h(x-y) = h(0) \Longrightarrow h(x) = h(y)$,
(3) $h(x) \subset h(y) \Longrightarrow h(x-y) = h(x) = h(y-x)$

for all $x, y \in V$.

Definition 1.46 A hesitant fuzzy set h, i.e., a map $h : L \to H([0, 1])$, is called a *hesitant fuzzy Lie subalgebra* of L over a field $\mathbb{F}$ if it is a hesitant fuzzy subspace of L such that

(iii) $h([x, y]) \geq h(x) \cap h(y)$

hold for all $x, y \in L$ and $\alpha \in \mathbb{F}$.

Example 1.8 The real vector space $\Re^3$ with $[x, y] = x \times y$, where $x, y \in \Re^3$, is a real Lie algebra. Define a hesitant fuzzy set h on $\Re^3$ by

$$h(x) = \begin{cases} \{0.9, 0.8, 0.6\}, & \text{if } x = (0, 0, 0), \\ \{0.6, 0.4, 0.3\}, & \text{if } x = (c, 0, 0), \quad c \neq 0, \\ \{0.1, 0.2, 0.2\}, & \text{otherwise.} \end{cases}$$

By direct calculations, it is easy to see that h is a hesitant fuzzy Lie subalgebra.

Definition 1.47 A hesitant fuzzy set $h : L \to [0, 1]$ is called a *hesitant fuzzy Lie ideal* of L if

(1) $h(x+y) \supseteq h(x) \cap h(y)$,
(2) $h(\alpha x) \supseteq h(x)$,
(3) $h([x, y]) \supseteq h(x)$

hold for all $x, y \in L$ and $\alpha \in \mathbb{F}$.

Proposition 1.9 *Every hesitant fuzzy Lie ideal is a hesitant fuzzy Lie subalgebra.*

The converse of Proposition 1.9 is not true, in general. The hesitant fuzzy set h defined in Example 1.8 is a hesitant fuzzy Lie subalgebra, but it is not a hesitant fuzzy Lie ideal.
The following Lemma is obvious.

Lemma 1.9 *Let h be a hesitant fuzzy Lie ideal of L, then*

(1) $h(0) \supseteq h(x)$,
(2) $h([x, y])) \supseteq h(x) \cup h(y)$,
(3) $h([x, y]) = h(-[y, x]) = h([y, x])$,
(4) $h(x - y) = h(0) \Rightarrow h(x) = h(y)$,
(5) $h(x - y) = h(x) = h(y - x)$, *if* $h(x) \subset h(y)$ *for all* $x, y \in L$.

Theorem 1.39 *Let* h_1 *and* h_2 *be two hesitant fuzzy Lie ideal of L. Then,* $h_1 \bigcap h_2 : L \to H([0, 1])$ *and* $h_1 + h_2 : L \to H([0, 1])$ *are hesitant fuzzy Lie ideals of L.*

Definition 1.48 Let L_1 and L_2 be two Lie algebras and f a function of L_1 into L_2. If h is a hesitant fuzzy set in L_2, then the *pre-image* of h under f is the hesitant fuzzy set in L_1 defined by

$$f^{-1}(h)(x) = h(f(x)) \qquad \forall\, x \in L_1.$$

Equivalently, if h is a hesitant fuzzy set in $f(L_1)$, then the *pre-image* of h under f is the hesitant fuzzy set h_1 in L_1 defined by

$$h_1(x) = h(f(x)) \qquad \forall\, x \in L_1.$$

Theorem 1.40 *Let* $f : L_1 \to L_2$ *be an epimorphism of Lie algebras. If* h_1 *is a hesitant fuzzy Lie ideal of* L_2 *and h is the pre-image of* h_1 *under* f*. Then, h is a hesitant fuzzy Lie ideal of* L_1.

Proof For any $x, y \in L_1$ and $\alpha \in \mathbb{F}$,

$$\begin{aligned} h(x + y) &= h_1(f(x + y)) = h_1(f(x) + f(y)) \\ &\supseteq h_1(f(x)) \cap h_1(f(y)) = h(x) \cap h(y), \\ h(\alpha x) &= h_1(f(\alpha x)) = h_1(\alpha f(x)) \supseteq h_1(f(x)) = h(x), \\ h([x, y]) &= h_1(f([x, y])) \supseteq h_1(f(x)) = h(x). \end{aligned}$$

Hence, h is a hesitant fuzzy Lie ideal of L_1.

Theorem 1.41 *Let* $f : L_1 \to L_2$ *be an onto homomorphism of Lie algebras. If h is a hesitant fuzzy Lie ideal of* L_2*, then* $f^{-1}(h)$ *is a hesitant fuzzy Lie ideal of* L_1.

Proof For any $x, y \in L_1$ and $\alpha \in \mathbb{F}$,

$$\begin{aligned} f^{-1}(h)(x + y) &= h(f(x + y)) = h(f(x) + f(y)) \\ &\supseteq h(f(x)) \cap h(f(y)) = f^{-1}(h)(x) \cap f^{-1}(h)(y), \\ f^{-1}(h)(\alpha x) &= h(f(\alpha x)) = h(\alpha f(x)) \supseteq h(f(x)) = f^{-1}(h)(x) \\ f^{-1}(h)([x, y]) &= h(f([x, y])) \supseteq h(f(x)) = f^{-1}(h)(x). \end{aligned}$$

Hence, h is a hesitant fuzzy Lie ideal of L_1.

Definition 1.49 Let L_1 and L_2 be two sets, and let $f : L_1 \to L_2$ be any function. A hesitant fuzzy set h is called *f-invariant* if and only if for $x, y \in L_1$ $f(x) = f(y)$ implies $h(x) = h(y)$.

Theorem 1.42 *Let $f : L_1 \to L_2$ be an epimorphism of Lie algebras. Then, h is an f-invariant hesitant fuzzy Lie ideal of L_1 if and only if $f(h)$ is a hesitant fuzzy Lie ideal of L_2.*

Proof Let $x, y \in L_2$ and $\alpha \in \mathbb{F}$. Then, there exist $a, b \in L_1$ such that $f(a) = x$, $f(b) = y$, $x + y = f(a + b)$ and $\alpha x = \alpha f(a)$. Since h is f-invariant,

$$
\begin{aligned}
f(h)(x + y) &= h(a + b) \supseteq h(a) \cap (b) = f(h)(x) \cap f(h)(y),\\
f(h)(\alpha x) &= h(\alpha a) \supseteq h(a) = f(h)(x),\\
f(h)([x, y]) &= h([a, b]) = [h(a), h(b)] \supseteq h(a) = f(h)(x).
\end{aligned}
$$

Hence, $f(h)$ is a hesitant fuzzy Lie ideal of L_2.

Conversely, if $f(h)$ is a hesitant fuzzy Lie ideal of L_2, then for any $x \in L_1$,

$$
\begin{aligned}
f^{-1}(f(h))(x) &= f(h)(f(x)) = \bigcup\{h(t) \mid t \in L_1, f(t) = f(x)\}\\
&= \bigcup\{h(t) \mid t \in L_1, h(t) = h(x)\} = h(x).
\end{aligned}
$$

Hence, $f^{-1}(f(h)) = h$ is a hesitant fuzzy Lie ideal by Theorem 1.41.

Chapter 2
Intuitionistic Fuzzy Lie Ideals

In this chapter, we present certain concepts, including intuitionistic fuzzy Lie subalgebras, Lie homomorphisms, intuitionistic fuzzy Lie ideals, special types of intuitionistic fuzzy Lie ideals, intuitionistic (S, T)-fuzzy Lie ideals, nilpotency of intuitionistic (S, T)-fuzzy Lie ideals, and intuitionistic (S, T)-fuzzy Killing form.

2.1 Introduction

Atanassov [29] introduced the concept of *intuitionistic fuzzy sets* in 1983 as a generalization of fuzzy sets. Atanassov added a new component (which determines the degree of nonmembership) in the definition of fuzzy set. The fuzzy sets give the degree of membership of an element in a given set (and the nonmembership degree equals one minus the degree of membership), while intuitionistic fuzzy sets give both a degree of membership and a degree of nonmembership which are more-or-less independent from each other, and the only requirement is that the sum of these two degrees is not greater than 1. Intuitionistic fuzzy sets have been applied in a wide variety of fields including computer science, engineering, mathematics, medicine, chemistry, and economics. Intuitionistic fuzzy sets are also defined by Takeuti and Titanti in [124]. Takeuti and Titanti considered intuitionistic fuzzy logic in the narrow sense and derived a set theory from logic which they called intuitionistic fuzzy set theory.

Definition 2.1 A mapping $A = (\mu_A, \lambda_A) : X \to [0, 1] \times [0, 1]$ is called an *intuitionistic fuzzy set* on X if $\mu_A(x) + \lambda_A(x) \leq 1$ for all $x \in X$, where the mappings $\mu_A : X \to [0, 1]$ and $\lambda_A : X \to [0, 1]$ denote the *degree of membership* (namely $\mu_A(x)$) and the *degree of nonmembership* (namely $\lambda_A(x)$) of each element $x \in X$ to A, respectively.

An intuitionistic fuzzy set A in X can be represented as an object of the form

$$A = (\mu_A, \lambda_A) = \{(x, \mu_A(x), \lambda_A(x)) \mid x \in X\},$$

M. Akram, *Fuzzy Lie Algebras*, Infosys Science Foundation Series,
https://doi.org/10.1007/978-981-13-3221-0_2

where the functions

$$\mu_A : X \to [0, 1]$$

and

$$\lambda_A : X \to [0, 1]$$

denote the *degree of membership* (namely $\mu_A(x)$) and the *degree of nonmembership* (namely $\lambda_A(x)$) of the element $x \in X$, respectively, and for all $x \in X$

$$0 \leq \mu_A(x) + \lambda_A(x) \leq 1.$$

Obviously, each fuzzy set may be written as

$$A = \{(x, \mu_A(x), 1 - \mu_A(x)) \mid x \in X\}.$$

The value

$$\pi_A(x) = 1 - \mu_A(x) - \lambda_A(x) \tag{2.1}$$

is called *uncertainty (intuitionistic index)* of the elements $x \in X$ to the intuitionistic fuzzy sets A. It represents hesitancy degree of x to A.

Clearly, in the case of ordinary fuzzy set, $\pi_A(x) = 0$ for all $x \in X$.

Geometrical Interpretations of an Intuitionistic Fuzzy Set (IFS) [31]

A geometrical interpretation of an intuitionistic fuzzy set is shown in Fig. 2.1. Atanassov considers a universe X and subset F in the Euclidean plane with the Cartesian coordinates.

This geometrical interpretation can be used as an example when considering a situation at the beginning of negotiations (applications of intuitionistic fuzzy sets for group decision making, negotiations, and other real situations are presented in Fig. 2.2). Each expert i is represented as a point having coordinates $\langle \mu_i, \lambda_i, \pi_i \rangle$. Expert $A : \langle 1, 0, 0 \rangle$—fully accepts a discussed idea. Expert $B : \langle 0, 1, 0 \rangle$—fully rejects it. The experts placed on the segment *AB* fixed their point of view (their hesitation margins equal zero for segment *AB*, so each expert is convinced to the extent μ_i, are against the extent λ_i and $\mu_i + \lambda_i = 1$; segment *AB* represents a fuzzy set). Expert $C : \langle 0, 0, 1 \rangle$ is absolutely hesitant, i.e., undecided, he or she is the most open to the influence of the arguments presented. A line parallel to *AB* describes a set of experts with the same level of hesitancy. For example, in Fig. 2.2, two sets are presented with intuitionistic indices equal to π_m and π_n, where $\pi_n > \pi_m$. In other words, Fig. 2.2 (the triangle *ABC*) is an orthogonal projection of the real situation (the triangle *ABD*) presented in Fig. 2.3.

An element of an intuitionistic fuzzy sets has three coordinates $\langle \mu_i, \lambda_i, \pi_i \rangle$; hence, the most natural representation of an intuitionistic fuzzy set is to draw a cube (with

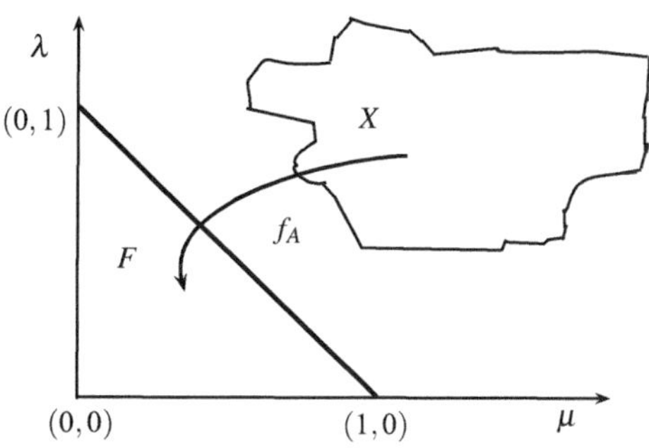

Fig. 2.1 A geometrical interpretation of an IFS

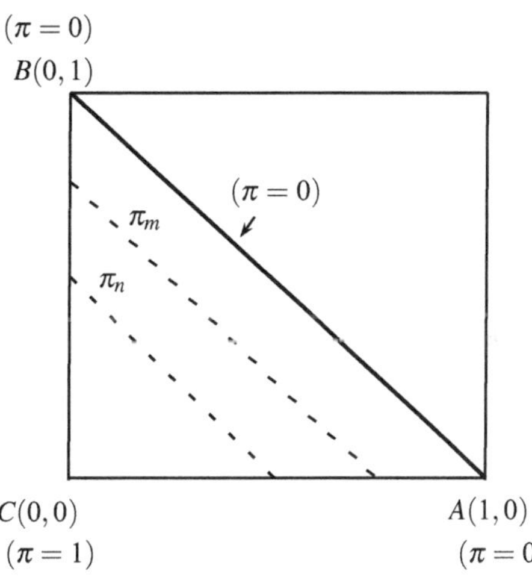

Fig. 2.2 An orthogonal projection of the real (three dimensions) representation (triangle ABD in Fig. 2.3) of an IFS

edge length equal to 1) and because of Equation (2.1), the triangle *ABD* (Fig. 2.3) represents an intuitionistic fuzzy set. As before (Fig. 2.2), the triangle *ABC* is the orthogonal projection of *ABD*.

Definition 2.2 We use $0_\sim$ and $1_\sim$ to denote the *intuitionistic fuzzy empty set* and the *intuitionistic fuzzy whole set* in a set X such that $0_\sim(x) = (0, 1)$ and $1_\sim(x) = (1, 0)$, for each $x \in X$, respectively.

Definition 2.3 For every two intuitionistic fuzzy sets $A = (\mu_A, \lambda_A)$ and $B = (\mu_B, \lambda_B)$ in X, the following relations and operations hold:

- $\overline{A} = (\lambda_A, \ \ \mu_A)$,
- $A \subseteq B \longleftrightarrow \mu_A(x) \le \mu_B(x)$ and $\lambda_A(x) \ge \lambda_B(x)$ for all $x \in X$,
- $A = B \longleftrightarrow A \subseteq B$ and $B \subseteq A$,
- $A + B = (\mu_A + \mu_B - \mu_A.\mu_B, \ \ \lambda_A.\lambda_B)$,
- $A \cdot B = (\mu_A.\mu_B, \ \ \lambda_A + \lambda_B - \lambda_A.\lambda_B)$,
- $A \bigcap B = (\mu_A \cap \mu_B, \ \ \lambda_A \cup \lambda_B)$,
- $A \bigcup B = (\mu_A \cup \mu_B, \ \ \lambda_A \cap \lambda_B)$,

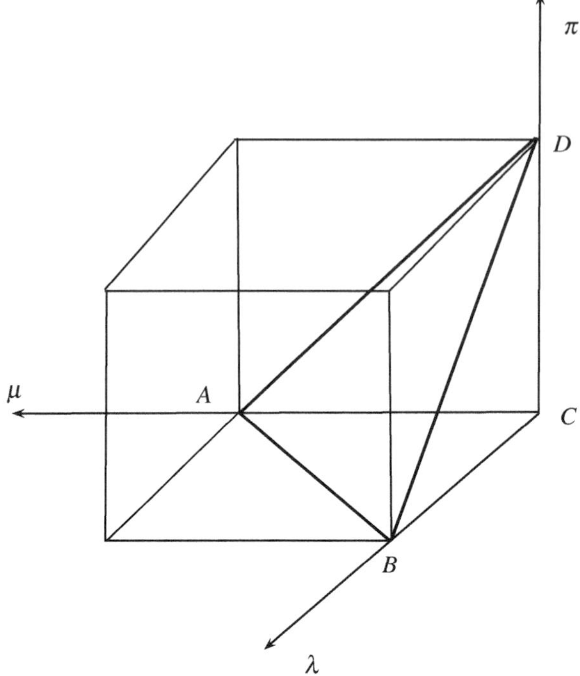

Fig. 2.3 A three-dimensional representation of an IFS

- $\Box A = (\mu_A, \ \overline{\mu}_A)$,
- $\Diamond A = (\overline{\lambda}_A, \ \lambda_A)$.

Remark 2.1 The following two laws do not hold in intuitionistic fuzzy sets:

(1) The law of excluded middle is not valid in intuitionistic fuzzy set theory, that is,

$$A \cup \overline{A} \neq 1_{\sim}$$

(2) The law of contradiction is not valid in intuitionistic fuzzy set theory, that is,

$$A \cap \overline{A} \neq 0_{\sim}$$

Definition 2.4 For $s, t \in [0, 1]$, the set $U(\mu_A, s) = \{x \in X \mid \mu_A(x) \geq s\}$ is called *upper level* of μ_A. The set $L(\lambda_A, t) = \{x \in X \mid \lambda_A(x) \leq t\}$ is called *lower level* of λ_A.

Definition 2.5 Let $A = (\mu_A, \lambda_A)$ be an intuitionistic fuzzy set on X. For $s, t \in [0, 1]$ with $s + t \leq 1$,

(i) the set $A^{(s,t)} := \{x \in X \mid s \leq \mu_A(x), \ \lambda_A(x) \leq t\}$ is called an *(s, t)-level subset* of A.

The set of all $(s, t) \in \mathrm{Im}(\mu_A) \times \mathrm{Im}(\lambda_A)$ such that $s + t \leq 1$ is called the *image of* $A = (\mu_A, \lambda_A)$.

(ii) the set $A^{(s,t)} := \{x \in X \mid s < \mu_A(x),\ \lambda_A(x) < t\}$ is called a strong (s, t)-*level subset* of A.

Note that

$$\begin{aligned} A^{(s,t)} &= \{x \in X \mid \mu_A(x) \geq s,\ \ \lambda_A(x) \leq t\} \\ &= \{x \in X \mid \mu_A(x) \geq s\} \cap \{x \in X \mid \lambda_A(x) \leq t\} \\ &= U(\mu_A, s) \cap L(\lambda_A, t). \end{aligned}$$

Definition 2.6 Let X and Y be two universes and let

$$A = \{(x, \mu_A(x), \lambda_A(x)) \mid x \in X\},$$

$$B = \{(x, \mu_B(x), \lambda_B(x)) \mid x \in Y\}$$

be two intuitionistic fuzzy sets over X and Y, respectively. Then we define only two versions of Cartesian product of the intuitionistic fuzzy sets.

(i) $A \times_1 B = \{< (x, y), \mu_A(x).\mu_B(y), \lambda_A(x).\lambda_B(y) > |x \in X\ \&\ y \in Y\}$.
(ii) $A \times_2 B = \{<(x, y), \min(\mu_A(x), \mu_B(y)), \max(\lambda_A(x), \lambda_B(y)) > |x \in X\ \&\ y \in Y\}$.

Definition 2.7 An intuitionistic fuzzy relation $R = (\mu_R(x, y), \lambda_R(x, y))$ in a universe $X \times Y$ ($R(X \to Y)$, for short) is an intuitionistic fuzzy set of the form

$$R = \{< (x, y), \mu_A(x, y), \nu_A(x, y) > |(x, y) \in X \times Y\},$$

where $\mu_A : X \times Y \to [0, 1]$ and $\lambda_A : X \times Y \to [0, 1]$. The intuitionistic fuzzy relation R satisfies $\mu_R(x, y) + \lambda_R(x, y) \leq 1$ for all $x, y \in X$.

Definition 2.8 Let R be an intuitionistic fuzzy relation on universe X. Then R is called *an intuitionistic fuzzy equivalence relation* on X if it satisfies the following conditions:

(a) R is intuitionistic fuzzy reflexive, i.e., $R(x, x) = (1, 0)$ for each $x \in X$,
(b) R is intuitionistic fuzzy symmetric, i.e., $R(x, y) = R(y, x)$ for any $x, y \in X$,
(c) R is intuitionistic fuzzy transitive, i.e., $R(x, z) \geq \bigvee_y (R(x, y) \bigwedge R(y, z))$.

Definition 2.9 Let $Q(X \to Y)$ and $R(Y \to Z)$ be two intuitionistic fuzzy relations. The *max-min-max composition* $R \circ Q(X \to Z)$ is the intuitionistic fuzzy relation defined by the membership function

$$\mu_{R \circ Q}(x, z) = \bigvee_y (\mu_Q(x, y) \wedge \mu_R(y, z))$$

and the nonmembership function

$$\lambda_{R\circ Q}(x, z) = \bigwedge_y (\lambda_Q(x, y) \vee \lambda_R(y, z))$$

for all $(x, z) \in X \times Z$ and for all $y \in Y$.

Remark 2.2 Let R, R_1, and R_2 be intuitionistic fuzzy relations on a nonempty set X.

- If R is symmetric, then so is R^{-1}.
- R is symmetric if and only if $R^{-1} = R$.
- If R_1 and R_2 are symmetric relations on X, then $R_1 \cup R_2$, $R_1 \cap R_2$, R_1^C, $R_1 + R_2$, $R_1 \cdot R_2$ are symmetric.
- If R_1 and R_2 are reflexive relations on X, then $R_1 \cup R_2$, $R_1 \cap R_2$, R_1^C, $R_1 + R_2$, $R_1 \cdot R_2$ are reflexive.
- If R_1 and R_2 are transitive relations on X, then $R_1 \cup R_2$, $R_1 \cap R_2$, R_1^C, $R_1 + R_2$, $R_1 \cdot R_2$ are transitive.
- If R_1 and R_2 are reflexive (symmetric, transitive) relations, then their composition $R_1 \circ R_2$ may not be reflexive (symmetric, transitive).

Definition 2.10 A *t-norm* is a mapping $T : [0, 1] \times [0, 1] \to [0, 1]$ such that

(T_1) $T(x, 1) = x$,
(T_2) $T(x, y) = T(y, x)$,
(T_3) $T(x, T(y, z)) = T(T(x, y), z)$,
(T_4) $T(x, y) \leqslant T(x, z)$ whenever $y \leqslant z$,

where $x, y, z \in [0, 1]$. Replacing 1 by 0 in condition (T_1), we obtain the concept of *s-norm* S.

2.2 Intuitionistic Fuzzy Lie Subalgebras

Definition 2.11 An intuitionistic fuzzy set $A = (\mu_A, \lambda_A)$ on Lie algebra L is called an *intuitionistic fuzzy Lie subalgebra* if the following conditions are satisfied:

(1) $\mu_A(x + y) \geq \min(\mu_A(x), \mu_A(y))$ and $\lambda_A(x + y) \leq \max(\lambda_A(x), \lambda_A(y))$,
(2) $\mu_A(\alpha x) \geq \mu_A(x)$ and $\lambda_A(\alpha x) \leq \lambda_A(x)$,
(3) $\mu_A([x, y]) \geq \min\{\mu_A(x), \mu_A(y)\}$ and $\lambda_A([x, y]) \leq \max\{\lambda_A(x), \lambda_A(y)\}$

for all $x, y \in L$ and $\alpha \in \mathbb{F}$.

Definition 2.12 An intuitionistic fuzzy set $A = (\mu_A, \lambda_A)$ on L is called an *intuitionistic fuzzy Lie ideal* if it satisfies the conditions (1), (2) and the following additional condition:

(4) $\mu_A([x, y]) \geq \mu_A(x)$ and $\lambda_A([x, y]) \leq \lambda_A(x)$

for all $x, y \in L$.

From (2), it follows that:

(5) $\mu_A(0) \geq \mu_A(x), \quad \lambda_A(0) \leq \lambda_A(x),$
(6) $\mu_A(-x) \geq \mu_A(x), \quad \lambda_A(-x) \leq \lambda_A(x).$

Example 2.1 Let $\mathfrak{R}^3 = \{(x, y, z) : x, y, z \in \mathbb{R}\}$ be the set of all three-dimensional real vectors. Then $\mathfrak{R}^3$ with the bracket $[\cdot, \cdot]$ defined as the usual cross product, i.e., $[x, y] = x \times y$, forms a real Lie algebra. We define an intuitionistic fuzzy set $A = (\mu_A, \lambda_A) : \mathfrak{R}^3 \to [0, 1] \times [0, 1]$ by

$$\mu_A(x, y, z) = \begin{cases} t_1 \text{ if } x = y = z = 0, \\ t_2 \text{ otherwise,} \end{cases} \qquad \lambda_A(x, y, z) = \begin{cases} t_2 \text{ if } x = y = z = 0, \\ t_1 \text{ otherwise,} \end{cases}$$

where $t_1 > t_2$ and $t_1, t_2 \in [0, 1]$. By routine computations, we can verify that the above intuitionistic fuzzy set A is an intuitionistic fuzzy Lie subalgebra and Lie ideal of the Lie algebra $\mathfrak{R}^3$.

Proposition 2.1 *Every intuitionistic fuzzy Lie ideal is an intuitionistic fuzzy Lie subalgebra.*

We note here that the converse of Proposition 2.1 does not hold in general as it can be seen in the following example.

Example 2.2 Consider $\mathbb{F} = \mathbb{R}$. Let $L = \mathfrak{R}^3 = \{(x, y, z) : x, y, z \in \mathbb{R}\}$ be the set of all three-dimensional real vectors which form a Lie algebra and define

$$\mathfrak{R}^3 \times \mathfrak{R}^3 \to \mathfrak{R}^3$$

$$[x, y] \to x \times y,$$

where $\times$ is the usual cross product. We define an intuitionistic fuzzy set $A = (\mu_A, \lambda_A) : \mathfrak{R}^3 \to [0, 1] \times [0, 1]$ by

$$\mu_A(x, y, z) = \begin{cases} 1 & \text{if } x = y = z = 0, \\ 0.5 & \text{if } x \neq 0, y = z = 0, \\ 0 & \text{otherwise,} \end{cases} \quad \lambda_A(x, y, z) = \begin{cases} 0 & \text{if } x = y = z = 0, \\ 0.3 & \text{if } x \neq 0, y = z = 0, \\ 1 & \text{otherwise.} \end{cases}$$

Then $A = (\mu_A, \lambda_A)$ is an intuitionistic fuzzy Lie subalgebra of L but $A = (\mu_A, \lambda_A)$ is not an intuitionistic fuzzy Lie ideal of L since

$$\mu_A([(1, 0, 0)\ (1, 1, 1)]) = \mu_A(0, -1, 1) = 0,$$

$$\lambda_A([(1, 0, 0)\ (1, 1, 1)]) = \lambda_A(0, -1, 1) = 1,$$

$$\mu_A(1,0,0) = 0.5, \quad \lambda_A(1,0,0) = 0.3.$$

That is,

$$\mu_A([(1,0,0)\ (1,1,1)]) \ngeq \mu_A(1,0,0),$$

$$\lambda_A([(1,0,0)\ (1,1,1)]) \nleq \lambda_A(1,0,0).$$

Theorem 2.1 *Let $A = (\mu_A, \lambda_A)$ be an intuitionistic fuzzy Lie subalgebra in a Lie algebra L. Then $A = (\mu_A, \lambda_A)$ is an intuitionistic fuzzy Lie subalgebra of L if and only if the nonempty upper s-level cut $U(\mu_A, s) = \{x \in L \mid \mu_A(x) \geq s\}$ and the nonempty lower t-level cut $L(\lambda_A, t) = \{x \in L \mid \lambda_A(x) \leq t\}$ are Lie subalgebras of L, for all s, t $\in [0, 1]$.*

Proof Assume that $A = (\mu_A, \lambda_A)$ is an intuitionistic fuzzy Lie subalgebra of L and let $s \in [0, 1]$ be such that $U(\mu_A, s) \neq \emptyset$. Let $x, y \in L$ be such that $x \in U(\mu_A, s)$, and $y \in U(\mu_A, s)$. Then $\mu_A(x) \geq s$ and $\mu_A(y) \geq s$. It follows that

$$\mu_A(x+y) \geq \min(\mu_A(x), \mu_A(y)) \geq s,$$

$$\mu_A(\alpha x) \geq \mu_A(x) \geq s,$$

$$\mu_A([x, y]) \geq \min(\mu_A(x), \mu_A(y)) \geq s$$

and hence, $x+y \in U(\mu_A, s)$, $\alpha x \in U(\mu_A, s)$, and $[x, y] \in U(\mu_A, s)$. Thus, $U(\mu_A, s)$ forms a Lie subalgebra of L. For the case $L(\lambda_A, t)$, the proof is similar.

Conversely, suppose that $U(\mu_A, s) \neq \emptyset$ is a Lie subalgebra of L for every $s \in [0, 1]$. Assume that

$$\mu_A(x+y) < \min\{\mu_A(x), \mu_A(y)\}$$

for some $x, y \in L$. Now, taking

$$s_0 := \frac{1}{2}\{\mu_A(x+y) + \min\{\mu_A(x) + \mu_A(y)\}\},$$

then we have

$$\mu_A(x+y) < s_0 < \min\{\mu_A(x), \mu_A(y)\}.$$

and hence, $x+y \notin U(\mu_A, s)$, $x \in U(\mu_A, s)$, and $y \in U(\mu_A, s)$. However, this is clearly a contradiction. Therefore,

$$\mu_A(x+y) \geq \min\{\mu_A(x), \mu_A(y)\}$$

for all $x, y \in L$. Similarly, we can show that

$$\mu_A(\alpha x) \geq \mu_A(x),$$

$$\mu_A([x, y]) \geq \min(\mu_A(x), \mu_A(y)).$$

Hence, $U(\mu_A, s)$ is a fuzzy Lie subalgebra of L. For the case $L(\lambda_A, t)$, the proof is similar.

Definition 2.13 Let $A = (\mu_A, \lambda_A)$ and $B = (\mu_B, \lambda_B)$ be two intuitionistic fuzzy sets of L. We define the *sup-min product* $[\mu_A \mu_B]$ of μ_A and μ_B and the *inf-max product* $[\lambda_A \lambda_B]$ of λ_A and λ_B as follows:

$$[\mu_A \mu_B](x) = \begin{cases} \sup_{x=[yz]}\{\min(\mu_A(y), \mu_B(z))\} \\ 0, \ \text{if x} \neq [\text{yz}], \end{cases}$$

$$[\lambda_B \lambda_B](x) = \begin{cases} \inf_{x=[yz]}\{\max(\lambda_A(y), \lambda_B(z))\} \\ 1, \ \text{if x} \neq [\text{yz}]. \end{cases}$$

for all $x, y, z \in L$.

Let $A = (\mu_A, \lambda_A)$ and $B = (\mu_B, \lambda_B)$ be intuitionistic fuzzy Lie subalgebras of the Lie algebra L. Then $[AB]$ may not be an intuitionistic fuzzy Lie subalgebra of L as this can be seen in the following counter example:

Example 2.3 Let $\{e_1, e_2, \ldots, e_8\}$ be a basis of a vector space over a field $\mathbb{F}$. Then, it is not difficult to see that, by putting:

$$[e_1, e_2] = e_5, \quad [e_1, e_3] = e_6, \quad [e_1, e_4] = e_7, \quad [e_1, e_5] = -e_8,$$

$$[e_2, e_3] = e_8, \quad [e_2, e_4] = e_6, \quad [e_2, e_6] = -e_7, \quad [e_3, e_4] = -e_5,$$

$$[e_3, e_5] = -e_7, \quad [e_4, e_6] = -e_8, \quad [e_i, e_j] = -[e_j, e_i]$$

and $[e_i, e_j] = 0$ for all $i \leq j$, we can obtain a Lie algebra over a field $\mathbb{F}$. The following fuzzy sets

$$\mu_A(x) := \begin{cases} 1 \text{ if } x \in \{0, e_1, e_5, e_6, e_7, e_8\}, \\ 0 \text{ otherwise,} \end{cases} \qquad \lambda_A(x) := \begin{cases} 0 \text{ if } x \in \{0, e_1, e_5, e_6, e_7, e_8\}, \\ 1 \text{ otherwise,} \end{cases}$$

$$\mu_B(x) := \begin{cases} 1 \ \text{ if } x = 0, \\ 0.5 \text{ if } x \in \{e_2, e_5, e_6, e_7, e_8\}, \\ 0 \ \text{ otherwise,} \end{cases} \quad \lambda_B(x) := \begin{cases} 0 \ \text{ if } x = 0, \\ 0.3 \text{ if } x \in \{e_2, e_5, e_6, e_7, e_8\}, \\ 1 \ \text{ otherwise,} \end{cases}$$

are clearly fuzzy Lie subalgebras of a Lie algebra L. Thus, $A = (\mu_A, \lambda_A)$ and $B = (\mu_B, \lambda_B)$ are intuitionistic fuzzy Lie subalgebras of L because the level Lie subalgebras

$$U(\mu_A, 1) =< e_1, e_5, e_6, e_7, e_8 >, \quad L(\lambda_A, 0) =< e_1, e_5, e_6, e_7, e_8 >$$

$$U(\mu_B, 0.5) =< e_2, e_5, e_6, e_7, e_8 >, \quad L(\lambda_B, 0.3) =< e_2, e_5, e_6, e_7, e_8 >$$

are Lie subalgebras of L. But $[AB]$ is not an intuitionistic fuzzy Lie subalgebra because the following condition does not hold:

$$[AB](e_7 + e_8) \geq \min\{[AB](e_7), [AB](e_8)\}.$$

$$(1) \quad [\mu_A\mu_B](e_7) = \sup \begin{cases} \min\{\mu_A(e_1), \mu_B(e_4)\} = 0, & e_7 = [e_1, e_4], \\ \min\{\mu_A(e_2), \mu_B(e_6)\} = 0, & e_7 = -[e_2, e_6], \\ \min\{\mu_A(e_3), \mu_B(e_5)\} = 0, & e_7 = -[e_3, e_5], \\ \min\{\mu_A(e_4), \mu_B(e_1)\} = 0, & e_7 = -[e_4, e_1], \\ \min\{\mu_A(e_6), \mu_B(e_2)\} = 0.5, & e_7 = [e_6, e_2], \\ \min\{\mu_A(e_5), \mu_B(e_3)\} = 0, & e_7 = [e_5, e_3]. \end{cases}$$

Thus, $[\mu_A\mu_B](e_7) = 0.5$.
(2) By using similar arguments, we can show that $[\mu_A\mu_B](e_8) = 0.5$.
(3) $[\mu_A\mu_B](e_7 + e_8) = \sup\{(i) - (vi)\}$

(i) if $e_7 + e_8 = [e_1(e_4 - e_5)]$, then $\min\{\mu_A(e_1), \mu_B(e_4 - e_5)\} = \min\{\mu_A(e_1), \mu_B(e_4), \mu_B(e_5)\} = 0$, since $\mu_B(e_4) = 0$, and if $e_7 + e_8 = [(e_5 - e_4)e_1]$, then $\min\{\mu_A(e_5 - e_4), \mu_B(e_1)\} = \min\{\mu_A(e_5), \mu_B(e_4), \mu_B(e_1)\} = 0$, since $\mu_A(e_4) = 0$.

By using similar method, we can also obtain the following numerical results:
(ii) If $e_7 + e_8 = [e_2(e_3 - e_6)]$, then $\min(\mu_A(e_2), \mu_B(e_3 - e_6)) = 0$.
(iii) If $e_7 + e_8 = [e_3(-e_2 - e_5)]$, then $\min(\mu_A(e_3), \mu_B(e_2 - e_5)) = 0$.
(iv) If $e_7 + e_8 = [e_4(-e_1 - e_6)]$, then $\min(\mu_A(e_4), \mu_B(-e_3 - e_1)) = 0$.
(v) If $e_7 + e_8 = [e_5(-e_3 - e_1)]$, then $\min(\mu_A(e_5), \mu_B(-e_3 - e_1)) = 0$.
(vi) If $e_7 + e_8 = [e_6(-e_2 - e_4)]$, then $\min(\mu_A(e_6), \mu_B(-e_2 - e_4)) = 0$.

Thus, $[\mu_A\mu_B](e_7 + e_8) = \sup\{0, 0, 0, 0, 0, 0\} = 0$.
Hence, we have proved that

$$[\mu_A\mu_B](e_7 + e_8) \ngeq \min\{[\mu_A\mu_B](e_7), [\mu_A\mu_B](e_8)\}.$$

The verification of

$$[\lambda_A\lambda_B](e_7 + e_8) \nleq \max\{[\lambda_A\lambda_B](e_7), [\lambda_A\lambda_B](e_8)\}$$

is similar.

We now refine the product of two intuitionistic fuzzy Lie subalgebras A and B of L to an extended form.

Definition 2.14 Let $A = (\mu_A, \lambda_A)$ and $B = (\mu_B, \lambda_B)$ be two intuitionistic fuzzy sets of L. Then, we define the *sup-min product* $\ll \mu_A\mu_B \gg$ of μ_A and μ_B, and the *inf-max product* $\ll \lambda_A\lambda_B \gg$ of λ_A and λ_B as follows, for all $x, y, z \in L$

$$\ll \mu_A\mu_B \gg (x) = \begin{cases} \sup_{x=\sum_{i=1}^{n}[x_iy_i]}\{\min_{i\in\mathbb{N}}\{\min(\mu_A(x_i), \mu_B(y_i))\}\} \\ 0, \quad \text{if } x \neq \sum_{i=1}^{n}[x_iy_i], \end{cases}$$

$$\ll \lambda_A\lambda_B \gg (x) = \begin{cases} \inf_{x=\sum_{i=1}^{n}[x_iy_i]}\{\max_{i\in\mathbb{N}}\{\max(\mu_A(x_i), \mu_B(y_i))\}\} \\ 1, \quad \text{if } x \neq \sum_{i=1}^{n}[x_iy_i]. \end{cases}$$

From the definitions of $[AB]$ and $\ll AB \gg$, we can easily see that $[AB] \subseteq \ll AB \gg$ and $[AB] \neq \ll AB \gg$ hold generally even if A and B are both intuitionistic fuzzy Lie subalgebras of L, and in this case, $\ll AB \gg$ is also an intuitionistic fuzzy Lie subalgebra of L.

We now formulate the following theorem.

Theorem 2.2 *Let $A = (\mu_A, \lambda_A)$ be an intuitionistic fuzzy Lie subalgebra of Lie algebra L. Define a binary relation $\sim$ on L by $x \sim y$ if and only if $\mu_A(x - y) = \mu_A(0)$, $\lambda_A(x - y) = \lambda_A(0)$ for all $x, y \in L$. Then $\sim$ is a congruence relation on L.*

Proof We first prove that "$\sim$" is an equivalent relation. We only need to show the transitivity of "$\sim$" because the reflectivity and symmetricity of "$\sim$" hold trivially. Let $x, y, z \in L$. If $x \sim y$ and $y \sim z$, then $\mu_A(x - y) = \mu_A(0)$, $\mu_A(y - z) = \mu_A(0)$, and $\lambda_A(x - y) = \lambda_A(0)$, $\lambda_A(y - z) = \lambda_A(0)$. Hence, it follows that

$$\mu_A(x - z) = \mu_A(x - y + y - z) \geq \min(\mu_A(x - y), \mu_A(y - z)) = \mu_A(0),$$
$$\lambda_A(x - z) = \lambda_A(x - y + y - z) \leq \max(\lambda_A(x - y), \lambda_A(y - z)) = \lambda_A(0).$$

Consequently, $x \sim z$. We now verify that "$\sim$" is a congruence relation on L. For this purpose, we let $x \sim y$ and $y \sim z$. Then $\mu_A(x - y) = \mu_A(0)$, $\mu_A(y - z) = \mu_A(0)$, $\lambda_A(x - y) = \lambda_A(0)$, and $\lambda_A(y - z) = \lambda_A(0)$. Now, for $x_1, x_2, y_1, y_2 \in L$, we have

$$\begin{aligned} \mu_A((x_1 + x_2) - (y_1 + y_2)) &= \mu_A((x_1 - y_1) + (x_2 - y_2)) \\ &\geq \min(\mu_A(x_1 - y_1), \mu_A(x_2 - y_2)) = \mu_A(0), \\ \lambda_A((x_1 + x_2) - (y_1 + y_2)) &= \lambda_A((x_1 - y_1) + (x_2 - y_2)) \\ &\leq \max(\lambda_A(x_1 - y_1), \lambda_A(x_2 - y_2)) = \lambda_A(0), \end{aligned}$$

$$\begin{aligned} \mu_A(\alpha x_1 - \alpha y_1) &= \mu_A(\alpha(x_1 - y_1)) \geq \mu_A(x_1 - y_1) = \mu_A(0), \\ \lambda_A(\alpha x_1 - \alpha y_1) &= \lambda_A(\alpha(x_1 - y_1)) \leq \lambda_A(x_1 - y_1) = \lambda_A(0), \\ \mu_A([x_1, x_2] - [y_1, y_2]) &= \mu_A([x_1 - y_1], [x_2 - y_2]) \\ &\geq \min\{\mu_A(x_1 - y_1), \mu_A(x_2 - y_2)\} = \mu_A(0), \\ \lambda_A([x_1, x_2] - [y_1, y_2]) &= \lambda_A([x_1 - y_1], [x_2 - y_2]) \\ &\leq \max\{\lambda_A(x_1 - y_1), \lambda_A(x_2 - y_2)\} = \lambda_A(0). \end{aligned}$$

That is, $x_1 + x_2 \sim y_1 + y_2$, $\alpha x_1 \sim \alpha y_1$ and $[x_1, x_2] \sim [y_1, y_2]$. Thus, "$\sim$" is indeed a congruence relation on L.

Definition 2.15 Let L be a nonempty set. Then we call a complex mapping $A = (\mu_A, \lambda_A) : L \times L \to [0, 1] \times [0, 1]$ an *intuitionistic fuzzy relation* on L if $\mu_A(x, y) + \lambda_A(x, y) \leq 1$, for all $(x, y) \in L \times L$.

Definition 2.16 Let $A = (\mu_A, \lambda_A)$ and $B = (\mu_B, \lambda_B)$ be intuitionistic fuzzy sets on a set L. If $A = (\mu_A, \lambda_A)$ is an intuitionistic fuzzy relation on a set L, then $A = (\mu_A, \lambda_A)$ is said to be an *intuitionistic fuzzy relation* on $B = (\mu_B, \lambda_B)$ if $\mu_A(x, y) \leq \min(\mu_B(x), \mu_B(y))$ and $\lambda_A(x, y) \geq \max(\lambda_B(x), \lambda_B(y))$, for all $x, y \in L$.

Definition 2.17 Let $A = (\mu_A, \lambda_A)$ and $B = (\mu_B, \lambda_B)$ be two intuitionistic fuzzy sets on a set L. Then the *generalized Cartesian product* $A \times B$ is defined as follows:

$$\begin{aligned} A \times B &= (\mu_A, \lambda_A) \times (\mu_B, \lambda_B) \\ &= (\mu_A \times \mu_B, \lambda_A \times \lambda_B), \end{aligned}$$

where $(\mu_A \times \mu_B)(x, y) = \min(\mu_A(x), \mu_B(y))$ and $(\lambda_A \times \lambda_B)(x, y) = \max(\lambda_A(x), \lambda_B(y))$.

We note that the generalized Cartesian product $A \times B$ is always an intuitionistic fuzzy set in $L \times L$ if

$$\min(\mu_A(x), \mu_B(y)) + \max(\lambda_A(x), \lambda_B(y)) \leq 1.$$

The proof of the following proposition is trivial.

Proposition 2.2 *Let $A = (\mu_A, \lambda_A)$ and $B = (\mu_B, \lambda_B)$ be intuitionistic fuzzy sets on a set L. Then*

(i) $A \times B$ is an intuitionistic fuzzy relation on L,
(ii) $U(\mu_A \times \mu_B, t) = U(\mu_A, t) \times U(\mu_B, t)$ and $L(\lambda_A \times \lambda_B, t) = L(\lambda_A, t) \times L(\lambda_B, t)$ for all $t \in [0, 1]$.

Theorem 2.3 *Let $A = (\mu_A, \lambda_A)$ and $B = (\mu_B, \lambda_B)$ be two intuitionistic fuzzy Lie subalgebras of a Lie algebra L. Then $A \times B$ is an intuitionistic fuzzy Lie subalgebra of $L \times L$.*

Proof Let $x = (x_1, x_2)$ and $y = (y_1, y_2) \in L \times L$. Then

$$\begin{aligned} (\mu_A \times \mu_B)(x + y) &= (\mu_A \times \mu_B)((x_1, x_2) + (y_1, y_2)) \\ &= (\mu_A \times \mu_B)(x_1 + y_1, x_2 + y_2) \\ &= \min(\mu_A(x_1 + y_1), \mu_B(x_2 + y_2)) \\ &\geq \min(\min(\mu_A(x_1), \mu_A(y_1)), \min(\mu_B(x_2), \mu_B(y_2))) \\ &= \min(\min(\mu_A(x_1), \mu_B(x_2)), \min(\mu_A(y_1), \mu_B(y_2))) \\ &= \min((\mu_A \times \mu_B)(x_1, x_2)), (\mu_A \times \mu_B)(y_1, y_2)) \\ &= \min((\mu_A \times \mu_B)(x), (\mu_A \times \mu_B)(y)), \end{aligned}$$

$$\begin{aligned}(\lambda_A \times \lambda_B)(x+y) &= (\lambda_A \times \lambda_B)((x_1, x_2) + (y_1, y_2))\\ &= (\lambda_A \times \lambda_B)(x_1 + y_1, x_2 + y_2)\\ &= \max(\lambda_A(x_1 + y_1), \lambda_B(x_2 + y_2))\\ &\le \max(\max(\lambda_A(x_1), \lambda_A(y_1)), \max(\lambda_B(x_2), \lambda_B(y_2)))\\ &= \max(\max(\mu_A(x_1), \lambda_B(x_2)), \max(\lambda_A(y_1), \lambda_B(y_2)))\\ &= \max((\lambda_A \times \lambda_B)(x_1, x_2)), (\lambda_A \times \lambda_B)(y_1, y_2))\\ &= \max((\lambda_A \times \lambda_B)(x), (\lambda_A \times \lambda_B)(y)),\end{aligned}$$

$$\begin{aligned}(\mu_A \times \mu_B)(\alpha x) &= (\mu_A \times \mu_B)(\alpha(x_1, x_2)) = (\mu_A \times \mu_B)(\alpha x_1, \alpha x_2)\\ &= \min(\mu_A(\alpha x_1), \mu_B(\alpha x_2)) \ge \min(\mu_A(x_1), \mu_B(x_2))\\ &= (\mu_A \times \mu_B)(x_1, x_2) = (\mu_A \times \mu_B)(x),\end{aligned}$$

$$\begin{aligned}(\lambda_A \times \lambda_B)(\alpha x) &= (\lambda_A \times \lambda_B)(\alpha(x_1, x_2)) = (\lambda_A \times \lambda_B)(\alpha x_1, \alpha x_2)\\ &= \max(\lambda_A(\alpha x_1), \lambda_B(\alpha x_2)) \le \max(\lambda_A(x_1), \lambda_B(x_2))\\ &= (\lambda_A \times \lambda_B)(x_1, x_2) = (\lambda_A \times \lambda_B)(x),\end{aligned}$$

$$\begin{aligned}(\mu_A \times \mu_B)([x, y]) &= (\mu_A \times \mu_B)([(x_1, x_2), (y_1, y_2)])\\ &\ge \min(\min(\mu_A(x_1), \mu_B(x_2)), \min(\mu_A(y_1), \mu_B(y_2)))\\ &= \min((\mu_A \times \mu_B)(x_1, x_2), (\mu_A \times \mu_B)(y_1, y_2))\\ &= \min((\mu_A \times \mu_B)(x), (\mu_A \times \mu_B)(y)),\end{aligned}$$

$$\begin{aligned}(\lambda_A \times \lambda_B)([x, y]) &= (\lambda_A \times \lambda_B)([(x_1, x_2), (y_1, y_2)])\\ &\le \max(\max(\mu_A(x_1), \lambda_B(x_2)), \max(\lambda_A(y_1), \lambda_B(y_2)))\\ &= \max((\lambda_A \times \lambda_B)(x_1, x_2), (\lambda_A \times \lambda_B)(y_1, y_2))\\ &= \max((\lambda_A \times \lambda_B)(x), (\lambda_A \times \lambda_B)(y)).\end{aligned}$$

This shows that $A \times B$ is an intuitionistic fuzzy Lie subalgebra of $L \times L$.

2.3 Lie Homomorphism of Intuitionistic Fuzzy Lie Subalgebras

For the Lie algebras L_1 and L_2, it can be easily observed that if $f : L_1 \to L_2$ is a Lie homomorphism and A is an intuitionistic fuzzy Lie subalgebra of L_2, then the intuitionistic fuzzy set $f^{-1}(A)$ of L_1 is also an intuitionistic fuzzy Lie subalgebra.

Definition 2.18 Let L_1 and L_2 be two Lie algebras. Then, a Lie homomorphism $f : L_1 \to L_2$ is said to have a natural extension $f : J^{L_1} \to J^{L_2}$ defined by for all $A = (\mu_A, \lambda_A) \in J^{L_1}$, $y \in L_2$:

$$f(\mu_A)(y) = \sup\{\mu_A(x) : x \in f^{-1}(y)\}$$

$$f(\lambda_A)(y) = \inf\{\lambda_A(x) : x \in f^{-1}(y)\}.$$

We now call these sets the homomorphic images of the intuitionistic fuzzy set $A = (\mu_A, \lambda_A)$.

We now formulate the following theorems:

Theorem 2.4 *The homomorphic image of an intuitionistic fuzzy Lie subalgebra is still an intuitionistic fuzzy Lie subalgebra of its co-domain.*

Proof Let $y_1, y_2 \in L_2$. Then

$$\{x \mid x \in f^{-1}(y_1 + y_2)\} \supseteq \{x_1 + x_2 \mid x_1 \in f^{-1}(y_1) \text{ and } x_2 \in f^{-1}(y_2)\}.$$

Now, we have

$$\begin{aligned} f(\mu_A)(y_1 + y_2) &= \sup\{\mu_A(x) \mid x \in f^{-1}(y_1 + y_2)\} \\ &\geq \{\mu_A(x_1 + x_2) \mid x_1 \in f^{-1}(y_1) \text{ and } x_2 \in f^{-1}(y_2)\} \\ &\geq \sup\{\min\{\mu_A(x_1), \mu_A(x_2)\} \mid x_1 \in f^{-1}(y_1) \text{ and } x_2 \in f^{-1}(y_2)\} \\ &= \min\{\sup\{\mu_A(x_1) \mid x_1 \in f^{-1}(y_1)\}, \sup\{\mu_A(x_2) \mid x_2 \in f^{-1}(y_2)\}\} \\ &= \min\{f(\mu_A)(y_1), f(\mu_A)(y_2)\}. \end{aligned}$$

For $y \in L_2$ and $\alpha \in \mathbb{F}$, we have

$$\{x \mid x \in f^{-1}(\alpha y)\} \supseteq \{\alpha x \mid x \in f^{-1}(y)\}.$$

$$\begin{aligned} f(\mu_A)(\alpha y) &= \sup\{\mu_A(\alpha x) \mid x \in f^{-1}(y)\} \\ &\geq \{\mu_A(\alpha x) \mid x \in f^{-1}(\alpha y)\} \\ &\geq \sup\{\mu_A(x) \mid x \in f^{-1}(y)\} \\ &= f(\mu_A)(y). \end{aligned}$$

If $y_1, y_2 \in L_2$, then

$$\{x \mid x \in f^{-1}([y_1, y_2])\} \supseteq \{[x_1, x_2] | x_1 \in f^{-1}(y_1),\ x_2 \in f^{-1}(y_2)\}.$$

Now

$$\begin{aligned} f(\mu_A)([y_1, y_2]) &= \sup\{\mu_A(x) \mid x \in f^{-1}([y_1, y_2])\} \\ &\geq \{\mu_A([x_1, x_2]) \mid x_1 \in f^{-1}(y_1) \text{ and } x_2 \in f^{-1}(y_2)\} \\ &\geq \sup\{\min\{\mu_A(x_1), \mu_A(x_2)\} \mid x_1 \in f^{-1}(y_1) \text{ and } x_2 \in f^{-1}(y_2)\} \\ &= \min\{\sup\{\mu_A(x_1) \mid x_1 \in f^{-1}(y_1)\}, \sup\{\mu_A(x_2) \mid x_2 \in f^{-1}(y_2)\}\} \\ &= \min\{f(\mu_A)(y_1), f(\mu_A)(y_2)\}. \end{aligned}$$

Thus, $f(\mu_A)$ is a fuzzy Lie algebra of L_2. In the same manner, we can prove that $f(\lambda_A)$ is a fuzzy Lie subalgebra of L_2. Hence, $f(A) = (f(\mu_A), f(\lambda_A))$ is an intuitionistic fuzzy Lie subalgebra of L_2.

Theorem 2.5 *Let $f : L_1 \to L_2$ be a surjective Lie homomorphism. If A and B are intuitionistic fuzzy Lie subalgebras of L_1, then $f(\ll AB \gg) = \ll f(A)f(B) \gg$.*

Proof Assume that $f(\ll AB \gg) < \ll f(A)f(B) \gg$. Now, we choose a number $t \in [0, 1]$ such that $f(\ll AB \gg)(x) < t < \ll f(A)f(B) \gg (x)$. Then, there exist $y_i, z_i \in L_2$ such that $x = \sum_{i=1}^{n}[y_i z_i]$ with $f(A)(y_i) > t$ and $f(B)(z_i) > t$. Since f is surjective, there exists $y \in L_1$ such that $f(y) = x$ and $y = \sum_{i=1}^{n}[a_i b_i]$ for some $a_i \in f^{-1}(y_i), b_i \in f^{-1}(z_i)$ with $f(a_i) = y_i, f(b_i) = z_i, A(a_i) > t$, and $B(b_i) > t$. Since

$$f\left(\sum_{i=1}^{n}[a_i b_i]\right) = \sum_{i=1}^{n} f([a_i b_i]) = \sum_{i=1}^{n}[f(a_i)f(b_i)] = \sum_{i=1}^{n}[y_i z_i] = x,$$

$f(\ll AB \gg)(x) > t$. This is a contradiction. Similarly, for the case $f(\ll AB \gg) > \ll f(A)f(B) \gg$, we can also obtain a contradiction. Hence, $f(\ll AB \gg) = \ll f(A)f(B) \gg$.

Definition 2.19 Let $A = (\mu_A, \lambda_A)$ and $B = (\mu_B, \lambda_B)$ be intuitionistic fuzzy subal gebras of L. Then A is said to be of the *same type* of B if there exists $f \in Aut(L)$ such that $A = B \circ f$, i.e., $\mu_A(x) = \mu_B(f(x)), \lambda_A(x) = \lambda_B(f(x))$ for all $x \in L$.

Theorem 2.6 *Let $A = (\mu_A, \lambda_A)$ and $B = (\mu_B, \lambda_B)$ be two intuitionistic fuzzy subalgebras of L. Then A is an intuitionistic fuzzy subalgebra having the same type of B if and only if A is isomorphic to B.*

Proof We only need to prove the necessity part because the sufficiency part is trivial. Let $A = (\mu_A, \lambda_A)$ be an intuitionistic fuzzy subalgebra having the same type of $B = (\mu_B, \lambda_B)$. Then there exists $\phi \in Aut(L)$ such that

$$\mu_A(x) = \mu_B(\phi(x)), \ \lambda_A(x) = \lambda_B(\phi(x)) \ \ \forall x \in L.$$

Let $f : A(L) \to B(L)$ be a mapping defined by $f(A(x)) = B(\phi(x))$ for all $x \in L$, that is,

$$f(\mu_A(x)) = \mu_B(\phi(x)), \ f(\lambda_A(x)) = \lambda_B(\phi(x)) \ \ \forall x \in L.$$

Then, it is clear that f is surjective. Also, f is injective because if $f(\mu_A(x)) = f(\mu_A(y))$ for all $x, y \in L$, then $\mu_B(\phi(x)) = \mu_B(\phi(y))$, and hence, $\mu_A(x) = \mu_B(y)$. Likewise, we have $f(\lambda_A(x)) = f(\lambda_A(y)) \Longrightarrow \lambda_A(x) = \lambda_B(y)$ for all $x \in L$. Finally, f is a homomorphism because for $x, y \in L$,

$$f(\mu_A(x + y)) = \mu_B(\phi(x + y)) = \mu_B(\phi(x) + \phi(y)),$$

$$f(\lambda_A(x + y)) = \lambda_B(\phi(x + y)) = \lambda_B(\phi(x) + \phi(y)),$$

$$f(\mu_A(\alpha x)) = \mu_B(\phi(\alpha x)) = \alpha\mu_B(\phi(x)),$$

$$f(\lambda_A(\alpha x)) = \lambda_B(\phi(\alpha x)) = \alpha\lambda_B(\phi(x)),$$

$$f(\mu_A([x, y])) = \mu_B(\phi([x, y])) = \mu_B([\phi(x), \phi(y)]),$$

$$f(\lambda_A([x, y])) = \lambda_B(\phi([x, y])) = \lambda_B([\phi(x), \phi(y)]).$$

Hence, $A = (\mu_A, \lambda_A)$ is isomorphic to $B = (\mu_B, \lambda_B)$. This completes the proof.

2.4 Intuitionistic Fuzzy Lie Ideals

Definition 2.20 An intuitionistic fuzzy set $A = (\mu_A, \lambda_A)$ on L is called an *intuitionistic fuzzy Lie ideal* if the following conditions are satisfied:

1. $\mu_A(x + y) \geq \min\{\mu_A(x), \mu_A(y)\}$,
2. $\mu_A(\alpha x) \geq \mu_A(x)$,
3. $\mu_A([x, y]) \geq \mu_A(x)$,
4. $\lambda_A(x + y) \leq \max\{\lambda_A(x), \lambda_A(y)\}$,
5. $\lambda_A(\alpha x) \leq \lambda_A(x)$,
6. $\lambda_A([x, y]) \leq \lambda_A(x)$

for all $x, y \in L$ and $\alpha \in \mathbb{F}$.

Example 2.4 Let $\Re^2 = \{(x, y) : x, y \in \mathbb{R}\}$ be the set of all two-dimensional real vectors. Then $\Re^2$ with the bracket $[\cdot, \cdot]$ defined as usual cross product, i.e., $[x, y] = x \times y$, is a real Lie algebra. Putting

$$\mu_A(x, y) = \begin{cases} 1 \text{ if } x = y = 0, \\ 0 \text{ otherwise,} \end{cases} \qquad \lambda_A(x, y) = \begin{cases} 0 \text{ if } x = y = 0, \\ 1 \text{ otherwise,} \end{cases}$$

We obtain an intuitionistic fuzzy set $A = (\mu_A, \lambda_A)$. By routine computations, we can check that it is an intuitionistic fuzzy Lie ideal of a Lie algebra L.

The following propositions are obvious.

Proposition 2.3 *An intuitionistic fuzzy set $A = (\mu_A, \lambda_A)$ is an intuitionistic fuzzy Lie ideal of L if and only if $\Box A$ and $\Diamond A$ are intuitionistic fuzzy Lie ideals of L.*

Proposition 2.4 *If A is an intuitionistic fuzzy Lie ideal of L, then*

1. $\mu_A(0) \geq \mu_A(x), \quad \lambda_A(0) \leq \lambda_A(x)$,
2. $\mu_A([x, y]) \geq \max\{\mu_A(x), \mu_A(y)\}$,
3. $\lambda_A([x, y]) \leq \min\{\lambda_A(x), \lambda_A(y)\}$,

4. $\mu_A([x, y]) = \mu_A(-[y, x]) = \mu_A([y, x])$,
5. $\lambda_A([x, y]) = \lambda_A(-[y, x]) = \lambda_A([y, x])$

for all $x, y \in L$.

Proposition 2.5 *If* $\{A_i \mid i \in I\}$ *is a family of intuitionistic fuzzy Lie ideals of* L*, then* $\bigcap A_i = (\bigwedge \mu_{A_i}, \bigvee \lambda_{A_i})$ *is an intuitionistic fuzzy Lie ideal of* L*, where*

$$\bigwedge \mu_{A_i}(x) = \inf\{\mu_{A_i}(x) \mid i \in I,\ x \in L\},$$

$$\bigvee \lambda_{A_i}(x) = \sup\{\lambda_{A_i}(x) \mid i \in I,\ x \in L\}.$$

Note that union $A \bigcup B$ of two intuitionistic fuzzy Lie ideals of a Lie algebra L is not an intuitionistic fuzzy Lie ideal, in general as seen in the following example.

Example 2.5 Let $\{e_1, e_2, \ldots, e_8\}$ be a basis of a vector space over a field $\mathbb{F}$. It is not difficult to see that putting:

$$[e_1, e_2] = e_5,\ \ [e_1, e_3] = e_6,\ \ [e_1, e_4] = e_7,\ \ [e_1, e_5] = -e_8,$$

$$[e_2, e_3] = e_8,\ \ [e_2, e_4] = e_6,\ \ [e_2, e_6] = -e_7,\ \ [e_3, e_4] = -e_5,$$

$$[e_3, e_5] = -e_7,\ \ [e_4, e_6] = -e_8,\ \ [e_i, e_j] = -[e_j, e_i]$$

and $[e_i, e_j] = 0$ for all $i \leq j$, we obtain a Lie algebra over a field $\mathbb{F}$.

The following two fuzzy sets

$$\mu_A(x) := \begin{cases} 1 & \text{if } x \in \{0, e_8\}, \\ 0.7 & \text{if } x = e_7, \\ 0 & \text{otherwise}, \end{cases} \qquad \mu_B(x) := \begin{cases} 1 & \text{if } x \in \{0, e_7\}, \\ 0.5 & \text{if } x = e_8, \\ 0 & \text{otherwise}, \end{cases}$$

are fuzzy Lie ideals of a Lie algebra L. By Proposition 2.3, an intuitionistic fuzzy sets $A = (\mu_A, \overline{\mu}_A)$, $B = (\mu_B, \overline{\mu}_B)$ are intuitionistic fuzzy Lie ideals of L, but $A \bigcup B$ is not an intuitionistic fuzzy Lie ideal. Indeed,

$$\begin{aligned}(\mu_A \cup \mu_B)(e_7 + e_8) &= \max\{\mu_A(e_7 + e_8),\ \mu_B(e_7 + e_8)\} \\ &\geq \max\{\min\{\mu_A(e_7),\ \mu_A(e_8)\}, \min\{\mu_B(e_7),\ \mu_B(e_8)\}\} = 0.7,\end{aligned}$$

and

$$\begin{aligned}&\min\{(\mu_A \cup \mu_B)(e_7),\ (\mu_A \cup \mu_B)(e_8)\} \\ &\quad = \min\{\max\{\mu_A(e_7),\ \mu_B(e_7)\},\ \max\{\mu_A(e_8),\ \mu_B(e_8)\}\} = 1.\end{aligned}$$

This proves that the axiom (1) is not satisfied. Hence, $A \bigcup B$ is not an intuitionistic fuzzy Lie ideal of L.

By a simple verification of the corresponding axioms, we can see that the following theorem is true.

Theorem 2.7 *If $A = (\mu_A, \lambda_A)$ is an intuitionistic fuzzy Lie ideal of a Lie algebra L, then the level subsets $U(\mu_A, s) = \{x \in L \mid \mu_A(x) \geq s\}$ and $L(\lambda_A, s) = \{x \in L \mid \lambda_A(x) \leq s\}$ are Lie ideals of L for every $s \in \mathrm{Im}(\mu_A) \cap \mathrm{Im}(\lambda_A) \subseteq [0, 1]$, where $\mathrm{Im}(\mu_A)$ and $\mathrm{Im}(\lambda_A)$ are sets of values of μ_A and λ_A, respectively.*

Theorem 2.8 *If all nonempty level subsets $U(\mu_A, s)$ and $L(\lambda_A, s)$ of an intuitionistic fuzzy set $A = (\mu_A, \lambda_A)$ are Lie ideals of a Lie algebra L, then A is an intuitionistic fuzzy Lie ideal of L.*

Proof Let $s \in [0, 1]$. Suppose that $U(\mu_A, s) \neq \emptyset$ and $L(\lambda_A, s) \neq \emptyset$ are Lie ideals of L. We must show that $A = (\mu_A, \lambda_A)$ satisfies the conditions (1) – (6) from the Definition 2.20.

If the condition (1) is false, then there exist $x, y \in L$ such that

$$\mu_A(x + y) < \min\{\mu_A(x), \mu_A(y)\}.$$

Taking

$$s_0 := \frac{1}{2}\{\mu_A(x + y) + \min\{\mu_A(x) + \mu_A(y)\}\},$$

we have

$$\mu_A(x + y) < s_0 < \min\{\mu_A(x), \mu_A(y)\}.$$

It follows that $x + y \notin U(\mu_A, s)$ and $x, y \in U(\mu_A, s)$, which is a contradiction. Hence, the condition (1) is true. The proof of other conditions is similar.

Theorem 2.9 *An intuitionistic fuzzy set $A = (\mu_A, \lambda_A)$ of L is an intuitionistic fuzzy Lie ideal of L if and only if $L_A^{(s,t)}$ is a Lie ideal of L for every $(s, t) \in \mathrm{Im}(\mu_A) \times \mathrm{Im}(\lambda_A)$ with $s + t \leq 1$.*

Proof If $A = (\mu_A, \lambda_A)$ is an intuitionistic fuzzy Lie ideal of L, then according to Theorem 2.7, all nonempty level subsets $U(\mu_A, s)$ and $L(\lambda_A, t)$ are Lie ideals of L. So, $L_A^{(s,t)} = U(\mu_A, s) \cap L(\lambda_A, t)$ is a Lie ideal of L.

Conversely, let $L_A^{(s,t)}$ be a Lie ideal of L and let $A = (\mu_A, \lambda_A)$ be an intuitionistic fuzzy set on L. Consider $x, y \in L$ such that $A(x) = (s_1, t_1)$ and $A(y) = (s_2, t_2)$, that is, $\mu_A(x) = s_1$, $\lambda_A(x) = t_1$, $\mu_A(y) = s_2$, and $\lambda_A(y) = t_2$. Without loss of generality, we can assume that $(s_1, t_1) \leq (s_2, t_2)$, i.e., $s_1 \leq s_2$ and $t_2 \leq t_1$. Then $L_A^{(s_2,t_2)} \subseteq L_A^{(s_1,t_1)}$, i.e., $x, y \in L_A^{(s_1,t_1)}$, which implies $x + y, \alpha x, [x, y] \in L_A^{(s_1,t_1)}$ because $L_A^{(s_1,t_1)}$ is a Lie ideal of L. Thus,

$$\mu(x + y) \geq s_1 = \min\{\mu_A(x), \mu_A(y)\},$$

$$\lambda_A(x + y) \leq t_1 = \max\{\lambda_A(x)), \lambda_A(y)\},$$

$$\mu_A(\alpha x) \geq s_1 = \mu_A(x), \quad \lambda_A(\alpha x) \leq t_1 = \lambda_A(x),$$

$$\mu_A([x, y]) \geq s_1 = \mu_A(x), \quad \lambda_A([x, y]) \leq t_1 = \lambda_A(x).$$

Hence, $A = (\mu_A, \lambda_A)$ is an intuitionistic fuzzy Lie ideal of L.

Proposition 2.6 *Let $A = (\mu_A, \lambda_A)$ be an intuitionistic fuzzy Lie ideal of L and $(s_1, t_1), (s_2, t_2) \in \mathrm{Im}(\mu) \times \mathrm{Im}(\lambda)$ with $s_i + t_i \leq 1$ for $i = 1, 2$. Then $L_A^{(s_1,t_1)} = L_A^{(s_2,t_2)}$ if and only if $(s_1, t_1) = (s_2, t_2)$.*

Proof If $(s_1, t_1) = (s_2, t_2)$, then clearly $L_A^{(s_1,t_1)} = L_A^{(s_2,t_2)}$. Assume that $L_A^{(s_1,t_1)} = L_A^{(s_2,t_2)}$. Since $(s_1, t_1) \in \mathrm{Im}(\mu) \times \mathrm{Im}(\lambda)$, there exists $x \in L$ such that $\mu(x) = s_1$ and $\lambda(x) = t_1$. It follows that $x \in L_A^{(s_1,t_1)} = L_A^{(s_2,t_2)}$ so that $s_1 = \mu(x) \geq s_2$ and $t_1 = \lambda(x) \leq t_2$. Similarly, we have $s_1 \leq s_2$ and $t_1 \geq t_2$. Hence, $(s_1, t_1) = (s_2, t_2)$.

Theorem 2.10 *Let $G_0 \subset G_1 \subset G_2 \subset \ldots G_n = L$ be a chain of Lie ideals of a Lie algebra L. Then there exists an intuitionistic fuzzy Lie ideal μ_A of L for which level subsets $U(\mu_A, s)$ and $L(\mu_A, t)$ coincide with this chain.*

Proof Let $\{s_k \mid k = 0, 1, \ldots, n\}$ and $\{t_k \mid k = 0, 1, \ldots, n\}$ be finite decreasing and increasing sequences in $[0, 1]$ such that $s_i + t_i \leq 1$, for $i = 0, 1, \ldots, n$. Let $A = (\mu_A, \lambda_A)$ be an intuitionistic fuzzy set in L defined by $\mu_A(G_0) = s_0$, $\lambda_A(G_0) = t_0$, $\mu_A(G_k \setminus G_{k-1}) = s_k$, and $\lambda_A(G_k \setminus G_{k-1}) = t_k$ for $0 < k \leq n$. Let $x, y \in L$. If $x, y \in G_k \setminus G_{k-1}$, then $x + y, \alpha x, [x, y] \in G_k$ and

$$\mu_A(x + y) \geq s_k = \min\{\mu_A(x), \mu_A(y)\},$$

$$\lambda_A(x + y) \leq t_k = \max\{\lambda_A(x)), \lambda_A(y)\},$$

$$\mu_A(\alpha x) \geq s_k = \mu_A(x), \quad \lambda_A(\alpha x) \leq t_k = \lambda_A(x),$$

$$\mu_A([x, y]) \geq s_k = \mu_A(x), \quad \lambda_A([x, y]) \leq t_k = \lambda_A(x).$$

For $i > j$, if $x \in G_i \setminus G_{i-1}$ and $y \in G_j \setminus G_{j-1}$, then $\mu_A(x) = s_i = \mu_A(y)$, $\lambda_A(x) = t_j = \lambda_A(y)$ and $x + y, \alpha x, [x, y] \in G_i$. Thus,

$$\mu_A(x + y) \geq s_i = \min\{\mu_A(x), \mu_A(y)\},$$

$$\lambda_A(x + y) \leq t_j = \max\{\lambda_A(x)), \lambda_A(y)\},$$

$$\mu_A(\alpha x) \geq s_i = \mu_A(x), \quad \lambda_A(\alpha x) \leq t_j = \lambda_A(x),$$

$$\mu_A([x, y]) \geq s_i = \mu_A(x), \quad \lambda_A([x, y]) \leq t_j = \lambda_A(x).$$

So, $A = (\mu_A, \lambda_A)$ is an intuitionistic fuzzy Lie ideal of a Lie algebra L and all its nonempty level subsets are Lie ideals.

Since $\mathrm{Im}(\mu_A) = \{s_0, s_1, \ldots, s_n\}$, $\mathrm{Im}(\lambda_A) = \{t_0, t_1, \ldots, t_n\}$, level subsets of A form chains:

$$U(\mu_A, s_0) \subset U(\mu_A, s_1) \subset \ldots \subset U(\mu_A, s_n) = L$$

and

$$L(\lambda_A, t_0) \subset L(\lambda_A, t_1) \subset \ldots \subset L(\lambda_A, t_n) = L,$$

respectively. Indeed,

$$U(\mu_A, s_0) = \{x \in L \mid \mu_A(x) \geq s_0\} = G_0,$$

$$L(\mu_A, t_0) = \{x \in L \mid \lambda_A(x) \leq t_0\} = G_0.$$

We now prove that

$$U(\mu_A, s_k) = G_k = L(\lambda_A, t_k) \text{ for } 0 < k \leq n.$$

Clearly, $G_k \subseteq U(\mu_k, s_k)$ and $G_k \subseteq L(\lambda_A, t_k)$. If $x \in U(\mu_A, s_k)$, then $\mu_A(x) \geq s_k$ and so $x \notin G_i$ for $i > k$. Hence,

$$\mu_A(x) \in \{s_0, s_1, \ldots, s_k\},$$

which implies $x \in G_i$ for some $i \leq k$. Since $G_i \subseteq G_k$, it follows that $x \in G_k$. Consequently, $U(\mu_A, s_k) = G_k$ for some $0 < k \leq n$. Now if $y \in L(\lambda_A, t_k)$, then $\lambda_A(x) \leq t_k$ and so $y \notin G_i$ for $j \leq k$. Thus,

$$\lambda_A(x) \in \{t_0, t_1, \ldots, t_k\},$$

which implies $x \in G_j$ for some $j \leq k$. Since $G_j \subseteq G_k$, it follows that $y \in G_k$. Consequently, $L(\lambda_A, t_k) = G_k$ for some $0 < k \leq n$. This completes the proof.

Theorem 2.11 *Let* $\{C_s \mid s \in \Lambda \subseteq [0, \frac{1}{2}]\}$ *be a collection of Lie ideals of a Lie algebra L such that* $L = \bigcup_{s \in \Lambda} C_s$, *and for every* $s, t \in \Lambda$, $s < t$ *if and only if* $C_t \subset C_s$. *Then an intuitionistic fuzzy set* $A = (\mu, \lambda)$ *defined by*

$$\mu(x) = \sup\{s \in \Lambda \mid x \in C_s\} \text{ and } \lambda(x) = \inf\{s \in \Lambda \mid x \in C_s\}$$

is an intuitionistic fuzzy Lie ideal of L.

Theorem 2.12 *If* $A = (\mu_A, \lambda_A)$ *is an intuitionistic fuzzy Lie ideal of a Lie algebra L, then*

$$\mu_A(x) = \sup\{s \in [0, 1] \mid x \in U(\mu_A, s)\},$$

$$\lambda_A(x) = \inf\{t \in [0, 1] \mid x \in L(\lambda_A, t)\}$$

for every $x \in L$.

Definition 2.21 For any $t \in [0, 1]$, we define relation $\mathscr{R}^t$ on the intuitionistic fuzzy Lie ideal over L (briefly, $IFI(L)$) as follows:

$$(A, B) \in \mathscr{R}^t \quad \longleftrightarrow \quad L_A^{(t,t)} = L_B^{(t,t)}.$$

Then the relation $\mathscr{R}^t$ is an *equivalence relation* on $IFI(L)$.

Theorem 2.13 *For any $t \in (0, 1)$, the map $\varphi_t : IFI(L) \to I(L) \cup \{\emptyset\}$ defined by $\varphi_t(A) = L_A^{(t,t)}$ is surjective, where $I(L)$ denote the family of all Lie ideals of L.*

Proof Let $t \in (0, 1)$. Then $\varphi_t(0_\sim) = L_A^{(t,t)} = U(0, t) \cap L(1, t) = \emptyset$. For any $H \in IFI(L)$, there exists $H_\sim = (\chi_H, \overline{\chi}_H) \in IFI(L)$ such that $\varphi_t(H_\sim) = L_{H_\sim}^{(t,t)} = U(\chi_H, t) \cap L(\overline{\chi}_H, t) = H$. So, φ_t is surjective.

Theorem 2.14 *For any $t \in (0, 1)$, the quotient set $IFI(L)/\mathscr{R}^t$ is equipotent to $I(L) \cup \{\emptyset\}$.*

Proof Let $t \in (0, 1)$ and let $\varphi_t^* : IFI(L)/\mathscr{R}^t \to I(L) \cup \{\emptyset\}$ be a map defined by $\varphi_t^*([A]_{\mathscr{R}^t}) = \varphi_t(A)$ for all $[A]_{\mathscr{R}^t} \in IFI(L)/\mathscr{R}^t$. If $\varphi_t^*([A]_{\mathscr{R}^t}) = \varphi_t^*([B]_{\mathscr{R}^t})$ for any $[A]_{\mathscr{R}^t}$, $[B]_{\mathscr{R}^t} \in IFI(L)/\mathscr{R}^t$, then $L_A^{(t,t)} = L_B^{(t,t)}$, i.e., $(A, B) \in \mathscr{R}^t$. It follows that $[A]_{\mathscr{R}^t} = [B]_{\mathscr{R}^t}$ so that φ_t^* is injective.

Moreover, $\varphi_t^*([0_\sim]_{\mathscr{R}^t}) = \varphi_t(0_\sim) = L_{0_\sim}^{(t,t)} = \emptyset$. For any $H \in I(L)$, we have $H_\sim = (\chi_H, \overline{\chi}_H) \in IFI(L)$ and

$$\varphi_t^*([H_\sim]_{\mathscr{R}^t}) = \varphi_t(H_\sim) = L_{H_\sim}^{(t,t)} = U(\chi_H; t) \cap L(\overline{\chi}_H; t) = H.$$

This proves that φ_t^* is surjective.

Definition 2.22 Let f be a map from a set L_1 to a set L_2. If $A = (\mu_A, \lambda_A)$ and $B = (\mu_B, \lambda_B)$ are intuitionistic fuzzy sets in L_1 and L_2 respectively, then the pre-image of B under f, denoted by $f^{-1}(B)$, is an intuitionistic fuzzy set defined by

$$f^{-1}(B) = (f^{-1}(\mu_B), f^{-1}(\lambda_B)).$$

Theorem 2.15 *Let $f : L_1 \to L_2$ be an onto homomorphism of Lie algebras. If $B = (\mu_B, \lambda_B)$ is an intuitionistic fuzzy Lie ideal of L_2, then the pre-image $f^{-1}(B) = (f^{-1}(\mu_B), f^{-1}(\lambda_B))$ of B under f is an intuitionistic fuzzy Lie ideal of L_1.*

Proof Assume that $B = (\mu_B, \lambda_B)$ is an intuitionistic fuzzy Lie ideal of L_2. Let $x, y \in L_1$ and $\alpha \in \mathbb{F}$. Then

$$\begin{aligned} f^{-1}(\mu_B)(x + y) &= \mu_B(f(x + y)) \\ &= \mu_B(f(x) + f(y)) \\ &\geq \min\{\mu_B(f(x)), \mu_B(f(y))\} \\ &= \min\{f^{-1}(\mu_B(x)), f^{-1}(\mu_B(y))\}, \end{aligned}$$

$$\begin{aligned} f^{-1}(\lambda_B)(x+y) &= \lambda_B(f(x+y)) \\ &= \lambda_B(f(x)+f(y)) \\ &\leq \max\{\lambda_B(f(x)), \lambda_B(f(y))\} \\ &= \max\{f^{-1}(\lambda_B(x)), f^{-1}(\lambda_B(y))\}, \end{aligned}$$

$$\begin{aligned} f^{-1}(\mu_B)(\alpha x) &= \mu_B(f(\alpha x)) = \mu_B(\alpha f(x)) \\ &\geq \mu_B(f(x)) = f^{-1}(\mu_B(x)), \end{aligned}$$

$$\begin{aligned} f^{-1}(\lambda_B)(\alpha x) &= \lambda_B(f(\alpha x)) = \lambda_B(\alpha f(x)) \\ &\leq \lambda_B(f(x)) = f^{-1}(\lambda_B(x)), \end{aligned}$$

$$\begin{aligned} f^{-1}(\mu_B)([x,y]) &= \mu_B(f([x,y])) = \mu_B([f(x), f(y)]) \\ &\geq \mu_B(f(x)) = f^{-1}(\mu_B(x)), \end{aligned}$$

$$\begin{aligned} f^{-1}(\lambda_B)([x,y]) &= \lambda_B(f([x,y])) = \lambda_B([f(x), f(y)]) \\ &\leq \lambda_B(f(x)) = f^{-1}(\lambda_B(x)). \end{aligned}$$

Hence, $f^{-1}(B) = (f^{-1}(\mu_B), f^{-1}(\lambda_B))$ is an intuitionistic fuzzy Lie ideal of L_1.

Theorem 2.16 *Let $f : L_1 \to L_2$ be an epimorphism of Lie algebras. If $A = (\mu_A, \lambda_A)$ is an intuitionistic fuzzy Lie ideal of L_2, then $f^{-1}(A^c) = (f^{-1}(A))^c$.*

Proof Let $A = (\mu_A, \lambda_A)$ is an intuitionistic fuzzy set in L_2. Then for $x \in L_1$,

$$\begin{aligned} f^{-1}(\mu_A^c)(x) &= \mu_A^c(f(x)) = 1 - \mu_A(f(x)) \\ &= 1 - f^{-1}(\mu_A)(x) = (f^{-1}(\mu_A))^c(x), \end{aligned}$$

$$f^{-1}(\lambda_A^c)(x) = (f^{-1}(\lambda_A))^c(x).$$

That is,

$$f^{-1}(\mu_A^c) = (f^{-1}(\mu_A))^c, \quad f^{-1}(\lambda_A^c) = (f^{-1}(\lambda_A))^c.$$

Hence,

$$f^{-1}(A^c) = f^{-1}(\mu_A^c, \lambda_A^c) = (f^{-1}(\mu_A^c), f^{-1}(\lambda_A^c)) = ((f^{-1}(\mu_A))^c, (f^{-1}(\lambda_A))^c) = (f^{-1}(A))^c.$$

Theorem 2.17 *Let $f : L_1 \to L_2$ be an epimorphism of Lie algebras. If $A = (\mu_A, \lambda_A)$ is an intuitionistic fuzzy Lie ideal of L_2 and $B = (\mu_B, \lambda_B)$ is the pre-image of $A = (\mu_A, \lambda_A)$ under f, then $B = (\mu_B, \lambda_B)$ is an intuitionistic fuzzy Lie ideal of L_1.*

Proof For any $x, y \in L_1$ and $\alpha \in \mathbb{F}$, we have

$$\begin{aligned} \mu_B(x+y) &= \mu_A(f(x+y)) = \mu_A(f(x)+f(y)) \\ &\geq \min\{\mu_A(f(x)), \mu_A(f(y))\} \end{aligned}$$

$$= \min\{\mu_B(x), \mu_B(y)\},$$

$$\begin{aligned}\lambda_B(x+y) &= \lambda_A(f(x+y)) = \lambda_A(f(x)+f(y)) \\ &\leq \max\{\lambda_A(f(x)), \lambda_A(f(y))\} \\ &= \max\{\mu_B(x), \mu_B(y)\},\end{aligned}$$

$$\begin{aligned}\mu_B(\alpha x) &= \mu_A(f(\alpha x)) = \mu_A(\alpha f(x)) \\ &\geq \mu_A(f(x)) = \mu_B(x),\end{aligned}$$

$$\begin{aligned}\lambda_B(\alpha x) &= \lambda_A(f(\alpha x)) = \lambda_A(\alpha f(x)) \\ &\leq \lambda_A(f(x)) = \lambda_B(x),\end{aligned}$$

$$\begin{aligned}\mu_B([x,y]) &= \mu_A(f([x,y])) = \mu_A([f(x), f(y)]) \\ &\geq \mu_A(f(x)) = \mu_B(x),\end{aligned}$$

$$\begin{aligned}\lambda_B([x,y]) &= \lambda_A(f([x,y])) = \lambda_A([f(x), f(y)]) \\ &\leq \lambda_A(f(x)) = \lambda_B(x).\end{aligned}$$

Hence, $B = (\mu_B, \lambda_B)$ is an intuitionistic fuzzy Lie ideal of L_1.

Definition 2.23 Let L_1 and L_2 be two Lie algebras and f be a mapping of L_1 into L_2. If $A = (\mu_A, \lambda_A)$ is an intuitionistic fuzzy set of L_1, then the image of $A = (\mu_A, \lambda_A)$ under f is the intuitionistic fuzzy set in L_2 defined by

$$f(\mu_A)(y) = \begin{cases} \sup_{x \in f^{-1}(y)} \mu_A(x), & \text{if } f^{-1}(y) \neq \emptyset, \\ 0, & \text{otherwise,} \end{cases}$$

$$f(\lambda_A)(y) = \begin{cases} \inf_{x \in f^{-1}(y)} \lambda_A(x), & \text{if } f^{-1}(y) \neq \emptyset, \\ 1, & \text{otherwise.} \end{cases}$$

for each $y \in L_2$.

Theorem 2.18 *Let $f: L_1 \to L_2$ be an epimorphism of Lie algebras. If $A = (\mu_A, \lambda_A)$ is an intuitionistic fuzzy Lie ideal of L_1, then $f(A)$ is an intuitionistic fuzzy Lie ideal of L_2.*

Proof Let $y_1, y_2 \in L_2$, then

$$\{x \mid x \in f^{-1}(y_1+y_2)\} \supseteq \{x_1+x_2 \mid x_1 \in f^{-1}(y_1) \textit{ and } x_2 \in f^{-1}(y_2)\}.$$

Now for $n > 1$, we get

$$\begin{aligned}f(\mu_A)(y_1+y_2) &= \sup\{\mu_A(x) \mid x \in f^{-1}(y_1+y_2)\} \\ &\geq \{\mu_A(x_1+x_2) \mid x_1 \in f^{-1}(y_1) \textit{ and } x_2 \in f^{-1}(y_2)\} \\ &\geq \sup\{\min\{\mu_A(x_1), \mu_A(x_2)\} \mid x_1 \in f^{-1}(y_1) \textit{ and } x_2 \in f^{-1}(y_2)\} \\ &= \min\{\sup\{\mu_A(x_1) \mid x_1 \in f^{-1}(y_1)\}, \sup\{\mu_A(x_2) \mid x_2 \in f^{-1}(y_2)\}\}\end{aligned}$$

$$= \min\{f(\mu_A)(y_1), f(\mu_A)(y_2)\}.$$

Let $y \in L_2$ and $\alpha \in \mathbb{F}$, then

$$\{x \mid x \in f^{-1}(\alpha y)\} \supseteq \{\alpha x, \mid x \in f^{-1}(y)\}.$$

$$\begin{aligned} f(\mu_A)(\alpha y) &= \sup\{\mu_A(\alpha x) \mid x \in f^{-1}(y)\} \\ &\geq \{\mu_A(\alpha x) \mid x \in f^{-1}(\alpha y)\} \\ &\geq \sup\{\mu_A(x) \mid x \in f^{-1}(y)\} \\ &= f(\mu_A)(y). \end{aligned}$$

Let $y_1, y_2 \in L_2$, then

$$\{x \mid x \in f^{-1}([y_1, y_2])\} \supseteq \{[x_1, x_2] \mid x_1 \in f^{-1}(y_1),\ x_2 \in f^{-1}(y_2)\}.$$

Now

$$\begin{aligned} f(\mu_A)([y_1, y_2]) &= \sup\{\mu_A(x) \mid x \in f^{-1}([y_1, y_2])\} \\ &\geq \{\mu_A([x_1, x_2]) \mid x_1 \in f^{-1}(y_1),\ x_2 \in f^{-1}(y_2)\} \\ &\geq \sup\{\mu_A(x_1) \mid x_1 \in f^{-1}(y_1)\} \\ &= f(\mu_A)(y_1). \end{aligned}$$

Thus, $f(\mu_A)$ is a fuzzy Lie ideal of L_2. In the same manner, we can prove that $f(\lambda_A)$ is a fuzzy Lie ideal of L_2. Hence, $f(A) = (f(\mu_A), f(\lambda_A))$ is an intuitionistic fuzzy Lie ideal of L_2.

Definition 2.24 Let $f : L_1 \to L_2$ be a homomorphism of Lie algebras. For any intuitionistic fuzzy set $A = (\mu_A, \lambda_A)$ in a Lie algebra L_2, we define an intuitionistic fuzzy set $A^f = (\mu_A^f, \lambda_A^f)$ in L_2 by

$$\mu_A^f(x) = \mu_A(f(x)), \qquad \lambda_A^f(x) = \lambda_A(f(x))$$

for all $x \in L_1$. Clearly, $A^f(x_1) = A^f(x_2) = A(x)$ for all $x_1, x_2 \in f^{-1}(x)$.

Lemma 2.1 *Let $f : L_1 \to L_2$ be a homomorphism of Lie algebras. If $A = (\mu_A, \lambda_A)$ is an intuitionistic fuzzy Lie ideal of L_2, then $A^f = (\mu_A^f, \lambda_A^f)$ is an intuitionistic fuzzy Lie ideal of L_1.*

Proof Let $x, y \in L_1$ and $\alpha \in \mathbb{F}$. Then

$$\mu_A^f(x+y) = \mu_A(f(x+y)) = \mu_A(f(x)+f(y)) \geq \min\{\mu_A(f(x)), \mu_A(f(y))\} = \min\{\mu_A^f(x), \mu_A^f(y)\},$$

$$\lambda_A^f(x+y) = \lambda_A(f(x+y)) = \lambda_A(f(x)+f(y)) \leq \max\{\lambda_A(f(x)), \lambda_A(f(y))\} = \max\{\lambda_A^f(x), \lambda_A^f(y)\},$$

$$\mu_A^f(\alpha x) = \mu_A(f(\alpha x)) = \mu_A(\alpha f(x)) \geq \mu_A(f(x)) = \alpha_A^f(x),$$

$$\lambda_A^f(\alpha x) = \lambda_A(\alpha f(x)) \leq \lambda_A(f(x)) = \lambda_A^f(x).$$

Similarly,

$$\mu_A^f([x, y]) = \mu_A(f([x, y])) = \mu_A([f(x), f(y)]) \geq \mu_A(f(x)) = \mu_A^f(x),$$

$$\lambda_A^f([x, y]) = \lambda_A([f(x, y)]) = \lambda_A([f(x), f(y)]) \leq \lambda_A(f(x)) = \lambda_A^f(x).$$

This proves that $A^f = (\mu_A^f, \lambda_A^f)$ is an intuitionistic fuzzy Lie ideal of L_1.

We now characterize the intuitionistic fuzzy Lie ideals of Lie algebras.

Theorem 2.19 *Let $f : L_1 \to L_2$ be an epimorphism of Lie algebras. Then $A^f = (\mu_A^f, \lambda_A^f)$ is an intuitionistic fuzzy Lie ideal of L_1 if and only if $A = (\mu_A, \lambda_A)$ is an intuitionistic fuzzy Lie ideal of L_2.*

Proof The sufficiency follows from Lemma 2.1. In proving the necessity, since f is a surjective mapping, for any $x, y \in L_2$ there are $x_1, y_1 \in L_1$ such that $x = f(x_1)$, $y = f(y_1)$. Thus, $\mu_A(x) = \mu_A^f(x_1)$, $\mu_A(y) = \mu_A^f(y_1)$, $\lambda_A(x) = \lambda_A^f(x_1)$, $\lambda_A(y) = \lambda_A^f(y_1)$, whence

$$\begin{aligned} \mu_A(x+y) &= \mu_A(f(x_1)+f(y_1)) = \mu_A(f(x_1+y_1)) = \mu_A^f(x_1+y_1) \\ &\geq \min\{\mu_A^f(x_1), \mu_A^f(y_1)\} = \min\{\mu_A(x), \mu_A(y)\} \\ \lambda_A(x+y) &= \lambda_A(f(x_1)+f(y_1)) = \lambda_A(f(x_1+y_1)) = \lambda_A^f(x_1+y_1) \\ &\leq \max\{\lambda_A^f(x_1), \lambda_A^f(y_1)\} = \max\{\lambda_A(x), \lambda_A(y)\}, \end{aligned}$$

$$\mu_A(\alpha x) = \mu_A(\alpha f(x_1)) = \mu_A(f(\alpha x_1)) = \mu_A^f(\alpha x_1) \geq \mu_A^f(x_1) = \mu_A(x),$$

$$\lambda_A(\alpha x) = \lambda_A(\alpha f(x_1)) = \lambda_A(f(\alpha x_1)) = \lambda_A^f(\alpha x_1) \leq \lambda_A^f(x_1) = \lambda_A(x).$$

Similarly,

$$\begin{aligned} \mu_A([x, y]) &= \mu_A([f(x_1), f(y_1)]) = \mu_A(f([x_1, y_1])) = \mu_A^f([x_1, y_1]) \\ &\geq \mu_A^f(x_1) = \mu_A(x), \\ \lambda_A([x, y]) &= \lambda_A([f(x_1), f(y_1)]) = \lambda_A(f([x_1, y_1])) = \lambda_A^f([x_1, y_1]) \\ &\leq \lambda_A^f(x_1) = \lambda_A(x). \end{aligned}$$

This proves that $A = (\mu_A, \lambda_A)$ is an intuitionistic fuzzy Lie ideal of L_2.

2.5 Special Types of Intuitionistic Fuzzy Lie Ideals

Definition 2.25 Let $A = (\mu_A, \lambda_A)$ be an intuitionistic fuzzy Lie ideal in L. Define inductively a sequence of intuitionistic fuzzy Lie ideals in L by

$$A^0 = A,\ A^1 = [A^0, A^0],\ A^2 = [A^1, A^1], \ldots, A^n = [A^{n-1}, A^{n-1}].$$

A^n is called the *nth derived intuitionistic fuzzy Lie ideal* of L. A series

$$A^0 \supseteq A^1 \supseteq A^2 \supseteq \cdots \supseteq A^n \supseteq \cdots$$

is called *derived series* of an intuitionistic fuzzy Lie ideal A in L.

Definition 2.26 An intuitionistic fuzzy Lie ideal A in L is called a *solvable intuitionistic fuzzy Lie ideal* if there exists a positive integer n such that

$$A^0 \supseteq A^1 \supseteq A^2 \supseteq \cdots \supseteq A^n = 0_{\sim}.$$

Theorem 2.20 *Homomorphic images of solvable intuitionistic fuzzy Lie ideals are solvable intuitionistic fuzzy Lie ideals.*

Proof Let $f : L_1 \to L_2$ be a homomorphism of Lie algebras. Suppose that $A = (\mu_A, \lambda_A)$ is an intuitionistic fuzzy Lie ideal in L_1. We prove by induction on n that $f(A^n) \supseteq [f(A)]^n$, where n is any positive integer. First we claim that $f([A, A]) \supseteq [f(A), f(A)]$. Let $y \in L_2$, then

$$\begin{aligned}
f(\ll \mu_A, \mu_A \gg)(y) &= \sup\{\ll \mu_A, \mu_A \gg (x) \mid f(x) = y\}\\
&= \sup\{\sup\{\min(\mu_A(a), \mu_A(b)) \mid a, b \in L_1, [a, b] = x, f(x) = y\}\}\\
&= \sup\{\min(\mu_A(a), \mu_A(b)) \mid a, b \in L_1, [a, b] = x, f(x) = y\}\\
&= \sup\{\min(\mu_A(a), \mu_A(b)) \mid a, b \in L_1, [f(a), f(b)] = x\}\\
&= \sup\{\min(\mu_A(a), \mu_A(b)) \mid a, b \in L_1, f(a) = u, f(b) = v, [u, v] = y\}\\
&\geq \sup\{\min(\sup_{a \in f^{-1}(u)} \mu_A(a), \sup_{b \in f^{-1}(v)} \mu_A(b)) \mid [u, v] = y\}\\
&= \sup\{\min(f(\mu_A)(u), f(\mu_A)(v)) \mid [u, v] = y\}\\
&= \ll f(\mu_A), f(\mu_A) \gg (y),
\end{aligned}$$

$$\begin{aligned}
f(\ll \lambda_A, \lambda_A \gg)(y) &= \inf\{\ll \lambda_A, \lambda_A \gg)(x) \mid f(x) = y\}\\
&= \inf\{\inf\{\max(\lambda_A(a), \lambda_A(b)) \mid a, b \in L_1, [a, b] = x, f(x) = y\}\}\\
&= \inf\{\max(\lambda_A(a), \lambda_A(b)) \mid a, b \in L_1, [a, b] = x, f(x) = y\}\\
&= \inf\{\max(\lambda_A(a), \lambda_A(b)) \mid a, b \in L_1, [f(a), f(b)] = x\}\\
&= \inf\{\max(\lambda_A(a), \lambda_A(b)) \mid a, b \in L_1, f(a) = u, f(b) = v, [u, v] = y\}\\
&\leq \inf\{\max(\inf_{a \in f^{-1}(u)} \lambda_A(a), \inf_{b \in f^{-1}(v)} \lambda_A(b)) \mid [u, v] = y\}\\
&= \inf\{\max(f(\lambda_A)(u), f(\lambda_A)(v)) \mid [u, v] = y\}\\
&= \ll f(\lambda_A), f(\lambda_A) \gg (y).
\end{aligned}$$

Thus,

$$f([A,A]) \supseteq f(\ll A, A \gg) \supseteq \ll f(A), f(A) \gg = [f(A), f(A)].$$

Now for $n > 1$, we get

$$\begin{aligned} f(A^n) &= f([A^{n-1}, A^{n-1}]) \supseteq [f(A^{n-1}), f(A^{n-1})] \\ &\supseteq [(f(A))^{n-1}, (f(A))^{n-1}] = (f(A))^n. \end{aligned}$$

This completes the proof.

Definition 2.27 Let $A = (\mu_A, \lambda_A)$ be an intuitionistic fuzzy Lie ideal in L. We define inductively a sequence of intuitionistic fuzzy Lie ideals in L by

$$A_0 = A,\ A_1 = [A, A_0],\ A_2 = [A, A_1], \ldots, , A_n = [A, A_{n-1}].$$

A series

$$A_0 \supseteq A_1 \supseteq A_2 \supseteq \cdots \supseteq A_n \supseteq \cdots$$

is called *descending central series* of an intuitionistic fuzzy Lie ideal A in L.

Definition 2.28 An intuitionistic fuzzy Lie ideal A in L is called a *nilpotent intuitionistic fuzzy Lie ideal*, if there exists a positive integer n such that

$$A_0 \supseteq A_1 \supseteq A_2 \supseteq \cdots \supseteq A_n = 0_{\sim}.$$

Theorem 2.21 *Homomorphic image of a nilpotent intuitionistic fuzzy Lie ideal is a nilpotent intuitionistic fuzzy Lie ideal.*

Proof Straightforward.

Theorem 2.22 *Let J be a Lie ideal of a Lie algebra L. If $A = (\mu_A, \lambda_A)$ is an intuitionistic fuzzy Lie ideal of L, then the intuitionistic fuzzy set $\overline{A} = (\overline{\mu}_A, \overline{\lambda}_A)$ of L/J defined by*

$$\overline{\mu}_A(a + J) = \sup_{x \in J} \mu_A(a + x),$$

$$\overline{\lambda}_A(a + J) = \inf_{x \in J} \lambda_A(a + x)$$

is an intuitionistic fuzzy Lie ideal of the quotient Lie algebra L/J of L with respect to J.

Proof Clearly, $\overline{A}$ is well defined. Let $x + J, y + J \in L/J$, then

$$\begin{aligned} \overline{\mu}_A((x + J) + (y + J)) &= \overline{\mu}_A((x + y) + J) \\ &= \sup_{z \in J} \mu_A((x + y) + z) \end{aligned}$$

$$\begin{aligned}
&= \sup_{z=s+t\in J} \mu_A((x+y)+(s+t)) \\
&\geq \sup_{s,\ t\in J} \min\{\mu_A(x+s), \mu_A(y+t)\} \\
&= \min\{\sup_{s\in J} \mu_A(x+s), \sup_{t\in J} \mu_A(y+t)\} \\
&= \min\{\overline{\mu}_A(x+J), \overline{\mu}_A(y+J)\},
\end{aligned}$$

$$\begin{aligned}
\overline{\mu}_A(\alpha(x+J)) &= \overline{\mu}_A(\alpha x+J) = \sup_{z\in J} \mu(\alpha x+z) \\
&\geq \sup_{z\in J} \mu(x+z) = \overline{\mu}_A(x+J), \\
\overline{\mu}_A([x+J, y+J]) &= \overline{\mu}_A([x,y]+J) = \sup_{z\in J} \mu_A([x,y]+z) \\
&\geq \sup_{z\in J} \mu_A(x+z) = \overline{\mu}_A(x+J).
\end{aligned}$$

Thus, $\overline{\mu}_A$ is an intuitionistic fuzzy Lie ideal of L/J. In a similar way, we can verify that $\overline{\lambda}_A$ is an intuitionistic fuzzy Lie ideal of L/J. Hence, $\overline{A} = (\overline{\mu}_A, \overline{\lambda}_A)$ is an intuitionistic fuzzy Lie ideal of L/J.

Theorem 2.23 *Let J be a Lie ideal of a Lie algebra L. Then there is a one-to-one correspondence between the set of intuitionistic fuzzy Lie ideals $A = (\mu_A, \lambda_A)$ of L such that $A(0) = A(s)$ for all $s \in J$ and the set of all intuitionistic fuzzy Lie ideals $\overline{A} = (\overline{\mu}_A, \overline{\lambda}_A)$ of L/J.*

Proof Let $A = (\mu_A, \lambda_A)$ be an intuitionistic fuzzy Lie ideal of L. Using Theorem 2.22, we prove that $\overline{\mu}_A$ and $\overline{\lambda}_A$ defined by

$$\overline{\mu}_A(a+J) = \sup_{x\in J} \mu_A(a+x)$$

$$\overline{\lambda}_A(a+J) = \inf_{x\in J} \lambda_A(a+x)$$

are intuitionistic fuzzy Lie ideals of L/J. Since $\mu_A(0) = \mu_A(s)$, $\lambda_A(0) = \lambda_A(s)$ for all $s \in J$,

$$\mu_A(a+s) \geq \min(\mu_A(a), \mu_A(s)) = \mu_A(a),$$

$$\lambda_A(a+s) \leq \max(\lambda_A(a), \lambda_A(s)) = \lambda_A(a).$$

Again,

$$\begin{aligned}
\mu_A(a) &= \mu_A(a+s-s) \\
&\geq \min(\mu_A(a+s), \mu_A(s)) \\
&= \mu_A(a+s).
\end{aligned}$$

$$\begin{aligned}\lambda_A(a) &= \lambda_A(a+s-s)\\ &\leq \max(\lambda_A(a+s), \lambda_A(s))\\ &= \lambda_A(a+s).\end{aligned}$$

Thus, $A(a+s) = A(a)$ for all $s \in J$. Hence, the correspondence $A \mapsto \overline{A}$ is one to one. Let $\overline{A}$ be an intuitionistic fuzzy Lie ideal of L/J and define an intuitionistic fuzzy set $A = (\mu_A, \lambda_A)$ in L by $\mu_A(a) = \overline{\mu}_A(a+J)$, $\lambda_A(a) = \overline{\lambda}_A(a+J)$ for all $a \in J$.

For $x, y \in L$, we have

$$\begin{aligned}\mu_A(x+y) &= \overline{\mu}_A((x+y)+J)\\ &= \overline{\mu}_A((x+J)+(y+J))\\ &\geq \min\{\overline{\mu}_A(x+J), \overline{\mu}_A(y+J)\}\\ &= \min\{\mu_A(x), \mu_A(y)\},\end{aligned}$$

$$\mu_A(\alpha x) = \overline{\mu}_A(\alpha x+J) \geq \overline{\mu}_A(x+J) = \mu_A(x),$$

$$\begin{aligned}\mu_A([x,y]) &= \overline{\mu}_A([x,y]+J) = \overline{\mu}_A([x+J, y+J])\\ &\geq \overline{\mu}_A(x+J) = \mu_A(x).\end{aligned}$$

Thus, μ_A is an intuitionistic fuzzy Lie ideal of L. In a similar way, we can verify that λ_A is an intuitionistic fuzzy Lie ideal of L. Hence, $A = (\mu_A, \lambda_A)$ is an intuitionistic fuzzy Lie ideal of L. Note that $\mu_A(z) = \overline{\mu}_A(z+J) = \mu_A(J)$, $\lambda_A(z) = \overline{\lambda}_A(z+J) = \lambda_A(J)$ for all $z \in J$, which shows that $A(z) = A(0)$ for all $z \in J$. This ends the proof.

2.6 Intuitionistic (S, T)-Fuzzy Lie Ideals

Definition 2.29 An intuitionistic fuzzy set $A = (\mu_A, \lambda_A)$ on L is called an *intuitionistic fuzzy Lie ideal* of L with respect to the t-norm T and the s-norm S (shortly, intuitionistic (S, T)-fuzzy Lie ideals of L) if

1. $\mu_A(x+y) \geq T(\mu_A(x), \mu_A(y))$ and $\lambda_A(x+y) \leq S(\lambda_A(x), \lambda_A(y))$,
2. $\mu_A(\alpha x) \geqslant \mu_A(x)$ and $\lambda_A(\alpha x) \leq \lambda_A(x)$,
3. $\mu_A([x,y]) \geq \mu_A(x)$ and $\lambda_A([x,y]) \leq \lambda_A(x)$ hold for all $x, y \in L$ and $\alpha \in \mathbb{F}$.
   From (2) it follows that
4. $\mu_A(0) \geq \mu_A(x)$ and $\lambda_A(0) \leq \lambda_A(x)$,
5. $\mu_A(-x) = \mu_A(x)$ and $\lambda_A(-x) = \lambda_A(x)$

for all $x \in L$.

Example 2.6 Let $\Re^2 = \{(x, y) \mid x, y \in \mathbb{R}\}$ be the set of all two-dimensional real vectors. Then $\Re^2$ with the bracket $[\ ,\ ]$ defined as usual cross product, i.e., $[x, y] = x \times y$, is a real Lie algebra. We define an intuitionistic fuzzy set $A = (\mu_A, \lambda_A) : L \to [0, 1] \times [0, 1]$ as follows:

$$\mu_A(x, y) = \begin{cases} m_1 \text{ if } x = y = 0, \\ m_2 \text{ otherwise,} \end{cases} \qquad \lambda_A(x, y) = \begin{cases} m_2 \text{ if } x = y = 0, \\ m_1 \text{ otherwise,} \end{cases}$$

where $m_1 > m_2$ and $m_1, m_2 \in [0, 1]$. Let T be a t-norm which is defined by $T(x, y) = \max\{x + y - 1, 0\}$ and S an s-norm which is defined by $S(x, y) = \min\{x + y, 1\}$ for all $x, y \in [0, 1]$. Then by routine computations, we see that $A = (\mu_A, \lambda_A)$ is an intuitionistic (S, T)-fuzzy Lie ideal of L.

The following proposition is obvious.

Proposition 2.7 *If A is an intuitionistic (S, T)-fuzzy Lie ideal of L, then*

(i) $\mu_A([x, y]) \geq S(\mu_A(x), \mu_A(y))$,
(ii) $\lambda_A([x, y]) \leq T(\lambda_A(x), \lambda_A(y))$

for all $x, y \in L$. □

Theorem 2.24 *Let $G_0 \subset G_1 \subset G_2 \subset \ldots G_n = L$ be a chain of Lie ideals of a Lie algebra L. Then there exists an intuitionistic (S, T)-fuzzy Lie ideal A of L for which level subsets $U(\mu_A, s)$ and $L(\lambda_A, t)$ coincide with this chain.*

Proof Let $\{s_k \mid k = 0, 1, \ldots, n\}$ and $\{t_k \mid k = 0, 1, \ldots, n\}$ be finite decreasing and increasing sequences in $[0, 1]$ such that $s_i + t_i \leq 1$, for $i = 0, 1, \ldots, n$. Let $A = (\mu_A, \lambda_A)$ be an intuitionistic fuzzy set in L defined by $\mu_A(G_0) = s_0$, $\lambda_A(G_0) = t_0$, $\mu_A(G_k \setminus G_{k-1}) = s_k$, and $\lambda_A(G_k \setminus G_{k-1}) = t_k$ for $0 < k \leq n$. Let $x, y \in L$. If $x, y \in G_k \setminus G_{k-1}$, then $x + y, \alpha x, [x, y] \in G_k$ and

$$\mu_A(x + y) \geq slants_k = T(\mu_A(x), \mu_A(y)),$$

$$\lambda_A(x + y) \leq t_k = S(\lambda_A(x)), \lambda_A(y)),$$

$$\mu_A(\alpha x) \geq s_k = \mu_A(x), \quad \lambda_A(\alpha x) \leq t_k = \lambda_A(x),$$

$$\mu_A([x, y]) \geq s_k = \mu_A(x), \quad \lambda_A([x, y]) \leq t_k = \lambda_A(x).$$

For $i > j$, if $x \in G_i \setminus G_{i-1}$ and $y \in G_j \setminus G_{j-1}$, then $\mu_A(x) = s_i = \mu_A(y)$, $\lambda_A(x) = t_j = \lambda_A(y)$ and $x + y, \alpha x, [x, y] \in G_i$. Thus,

$$\mu_A(x + y) \geq s_i = T(\mu_A(x), \mu_A(y)),$$

$$\lambda_A(x + y) \leq t_j = S(\lambda_A(x)), \lambda_A(y)),$$

$$\mu_A(\alpha x) \geq s_i = \mu_A(x), \quad \lambda_A(\alpha x) \leq t_j = \lambda_A(x),$$

$$\mu_A([x, y]) \geq s_i = \mu_A(x), \quad \lambda_A([x, y]) \leq t_j = \lambda_A(x).$$

So, $A = (\mu_A, \lambda_A)$ is an intuitionistic (S, T)-fuzzy Lie ideal of a Lie algebra L and all its nonempty level subsets are Lie ideals. Since $\mathrm{Im}(\mu_A) = \{s_0, s_1, \ldots, s_n\}$, $\mathrm{Im}(\lambda_A) = \{t_0, t_1, \ldots, t_n\}$, level subsets of A form chains:

$$U(\mu_A, s_0) \subset U(\mu_A, s_1) \subset \ldots \subset U(\mu_A, s_n) = L$$

and

$$L(\lambda_A, t_0) \subset L(\lambda_A, t_1) \subset \ldots \subset L(\lambda_A, t_n) = L,$$

respectively. Indeed,

$$U(\mu_A, s_0) = \{x \in L \mid \mu_A(x) \geq s_0\} = G_0,$$

$$L(\lambda_A, t_0) = \{x \in L \mid \lambda_A(x) \leq t_0\} = G_0.$$

We now prove that

$$U(\mu_A, s_k) = G_k = L(\lambda_A, t_k) \text{ for } 0 < k \leq n.$$

Clearly, $G_k \subseteq U(\mu_k, s_k)$ and $G_k \subseteq L(\lambda_A, t_k)$. If $x \in U(\mu_A, s_k)$, then $\mu_A(x) \geq s_k$ and so $x \notin G_i$ for $i > k$. Hence,

$$\mu_A(x) \in \{s_0, s_1, \ldots, s_k\},$$

which implies $x \in G_i$ for some $i \leq k$. Since $G_i \subseteq G_k$, it follows that $x \in G_k$. Consequently, $U(\mu_A, s_k) = G_k$ for some $0 < k \leq n$. Now if $x \subset L(\lambda_A, t_k)$, then $\lambda_A(x) \leq t_k$ and so $x \notin G_i$ for $j \leq k$. Thus,

$$\lambda_A(x) \in \{t_0, t_1, \ldots, t_k\},$$

which implies $x \in G_j$ for some $j \leq k$. Since $G_j \subseteq G_k$, it follows that $x \in G_k$. Consequently, $L(\lambda_A, t_k) = G_k$ for some $0 < k \leq n$. This completes the proof.

Definition 2.30 Let $f : L_1 \to L_2$ be a homomorphism of Lie algebras. Let $A = (\mu_A, \lambda_A)$ be an intuitionistic fuzzy set of L_2. Then we can define an intuitionistic fuzzy set $f^{-1}(A)$ of L_1 by

$$f^{-1}(A)(x) = A(f(x)) = (\mu_A(f(x)), \lambda_A(f(x))) \ \forall\, x \in L_1.$$

Proposition 2.8 *Let $f : L_1 \to L_2$ be an epimorphism of Lie algebras. Then A is an intuitionistic (S, T)-fuzzy Lie ideal of L_2 if and only if $f^{-1}(A)$ is an intuitionistic (S, T)-fuzzy Lie ideal of L_1.*

Proof Straightforward.

Definition 2.31 Let $f : L_1 \to L_2$ be a homomorphism of Lie algebras. Let $A = (\mu_A, \lambda_A)$ be an intuitionistic fuzzy set of L_1. Then intuitionistic fuzzy set $f(A) = (f(\mu_A), f(\lambda_A))$ in L_2 is defined by

$$f(\mu_A)(y) = \begin{cases} \sup\{\mu_A(t) \mid t \in L_1, f(t) = y\}, & \text{if } f^{-1}(y) \neq \emptyset, \\ 0, & \text{otherwise,} \end{cases}$$

$$f(\lambda_A)(y) = \begin{cases} \inf\{\lambda_A(t) \mid t \in L_1, f(t) = y\}, & \text{if } f^{-1}(y) \neq \emptyset, \\ 1, & \text{otherwise.} \end{cases}$$

Definition 2.32 Let L_1 and L_2 be any sets and $f : L_1 \to L_2$ any function. Then, we call an intuitionistic fuzzy set $A = (\mu_A, \lambda_A)$ of L_1 *f-invariant* if $f(x) = f(y)$ implies $A(x) = A(y)$, i.e., $\mu_A(x) = \mu_A(y)$, $\lambda_A(x) = \lambda_A(y)$ for $x, y \in L_1$.

Theorem 2.25 *Let $f : L_1 \to L_2$ be an epimorphism of Lie algebras. Then $A = (\mu_A, \lambda_A)$ is an f-invariant intuitionistic (S, T)-fuzzy Lie ideal of L_1 if and only if $f(A)$ is an intuitionistic (S, T)-fuzzy Lie ideal of L_2.*

Proof Let $x, y \in L_2$. Then there exist $a, b \in L_1$ such that $f(a) = x$, $f(b) = y$, and $x + y = f(a + b)$, $\alpha x = \alpha f(a)$. Since A is f-invariant, by straightforward verification, we have

$$f(\mu_A)(x + y) = \mu_A(a + b) \geq T(\mu_A(a), \mu_A(b)) = T(f(\mu_A)(x), f(\mu_A)(y)),$$
$$f(\lambda_A)(x + y) = \lambda_A(a + b) \leq S(\lambda_A(a), \lambda_A(b)) = S(f(\lambda_A)(x), f(\lambda_A)(y)),$$
$$f(\mu_A)(\alpha x) = \mu_A(\alpha a) \geq \mu_A(a) = f(\mu_A)(x),$$
$$f(\lambda_A)(\alpha x) = \lambda_A(\alpha a) \leq \lambda_A(a) = f(\lambda_A)(x),$$
$$f(\mu_A)([x, y]) = \mu_A([a, b]) = [\mu_A(a), \mu_A(b)] \geq \mu_A(a) = f(\mu_A)(x),$$
$$f(\lambda_A)([x, y]) = \lambda_A([a, b]) = [\lambda_A(a), \lambda_A(b)] \leq \lambda_A(a) = f(\lambda_A)(x).$$

Hence, $f(A)$ is an intuitionistic (S, T)-fuzzy Lie ideal of L_2.

Conversely, if $f(A)$ is an intuitionistic (S, T)-fuzzy Lie ideal of L_2, then for any $x \in L_1$

$$\begin{aligned} f^{-1}(f(\mu_A))(x) &= f(\mu_A)(f(x)) = \sup\{\mu_A(t) \mid t \in L_1, f(t) = f(x)\} \\ &= \sup\{\mu_A(t) \mid t \in L_1, \mu_A(t) = \mu_A(x)\} = \mu_A(x), \end{aligned}$$

$$\begin{aligned} f^{-1}(f(\lambda_A))(x) &= f(\lambda_A)(f(x)) = \inf\{\lambda_A(t) \mid t \in L_1, f(t) = f(x)\} \\ &= \inf\{\lambda_A(t) \mid t \in L_1, \lambda_A(t) = \lambda_A(x)\} = \lambda_A(x). \end{aligned}$$

Hence, $f^{-1}(f(A)) = A$ is an intuitionistic (S, T)-fuzzy Lie ideal.

Lemma 2.2 *Let $A = (\mu_A, \lambda_A)$ be an intuitionistic (S, T)-fuzzy Lie ideal of a Lie algebra L and let $x \in L$. Then $\mu_A(x) = t$, $\lambda_A(x) = s$ if and only if $x \in U(\mu_A, t)$, $x \notin U(\mu_A, s)$ and $x \in L(\lambda_A, s)$, $x \notin L(\lambda_A, t)$, for all $s > t$.*

Proof Straightforward.

Definition 2.33 An intuitionistic (S, T)-fuzzy Lie ideal $A = (\mu_A, \lambda_A)$ of a Lie algebra L is called *characteristic* if $\mu_A(f(x)) = \mu_A(x)$ and $\lambda_A(f(x)) = \lambda_A(x)$ for all $x \in L$ and $f \in Aut(L)$.

Theorem 2.26 *An intuitionistic (S, T)-fuzzy Lie ideal is characteristic if and only if each of its level set is a characteristic Lie ideal.*

Proof Let an intuitionistic (S, T)-fuzzy Lie ideal $A = (\mu_A, \lambda_A)$ be characteristic, $t \in Im(\mu_A), f \in Aut(L)$, $x \in U(\mu_A, t)$. Then $\mu_A(f(x)) = \mu_A(x) \geq t$, which means that $f(x) \in U(\mu_A, t)$. Thus, $f(U(\mu_A, t)) \subseteq U(\mu_A, t)$. Since for each $x \in U(\mu_A, t)$ there exists $y \in L$ such that $f(y) = x$ we have $\mu_A(y) = \mu_A(f(y)) = \mu_A(x) \geq t$, whence we conclude $y \in U(\mu_A, t)$. Consequently $x = f(y) \in f(U(\mu_A, t))$. Hence, $f(U(\mu_A, t)) = U(\mu_A, t)$. Similarly, $f(L(\lambda_A, s)) = L(\lambda_A, s)$. This proves that $U(\mu_A, t)$ and $L(\lambda_A, s)$ are characteristic.

Conversely, if all levels of $A = (\mu_A, \lambda_A)$ are characteristic Lie ideals of L, then for $x \in L$, $f \in Aut(L)$, and $\mu_A(x) = t < s = \lambda_A(x)$, by Lemma 2.2, we have $x \in U(\mu_A, t), x \notin U(\mu_A, s)$, and $x \in L(\lambda_A, s), x \notin L(\lambda_A, t)$. Thus, $f(x) \in f(U(\mu_A, t)) = U(\mu_A, t)$ and $f(x) \in f(L(\lambda_A, s)) = L(\lambda_A, s)$, i.e., $\mu_A(f(x)) \geq t$ and $\lambda_A(f(x)) \leq s$. For $\mu_A(f(x)) = t_1 > t$, $\lambda_A(f(x)) = s_1 < s$, we have $f(x) \in U(\mu_A, t_1) = f(U(\mu_A, t_1))$, $f(x) \in L(\lambda_A, s_1) = f(L(\lambda_A, s_1))$, whence $x \in U(\mu_A, t_1), x \in L(\mu_A, s_1)$. This is a contradiction. Thus, $\mu_A(f(x)) = \mu_A(x)$ and $\lambda_A(f(x)) = \lambda_A(x)$. So, $A = (\mu_A, \lambda_A)$ is characteristic.

We sate the following theorem without proof.

Theorem 2.27 *Let $\{C_s \mid s \in \Lambda \subseteq [0, \frac{1}{2}]\}$ be a collection of Lie ideals of a Lie algebra L such that $L = \bigcup_{s \in \Lambda} C_s$, and for every $s, t \in \Lambda$, $s < t$ if and only if $C_t \subset C_s$. Then an intuitionistic fuzzy set $A = (\mu_A, \lambda_A)$ defined by*

$$\mu_A(x) = \sup\{s \in \Lambda \mid x \in C_s\} \quad \textit{and} \quad \lambda_A(x) = \inf\{s \in \Lambda \mid x \in C_s\}$$

is an intuitionistic (S, T)-fuzzy Lie ideal of L. □

Theorem 2.28 *Let $A = (\mu_A, \lambda_A)$ be an intuitionistic (S, T)-fuzzy Lie ideal of Lie algebra L. Define a binary relation $\sim$ on L by*

$$x \sim y \longleftrightarrow \mu_A(x - y) = \mu_A(0) \textit{ and } \lambda_A(x - y) = \lambda_A(0).$$

Then $\sim$ is a congruence on L.

Proof The reflexivity and symmetry are obvious. To prove transitivity, let $x \sim y$ and $y \sim z$. Then $\mu_A(x - y) = \mu_A(0)$, $\mu_A(y - z) = \mu_A(0)$ and $\lambda_A(x - y) = \lambda_A(0)$, $\lambda_A(y - z) = \lambda_A(0)$. Thus,

$$\mu_A(x - z) = \mu_A(x - y + y - z) \geq T(\mu_A(x - y), \mu_A(y - z)) = \mu_A(0),$$
$$\lambda_A(x - z) = \lambda_A(x - y + y - z) \leq S(\lambda_A(x - y), \lambda_A(y - z)) = \lambda(0),$$

whence, we conclude $x \sim z$.

If $x_1 \sim y_1$ and $x_2 \sim y_2$, then

$$\begin{aligned}\mu_A((x_1+x_2)-(y_1+y_2)) &= \mu_A((x_1-y_1)+(x_2-y_2))\\ &\geq T(\mu_A(x_1-y_1),\mu_A(x_2-y_2)) = \mu_A(0),\\ \lambda_A((x_1+x_2)-(y_1+y_2)) &= \lambda_A((x_1-y_1)+(x_2-y_2))\\ &\leq S(\lambda_A(x_1-y_1),\lambda_A(x_2-y_2)) = \lambda_A(0),\end{aligned}$$

$$\begin{aligned}\mu_A(\alpha x_1-\alpha y_1) &= \mu_A(\alpha(x_1-y_1)) \geq \mu_A(x_1-y_1) = \mu(0),\\ \lambda_A(\alpha x_1-\alpha y_1) &= \lambda_A(\alpha(x_1-y_1)) \leq \lambda_A(x_1-y_1) = \lambda_A(0),\\ \mu_A([x_1,x_2]-[y_1,y_2]) &= \mu_A([x_1-y_1],[x_2-y_2]) \geq \mu_A(x_1-y_1) = \mu_A(0),\\ \lambda_A([x_1,x_2]-[y_1,y_2]) &= \lambda_A([x_1-y_1],[x_2-y_2]) \leq \lambda_A(x_1-y_1) = \lambda_A(0).\end{aligned}$$

Now, applying (4), it is easy to see that $x_1+x_2 \sim y_1+y_2$, $\alpha x_1 \sim \alpha y_1$, and $[x_1,x_2] \sim [y_1,y_2]$. So, $\sim$ is a congruence.

2.7 Nilpotency of Intuitionistic (*S*, *T*)-Fuzzy Lie Ideals

Definition 2.34 Let $A=(\mu_A,\lambda_A)\in J^L$, an intuitionistic fuzzy subspace of L generated by A will be denoted by $[A]$. It is the intersection of all intuitionistic fuzzy subspaces of L containing A. For all $x\in L$, we define:

$$[\mu_A](x) = \sup\{\min \mu_A(x_i) \mid x = \sum \alpha_i x_i,\ \alpha_i \in \mathbb{F}, x_i \in L\},$$

$$[\lambda_A](x) = \inf\{\max \lambda_A(x_i) \mid x = \sum \alpha_i x_i,\ \alpha_i \in \mathbb{F}, x_i \in L\}.$$

Definition 2.35 Let $f : L_1 \to L_2$ be a homomorphism of Lie algebras which has an extension $f : J^{L_1} \to J^{L_2}$ defined by:

$$f(\mu_A)(y) = \sup\{\mu_A(x), x \in f^{-1}(y)\},$$

$$f(\lambda_A)(y) = \inf\{\lambda_A(x), x \in f^{-1}(y)\},$$

for all $A=(\mu_A,\lambda_A)\in J^{L_1}$, $y\in L_2$. Then $f(A)$ is called the *homomorphic image* of A.

The following two propositions are obvious.

Proposition 2.9 *Let $f : L_1 \to L_2$ be a homomorphism of Lie algebras and let $A = (\mu_A,\lambda_A)$ be an intuitionistic (S,T)-fuzzy Lie ideal of L_1. Then*
(i) $f(A)$ is an intuitionistic (S,T)-fuzzy Lie ideal of L_2,
(ii) $f([A]) \supseteq [f(A)]$.

Proposition 2.10 *If A and B are intuitionistic (S,T)-fuzzy Lie ideals in L, then $[A,B]$ is an intuitionistic (S,T)-fuzzy Lie ideal of L.*

Theorem 2.1 *Let A_1, A_2, B_1, B_2 be intuitionistic (S, T)-fuzzy Lie ideals in L such that $A_1 \subseteq A_2$ and $B_1 \subseteq B_2$, then $[A_1, B_1] \subseteq [A_2, B_2]$.*

Proof Indeed,

$$\begin{aligned} \ll \mu_{A_1}, \mu_{B_1} \gg (x) &= \sup\{T(\mu_{A_1}(a), \mu_{B_1}(b)) \mid a, b \in L_1, [a, b] = x\} \\ &\geq \sup\{T(\mu_{A_2}(a), \mu_{B_2}(b)) \mid a, b \in L_1, [a, b] = x\} \\ &= \ll \mu_{A_2}, \mu_{B_2} \gg (x), \\ \ll \lambda_{A_1}, \lambda_{B_1} \gg (x) &= \inf\{S(\lambda_{A_1}(a), \lambda_{B_1}(b)) \mid a, b \in L_1, [a, b] = x\} \\ &\leq \inf\{S(\lambda_{A_2}(a), \lambda_{B_2}(b)) \mid a, b \in L_1, [a, b] = x\} \\ &= \ll \lambda_{A_2}, \lambda_{B_2} \gg (x). \end{aligned}$$

Hence, $[A_1, B_1] \subseteq [A_2, B_2]$.

Let $A = (\mu_A, \lambda_A)$ be an intuitionistic (S, T)-fuzzy Lie ideal in L. Putting

$$A^0 = A,\ A^1 = [A, A_0],\ A^2 = [A, A_1],\ \ldots,\ A^n = [A, A^{n-1}]$$

we obtain a descending series of an intuitionistic (S, T)-fuzzy Lie ideals

$$A^0 \supseteq A^1 \supseteq A^2 \supseteq \ldots \supseteq A^n \supseteq \ldots$$

and a series of intuitionistic fuzzy sets $B^n = (\mu_B^n, \lambda_B^n)$ such that

$$\mu_B^n = \sup\{\mu_A^n(x) \mid 0 \neq x \in L\}, \quad \lambda_B^n = \inf\{\lambda_A^n(x) \mid 0 \neq x \in L\}.$$

Definition 2.36 An intuitionistic (S, T)-fuzzy Lie ideal $A = (\mu_A, \lambda_A)$ is called *nilpotent* if there exists a positive integer n such that $B^n = 0_\sim$.

Theorem 2.2 *A homomorphic image of a nilpotent intuitionistic (S, T)-fuzzy Lie ideal is a nilpotent intuitionistic (S, T)-fuzzy Lie ideal.*

Proof Let $f : L_1 \to L_2$ be a homomorphism of Lie algebras and let $A = (\mu_A, \lambda_A)$ be a nilpotent intuitionistic (S, T)-fuzzy Lie ideal in L_1. Assume that $f(A) = B$. We prove by induction that $f(A^n) \supseteq B^n$ for every natural n. First we claim that $f([A, A]) \supseteq [f(A), f(A)] = [B, B]$. Let $y \in L_2$, then

$$\begin{aligned} f(\ll \mu_A, \mu_A \gg)(y) &= \sup\{\ll \mu_A, \mu_A \gg)(x) \mid f(x) = y\} \\ &= \sup\{\sup\{T(\mu_A(a), \mu_A(b)) \mid a, b \in L_1, [a, b] = x, f(x) = y\}\} \\ &= \sup\{T(\mu_A(a), \mu_A(b)) \mid a, b \in L_1, [a, b] = x, f(x) = y\} \\ &= \sup\{T(\mu_A(a), \mu_A(b)) \mid a, b \in L_1, [f(a), f(b)] = x\} \\ &= \sup\{T(\mu_A(a), \mu_A(b)) \mid a, b \in L_1, f(a) = u, f(b) = v, [u, v] = y\} \end{aligned}$$

$$\geq \sup\{T(\sup_{a\in f^{-1}(u)} \mu_A(a), \sup_{b\in f^{-1}(v)} \mu_A(b)) \mid [u, v] = y\}$$
$$= \sup\{T(f(\mu_A)(u), f(\mu_A)(v)) \mid [u, v] = y\} = \ll f(\mu_A), f(\mu_A) \gg (y),$$

$$\begin{aligned}
f(\ll \lambda_A, \lambda_A \gg)(y) &= \inf\{\ll \lambda_A, \lambda_A \gg)(x) \mid f(x) = y\} \\
&= \inf\{\inf\{S(\lambda_A(a), \lambda_A(b)) \mid a, b \in L_1, [a, b] = x, f(x) = y\}\} \\
&= \inf\{S(\lambda_A(a), \lambda_A(b)) \mid a, b \in L_1, [a, b] = x, f(x) = y\} \\
&= \inf\{S(\lambda_A(a), \lambda_A(b)) \mid a, b \in L_1, [f(a), f(b)] = x\} \\
&= \inf\{S(\lambda_A(a), \lambda_A(b)) \mid a, b \in L_1, f(a) = u, f(b) = v, [u, v] = y\} \\
&\leq \inf\{S(\inf_{a\in f^{-1}(u)} \lambda_A(a), \inf_{b\in f^{-1}(v)} \lambda_A(b)) \mid [u, v] = y\} \\
&= \inf\{S(f(\lambda_A)(u), f(\lambda_A)(v)) \mid [u, v] = y\} = \ll f(\lambda_A), f(\lambda_A) \gg (y).
\end{aligned}$$

Thus,

$$f([A, A]) \supseteq f(\ll A, A \gg) \supseteq \ll f(A), f(A) \gg = [f(A), f(A)].$$

For $n > 1$, we get

$$f(A^n) = f([A, A^{n-1}]) \supseteq [f(A), f(A^{n-1})] \supseteq [B, B^{n-1}] = B^n.$$

Let m be a positive integer such that $A^m = 0_\sim$. Then for $0 \neq y \in L_2$ we have

$$\mu_B^m(y) \leq f(\mu_A^m)(y) = f(0)(y) = \sup\{0(a) \mid f(x) = y\} = 0,$$

$$\lambda_B^m(y) \geq f(\lambda_A^m)(y) = f(1)(y) = \inf\{1(a) \mid f(x) = y\} = 1.$$

Thus, $B^m = 0_\sim$. This completes the proof.

Let $A = (\mu_A, \lambda_A)$ be an intuitionistic (S, T)-fuzzy Lie ideal in L. Putting

$$A^{(0)} = A,\ A^{(1)} = [A^{(0)}, A^{(0)}],\ A^{(2)} = [A^{(1)}, A^{(1)}], \ldots, A^{(n)} = [A^{(n-1)}, A^{(n-1)}]$$

we obtain series

$$A^{(0)} \subseteq A^{(1)} \subseteq A^{(2)} \subseteq \ldots \subseteq A^{(n)} \subseteq \ldots$$

of intuitionistic (S, T)-fuzzy Lie ideals and a series of intuitionistic fuzzy sets $B^{(n)} = (\mu_B^{(n)}, \lambda_B^{(n)})$ such that

$$\mu_B^{(n)} = \sup\{\mu_A^{(n)}(x) \mid 0 \neq x \in L\},\ \ \lambda_B^{(n)} = \inf\{\lambda_A^{(n)}(x) \mid 0 \neq x \in L\}.$$

Definition 2.37 An intuitionistic (S, T)-fuzzy Lie ideal $A = (\mu_A, \lambda_A)$ is called *solvable* if there exists a positive integer n such that $B^{(n)} = 0_\sim$.

Theorem 2.3 *A nilpotent intuitionistic (S, T)-fuzzy Lie ideal is solvable.*

Proof It is enough to prove that $A^{(n)} \subseteq A^n$ for all positive integers n. We prove it by induction on n and by the use of Theorem 2.1:

$$A^{(1)} = [A, A] = A^1, \quad A^{(2)} = [A^{(1)}, A^{(1)}] \subseteq [A, A^{(1)}] = A^2.$$

$$A^{(n)} = [A^{(n-1)}, A^{(n-1)}] \subseteq [A, A^{(n-1)}] \subseteq [A, A^{(n-1)}] = A^n.$$

This completes the proof.

Definition 2.38 Let $A = (\mu_A, \lambda_A)$ and $B = (\mu_B, \lambda_B)$ be two intuitionistic (S, T)-fuzzy Lie ideals of a Lie algebra L. The sum $A \oplus B$ is called a *direct sum* if $A \cap B = 0_\sim$.

Theorem 2.4 *The direct sum of two nilpotent intuitionistic (S, T)-fuzzy Lie ideals is also a nilpotent intuitionistic (S, T)-fuzzy Lie ideal.*

Proof Suppose that $A = (\mu_A, \lambda_A)$ and $B = (\mu_B, \lambda_B)$ are two intuitionistic (S, T)-fuzzy Lie ideals such that $A \cap B = 0_\sim$. We claim that $[A, B] = 0_\sim$. Let $x(\neq 0) \in L$, then

$$\ll \mu_A, \mu_B \gg (x) = \sup\{T(\mu_A(a), \mu_B(b)) \mid [a, b] = x\} \leq T(\mu_A(x), \mu_B(x)) = 0$$

and

$$\ll \lambda_A, \lambda_B \gg (x) = \inf\{S(\lambda_A(a), \lambda_B(b)) \mid [a, b] = x\} \geq S(\lambda_A(x), \lambda_B(x)) = 1.$$

This proves our claim. Thus, we obtain $[A^m, B^n] = 0_\sim$ for all positive integers m, n. Now we again claim that $(A \oplus B)^n \subseteq A^n \oplus B^n$ for positive integer n. We prove this claim by induction on n. For $n = 1$,

$$(A \oplus B)^1 = [A \oplus B, A \oplus B] \subseteq [A, A] \oplus [A, B] \oplus [B, A] \oplus [B, B] = A^1 \oplus B^1.$$

Now for $n > 1$,

$$\begin{aligned}(A \oplus B)^n &= [A \oplus B, (A \oplus B)^{n-1}] \subseteq [A \oplus B, A^{n-1} \oplus B^{n-1}] \\ &\subseteq [A, A^{n-1}] \oplus [A, B^{n-1}] \oplus [B, A^{n-1}] \oplus [B, B^{n-1}] = A^n \oplus B^n.\end{aligned}$$

Since there are two positive integers p and q such that $A^p = B^q = 0_\sim$, we have $(A \oplus B)^{p+q} \subseteq A^{p+q} \oplus B^{p+q} = 0_\sim$.

In a similar way, we can prove the following theorem.

Theorem 2.5 *The direct sum of two solvable intuitionistic (S, T)-fuzzy Lie ideals is a solvable intuitionistic (S, T)-fuzzy Lie ideal.*

Theorem 2.6 *Let $A = (\mu_A, \lambda_A)$ be an intuitionistic (S, T)-fuzzy Lie ideal in a Lie algebra L. Then $A^n \subseteq [A_n]$ for any $n > 0$, where an intuitionistic fuzzy subset $[A_n] = ([\mu_{A_n}], [\lambda_{A_n}])$ is defined by*

$$[\mu_{A_n}](x) = \sup\{\mu_A(a) \mid [x_1, [x_2, [\ldots, [x_n, a]\ldots]]] = x, \quad x_1, \ldots, x_n \in L\},$$
$$[\lambda_{A_n}](x) = \inf\{\lambda_A(a) \mid [x_1, [x_2, [\ldots, [x_n, a]\ldots]]] = x, \quad x_1, \ldots, x_n \in L\}.$$

Proof It is enough to prove that $\ll A, A^{n-1} \gg \subseteq [A_n]$. We prove it by induction on n. For n=1 and $x \in L$, we have

$$\begin{aligned}\ll \mu_A, \mu_A \gg (x) &= \sup\{T(\mu_A(a), \mu_A(b)) \mid [a, b] = x\}\\ &\geq \sup\{\mu_A(b) \mid [a, b] = x, a \in L\} = [\mu_{A_1}](x),\end{aligned}$$

$$\begin{aligned}\ll \lambda_A, \lambda_A \gg (x) &= \inf\{S(\mu_A(a), \mu_A(b)) \mid [a, b] = x\}\\ &\leq \inf\{\lambda_A(b) : [a, b] = x, a \in L\} = [\lambda_{A_1}](x).\end{aligned}$$

For $n > 1$,

$$\begin{aligned}&\ll \mu_A, \mu_A^{n-1} \gg (x) = \sup\{T(\mu_A(a), \mu_A^{n-1}(b)) \mid [a, b] = x\}\\ &= \sup\{T(\mu_A(a), [\mu_A(b), \mu_A^{n-2}(b)]) \mid [a, b] = x\}\\ &\geq \sup\{T(\mu_A(a), \sup\{\ll \mu_A, \mu_A^{n-2} \gg (b_i) \mid b = \textstyle\sum \alpha_i b_i\}) \mid [a, b] = x\}\\ &\geq \sup\{T(\mu_A(a), \sup\{[\mu_{A_{n-1}}](b_i) \mid b = \textstyle\sum \alpha_i b_i\}) \mid [a, b] = x\}\\ &\geq \sup\{T(\mu_A(a), [\mu_{A_{n-1}}](b_i)) \mid \textstyle\sum \alpha_i [a, b_i] = x\}\\ &\geq \sup\{T(\mu_A(a), \sup\{\mu_{A_{n-1}}(c_i) \mid b_i = \textstyle\sum \beta_i c_i\}) \mid \textstyle\sum \alpha_i [a, b_i] = x\}\\ &\geq \sup\{T(\mu_A(a), \mu_{A_{n-1}}(c_i)) \mid \textstyle\sum \gamma_i [a, c_i] = x\}\\ &\geq \sup\{T(\mu_A(a), \sup\{\mu_A(d_i)) \mid [x_1, [x_2, [\ldots, [x_{n-1}, d_i]\ldots]]] = c_i\} \mid \textstyle\sum \gamma_i [a, c_i] = x\}\\ &\geq \sup\{T(\mu_A(a), \mu_A(d_i)) \mid \textstyle\sum \gamma_i [a, [x_1, [x_2, [\ldots, [x_{n-1}, d_i]\ldots]]]] = x\}\\ &\geq \sup\{\mu_{A_n}(d_i) \mid \textstyle\sum \gamma_i [a, [x_1, [x_2, [\ldots, [x_{n-1}, d_i]\ldots]]]] = x\} \geq [\mu_{A_n}](x),\end{aligned}$$

$$\begin{aligned}&\ll \lambda_A, \lambda_A^{n-1} \gg (x) = \inf\{S(\lambda_A(a), \lambda_A^{n-1}(b)) \mid [a, b] = x\}\\ &= \inf\{S(\lambda_A(a), [\lambda_A(b), \lambda_A^{n-2}(b)]) \mid [a, b] = x\}\\ &\leq \inf\{S(\lambda_A(a), \inf\{\ll \lambda_A, \lambda_A^{n-2} \gg (b_i) \mid b = \textstyle\sum \alpha_i b_i\}) \mid [a, b] = x\}\\ &\leq \inf\{S(\lambda_A(a), \inf\{[\lambda_{A_{n-1}}](b_i) \mid b = \textstyle\sum \alpha_i b_i\}) \mid [a, b] = x\}\\ &\leq \inf\{S(\lambda_A(a), [\lambda_{A_{n-1}}](b_i)) \mid \textstyle\sum \alpha_i [a, b_i] = x\}\\ &\leq \inf\{S(\lambda_A(a), \inf\{\lambda_{A_{n-1}}(c_i) \mid b_i = \textstyle\sum \beta_i c_i\}) \mid \textstyle\sum \alpha_i [a, b_i] = x\}\\ &\leq \inf\{S(\lambda_A(a), \lambda_{A_{n-1}}(c_i)) \mid \textstyle\sum \gamma_i [a, c_i] = x\}\\ &\leq \inf\{S(\lambda_A(a), \inf\{\lambda_A(d_i)) \mid [x_1, [x_2, [\ldots, [x_{n-1}, d_i]\ldots]]] = c_i\} \mid \textstyle\sum \gamma_i [a, c_i] = x\}\\ &\leq \inf\{S(\lambda_A(a), \lambda_A(d_i)) \mid \textstyle\sum \gamma_i [a, [x_1, [x_2, [\ldots, [x_{n-1}, d_i]\ldots]]]] = x\}\\ &\leq \inf\{\lambda_{A_n}(d_i) \mid \textstyle\sum \gamma_i [a, [x_1, [x_2, [\ldots, [x_{n-1}, d_i]\ldots]]]] = x\} \leq [\lambda_{A_n}](x).\end{aligned}$$

This complete the proof.

Theorem 2.7 *If for an intuitionistic (S, T)-fuzzy Lie ideal $A = (\mu_A, \lambda_A)$, there exists a positive integer n such that*

$$(\overline{ad}x_1 \circ \overline{ad}x_2 \circ \cdots \circ \overline{ad}x_n)(\mu_A) = 0,$$

$$(\overline{ad}x_1 \circ \overline{ad}x_2 \circ \cdots \circ \overline{ad}x_n)(\lambda_A) = 1,$$

for all $x_1, \ldots, x_n \in L$, then A is nilpotent.

Proof For $x_1, \ldots, x_n \in L$ and $x(\neq 0) \in L$, we have

$$(\overline{ad}x_1 \circ \cdots \circ \overline{ad}x_n)(\mu_A)(x) = \sup\{\mu_A(a) \mid [x_1, [x_2, [\ldots, [x_n, a]\ldots]]] = x\} = 0,$$

$$(\overline{ad}x_1 \circ \cdots \circ \overline{ad}x_n)(\lambda_A)(x) = \inf\{\lambda_A(a) \mid [x_1, [x_2, [\ldots, [x_n, a]\ldots]]] = x\} = 1.$$

Thus, $[A_n] = 0_\sim$. From Theorem 2.6, it follows that $A^n = 0_\sim$. Hence, $A = (\mu_A, \lambda_A)$ is a nilpotent intuitionistic (S, T)-fuzzy Lie ideal.

2.8 Intuitionistic (S, T)-Fuzzy Killing Form

Killing form K can be naturally extended to $\overline{K} : J^{L\times L} \to J^{\mathbb{F}}$ defined by putting

$$\overline{K}(\mu_A)(\beta) = \sup\{\mu_A(x, y) \mid Tr(adx \circ ady) = \beta\},$$
$$\overline{K}(\lambda_A)(\beta) = \inf\{\lambda_A(x, y) \mid Tr(adx \circ ady) = \beta\}.$$

The Cartesian product of two intuitionistic (S, T)-fuzzy sets $A = (\mu_A, \lambda_A)$ and $B = (\mu_B, \lambda_B)$ is defined as

$$(\mu_A \times \mu_B)(x, y) = T(\mu_A(x), \mu_B(y)),$$
$$(\lambda_A \times \lambda_B)(x, y) = S(\lambda_A(x), \lambda_B(y)).$$

Similarly, we define

$$\overline{K}(\mu_A \times \mu_B)(\beta) = \sup\{T(\mu_A(x), \mu_B(y)) \mid Tr(adx \circ ady) = \beta\},$$
$$\overline{K}(\lambda_A \times \lambda_B)(\beta) = \inf\{S(\lambda_A(x), \lambda_B(y)) \mid Tr(adx \circ ady) = \beta\}.$$

Proposition 2.11 *Let $A = (\mu_A, \lambda_A)$ be an intuitionistic (S, T)-fuzzy Lie ideal of Lie algebra L. Then*

(i) $1_{\sim(x+y)} = 1_{\sim x} \oplus 1_{\sim y}$,

(ii) $1_{\sim(\alpha x)} = \alpha \odot 1_{\sim x}$

for all $x, y \in L$, $\alpha \in \mathbb{F}$.

Proof Straightforward.

Theorem 2.8 *Let* $A = (\mu_A, \lambda_A)$ *be an intuitionistic* (S, T)*-fuzzy Lie ideal of Lie algebra* L*. Then* $\overline{K}(\mu_A \times 1_{(\alpha x)}) = \alpha \odot \overline{K}(\mu_A \times 1_x)$ *and* $\overline{K}(\lambda_A \times 0_{(\alpha x)}) = \alpha \odot \overline{K}(\lambda_A \times 0_x)$ *for all* $x \in L$, $\alpha \in \mathbb{F}$.

Proof If $\alpha = 0$, then for $\beta = 0$ we have

$$\begin{aligned}
\overline{K}(\mu_A \times 1_0)(0) &= \sup\{T(\mu_A(x), 1_0(y)) \mid Tr(adx \circ ady) = 0\} \\
&\geq T(\mu_A(0), 1_0(0)) = 0, \\
\overline{K}(\lambda_A \times 0_0)(0) &= \inf\{S(\lambda_A(x), 0_0(y)) : Tr(adx \circ ady) = 0\} \\
&\leq S(\lambda_A(0), 0_0(0)) = 1.
\end{aligned}$$

For $\beta \neq 0$ $Tr(adx \circ ady) = \beta$ means that $x \neq 0$ and $y \neq 0$. So,

$$\begin{aligned}
\overline{K}(\mu_A \times 1_0)(\beta) &= \sup\{T(\mu_A(x), 1_0(y)) \mid Tr(adx \circ ady) = \beta\} = 0, \\
\overline{K}(\lambda_A \times 0_0)(\beta) &= \inf\{S(\lambda_A(x), 0_0(y)) \mid Tr(adx \circ ady) = \beta\} = 1.
\end{aligned}$$

If $\alpha \neq 0$, then for arbitrary β we obtain

$$\begin{aligned}
\overline{K}(\mu_A \times 1_{\alpha x})(\beta) &= \sup\{T(\mu_A(y), 1_{\alpha x}(z)) \mid Tr(ady \circ adz) = \beta\} \\
&= \sup\{T(\mu_A(y), \alpha \odot 1_x(z)) \mid Tr(ady \circ adz) = \beta\} \\
&= \sup\{T(\mu_A(y), 1_x(\alpha^{-1}z)) \mid \alpha Tr(ady \circ ad(\alpha^{-1}z)) = \beta\} \\
&= \sup\{T(\mu_A(y), 1_x(\alpha^{-1}z)) \mid Tr(ady \circ ad(\alpha^{-1}z)) = \alpha^{-1}\beta\} \\
&= \overline{K}(\mu_A \times 1_x)(\alpha^{-1}\beta) = \alpha \odot \overline{K}(\mu_A \times 1_x)(\beta),
\end{aligned}$$

$$\begin{aligned}
\overline{K}(\lambda_A \times 0_{\alpha x})(\beta) &= \inf\{S(\lambda_A(y), 0_{\alpha x}(z)) \mid Tr(ady \circ adz) = \beta\} \\
&= \inf\{S(\lambda_A(y), \alpha \odot 0_x(z)) \mid Tr(ady \circ adz) = \beta\} \\
&= \inf\{S(\lambda_A(y), 0_x(\alpha^{-1}z)) \mid \alpha Tr(ady \circ ad(\alpha^{-1}z)) = \beta\} \\
&= \inf\{S(\lambda_A(y), 0_x(\alpha^{-1}z)) \mid Tr(ady \circ ad(\alpha^{-1}z)) = \alpha^{-1}\beta\} \\
&= \overline{K}(\lambda_A \times 1_x)(\alpha^{-1}\beta) = \alpha \odot \overline{K}(\lambda_A \times 0_x)(\beta).
\end{aligned}$$

This completes the proof.

Theorem 2.9 *Let* $A = (\mu_A, \lambda_A)$ *be an intuitionistic* (S, T)*-fuzzy Lie ideal of a Lie algebra* L*. Then* $\overline{K}(\mu_A \times 1_{(x+y)}) = \overline{K}(\mu_A \times 1_x) \oplus \overline{K}(\mu_A \times 1_y)$ *and* $\overline{K}(\mu_A \times 0_{(x+y)}) = \overline{K}(\mu_A \times 0_x) \oplus \overline{K}(\mu_A \times 0_y)$ *for all* $x, y \in L$.

Proof Indeed,

$$
\begin{aligned}
\overline{K}(\mu_A \times 1_{(x+y)})(\beta) &= \sup\{T(\mu_A(z), 1_{x+y}(u)) \mid Tr(adz \circ adu) = \beta\} \\
&= \sup\{\mu_A(z) \mid Tr(adz \circ ad(x+y)) = \beta\} \\
&= \sup\{\mu_A(z) \mid Tr(adz \circ adx) + Tr(adz \circ ady) = \beta\}
\end{aligned}
$$

$$
\begin{aligned}
&= \sup\{T(\mu_A(z), T(1_x(v), 1_y(w))) \mid Tr(adz \circ adv) + Tr(adz \circ adw) = \beta\} \\
&= \sup\{T(\sup\{T(\mu_A(z), 1_x(v)) \mid Tr(adz \circ adv) = \beta_1\}, \\
&\qquad \sup\{T(\mu_A(z), 1_y(w)) \mid Tr(adz \circ adw) = \beta_2\} \mid \beta_1 + \beta_2 = \beta)\} \\
&= \sup\{T(\overline{K}(\mu_A \times 1_x)(\beta_1), \overline{K}(\mu_A \times 1_y)(\beta_2)) \mid \beta_1 + \beta_2 = \beta\} \\
&= \overline{K}(\mu_A \times 1_x) \oplus \overline{K}(\mu_A \times 1_y)(\beta),
\end{aligned}
$$

$$
\begin{aligned}
&\overline{K}(\lambda_A \times 0_{(x+y)})(\beta) = \inf\{S(\lambda_A(z), 0_{x+y}(u)) \mid Tr(adz \circ adu) = \beta\} \\
&= \inf\{\lambda_A(z) \mid Tr(adz \circ ad(x+y)) = \beta\} \\
&= \inf\{\lambda_A(z) \mid Tr(adz \circ adx) + Tr(adz \circ ady) + \beta\} \\
&= \inf\{S(\lambda_A(z), S(0_x(v), 0_y(w))) \mid Tr(adz \circ adv) + Tr(adz \circ adw) = \beta\} \\
&= \inf\{S(\inf\{S(\lambda_A(z), 0_x(v)) \mid Tr(adz \circ adv) = \beta_1\}, \\
&\qquad \inf\{S(\lambda_A(z), 0_y(w) \mid Tr(adz \circ adw) = \beta_2\} \mid \beta_1 + \beta_2 = \beta)\} \\
&= \inf\{S(\overline{K}(\lambda_A \times 0_x)(\beta_1), \overline{K}(\lambda_A \times 0_y)(\beta_2)) \mid \beta_1 + \beta_2 = \beta\} \\
&= \overline{K}(\lambda_A \times 0_x) \oplus \overline{K}(\lambda_A \times 0_y)(\beta).
\end{aligned}
$$

This completes the proof.

As a consequence of the above two theorems, we obtain

Corollary 2.1 *For each intuitionistic (S, T)-fuzzy Lie ideal $A = (\mu_A, \lambda_A)$ and all $x, y \in L$, $\alpha, \beta \in \mathbb{F}$, we have*

$$
\begin{aligned}
\overline{K}(\mu_A \times 1_{(\alpha x+\beta y)}) &= \alpha \odot \overline{K}(\mu_A \times 1_x) \oplus \beta \odot \overline{K}(\mu_A \times 1_y), \\
\overline{K}(\lambda_A \times 0_{(\alpha x+\beta y)}) &= \alpha \odot \overline{K}(\lambda_A \times 0_x) \oplus \beta \odot \overline{K}(\lambda_A \times 0_y).
\end{aligned}
$$

Chapter 3
Interval-Valued Fuzzy Lie Structures

In this chapter, we present properties of certain concepts, including interval-valued fuzzy Lie ideals, characterizations of Noetherian Lie algebras, quotient Lie algebras via interval-valued fuzzy Lie ideals, interval-valued intuitionistic fuzzy Lie ideals, fully invariant and characteristic interval-valued intuitionistic fuzzy Lie ideals, solvable, nilpotent interval-valued intuitionistic fuzzy Lie ideals, and interval-valued fuzzy Lie superalgebras.

3.1 Introduction

After introducing the concept of fuzzy sets, a number of new theories have been discussed for treating imprecision and uncertainty. Some of these theories are extensions of the fuzzy set theory. In 1975, Zadeh [141] introduced the notion of *interval-valued fuzzy sets* as an extension of fuzzy sets in which the values of the membership degrees are intervals of numbers instead of the numbers. Interval-valued fuzzy sets provide a more adequate description of uncertainty than traditional fuzzy sets. It is therefore important to use interval-valued fuzzy sets in applications, such as fuzzy control. One of the computationally most intensive parts of fuzzy control is defuzzification [102]. Since interval-valued fuzzy sets are widely studied and used, we describe briefly the papers of Gorzalczany [73, 74] on approximate reasoning, Roy and Biswas [117] on medical diagnosis, Turksen [127] on multivalued logic, and Mendel on intelligent control.

Definition 3.1 An *interval number* D is an interval $[a^-, a^+]$ with $0 \leq a^- \leq a^+ \leq 1$. The interval $[a, a]$ is identified with the number $a \in [0, 1]$. $D[0, 1]$ denotes the set of all interval numbers. For interval numbers $D_1 = [a_1^-, b_1^+]$ and $D_2 = [a_2^-, b_2^+]$, we define

- $\min\{D_1, D_2\} = \min\{[a_1^-, b_1^+], [a_2^-, b_2^+]\} = [\min\{a_1^-, a_2^-\}, \min\{b_1^+, b_2^+\}]$,
- $\max\{D_1, D_2\} = \max\{[a_1^-, b_1^+], [a_2^-, b_2^+]\} = [\max\{a_1^-, a_2^-\}, \max\{b_1^+, b_2^+\}]$,
- $D_1 + D_2 = [a_1^- + a_2^- - a_1^- \cdot a_2^-, b_1^+ + b_2^+ - b_1^+ \cdot b_2^+]$,

M. Akram, *Fuzzy Lie Algebras*, Infosys Science Foundation Series,
https://doi.org/10.1007/978-981-13-3221-0_3

- $D_1 \leq D_2 \Longleftrightarrow a_1^- \leq a_2^-$ and $b_1^+ \leq b_2^+$,
- $D_1 = D_2 \Longleftrightarrow a_1^- = a_2^-$ and $b_1^+ = b_2^+$,
- $D_1 < D_2 \Longleftrightarrow D_1 \leq D_2$ and $D_1 \neq D_2$,
- $kD = k[a_1^-, b_1^+] = [ka_1^-, kb_1^+]$, where $0 \leq k \leq 1$.

Similarly,

$$\sup_{i \in I}\{[a_i^-, b_i^+]\} = [\sup_{i \in I}\{a_i^-\}, \sup_{i \in I}\{b_i^+\}] \quad \text{and} \quad \inf_{i \in I}\{[a_i^-, b_i^+]\} = [\inf_{i \in I}\{a_i^-\}, \inf_{i \in I}\{b_i^+\}].$$

Clearly, $(D[0, 1], \leq, \vee, \wedge)$ is a complete lattice with $[0, 0]$ as the least element and $[1, 1]$ as the greatest element.

Definition 3.2 An interval-valued fuzzy set on X is a mapping $\widetilde{\mu} : X \to D[0, 1]$, where $D[0, 1]$ denotes the set of all interval numbers and $\widetilde{\mu} = [\mu^-, \mu^+]$. An interval-valued fuzzy set $\widetilde{\mu}$ on $X (\neq \emptyset)$ can be represented as an object of the form

$$\widetilde{\mu} = \{(x, [\mu^-(x), \mu^+(x)]) \mid x \in X\},$$

where $\mu^-(x) \leq \mu^+(x)$ for all $x \in X$.

Definition 3.3 If $\widetilde{\mu} = [\mu^-(x), \mu^+(x)]$, and $\widetilde{\nu} = [\nu^-(x), \nu^+(x)]$ are interval-valued fuzzy sets on X, then the following operations are defined:

- $\widetilde{\mu} \subseteq \widetilde{\nu} \Longleftrightarrow \mu^-(x) \leq \nu^-(x)$ and $\mu^+(x) \leq \nu^+(x)$ for all $x \in X$,
- $\widetilde{\mu} = \widetilde{\nu} \Longleftrightarrow \mu^-(x) = \nu^-(x)$ and $\mu^+(x) = \nu^+(x)$ for all $x \in X$,
- $\widetilde{\mu} \bigcup \widetilde{\nu} = \{< x, [\max\{\mu^-(x), \nu^-(x)\}, \max\{\mu^+(x), \nu^+(x)\}] >\mid x \in X\}$,
- $\widetilde{\mu} \bigcap \widetilde{\nu} = \{< x, [\min\{\mu^-(x), \nu^-(x)\}, \min\{\mu^+(x), \nu^+(x)\}] >\mid x \in X\}$,
- $\widetilde{\mu} \cdot \widetilde{\nu} = \{< x, [\mu^-(x) \cdot \nu^-(x), \mu^+(x) \cdot \nu^+(x)] >\mid x \in X\}$,
- $\widetilde{\mu} + \widetilde{\nu} = \{< x, [\mu^-(x) + \nu^-(x) - \mu^-(x) \cdot \nu^-(x), \mu^+(x) + \nu^+(x) - \mu^+(x) \cdot \nu^+(x)] >\mid x \in X\}$,
- $\widetilde{\mu}^c = \{< [x, 1 - \mu^-(x), 1 - \mu^+(x)] >\mid x \in X\}$.

Definition 3.4 Let $\widetilde{\mu}$ be an interval-valued fuzzy set of X and $[\delta_1, \delta_2] \in D[0, 1]$. Then, the *interval-valued level subset* $U(\widetilde{\mu}, [\delta_1, \delta_2])$ of X and *strong interval-valued level subset* $U_>(\widetilde{\mu}, [\delta_1, \delta_2])$ of X are defined as follows:

$$U(\widetilde{\mu}, [\delta_1, \delta_2]) = \{x \in X \mid \widetilde{\mu}(x) \geq [\delta_1, \delta_2]\},$$
$$U_>(\widetilde{\mu}, [\delta_1, \delta_2]) = \{x \in X \mid \widetilde{\mu}(x) > [\delta_1, \delta_2]\}.$$

Specially, we denote $\widetilde{\mu}^* = \{x \in X \mid \widetilde{\mu}(x) > [0, 0]\}$.

Definition 3.5 An interval-valued fuzzy set $\widetilde{\mu}$ in vector space V over $\mathbb{F}$ is called an *interval-valued fuzzy subspace* of V if

1. $\widetilde{\mu}(x + y) \geq \min\{\widetilde{\mu}(x), \widetilde{\mu}(y)\}$,
2. $\widetilde{\mu}(\alpha x) \geq \widetilde{\mu}(x)$

hold for all $x, y \in V$ and $\alpha \in \mathbb{F}$.

Definition 3.6 Let $\widetilde{\mu}$ and $\widetilde{\nu}$ be interval-valued fuzzy sets of a vector space V. We define the *sum* of $\widetilde{\mu}$ and $\widetilde{\nu}$ by

$$(\widetilde{\mu}+\widetilde{\nu})(x)=\sup_{x=a+b}\{\min\{\widetilde{\mu}(a),\widetilde{\nu}(b)\}\}.$$

Definition 3.7 Let $\widetilde{\mu}$ be an interval-valued fuzzy set of a vector space $V^{'}$ and f be a mapping from vector space V to $V^{'}$. Then, the *inverse image* of $\widetilde{\mu}$, denoted by $f^{-1}(\widetilde{\mu})$, is the interval-valued fuzzy set in V with the membership function given by $\widetilde{\mu}_{f^{-1}}(x)=\widetilde{\mu}(f(x))$, for all $x\in V$.

Definition 3.8 Let $\widetilde{\mu}$ be an interval-valued fuzzy set of a vector space V and f be a mapping from a vector space V to $V^{'}$. Then, the *image* of $\widetilde{\mu}$, denoted by $f(\widetilde{\mu})$, is the interval-valued fuzzy set in $V^{'}$ with the membership function defined by

$$\widetilde{\mu}_f(y)=\begin{cases}\sup\limits_{x\in f^{-1}(y)}\{\widetilde{\mu}(x)\} & y\in f(V),\\ [0,0] & y\notin f(V).\end{cases}$$

We state some results of interval-valued fuzzy subspaces of a vector space.

Lemma 3.1 *$\widetilde{\mu}$ is an interval-valued fuzzy subspace of vector space V if and only if μ^- and μ^+ are fuzzy subspaces of V.*

Lemma 3.2 *Let $\widetilde{\mu}$ and $\widetilde{\nu}$ be interval-valued fuzzy subspaces of a vector space V. Then, $\widetilde{\mu}+\widetilde{\nu}$ is also an interval-valued fuzzy subspace of V.*

Lemma 3.3 *Let $\widetilde{\mu}$ and $\widetilde{\nu}$ be interval-valued fuzzy subspaces of a vector space V. Then, $\widetilde{\mu}\cap\widetilde{\nu}$ is an interval-valued fuzzy subspace of V.*

Lemma 3.4 *Let $\widetilde{\mu}$ be an interval-valued fuzzy subspace of a vector space $V^{'}$ and f be a mapping from vector space V to $V^{'}$. Then, the inverse image $f^{-1}(\widetilde{\mu})$ is also an interval-valued fuzzy subspace of V.*

Lemma 3.5 *Let $\widetilde{\mu}$ be an interval-valued fuzzy subspace of a vector space V and f be a mapping from vector space V to $V^{'}$. Then, the image $f(\widetilde{\mu})$ is also an interval-valued fuzzy subspace of V.*

3.2 Interval-Valued Fuzzy Lie Ideals

Definition 3.9 An interval-valued fuzzy set $\widetilde{\mu}$ in a Lie algebra L is called an *interval-valued fuzzy Lie subalgebra* of L if

(1) $\widetilde{\mu}(x+y)\geq\min\{\widetilde{\mu}(x),\widetilde{\mu}(y)\}$,
(2) $\widetilde{\mu}(\alpha x)\geq\widetilde{\mu}(x)$,
(3) $\widetilde{\mu}([x,y])\geq\min\{\widetilde{\mu}(x),\widetilde{\mu}(y)\}$

hold for all $x,y\in L$ and $\alpha\in\mathbb{F}$.

Definition 3.10 An interval-valued fuzzy set $\widetilde{\mu}$ satisfying (1), (2), and

(4) $\widetilde{\mu}([x, y]) \geq \widetilde{\mu}(x)$

is called an *interval-valued fuzzy Lie ideal* of L.

From (2), it follows that

(5) $\widetilde{\mu}(0) \geq \widetilde{\mu}(x)$,
(6) $\widetilde{\mu}(-x) = \widetilde{\mu}(x)$

for all $x \in L$.

Example 3.1 The set $\Re^3$ with the operation $[x, y] = x \times y$, where $x, y \in \Re^3$, is a real Lie algebra. We define an interval-valued fuzzy set $\widetilde{\mu} : \Re^3 \to D[0, 1]$ by

$$\widetilde{\mu}(x) = \begin{cases} [s_1, s_2] \text{ if } x = (0, 0, 0), \\ [t_1, t_2] \text{ otherwise,} \end{cases}$$

where $[s_1, s_2] > [t_1, t_2]$ and $[s_1, s_2], [t_1, t_2] \in D[0, 1]$. By routine computations, we can see that it is an interval-valued fuzzy Lie subalgebra and a Lie ideal of $\Re^3$.

Proposition 3.1 *Every interval-valued fuzzy Lie ideal is an interval-valued fuzzy Lie subalgebra.*

The converse of Proposition 3.1 is not true in general.

Example 3.2 Let $\Re^3$ and [,] be as in the previous example. Putting

$$\widetilde{\mu}(x) = \begin{cases} [1, 1] & \text{if } x = (0, 0, 0), \\ [0.5, 0.5] & \text{if } x = (c, 0, 0),\ c \neq 0, \\ [0, 0] & \text{otherwise,} \end{cases}$$

we obtain an example of an interval-valued fuzzy Lie subalgebra which is not an interval-valued fuzzy Lie ideal. Indeed,

$$\widetilde{\mu}([(1, 0, 0), (1, 1, 1)]) = \widetilde{\mu}([(1, 0, 0) \times (1, 1, 1)]) = \widetilde{\mu}(0, -1, 1) = [0, 0],$$

$$\widetilde{\mu}(1, 0, 0) = [0.5, 0.5].$$

That is,

$$\widetilde{\mu}([(1, 0, 0), (1, 1, 1)]) \ngeq \widetilde{\mu}(1, 0, 0).$$

Theorem 3.1 *An interval-valued fuzzy set $\widetilde{\mu} = [\mu^-, \mu^+]$ in L is an interval-valued fuzzy Lie subalgebra (ideal) if and only if μ^- and μ^+ are fuzzy Lie subalgebras (ideals) of L.*

Proof Suppose that μ^- and μ^+ are fuzzy Lie ideals of L. Then,

$$\begin{aligned}\widetilde{\mu}(x+y) &= [\mu^-(x+y), \mu^+(x+y)]\\ &\geq [\min\{\mu^-(x), \mu^-(y)\}, \min\{\mu^+(x), \mu^+(y)\}]\\ &= [\min\{\mu^-(x), \mu^+(x)\}, \min\{\mu^-(y), \mu^+(y)\}]\\ &= \min\{\widetilde{\mu}(x), \widetilde{\mu}(y)\}\end{aligned}$$

for $x, y \in L$. The verification of (2), (3), and (4) is analogous. Hence, $\widetilde{\mu}$ is an interval-valued fuzzy Lie subalgebra (ideal) of L.

Conversely, assume that $\widetilde{\mu}$ is an interval-valued fuzzy Lie subalgebra (ideal) of L. Then,

$[\mu^-(x+y), \mu^+(x+y)] = \widetilde{\mu}(x+y) \geq \min\{\widetilde{\mu}(x), \widetilde{\mu}(y)\}$
$= \min\{[\mu^-(x)), \mu^+(x)], [\mu^-(y), \mu^+(y)]\} = [\min\{\mu^-(x), \mu^-(y)\}, \min\{\mu^+(x), \mu^+(y)\}]$

for $x, y \in L$. So,

$\mu^-(x+y) \geq \min\{\mu^-(x), \mu^-(y)\}$ and $\mu^+(x+y) \geq \min\{\mu^+(x), \mu^+(y)\}$.

In a similar way, we can verify (2), (3), and (4). This means that μ^- and μ^+ are fuzzy Lie subalgebras (ideals) of L.

The transfer principle for fuzzy sets described in [90] suggests the following theorem.

Theorem 3.2 *An interval-valued fuzzy set $\widetilde{\mu}$ of a Lie algebra L is an interval-valued fuzzy subalgebra (ideal) of L if and only if all nonempty interval-valued levels of $\widetilde{\mu}$ are interval-valued Lie subalgebras (ideals) of L.*

Proof Assume that $\widetilde{\mu}$ is an interval-valued fuzzy Lie subalgebra (ideal) of L, and let $[t_1, t_2] \in D[0, 1]$ be such that $U(\widetilde{\mu}, [t_1, t_2]) \neq \emptyset$. If $x \in U(\widetilde{\mu}, [t_1, t_2])$, and $y \in U(\widetilde{\mu}, [t_1, t_2])$, then $\widetilde{\mu}(x) \geq [t_1, t_2]$ and $\widetilde{\mu}(y) \geq [t_1, t_2]$. Hence,

$$\widetilde{\mu}(x+y) \geq \min\{\widetilde{\mu}(x), \widetilde{\mu}(y)\} \geq [t_1, t_2],$$

$$\widetilde{\mu}(\alpha x) \geq \widetilde{\mu}(x) \geq [t_1, t_2],$$

$$\widetilde{\mu}([x, y]) \geq \min\{\widetilde{\mu}(x), \widetilde{\mu}(y)\} \geq [t_1, t_2].$$

So, $x + y \in U(\widetilde{\mu}, [t_1, t_2])$, $\alpha x \in U(\widetilde{\mu}, [t_1, t_2])$, and $[x, y] \in U(\widetilde{\mu}, [t_1, t_2])$. This proves that $U(\widetilde{\mu}, [t_1, t_2])$ is a Lie subalgebra of L.

If $\widetilde{\mu}$ is an ideal, then

$$\widetilde{\mu}([x, y]) \geq \widetilde{\mu}(x) \geq [t_1, t_2].$$

Hence, in this case, $U(\widetilde{\mu}, [t_1, t_2])$ is a Lie ideal of L.

The converse statement is obvious.

Definition 3.11 Let $f : L_1 \to L_2$ be a homomorphism of Lie algebras. For any interval-valued fuzzy set $\widetilde{\mu}$ in a Lie algebra L_2, we define an interval-valued fuzzy set $\widetilde{\mu}^f$ in L_1 by $\widetilde{\mu}^f(x) = \widetilde{\mu}(f(x))$ for all $x \in L_1$.

Lemma 3.6 *Let $f : L_1 \to L_2$ be a homomorphism of Lie algebras. If $\widetilde{\mu}$ is an interval-valued fuzzy Lie ideal of L_2, then $\widetilde{\mu}^f$ is an interval-valued fuzzy Lie ideal of L_1.*

Proof Let $x, y \in L_1$ and $\alpha \in \mathbb{F}$. Then,

$$\begin{aligned}
\widetilde{\mu}^f(x+y) = \widetilde{\mu}(f(x+y)) &= \widetilde{\mu}(f(x)+f(y)) \geq \min\{\widetilde{\mu}(f(x)), \widetilde{\mu}(f(y))\} \\
&= \min\{\widetilde{\mu}^f(x), \widetilde{\mu}^f(y)\}, \\
\widetilde{\mu}^f(\alpha x) = \widetilde{\mu}(f(\alpha x)) &= \widetilde{\mu}(\alpha f(x)) \geq \widetilde{\mu}(f(x)) = \mu^f(x), \\
\widetilde{\mu}^f([x, y]) = \widetilde{\mu}(f([x, y])) &= \widetilde{\mu}([f(x), f(y)]) \geq \widetilde{\mu}(f(x)) = \widetilde{\mu}^f(x),
\end{aligned}$$

which proves that $\widetilde{\mu}^f$ is an interval-valued fuzzy Lie ideal of L_1.

Theorem 3.3 *Let $f : L_1 \to L_2$ be an epimorphism of Lie algebras. Then, $\widetilde{\mu}^f$ is an interval-valued fuzzy Lie ideal of L_1 if and only if $\widetilde{\mu}$ is an interval-valued fuzzy Lie ideal of L_2.*

Proof The sufficiency follows from Lemma 3.6. To prove the necessity, observe that f is surjective, so for any $x, y \in L_2$ there are $x_1, y_1 \in L_1$ such that $x = f(x_1)$, $y = f(y_1)$. Thus, $\widetilde{\mu}(x) = \widetilde{\mu}^f(x_1)$, $\widetilde{\mu}(y) = \widetilde{\mu}^f(y_1)$; whence

$$\begin{aligned}
\widetilde{\mu}(x+y) &= \widetilde{\mu}(f(x_1)+f(y_1)) = \widetilde{\mu}(f(x_1+y_1)) = \widetilde{\mu}^f(x_1+y_1) \\
&\geq \min\{\widetilde{\mu}^f(x_1), \widetilde{\mu}^f(y_1)\} = \min\{\widetilde{\mu}(x), \widetilde{\mu}(y)\},
\end{aligned}$$

$$\widetilde{\mu}(\alpha x) = \widetilde{\mu}(\alpha f(x_1)) = \widetilde{\mu}(f(\alpha x_1)) = \widetilde{\mu}^f(\alpha x_1) \geq \widetilde{\mu}^f(x_1) = \widetilde{\mu}(x),$$

$$\widetilde{\mu}([x, y]) = \widetilde{\mu}([f(x_1), f(y_1)]) = \widetilde{\mu}(f([x_1, y_1])) = \widetilde{\mu}^f([x_1, y_1]) \geq \widetilde{\mu}^f(x_1) = \widetilde{\mu}(x).$$

This proves that $\widetilde{\mu}$ is an interval-valued fuzzy Lie ideal of L_2.

Definition 3.12 Two interval-valued fuzzy Lie ideals $\widetilde{\mu}$ and $\widetilde{\lambda}$ of L have the *same type* if there exists $f \in Aut(L)$ such that $\widetilde{\mu}(x) = \widetilde{\lambda}(f(x))$ for all $x \in L$.

Theorem 3.4 *Let $\widetilde{\mu}$ and $\widetilde{\lambda}$ be interval-valued fuzzy Lie ideals of L. Then, the following are equivalent:*

(i) $\widetilde{\mu}$ and $\widetilde{\lambda}$ have the same type,
(ii) $\widetilde{\mu} \circ f = \widetilde{\lambda}$ for some $f \in Aut(L)$,
(iii) $g(\widetilde{\mu}) = \widetilde{\lambda}$ for some $g \in Aut(L)$,
(iv) $h(\widetilde{\lambda}) = \widetilde{\mu}$ for some $h \in Aut(L)$,
(v) $U(\widetilde{\mu}, [t_1, t_2]) = h(U(\widetilde{\lambda}; [t_1, t_2]))$ for some $h \in Aut(L)$ and all $[t_1, t_2] \in D[0, 1]$.

Proof $(i) \Rightarrow (ii)$. Proof follows immediately from the definition.

$(ii) \Rightarrow (iii)$. Suppose that $\widetilde{\mu} \circ f = \widetilde{\lambda}$ for some $f \in Aut(L)$. Then, $\widetilde{\mu}(f(x)) = \widetilde{\lambda}(x)$ and $f^{-1}(\widetilde{\mu})(x) = \sup_{y \in f(x)} \widetilde{\mu}(y) = \widetilde{\mu}(f(x)) = \widetilde{\lambda}(x)$ for all $x \in L$. If $g = f^{-1}$, then $g \in Aut(L)$ and $g(\widetilde{\mu}) = \widetilde{\lambda}$.

$(iii) \Rightarrow (iv)$. Suppose that $g(\widetilde{\mu}) = \widetilde{\lambda}$ for some $g \in Aut(L)$. Then, $\widetilde{\lambda}(x) = g(\widetilde{\mu}) = \sup_{y \in g^{-1}(x)} \widetilde{\mu}(y) = \widetilde{\mu}(g^{-1}(x))$. Hence, we have $g^{-1}(x) = \sup_{y \in g(x)} \widetilde{\lambda}(y) = \widetilde{\lambda}(g(y)) = \widetilde{\mu}(g^{-1}(g(x))) = \widetilde{\mu}(x)$ for all $x \in L$. If $h = g^{-1}$, then $h \in Aut(L)$ and $h(\widetilde{\lambda}) = \widetilde{\mu}$.

$(iv) \Rightarrow (v)$. If $h(\widetilde{\lambda}) = \widetilde{\mu}$ for some $h \in \mathrm{Aut}(L)$, then we obtain $\widetilde{\mu}(x) = h(\widetilde{\lambda})(x) = \sup_{y \in h^{-1}(x)} \widetilde{\lambda}(y) = \widetilde{\lambda}(h^{-1}(x))$ for all $x \in L$.

Let $[t_1, t_2] \in D[0, 1]$. We need to show that $U(\widetilde{\mu}, [t_1, t_2]) = h(U(\widetilde{\lambda}, [t_1, t_2]))$. So, if $x \in U(\widetilde{\mu}, [t_1, t_2])$, then $\widetilde{\lambda}(h^{-1}(x)) = \widetilde{\mu} \geq [t_1, t_2]]$ which implies $h^{-1}(x) \in U(\widetilde{\lambda}; [t_1, t_2])$, i.e., $x \in h(U(\widetilde{\lambda}; [t_1, t_2]))$. Thus, we obtain $U(\widetilde{\mu}, [t_1, t_2]) \subseteq h(U(\widetilde{\lambda}, [t_1, t_2]))$. On the other hand, let $x \in h(U(\widetilde{\lambda}, [t_1, t_2]))$. Then, $h^{-1}(x) \in U(\widetilde{\lambda}, [t_1, t_2])$ and so $\widetilde{\mu}(x) = \widetilde{\lambda}(h^{-1}(x)) \geq [t_1, t_2]$. It follows that $x \in U(\widetilde{\mu}, [t_1, t_2])$. Consequently, $h(U(\widetilde{\lambda}, [t_1, t_2])) \subseteq U(\widetilde{\mu}, [t_1, t_2])$ and (v) holds.

$(v) \Rightarrow (i)$. Suppose that $U(\widetilde{\mu}, [t_1, t_2]) = h(U(\widetilde{\lambda}, [t_1, t_2]))$ for some $h \in \mathrm{Aut}(L)$ and all $[t_1, t_2] \in D[0, 1]$. Let $\widetilde{\lambda}(h^{-1}(x)) = [s_1, s_2]$. Then, $h^{-1}(x) \in U(\widetilde{\lambda}; [s_1, s_2])$; hence, we get $x \in h(U(\widetilde{\lambda}, [s_1, s_2])) = U(\widetilde{\mu}; [s_1, s_2])$. Thus, $\widetilde{\mu}(x) \geq [s_1, s_2] = \widetilde{\lambda}(h^{-1}(x))$. Hence, $\widetilde{\mu}(x) = \widetilde{\lambda}(h^{-1}(x))$ for all $x \in L$, which proves that $\widetilde{\mu}$ and $\widetilde{\lambda}$ have the same type.

3.3 Characterizations of Noetherian Lie Algebras

Theorem 3.5 *A Lie algebra L is Noetherian if and only if the set of values of any interval-valued fuzzy Lie ideal is well ordered.*

Proof Suppose that $\widetilde{\mu}$ is an interval-valued fuzzy Lie ideal whose set of values is not well-ordered subset of $D[0, 1]$. Then, there exists a strictly decreasing sequence $\{[s_n, t_n]\}$ such that $[s_n, t_n] = \widetilde{\mu}(x_n)$ for some $x_n \in L$. Then, $B_1 \subset B_2 \subset B_3 \subset \ldots$, where $B_n := \{x \in L \mid \widetilde{\mu}(x) \geq [s_n, t_n]\}$, form a strictly ascending chain of Lie ideals of L, contradicting the assumption that L is Noetherian.

Conversely, suppose that the set of values of any interval-valued fuzzy Lie ideal of L is well ordered, but L is not Noetherian. Then, there exists a strictly ascending chain $A_1 \subset A_2 \subset A_3 \subset \ldots$ of Lie ideals of L. Suppose that $A = \bigcup_{k=1}^{\infty} A_k$ is a Lie ideal of L. Define an interval-valued fuzzy set $\widetilde{\mu}$ in L by putting

$$\widetilde{\mu}(x) := \begin{cases} \left[\frac{1}{k+1}, \frac{1}{k}\right] & \text{for } x \in A_k \backslash A_{k-1}, \\ [0, 0] & \text{for } x \notin A. \end{cases}$$

We claim that $\widetilde{\mu}$ is an interval-valued fuzzy Lie ideal of L.

Let $x, y \in L$. If $x, y \in A$, then there are m, n such that $x \in A_n \backslash A_{n-1}$, $y \in A_m \backslash A_{m-1}$. Obviously, $x + y \in A_k \backslash A_{k-1} \subset A_p$, where $k \leq p = \max\{m, n\}$. So, $\widetilde{\mu}(x) = \left[\frac{1}{n+1}, \frac{1}{n}\right]$, $\widetilde{\mu}(y) = [\frac{1}{m+1}, \frac{1}{m}]$ and

$$\widetilde{\mu}(x+y) = \left[\frac{1}{k+1}, \frac{1}{k}\right] \geq \left[\frac{1}{p+1}, \frac{1}{p}\right] = \min\{\widetilde{\mu}(x), \widetilde{\mu}(y)\}.$$

In the case $x \notin A$, $y \in A$, we have $y \in A_m \backslash A_{m-1}$ for some natural m. Hence, $\widetilde{\mu}(x) = [0, 0]$, $\widetilde{\mu}(y) = \left[\frac{1}{m+1}, \frac{1}{m}\right]$; consequently,

$$\widetilde{\mu}(x+y) \geq [0, 0] = \min\{\widetilde{\mu}(x), \widetilde{\mu}(y)\}.$$

The case $x \in A$, $y \notin A$ is analogous. The case $x \notin A$, $y \notin A$ is obvious. The verification of (2) and (4) is analogous. Thus, $\widetilde{\mu}$ is an interval-valued fuzzy Lie ideal of L. Consequently, $\widetilde{\mu}$ is an interval-valued fuzzy Lie ideal. Since the chain $A_1 \subset A_2 \subset A_3 \subset \ldots$ is not terminating, $\widetilde{\mu}$ has a strictly descending sequence of values. This contradicts that the value set of any interval-valued fuzzy Lie ideal is well ordered. This completes the proof.

We note that a set is well ordered if and only if it does not contain any infinite decreasing sequence.

Theorem 3.6 *Let $S = \{[s_1, t_1], [s_2, t_2], \ldots\} \cup \{[0, 0]\}$, where $\{[s_n, t_n]\}$ is a strictly decreasing sequence in $D[0, 1]$. Then, a Lie algebra L is Noetherian if and only if for each interval-valued fuzzy Lie ideal $\widetilde{\mu}$ of L, $Im(\widetilde{\mu}) \subseteq S$ implies that there exists a positive integer m such that $Im(\widetilde{\mu}) \subseteq \{[s_1, t_1], [s_2, t_2], \ldots, [s_m, t_m]\} \cup \{[0, 0]\}$.*

Proof If L is a Noetherian Lie algebra, then $Im(\widetilde{\mu})$ is a well-ordered subset of $D[0, 1]$.

Conversely, if the above condition is satisfied and L is not Noetherian, then there exists a strictly ascending chain $A_1 \subset A_2 \subset A_3 \subset \ldots$ of Lie ideals of L.

Define an interval-valued fuzzy set $\widetilde{\mu}$ by

$$\widetilde{\mu}(x) := \begin{cases} [s_1, t_1] \text{ if } x \in A_1, \\ [s_n, t_n] \text{ if } x \in A_n \backslash A_{n-1}, n = 2, 3, 4, \ldots \\ [0, 0] \quad \text{if } x \in G \backslash \bigcup_{n=1}^{\infty} A_n. \end{cases}$$

Let $x, y \in L$. If either x or y belongs to $G \backslash \bigcup_{n=1}^{\infty} A_n$, then either $\widetilde{\mu}(x) = [0, 0]$ or $\widetilde{\mu}(y) = [0, 0]$. Thus, $\widetilde{\mu}(x+y) \geq \min\{\widetilde{\mu}(x), \widetilde{\mu}(y)\}$.

If $x, y \in A_1$, then $x \in A_1$ and so $\widetilde{\mu}(x+y) = [s_1, t_1] \geq \min\{\widetilde{\mu}(x), \widetilde{\mu}(y)\}$.

If $x, y \in A_n \backslash A_{n-1}$, then $x \in A_n$ and $\widetilde{\mu}(x+y) \geq [s_n, t_n] = \min\{\widetilde{\mu}(x), \widetilde{\mu}(y)\}$. Assume that $x \in A_1$ and $y \in A_n \backslash A_{n-1}$ for $n = 2, 3, 4, \ldots$, then $x + y \in A_n$, and hence,

$$\widetilde{\mu}(x+y) \geq [s_n, t_n] = \min\{[s_1, t_1], [s_n, t_n]\} = \min\{\widetilde{\mu}(x), \widetilde{\mu}(y)\}.$$

Similarly for $x \in A_n \backslash A_{n-1}$ and $y \in A_1$ for $n = 2, 3, 4, \ldots$, we have

$$\widetilde{\mu}(x+y) \geq [s_n, t_n] = \min\{\widetilde{\mu}(x), \widetilde{\mu}(y)\}.$$

Hence, $\widetilde{\mu}$ is an interval-valued fuzzy Lie ideal of Lie algebra. This contradicts our assumption. The verification of (2) and (4) is analogous, and we omit the details. This completes the proof.

3.4 Quotient Lie Algebra via Interval-Valued Lie Ideals

Theorem 3.7 *Let J be a Lie ideal of a Lie algebra L. If $\widetilde{\mu}$ is an interval-valued Lie ideal of L, then an interval-valued fuzzy set $\overline{\widetilde{\mu}}$ defined by*

$$\overline{\widetilde{\mu}}(a+J) = \sup_{x \in J} \widetilde{\mu}(a+x)$$

is an interval-valued fuzzy Lie ideal of the quotient Lie algebra L/J.

Proof Clearly, $\overline{\widetilde{\mu}}$ is well defined. Let $x+J, y+J \in L/J$, then

$$\begin{aligned}\overline{\widetilde{\mu}}(x+J)+(y+J)) &= \overline{\widetilde{\mu}}_A((x+y)+J) = \sup_{z\in J} \widetilde{\mu}((x+y)+z)\\ &= \sup_{z=s+t\in J} \widetilde{\mu}((x+y)+(s+t))\\ &\geq \sup_{s,\ t\in J} \min\{\widetilde{\mu}(x+s), \widetilde{\mu}(y+t)\}\\ &= \min\{\sup_{s\in J} \widetilde{\mu}(x+s), \sup_{t\in J} \widetilde{\mu}(y+t)\}\\ &= \min\{\overline{\widetilde{\mu}}(x+J), \overline{\widetilde{\mu}}(y+J)\},\end{aligned}$$

$$\begin{aligned}\overline{\widetilde{\mu}}(\alpha(x+J)) = \overline{\widetilde{\mu}}(\alpha x+J) = \sup_{z\in J} \widetilde{\mu}(\alpha x+z) &\geq \sup_{z\in J} \widetilde{\mu}(x+z) = \overline{\widetilde{\mu}}(x+J),\\ \overline{\widetilde{\mu}}([x+J, y+J]) = \overline{\widetilde{\mu}}([x,y]+J) &= \sup_{z\in J} \widetilde{\mu}([x,y]+z)\\ &\geq \sup_{z\in J} \widetilde{\mu}(x+z) = \overline{\widetilde{\mu}}(x+J).\end{aligned}$$

Hence, $\overline{\widetilde{\mu}}$ is an interval-valued fuzzy Lie ideal of L/J.

Theorem 3.8 *Let $f : L_1 \to L_2$ be a homomorphism of a Lie algebra L_1 onto a Lie algebra L_2.*

(i) If $\widetilde{\mu}$ is an interval-valued fuzzy Lie ideal of L_1, then $f(\widetilde{\mu})$ is an interval-valued fuzzy Lie ideal of L_2.
(ii) If $\widetilde{\lambda}$ is an interval-valued fuzzy Lie ideal of L_2, then $f^{-1}(\widetilde{\lambda})$ is an interval-valued fuzzy Lie ideal of L_1.

Proof Straightforward.

For an interval-valued fuzzy Lie ideal $\widetilde{\mu}$ of a Lie algebra L, we define a binary relation $\sim$ by putting

$$x \sim y \Longleftrightarrow \widetilde{\mu}(x-y) = \widetilde{\mu}(0).$$

This relation is a congruence. The set of all its equivalence classes $\widetilde{\mu}[x]$ is denoted by $L/\widetilde{\mu}$. It is a Lie algebra under the following operations:

$$\widetilde{\mu}[x]+\widetilde{\mu}[y] = \widetilde{\mu}[x+y], \quad \alpha\widetilde{\mu}[x] = \widetilde{\mu}[\alpha x], \quad [\widetilde{\mu}[x], \widetilde{\mu}[y]] = \widetilde{\mu}[[x,y]],$$

where $x, y \in L, \alpha \in \mathbb{F}$.

Theorem 3.9 (First isomorphism theorem) *Let $f: L_1 \to L_2$ be an epimorphism of Lie algebras, and let $\widetilde{\mu}$ be an interval-valued fuzzy Lie ideal of L_2. Then, $L_1/f^{-1}(\widetilde{\mu}) \cong L_2/\widetilde{\mu}$.*

Proof Define a map $\theta: L_1/f^{-1}(\widetilde{\mu}) \to L_2/\widetilde{\mu}$ by $\theta(f^{-1}(\widetilde{\mu})[x]) = \widetilde{\mu}[f(x)]$.

θ is well defined since $f^{-1}(\widetilde{\mu})[x] = f^{-1}(\widetilde{\mu})[y]$ gives $f^{-1}(\widetilde{\mu})(x-y) = f^{-1}(\widetilde{\mu})(0)$. Whence $\widetilde{\mu}(f(x)-f(y)) = \widetilde{\mu}(f(0)) = \widetilde{\mu}(0)$. Thus, $\widetilde{\mu}[f(x)] = \widetilde{\mu}[f(y)]$.

θ is one to one because $\widetilde{\mu}[f(x)] = \widetilde{\mu}[f(y)]$ gives $\widetilde{\mu}(f(x)-f(y)) = \widetilde{\mu}(0)$, i.e., $\widetilde{\mu}(f(x)-f(y)) = \widetilde{\mu}(f(0))$, which proves $f^{-1}(\widetilde{\mu})(x-y) = f^{-1}(\widetilde{\mu})(0)$. Therefore, $f^{-1}(\widetilde{\mu})[x] = f^{-1}(\widetilde{\mu})[y]$.

Since f is an onto, θ is an onto. Finally, θ is a homomorphism because

$$\begin{aligned}\theta(f^{-1}(\widetilde{\mu})[x]+f^{-1}(\widetilde{\mu})[y]) &= \theta(f^{-1}(\widetilde{\mu})[x+y]) = \widetilde{\mu}[f(x+y)] = \widetilde{\mu}[f(x)+f(y)] \\ &= \widetilde{\mu}[f(x)] + \widetilde{\mu}[f(y)] = \theta(f^{-1}(\widetilde{\mu})[x]) + \theta(f^{-1}(\widetilde{\mu})[y]),\end{aligned}$$

$$\theta(\alpha f^{-1}(\widetilde{\mu})[x]) = \theta(f^{-1}(\widetilde{\mu})[\alpha x]) = \widetilde{\mu}[f(\alpha x)] = \alpha\widetilde{\mu}[f(x)] = \alpha\theta(f^{-1}(\widetilde{\mu})[x]),$$

$$\begin{aligned}\theta([f^{-1}(\widetilde{\mu})[x], f^{-1}(\widetilde{\mu})[y]]) &= \theta([f^{-1}(\widetilde{\mu})[x, y]]) = \widetilde{\mu}[f([x, y])] \\ &= \widetilde{\mu}[[f(x), f(y)]] = [\widetilde{\mu}[f(x)], \widetilde{\mu}[f(y)]] \\ &= [\theta(f^{-1}(\widetilde{\mu})[x]), \theta(f^{-1}(\widetilde{\mu})[y])].\end{aligned}$$

Hence, $L_1/f^{-1}(\widetilde{\mu}) \cong L_2/\widetilde{\mu}$.

We state the following isomorphism theorems without proofs.

Theorem 3.10 (Second isomorphism theorem) *Let $\widetilde{\mu}$ and $\widetilde{\lambda}$ be two interval-valued fuzzy subsets of the same Lie algebra. If $\widetilde{\mu}$ is a subalgebra and $\widetilde{\lambda}$ is a Lie ideal, then*

(1) $\widetilde{\lambda}$ is an interval-valued fuzzy Lie ideal of $\widetilde{\mu} + \widetilde{\lambda}$,

(2) $\widetilde{\mu} \cap \widetilde{\lambda}$ is an interval-valued fuzzy Lie ideal of $\widetilde{\mu}$,

(3) $(\widetilde{\mu} + \widetilde{\lambda})/\lambda \cong \widetilde{\mu}/(\widetilde{\mu} \cap \widetilde{\lambda})$.

Theorem 3.11 (Third isomorphism theorem) *Let $\widetilde{\mu}$ and $\widetilde{\lambda}$ be interval-valued fuzzy Lie ideals of the same Lie algebra such that $\widetilde{\mu} \leq \widetilde{\lambda}$. Then,*

(i) $\widetilde{\lambda}/\widetilde{\mu}$ is an interval-valued fuzzy Lie ideal of $L/\widetilde{\mu}$,

(ii) $(L/\widetilde{\mu})/(\widetilde{\lambda}/\widetilde{\mu}) \cong L/\widetilde{\lambda}$.

Theorem 3.12 (Interval-valued Zassenhaus lemma) *Let $\widetilde{\mu}$ and $\widetilde{\lambda}$ be interval-valued fuzzy subalgebras of a Lie algebra L, and let $\widetilde{\mu}_1$ and $\widetilde{\lambda}_1$ be interval-valued fuzzy Lie ideals of $\widetilde{\mu}$ and $\widetilde{\lambda}$, respectively. Then,*

(a) $\widetilde{\mu}_1 + (\widetilde{\mu} \cap \widetilde{\lambda}_1)$ is an interval-valued fuzzy Lie ideal of $\widetilde{\mu}_1 + (\widetilde{\mu} \cap \widetilde{\lambda})$,

(b) $\widetilde{\lambda}_1 + (\widetilde{\mu}_1 \cap \widetilde{\lambda})$ is an interval-valued fuzzy Lie ideal of $\widetilde{\lambda}_1 + (\widetilde{\mu} \cap \widetilde{\lambda})$,

(c) $(\widetilde{\mu}_1 + (\widetilde{\mu} \cap \widetilde{\lambda}))/(\widetilde{\mu}_1 + (\widetilde{\mu} \cap \widetilde{\lambda}_1)) \simeq (\widetilde{\lambda}_1 + (\widetilde{\mu} \cap \widetilde{\lambda}))/(\widetilde{\lambda}_1 + (\widetilde{\mu}_1 \cap \widetilde{\lambda}))$.

3.5 Interval-Valued Intuitionistic Fuzzy Lie Ideals

Atanassov and Gargov [32] introduced the notion of interval-valued intuitionistic fuzzy sets which is a generalization of both intuitionistic fuzzy sets and interval-valued fuzzy sets.

Definition 3.13 A mapping $\tilde{A} = (\widetilde{\mu}_{\tilde{A}}, \widetilde{\lambda}_{\tilde{A}}) : X \to D[0,1] \times D[0,1]$ is called an *interval-valued intuitionistic fuzzy set* in X if $\mu^+_{\tilde{A}}(x) + \lambda^+_{\tilde{A}}(x) \leq 1$ and $\mu^-_{\tilde{A}}(x) + \lambda^-_{\tilde{A}}(x) \leq 1$ for all $x \in X$, where the mappings $\widetilde{\mu}_{\tilde{A}} = [\mu^-_{\tilde{A}}, \mu^+_{\tilde{A}}] : X \to D[0,1]$ and $\widetilde{\lambda}_{\tilde{A}} = [\lambda^-_{\tilde{A}}, \lambda^+_{\tilde{A}}] : X \to D[0,1]$ denote the *degree of membership* (namely $\widetilde{\mu}_{\tilde{A}}(x)$) and the *degree of nonmembership* (namely $\widetilde{\lambda}_{\tilde{A}}(x)$) of each element $x \in X$ to $\tilde{A}$, respectively.

Definition 3.14 Let $\tilde{A} = (\widetilde{\mu}_{\tilde{A}}, \widetilde{\lambda}_{\tilde{A}})$ be an interval-valued intuitionistic fuzzy set on X, and let $\widetilde{\alpha}, \widetilde{\beta} \in D[0,1]$ be such that $\widetilde{\alpha} + \widetilde{\beta} \leq [1,1]$. Then, the set

$$\tilde{A}^{(\widetilde{\alpha},\widetilde{\beta})} := \{x \in X \mid \widetilde{\alpha} \leq \widetilde{\mu}_{\tilde{A}}(x),\ \widetilde{\lambda}_{\tilde{A}}(x) \leq \widetilde{\beta}\}$$

is called an $(\widetilde{\alpha}, \widetilde{\beta})$-*level subset* of $\tilde{A}$.

The set of all $(\widetilde{\alpha}, \widetilde{\beta}) \in \mathrm{Im}(\widetilde{\mu}_{\tilde{A}}) \times \mathrm{Im}(\widetilde{\lambda}_{\tilde{A}})$ such that $\widetilde{\alpha} + \widetilde{\beta} \leq [1,1]$ is called the *image of* $\tilde{A} = (\widetilde{\mu}_{\tilde{A}}, \widetilde{\lambda}_{\tilde{A}})$.

Note that

$$\begin{aligned}\tilde{A}^{(\widetilde{\alpha},\widetilde{\beta})} &= \{x \in X \mid \widetilde{\mu}_{\tilde{A}}(x) \geq \widetilde{\alpha},\ \widetilde{\lambda}_{\tilde{A}}(x) \leq \widetilde{\beta}\}\\ &= \{x \in X \mid \widetilde{\mu}_{\tilde{A}}(x) \geq \widetilde{\alpha}\} \cap \{x \in X \mid \widetilde{\lambda}_{\tilde{A}}(x) \leq \widetilde{\beta}\}\\ &= U(\widetilde{\mu}_{\tilde{A}}, \widetilde{\alpha}) \cap L(\widetilde{\lambda}_{\tilde{A}}, \widetilde{\beta}).\end{aligned}$$

Definition 3.15 We use $\widetilde{0}$ to denote the *interval-valued fuzzy empty set* and $\widetilde{1}$ to denote the *interval-valued fuzzy whole set* in a set X, and we define $\widetilde{0}(x) = [0,0]$ and $\widetilde{1}(x) = [1,1]$, for all $x \in X$.

Notation. For short, we may write $\widetilde{t} = [t_1, t_2], \widetilde{s} = [s_1, s_2], \widetilde{\alpha} = [\alpha_1, \alpha_2], \widetilde{s_1} = [s_3, s_4]$, $\widetilde{t_1} = [t_3, t_4]$, $\widetilde{\beta} = [\beta_1, \beta_2]$, $\widetilde{\alpha}_1 = [\alpha_3, \alpha_4]$, and $\widetilde{\beta}_1 = [\beta_3, \beta_4] \in D[0,1]$.

Definition 3.16 An interval-valued intuitionistic fuzzy set $\tilde{A} = (\widetilde{\mu}_{\tilde{A}}, \widetilde{\lambda}_{\tilde{A}})$ in L is called an *interval-valued intuitionistic fuzzy Lie ideal* (IIF, for short) of L if the following conditions are satisfied:

(1) $\widetilde{\mu}_{\tilde{A}}(x+y) \geq \min\{\widetilde{\mu}_{\tilde{A}}(x), \widetilde{\mu}_{\tilde{A}}(y)\}$,

(2) $\widetilde{\lambda}_{\tilde{A}}(x+y) \leq \max\{\widetilde{\lambda}_{\tilde{A}}(x), \widetilde{\lambda}_{\tilde{A}}(y)\}$,

(3) $\widetilde{\mu}_{\tilde{A}}(\alpha x) \geq \widetilde{\mu}_{\tilde{A}}(x),\quad \widetilde{\lambda}_{\tilde{A}}(\alpha x) \leq \widetilde{\lambda}_{\tilde{A}}(x)$,

(4) $\widetilde{\mu}_{\tilde{A}}([x,y]) \geq \widetilde{\mu}_{\tilde{A}}(x),\quad \widetilde{\lambda}_{\tilde{A}}([x,y]) \leq \widetilde{\lambda}_{\tilde{A}}(x)$

for all $x, y \in L$ and $\alpha \in \mathbb{F}$.

From (3), it follows that:

(5) $\widetilde{\mu}_{\tilde{A}}(0) \geq \widetilde{\mu}_{\tilde{A}}(x)$, $\widetilde{\lambda}_{\tilde{A}}(0) \leq \widetilde{\lambda}_{\tilde{A}}(x)$,
(6) $\widetilde{\mu}_{\tilde{A}}(-x) \geq \widetilde{\mu}_{\tilde{A}}(x)$, $\widetilde{\lambda}_{\tilde{A}}(-x) \leq \widetilde{\lambda}_{\tilde{A}}(x)$.

Example 3.3 Let $\Re^3 = \{(x, y, z) : x, y, z \in \mathbb{R}\}$ be the set of all three-dimensional real vectors. Then, $\Re^3$ with the bracket $[\cdot, \cdot]$ defined as usual cross product, i.e., $[x, y] = x \times y$, is a real Lie algebra. We define an interval-valued intuitionistic fuzzy set $\tilde{A} = (\widetilde{\mu}_{\tilde{A}}, \widetilde{\lambda}_{\tilde{A}}) : \Re^3 \to D[0, 1] \times D[0, 1]$ by

$$\widetilde{\mu_{\tilde{A}}}(x, y, z) = \begin{cases} \widetilde{s} \text{ if } x = y = z = 0, \\ \widetilde{t} \text{ otherwise,} \end{cases}$$

$$\widetilde{\lambda_{\tilde{A}}}(x, y, z) = \begin{cases} \widetilde{t} \text{ if } x = y = z = 0, \\ \widetilde{s} \text{ otherwise,} \end{cases}$$

where $\widetilde{s} > \widetilde{t}$ and $\widetilde{s}, \widetilde{t} \in D[0, 1]$. By routine calculations, it is easy to check that $\tilde{A} = (\widetilde{\mu}_{\tilde{A}}, \widetilde{\lambda}_{\tilde{A}})$ is an interval-valued intuitionistic fuzzy Lie ideal of a Lie algebra $\Re^3$.

The following lemma is trivial.

Lemma 3.7 *If $\tilde{A}$ is an interval-valued intuitionistic fuzzy Lie ideal of a Lie algebra L, then*

1. $\widetilde{\mu}_{\tilde{A}}([x, y]) \geq \max\{\widetilde{\mu}_{\tilde{A}}(x), \widetilde{\mu}_{\tilde{A}}(y)\}$,
2. $\widetilde{\lambda}_{\tilde{A}}([x, y]) \leq \min\{\widetilde{\lambda}_{\tilde{A}}(x), \widetilde{\lambda}_{\tilde{A}}(y)\}$,
3. $\widetilde{\mu}_{\tilde{A}}([x, y]) = \widetilde{\mu}_{\tilde{A}}(-[y, x]) = \widetilde{\mu}_{\tilde{A}}([y, x])$,
4. $\widetilde{\lambda}_{\tilde{A}}([x, y]) = \widetilde{\lambda}_{\tilde{A}}(-[y, x]) = \widetilde{\lambda}_{\tilde{A}}([y, x])$

for all $x, y \in L$.

Theorem 3.13 *If $\tilde{A} = (\widetilde{\mu}_{\tilde{A}}, \widetilde{\lambda}_{\tilde{A}})$ is an interval-valued intuitionistic fuzzy Lie ideal of a Lie algebra L, then the level subsets $U(\widetilde{\mu}_{\tilde{A}}, \widetilde{\alpha}) = \{x \in L \mid \widetilde{\mu}_{\tilde{A}}(x) \geq \widetilde{\alpha}\}$ and $L(\widetilde{\lambda}_{\tilde{A}}, \widetilde{\alpha}) = \{x \in L \mid \widetilde{\lambda}_{\tilde{A}}(x) \leq \widetilde{\alpha}\}$ are Lie ideals of L for every $\widetilde{\alpha} \in \mathrm{Im}(\widetilde{\mu}_{\tilde{A}}) \cap \mathrm{Im}(\widetilde{\lambda}_{\tilde{A}}) \subseteq D[0, 1]$, where $\mathrm{Im}(\widetilde{\mu}_{\tilde{A}})$ and $\mathrm{Im}(\widetilde{\lambda}_{\tilde{A}})$ are sets of values of $\widetilde{\mu}_{\tilde{A}}$ and $\widetilde{\lambda}_{\tilde{A}}$, respectively.*

Proof Let $\widetilde{\alpha} \in Im(\widetilde{\mu}_{\tilde{A}}) \cap Im(\widetilde{\lambda}_{\tilde{A}}) \subseteq D[0, 1]$, and let $x, y \in U(\widetilde{\mu}_{\tilde{A}}; \widetilde{\alpha})$ and $\alpha \in \mathbb{F}$. Then, $\widetilde{\mu}_{\tilde{A}}(x) \geq [s, t]$ and $\widetilde{\mu}_{\tilde{A}}(y) \geq \widetilde{\alpha}$. It follows that

$$\begin{aligned} \widetilde{\mu}_{\tilde{A}}(x + y) &\geq \min\{\widetilde{\mu}_{\tilde{A}}(x), \widetilde{\mu}_{\tilde{A}}(y)\} \geq \widetilde{\alpha}, \\ \widetilde{\mu}_{\tilde{A}}(\alpha x) &\geq \widetilde{\mu}_{\tilde{A}}(x) \geq \widetilde{\alpha}, \\ \widetilde{\mu}_{\tilde{A}}([x, y]) &\geq \widetilde{\mu}_{\tilde{A}}(x) \geq \widetilde{\alpha}, \end{aligned}$$

so that $x + y, \alpha x, [x, y] \in U(\mu_{\tilde{A}}, \widetilde{\alpha})$. Consequently, $U(\widetilde{\mu}_{\tilde{A}}, \widetilde{\alpha})$ is a Lie ideal of L. In the same manner, we can prove that $L(\widetilde{\lambda}_{\tilde{A}}, \widetilde{\alpha})$ is a Lie ideal of L. This completes the proof.

Theorem 3.14 *If all nonempty-level subsets $U(\widetilde{\mu}_{\tilde{A}}, \widetilde{\alpha})$ and $L(\widetilde{\lambda}_{\tilde{A}}, \widetilde{\alpha})$ of an interval-valued intuitionistic fuzzy set $\tilde{A} = (\widetilde{\mu}_{\tilde{A}}, \widetilde{\lambda}_{\tilde{A}})$ are Lie ideals of a Lie algebra L, then $\tilde{A}$ is an interval-valued intuitionistic fuzzy Lie ideal of L.*

Proof Let $\widetilde{\alpha} \in D[0, 1]$. Suppose that $U(\widetilde{\mu}_{\tilde{A}}, \widetilde{\alpha}) \neq \emptyset$ and $L(\widetilde{\lambda}_{\tilde{A}}, \widetilde{\alpha}) \neq \emptyset$ are Lie ideals of L. We must show that $\tilde{A} = (\widetilde{\mu}_{\tilde{A}}, \widetilde{\lambda}_{\tilde{A}})$ satisfies the conditions (1)–(4) from Definition 3.16. If the condition (1) is false, then there exist $x, y \in L$ such that

$$\widetilde{\mu}_{\tilde{A}}(x + y) < \min\{\widetilde{\mu}_{\tilde{A}}(x), \widetilde{\mu}_{\tilde{A}}(y)\}.$$

Taking

$$\widetilde{\alpha}_0 := \frac{1}{2}\{\widetilde{\mu}_{\tilde{A}}(x + y) + \min\{\widetilde{\mu}_{\tilde{A}}(x), \widetilde{\mu}_{\tilde{A}}(y)\}\},$$

we have

$$\widetilde{\mu}_{\tilde{A}}(x + y) < \widetilde{\alpha}_0 < \min\{\widetilde{\mu}_{\tilde{A}}(x), \widetilde{\mu}_{\tilde{A}}(y)\}.$$

It follows that $x + y \notin U(\widetilde{\mu}_{\tilde{A}}, \widetilde{\alpha})$ and $x, y \in U(\widetilde{\mu}_{\tilde{A}}, \widetilde{\alpha})$, a contradiction. Hence, the condition (1) is true. The verification is analogous for other conditions, and we omit the details.

Theorem 3.15 *An interval-valued intuitionistic fuzzy set $\tilde{A} = (\widetilde{\mu}_{\tilde{A}}, \widetilde{\lambda}_{\tilde{A}})$ of L is an interval-valued intuitionistic fuzzy Lie ideal of L if and only if $L_{\tilde{A}}^{(\widetilde{\alpha}, \widetilde{\beta})}$ is a Lie ideal of L for every $(\widetilde{\alpha}, \widetilde{\beta}) \in \mathrm{Im}(\widetilde{\mu}_{\tilde{A}}) \times \mathrm{Im}(\widetilde{\lambda}_{\tilde{A}})$ with $\widetilde{\alpha} + \widetilde{\beta} \leq [1, 1]$.*

Proof Straightforward.

Corollary 3.1 *An interval-valued intuitionistic fuzzy set $\tilde{A} = (\widetilde{\mu}_{\tilde{A}}, \widetilde{\lambda}_{\tilde{A}})$ is an interval-valued intuitionistic fuzzy Lie ideal of L if and only if for every $\widetilde{\alpha}, \widetilde{\beta} \in D[0, 1]$ such that $\widetilde{\alpha} + \widetilde{\beta} \leq [1, 1]$, all nonempty $U(\widetilde{\mu}_{\tilde{A}}, \widetilde{\alpha})$ and $L(\widetilde{\lambda}_{\tilde{A}}, \widetilde{\beta})$ are Lie ideals of L.*

Theorem 3.16 *Let $\{G_{[\alpha_1, \beta_1]} \mid [\alpha_1, \beta_1] \in \Omega \subseteq D[0, \frac{1}{2}]\}$ be a collection of Lie ideals of L such that $L = \bigcup G_{[\alpha_1, \beta_1]}$, and for every $[\alpha_1, \beta_1], [\alpha_2, \beta_2] \in \Omega$, $[\alpha_1, \beta_1] > [\alpha_2, \beta_2]$ if and only if $G_{[\alpha_1, \beta_1]} \subset G_{[\alpha_2, \beta_2]}$. Then, an interval-valued intuitionistic fuzzy set $(\widetilde{\mu}_{\tilde{A}}, \widetilde{\lambda}_{\tilde{A}})$ of L defined by*

$$\widetilde{\mu}_{\tilde{A}}(x) = \sup\{[\alpha_1, \beta_1] \in \Omega \mid x \in G_{[\alpha_1, \beta_1]}\}$$

$$\widetilde{\lambda}_{\tilde{A}}(x) = \inf\{[\alpha_1, \beta_1] \in \Omega \mid x \in G_{[\alpha_1, \beta_1]}\}$$

is an interval-valued intuitionistic fuzzy Lie ideal of L.

Proof According to Corollary 3.1, it suffices to show that for every $[\alpha_1, \beta_1]$, $[\alpha_2, \beta_2] \in D[0, 1]$, where $[\alpha_1, \beta_1] + [\alpha_2, \beta_2] \leq [1, 1]$, the nonempty sets $U(\widetilde{\mu}_{\tilde{A}}, [\alpha_1, \beta_1])$ and $L(\widetilde{\lambda}_{\tilde{A}}, [\alpha_2, \beta_2])$ are Lie ideals of L. To prove that $U(\widetilde{\mu}_{\tilde{A}}, [\alpha_1, \beta_1])$ is a Lie ideal, we consider two cases:

(1) $[\alpha_1, \beta_1] = \sup\{[\alpha_2, \beta_2] \in \Omega \mid [\alpha_2, \beta_2] < [\alpha_1, \beta_1]\} = \sup\{[\alpha_2, \beta_2] \in \Omega \mid G_{[\alpha_1,\beta_1]} \subset G_{[\alpha_2,\beta_2]}\}$,
(2) $[\alpha_1, \beta_1] \neq \sup\{[\alpha_2, \beta_2] \in \Omega \mid [\alpha_2, \beta_2] < [\alpha_1, \beta_1]\} = \sup\{[\alpha_2, \beta_2] \in \Omega \mid G_{[\alpha_1,\beta_1]} \subset G_{[\alpha_2,\beta_2]}\}$.

Case (1) implies that

$$\begin{aligned} x \in U(\widetilde{\mu}_{\tilde{A}}; [\alpha_1, \beta_1]) &\Leftrightarrow x \in G_w, \ \forall\, w < [\alpha_1, \beta_1] \\ &\Leftrightarrow x \in \bigcap\nolimits_{w<[\alpha_1,\beta_1]} G_w. \end{aligned}$$

Hence, $U(\widetilde{\mu}_{\tilde{A}}, [\alpha_1, \beta_1]) = \bigcap_{w<[\alpha_1,\beta_1]} G_w$, which is a Lie ideal of L.

For case (2), there exists $\varepsilon > 0$ such that $([\alpha_1, \beta_1] - \varepsilon, [\alpha_1, \beta_1]) \bigcap \Omega = \emptyset$. We claim that in this case $U(\widetilde{\mu}_{\tilde{A}}, [\alpha_1, \beta_1]) = \bigcup_{[\alpha_2,\beta_2]\geq[\alpha_1,\beta_1]} G_{[\alpha_2,\beta_2]}$. Indeed, if $x \in \bigcup_{[\alpha_2,\beta_2]\geq[\alpha_1,\beta_1]} G_{[\alpha_2,\beta_2]}$, then $x \in G_{[\alpha_2,\beta_2]}$ for some $[\alpha_2, \beta_2] \geq [\alpha_1, \beta_1]$, which gives $\widetilde{\mu}_{\tilde{A}}(x) \geq [\alpha_2, \beta_2] \geq [\alpha_1, \beta_1]$. Thus, $x \in U(\widetilde{\mu}_{\tilde{A}}, [\alpha_1, \beta_1])$, i.e., $\bigcup_{[\alpha_2,\beta_2]\geq[\alpha_1,\beta_1]} G_{[\alpha_2,\beta_2]} \subseteq U(\widetilde{\mu}_{\tilde{A}}, [\alpha_1, \beta_1])$. On the other hand, if $x \notin \bigcup_{[\alpha_2,\beta_2]\geq[\alpha_1,\beta_1]} G_{[\alpha_2,\beta_2]}$, then $x \notin G_{[\alpha_2,\beta_2]}$ for all $[\alpha_2, \beta_2] \geq [\alpha_1, \beta_1]$, which implies that $x \notin G_{[\alpha_2,\beta_2]}$ for all $[\alpha_2, \beta_2] > [\alpha_1, \beta_1] - \varepsilon$; i.e., if $x \in G_{[\alpha_2,\beta_2]}$, then $[\alpha_2, \beta_2] \leq [\alpha_1, \beta_1] - \varepsilon$. Thus, $\widetilde{\mu}_{\tilde{A}}(x) \leq [\alpha_1, \beta_1] - \varepsilon$. So $x \notin U(\widetilde{\mu}, [\alpha_1, \beta_1])$. Thus, $U(\widetilde{\mu}_{\tilde{A}}, [\alpha_1, \beta_1]) \subseteq \bigcup_{[\alpha_2,\beta_2]\geq[\alpha_1,\beta_1]} G_{[\alpha_2,\beta_2]}$. Hence, $U(\widetilde{\mu}_{\tilde{A}}, [\alpha_1, \beta_1]) = \bigcup_{[\alpha_2,\beta_2]\geq[\alpha_1,\beta_1]} G_{[\alpha_2,\beta_2]}$, which is a Lie ideal of L. For $L(\widetilde{\lambda}_{\tilde{A}}, [\alpha_2, \beta_2])$, the proof is similar.

Definition 3.17 For any $[s, t] \in D[0, 1]$, we define relation $\mathscr{R}^{[s,t]}$ on the interval-valued intuitionistic fuzzy Lie ideal over L (briefly, $IIFI(L)$) as follows:

$$(\tilde{A}, \tilde{B}) \in \mathscr{R}^{[s,t]} \quad \longleftrightarrow \quad L_{\tilde{A}}^{([s,t],[s,t])} = L_{\tilde{B}}^{([s,t],[s,t])}.$$

The relation $\mathscr{R}^{[s,t]}$ is an *equivalence relation* on $IIFI(L)$.

Theorem 3.17 *For any* $[s, t] \in D(0, 1)$, *the map* $\varphi_{[s,t]} : IIFI(L) \to I(L) \cup \{\emptyset\}$ *defined by* $\varphi_{[s,t]}(A) = L_{\tilde{A}}^{([s,t],[s,t])}$ *is surjective, where* $I(L)$ *denote the family of all Lie ideals of* L.

Proof Let $[s, t] \in D(0, 1)$. Then, $\varphi_{[s,t]}([0, 0], [1, 1]) = L_{\tilde{A}}^{([s,t],[s,t])} = U([0, 0], [s, t]) \cap L([1, 1], [s, t]) = \emptyset$. For any $H \in IIFI(L)$, there exists $\widetilde{H} = (\chi_H, \overline{\chi}_H) \in IIFI(L)$ such that $\varphi_t(\widetilde{H}) = L_{\tilde{A}}^{([s,t],[s,t])} = U(\chi_H, [s, t]) \cap L(\overline{\chi}_H, [s, t]) = H$. Hence, $\varphi_{[s,t]}$ is surjective.

Theorem 3.18 *Let* $\tilde{A} = (\widetilde{\mu}_{\tilde{A}}, \widetilde{\lambda}_{\tilde{A}})$ *be an interval-valued intuitionistic fuzzy Lie ideal of Lie algebra* L. *Define a binary relation* $\sim$ *on* L *by* $x \sim y$ *if and only if* $\widetilde{\mu}_{\tilde{A}}(x - y) = \widetilde{\mu}_{\tilde{A}}(0)$, $\widetilde{\lambda}_{\tilde{A}}(x - y) = \widetilde{\lambda}_{\tilde{A}}(0)$ *for all* $x, y \in L$. *Then,* $\sim$ *is a congruence relation on* L.

Proof We first prove that "$\sim$" is an equivalent relation. We only need to show the transitivity of "$\sim$" because the reflectivity and symmetry of "$\sim$" hold trivially. Let $x, y, z \in L$. If $x \sim y$ and $y \sim z$, then $\widetilde{\mu}_{\tilde{A}}(x - y) = \widetilde{\mu}_{\tilde{A}}(0)$, $\widetilde{\mu}_{\tilde{A}}(y - z) = \widetilde{\mu}_{\tilde{A}}(0)$ and $\widetilde{\lambda}_{\tilde{A}}(x - y) = \widetilde{\lambda}_{\tilde{A}}(0)$, $\widetilde{\lambda}_{\tilde{A}}(y - z) = \widetilde{\lambda}_{\tilde{A}}(0)$. Hence, it follows that

$$\begin{aligned}&\widetilde{\mu}_{\tilde{A}}(x-z)=\widetilde{\mu}_{\tilde{A}}(x-y+y-z)\\&\geq \min\{\widetilde{\mu}_{\tilde{A}}(x-y),\widetilde{\mu}_{\tilde{A}}(y-z)\}=\widetilde{\mu}_{\tilde{A}}(0),\\&\widetilde{\lambda}_{\tilde{A}}(x-z)=\widetilde{\lambda}_{\tilde{A}}(x-y+y-z)\\&\leq \max\{\widetilde{\lambda}_{\tilde{A}}(x-y),\widetilde{\lambda}_{\tilde{A}}(y-z)\}=\widetilde{\lambda}_{\tilde{A}}(0).\end{aligned}$$

Consequently, $x \sim z$. We now verify that "$\sim$" is a congruence relation on L. For this purpose, we let $x \sim y$ and $y \sim z$. Then, $\widetilde{\mu}_{\tilde{A}}(x-y)=\widetilde{\mu}_{\tilde{A}}(0)$, $\widetilde{\mu}_{\tilde{A}}(y-z)=\widetilde{\mu}_{\tilde{A}}(0)$, $\widetilde{\lambda}_{\tilde{A}}(x-y)=\widetilde{\lambda}_{\tilde{A}}(0)$, and $\widetilde{\lambda}_{\tilde{A}}(y-z)=\widetilde{\lambda}_{\tilde{A}}(0)$. Now, for $x_1, x_2, y_1, y_2 \in L$, we have

$$\begin{aligned}&\widetilde{\mu}_{\tilde{A}}((x_1+x_2)-(y_1+y_2))=\widetilde{\mu}_{\tilde{A}}((x_1-y_1)+(x_2-y_2))\\&\geq \min\{\widetilde{\mu}_{\tilde{A}}(x_1-y_1),\widetilde{\mu}_{\tilde{A}}(x_2-y_2)\}\\&=\widetilde{\mu}_{\tilde{A}}(0),\\&\widetilde{\mu}_{\tilde{A}}(\alpha x_1-\alpha y_1)=\widetilde{\mu}_{\tilde{A}}(\alpha(x_1-y_1))\\&\geq \widetilde{\mu}_{\tilde{A}}(x_1-y_1)=\widetilde{\mu}_{\tilde{A}}(0),\\&\widetilde{\mu}_{\tilde{A}}([x_1,x_2]-[y_1,y_2])=\widetilde{\mu}_{\tilde{A}}([x_1-y_1],[x_2-y_2])\\&\geq \widetilde{\mu}_{\tilde{A}}(x_1-y_1)=\widetilde{\mu}_{\tilde{A}}(0).\end{aligned}$$

For $\lambda_{\tilde{A}}$, the proof is analogous.

That is, $x_1+x_2 \sim y_1+y_2$, $\alpha x_1 \sim \alpha y_1$, and $[x_1,x_2] \sim [y_1,y_2]$. Thus, "$\sim$" is a congruence relation on L.

Definition 3.18 Let $\tilde{A}=(\widetilde{\mu}_{\tilde{A}},\widetilde{\lambda}_{\tilde{A}})$ and $\tilde{B}=(\widetilde{\mu}_{\tilde{B}},\widetilde{\lambda}_{\tilde{B}})$ be interval-valued intuitionistic fuzzy sets on a set L. Then, *generalized cartesian product* $\tilde{A}\times\tilde{B}$ is defined as follows: $\tilde{A}\times\tilde{B}=(\widetilde{\mu}_{\tilde{A}}\times\widetilde{\mu}_{\tilde{B}},\widetilde{\lambda}_{\tilde{A}}\times\widetilde{\lambda}_{\tilde{B}})$, where $(\widetilde{\mu}_{\tilde{A}}\times\widetilde{\mu}_{\tilde{B}})(x,y)=\min\{\widetilde{\mu}_{\tilde{A}}(x),\widetilde{\mu}_{\tilde{B}}(y)\}$ and $(\widetilde{\lambda}_{\tilde{A}}\times\widetilde{\lambda}_{\tilde{B}})(x,y)=\max\{\widetilde{\lambda}_{\tilde{A}}(x),\widetilde{\lambda}_{\tilde{B}}(y)\}$.

The following proposition is obvious.

Proposition 3.2 *Let $\tilde{A}=(\widetilde{\mu}_{\tilde{A}},\widetilde{\lambda}_{\tilde{A}})$ and $\tilde{B}=(\widetilde{\mu}_{\tilde{B}},\widetilde{\lambda}_{\tilde{B}})$ be interval-valued intuitionistic fuzzy sets on a set L. Then, for all $\widetilde{s}\in D[0,1]$.*

(i) $U(\widetilde{\mu}_{\tilde{A}}\times\widetilde{\mu}_{\tilde{B}},\widetilde{s})=U(\widetilde{\mu}_{\tilde{A}},\widetilde{s})\times U(\widetilde{\mu}_{\tilde{B}},\widetilde{s})$,
(ii) $L(\widetilde{\lambda}_{\tilde{A}}\times\widetilde{\lambda}_{\tilde{B}},\widetilde{s})=L(\widetilde{\lambda}_{\tilde{A}},\widetilde{s})\times L(\widetilde{\lambda}_{\tilde{B}},\widetilde{s})$.

Theorem 3.19 *Let $\tilde{A}=(\widetilde{\mu}_{\tilde{A}},\widetilde{\lambda}_{\tilde{A}})$ and $\tilde{B}=(\widetilde{\mu}_{\tilde{B}},\widetilde{\lambda}_{\tilde{B}})$ be two interval-valued intuitionistic fuzzy Lie ideals of a Lie algebras L. Then, $\tilde{A}\times B$ is an interval-valued intuitionistic fuzzy Lie ideal of $L\times L$.*

Proof We restrict our proof to the verification of the properties of $\widetilde{\mu}_{\tilde{A}}\times\widetilde{\mu}_{\tilde{B}}$. Let $x=(x_1,x_2)$ and $y=(y_1,y_2)\in L\times L$. Then,

$$\begin{aligned}&(\widetilde{\mu}_{\tilde{A}}\times\widetilde{\mu}_{\tilde{B}})(x+y)=(\widetilde{\mu}_{\tilde{A}}\times\widetilde{\mu}_{\tilde{B}})((x_1,x_2)+(y_1,y_2))\\&=(\widetilde{\mu}_{\tilde{A}}\times\widetilde{\mu}_{\tilde{B}})(x_1+y_1,x_2+y_2)\\&=\min\{\widetilde{\mu}_{\tilde{A}}(x_1+y_1),\widetilde{\mu}_{\tilde{B}}(x_2+y_2)\}\\&\geq \min\{\min\{\widetilde{\mu}_{\tilde{A}}(x_1),\widetilde{\mu}_{\tilde{A}}(y_1)\},\min\{\widetilde{\mu}_{\tilde{B}}(x_2),\widetilde{\mu}_{\tilde{B}}(y_2)\}\}\\&=\min\{\min\{\widetilde{\mu}_{\tilde{A}}(x_1),\widetilde{\mu}_{\tilde{B}}(x_2)\},\min\{\widetilde{\mu}_{\tilde{A}}(y_1),\widetilde{\mu}_{\tilde{B}}(y_2)\}\}\\&=\min\{(\widetilde{\mu}_{\tilde{A}}\times\widetilde{\mu}_{\tilde{B}})(x_1,x_2),(\widetilde{\mu}_{\tilde{A}}\times\widetilde{\mu}_{\tilde{B}})(y_1,y_2)\}\\&=\min\{(\widetilde{\mu}_{\tilde{A}}\times\widetilde{\mu}_{\tilde{B}})(x),(\widetilde{\mu}_{\tilde{A}}\times\widetilde{\mu}_{\tilde{B}})(y)\},\end{aligned}$$

$$\begin{aligned}
(\widetilde{\mu}_{\tilde{A}} \times \widetilde{\mu}_{\tilde{B}})(\alpha x) &= (\widetilde{\mu}_{\tilde{A}} \times \widetilde{\mu}_{\tilde{B}})(\alpha(x_1, x_2)) \\
&= (\widetilde{\mu}_{\tilde{A}} \times \widetilde{\mu}_{\tilde{B}})(\alpha x_1, \alpha x_2) \\
&= \min\{\widetilde{\mu}_{\tilde{A}}(\alpha x_1), \widetilde{\mu}_{\tilde{B}}(\alpha x_2)\} \geq \min\{\widetilde{\mu}_{\tilde{A}}(x_1), \widetilde{\mu}_{\tilde{B}}(x_2)\} \\
&= (\widetilde{\mu}_{\tilde{A}} \times \widetilde{\mu}_{\tilde{B}})(x_1, x_2) \\
&= (\widetilde{\mu}_{\tilde{A}} \times \widetilde{\mu}_{\tilde{B}})(x),
\end{aligned}$$

$$\begin{aligned}
(\widetilde{\mu}_{\tilde{A}} \times \widetilde{\mu}_{\tilde{B}})([x, y]) &= (\widetilde{\mu}_{\tilde{A}} \times \widetilde{\mu}_{\tilde{B}})([(x_1, x_2), (y_1, y_2)]) \\
&\geq \min\{(\widetilde{\mu}_{\tilde{A}} \times \widetilde{\mu}_{\tilde{B}})(x_1, x_2)) = \min\{(\widetilde{\mu}_{\tilde{A}} \times \widetilde{\mu}_{\tilde{B}})(x)).
\end{aligned}$$

In a similar way, we can verify the analogous properties of $\widetilde{\lambda}_{\tilde{A}} \times \widetilde{\lambda}_{\tilde{B}}$. Hence, $\tilde{A} \times \tilde{B}$ is an interval-valued intuitionistic fuzzy Lie ideal of $L \times L$.

3.6 Fully Invariant and Characteristic IIF Lie Ideals

Definition 3.19 An interval-valued intuitionistic fuzzy Lie ideal $\tilde{A} = (\widetilde{\mu}_A, \widetilde{\lambda}_A)$ of L is called a *fully invariant* if $\widetilde{\mu}^f_{\tilde{A}}(x) = \widetilde{\mu}_{\tilde{A}}(f(x)) \leq \widetilde{\mu}_{\tilde{A}}(x)$ and $\widetilde{\lambda}^f_{\tilde{A}}(x) = \widetilde{\lambda}_{\tilde{A}}(f(x)) \leq \widetilde{\lambda}_{\tilde{A}}(x)$ for all $x \in L$ and $f \in \text{End}(L)$.

Theorem 3.20 *If $\{\tilde{A}_i \mid i \in I\}$ is a family of interval-valued intuitionistic fuzzy fully invariant Lie ideals of L, then $\bigcap_{i \in I} \tilde{A}_i = (\bigwedge_{i \in I} \widetilde{\mu}_{\tilde{A}_i}, \bigvee_{i \in I} \widetilde{\lambda}_{\tilde{A}_i})$ is an interval-valued intuitionistic fuzzy fully invariant Lie ideal of L, where*

$$\bigwedge_{i \in I} \widetilde{\mu}_{\tilde{A}_i}(x) = \inf\{\widetilde{\mu}_{\tilde{A}_i}(x) \mid i \in I,\ x \in L\},$$

$$\bigvee_{i \in I} \widetilde{\lambda}_{\tilde{A}_i}(x) = \sup\{\widetilde{\lambda}_{\tilde{A}_i}(x) \mid i \in I,\ x \in L\}.$$

Proof Straightforward.

Theorem 3.21 *Let H be a nonempty subset of a Lie algebra L and $\tilde{A} = (\widetilde{\mu}_{\tilde{A}}, \widetilde{\lambda}_{\tilde{A}})$ an interval-valued intuitionistic fuzzy set defined by*

$$\widetilde{\mu}_{\tilde{A}}(x) = \begin{cases} [s_2, t_2] \text{ if } x \in H, \\ [s_1, t_1] \text{ otherwise}, \end{cases}$$

$$\widetilde{\lambda}_{\tilde{A}}(x) = \begin{cases} [u_2, v_2] \text{ if } x \in H, \\ [u_1, v_1] \text{ otherwise}, \end{cases}$$

where $[0, 0] \leq [s_1, t_1] < [s_2, t_2] \leq [1, 1]$, $[0, 0] \leq [u_2, v_2] < [u_1, v_1] \leq [1, 1]$, $[0, 0] \leq [s_i, t_i] + [u_i, v_i] \leq [1, 1]$ *for* $i = 1, 2$. *If H is an interval-valued intuitionistic fuzzy fully invariant Lie ideal of L, then $\tilde{A} = (\widetilde{\mu}_{\tilde{A}}, \widetilde{\lambda}_{\tilde{A}})$ is an interval-valued intuitionistic fuzzy fully invariant Lie ideal of L.*

Proof We can easily see that $\tilde{A} = (\widetilde{\mu}_{\tilde{A}}, \widetilde{\lambda}_{\tilde{A}})$ is an interval-valued intuitionistic fuzzy Lie ideal of L. Let $x \in L$ and $f \in \text{End}(L)$. If $x \in H$, then $f(x) \in f(H) \subseteq H$. Thus, we have

$$\widetilde{\mu}^f_{\tilde{A}}(x) = \widetilde{\mu}_{\tilde{A}}(f(x)) \leq \widetilde{\mu}_{\tilde{A}}(x) = [s_2, t_2],$$

$$\widetilde{\lambda}^f_{\tilde{A}}(x) = \widetilde{\lambda}_{\tilde{A}}(f(x)) \leq \widetilde{\lambda}_{\tilde{A}}(x) = [u_2, v_2].$$

For otherwise, we have

$$\widetilde{\mu}^f_{\tilde{A}}(x) = \widetilde{\mu}_{\tilde{A}}(f(x)) \leq \widetilde{\mu}_{\tilde{A}}(x) = [s_1, t_1],$$

$$\widetilde{\lambda}^f_{\tilde{A}}(x) = \widetilde{\lambda}_{\tilde{A}}(f(x)) \leq \widetilde{\lambda}_{\tilde{A}}(x) = [u_1, v_1].$$

Thus, we have verified that $\tilde{A} = (\widetilde{\mu}_{\tilde{A}}, \widetilde{\lambda}_{\tilde{A}})$ is an interval-valued intuitionistic fuzzy fully invariant Lie ideal of L.

Definition 3.20 An interval-valued intuitionistic fuzzy Lie ideal $\tilde{A} = (\widetilde{\mu}_{\tilde{A}}, \widetilde{\lambda}_{\tilde{A}})$ of L *has the same type* as an interval-valued intuitionistic fuzzy Lie ideal $\tilde{B} = (\widetilde{\mu}_{\tilde{B}}, \widetilde{\lambda}_{\tilde{B}})$ of L if there exists $f \in \text{End}(L)$ such that $\tilde{A} = B \circ f$, i.e., $\widetilde{\mu}_{\tilde{A}}(x) > \widetilde{\mu}_{\tilde{B}}(f(x)), \widetilde{\lambda}_{\tilde{A}}(x) \geq \widetilde{\lambda}_{\tilde{B}}(f(x))$ for all $x \in L$.

Theorem 3.22 *Interval-valued intuitionistic fuzzy Lie ideals of L have same type if and only if they are isomorphic.*

Proof We only need to prove the necessity part because the sufficiency part is obvious. If an interval-valued intuitionistic fuzzy Lie ideal $\tilde{A} = (\widetilde{\mu}_{\tilde{A}}, \widetilde{\lambda}_{\tilde{A}})$ of L has the same type as $\tilde{B} = (\widetilde{\mu}_{\tilde{B}}, \widetilde{\lambda}_{\tilde{B}})$, then there exists $\varphi \in \text{End}(L)$ such that

$$\widetilde{\mu}_{\tilde{A}}(x) \geq \widetilde{\mu}_{\tilde{B}}(\varphi(x)),\ \widetilde{\lambda}_{\tilde{A}}(x) \geq \widetilde{\lambda}_{\tilde{B}}(\varphi(x))\ \ \forall x \in L.$$

Let $f : A(L) \to B(L)$ be a mapping defined by $f(A(x)) = B(\varphi(x))$ for all $x \in L$, that is,

$$f(\widetilde{\mu}_{\tilde{A}}(x)) = \widetilde{\mu}_{\tilde{B}}(\varphi(x)),\ f(\widetilde{\lambda}_{\tilde{A}}(x)) \\ = \widetilde{\lambda}_{\tilde{B}}(\varphi(x))$$

for $x \in L$. Then, it is clear that f is a surjective homomorphism. Also, f is injective because $f(\widetilde{\mu}_{\tilde{A}}(x)) = f(\widetilde{\mu}_{\tilde{A}}(y))$ for all $x, y \in L$ implies $\widetilde{\mu}_{\tilde{B}}(\varphi(x)) = \widetilde{\mu}_{\tilde{B}}(\varphi(y))$. Whence $\widetilde{\mu}_{\tilde{A}}(x) = \widetilde{\mu}_{\tilde{B}}(y)$. Likewise, from $f(\widetilde{\lambda}_{\tilde{A}}(x)) = f(\widetilde{\lambda}_{\tilde{A}}(y))$, we conclude $\widetilde{\lambda}_{\tilde{A}}(x) = \widetilde{\lambda}_{\tilde{B}}(y)$ for all $x \in L$. Hence, $\tilde{A} = (\widetilde{\mu}_{\tilde{A}}, \widetilde{\lambda}_{\tilde{A}})$ is isomorphic to $\tilde{B} = (\widetilde{\mu}_{\tilde{B}}, \widetilde{\lambda}_{\tilde{B}})$. This completes the proof.

Definition 3.21 An interval-valued intuitionistic fuzzy Lie ideal $\tilde{A} = (\widetilde{\mu}_A, \widetilde{\lambda}_A)$ of L is called a *characteristic* if $\widetilde{\mu}_{\tilde{A}}(f(x)) = \widetilde{\mu}_{\tilde{A}}(x)$ and $\widetilde{\lambda}_{\tilde{A}}(f(x)) = \widetilde{\lambda}_{\tilde{A}}(x)$ for all $x \in L$ and $f \in \text{Aut}(L)$.

The following lemma is obvious.

Lemma 3.8 *Let $\tilde{A} = (\widetilde{\mu}_A, \widetilde{\lambda}_A)$ be an interval-valued intuitionistic fuzzy Lie ideal of L, and let $x \in L$. Then, $\widetilde{\mu}_{\tilde{A}}(x) = \widetilde{t}$, $\widetilde{\lambda}_{\tilde{A}}(x) = \widetilde{s}$ if and only if $x \in U(\widetilde{\mu}_{\tilde{A}}, \widetilde{t})$, $x \notin U(\widetilde{\mu}_{\tilde{A}}, \widetilde{s})$ and $x \in L(\widetilde{\lambda}_{\tilde{A}}, \widetilde{s})$, $x \notin L(\widetilde{\lambda}_{\tilde{A}}, \widetilde{t})$ for all $\widetilde{s} > \widetilde{t}$.*

Theorem 3.23 *An interval-valued intuitionistic fuzzy Lie ideal is characteristic if and only if each of its level set is a characteristic Lie ideal.*

Proof Let an interval-valued intuitionistic fuzzy Lie ideal $\tilde{A} = (\widetilde{\mu}_{\tilde{A}}, \widetilde{\lambda}_{\tilde{A}})$ be characteristic, $\widetilde{t} \in Im(\widetilde{\mu}_{\tilde{A}})$, $f \in \mathrm{Aut}(L)$, $x \in U(\widetilde{\mu}_{\tilde{A}}, \widetilde{t})$. Then, $\widetilde{\mu}_{\tilde{A}}(f(x)) = \widetilde{\mu}_{\tilde{A}}(x) \geq \widetilde{t}$, which means that $f(x) \in U(\widetilde{\mu}_{\tilde{A}}, \widetilde{t})$. Thus, $f(U(\widetilde{\mu}_{\tilde{A}}, \widetilde{t})) \subseteq U(\widetilde{\mu}_{\tilde{A}}, \widetilde{t})$. Since for each $x \in U(\widetilde{\mu}_{\tilde{A}}, \widetilde{t})$ there exists $y \in L$ such that $f(y) = x$, we have $\widetilde{\mu}_{\tilde{A}}(y) = \widetilde{\mu}_{\tilde{A}}(f(y)) = \widetilde{\mu}_{\tilde{A}}(x) \geq \widetilde{t}$; whence we conclude $y \in U(\widetilde{\mu}_{\tilde{A}}, \widetilde{t})$. Consequently, $x = f(y) \in f(U(\widetilde{\mu}_{\tilde{A}}, \widetilde{t}))$. Hence, $f(U(\widetilde{\mu}_{\tilde{A}}, \widetilde{t})) = U(\widetilde{\mu}_{\tilde{A}}, \widetilde{t})$. Similarly, $f(L(\widetilde{\lambda}_{\tilde{A}}, \widetilde{s})) = L(\widetilde{\lambda}_{\tilde{A}}, \widetilde{s})$. This proves that $U(\widetilde{\mu}_{\tilde{A}}, \widetilde{t})$ and $L(\widetilde{\lambda}_{\tilde{A}}, \widetilde{s})$ are characteristics.

Conversely, if all levels of $\tilde{A} = (\widetilde{\mu}_{\tilde{A}}, \widetilde{\lambda}_{\tilde{A}})$ are characteristic Lie ideals of L, then for $x \in L$, $f \in \mathrm{Aut}(L)$ and $\widetilde{\mu}_{\tilde{A}}(x) = \widetilde{t} < \widetilde{s} = \widetilde{\lambda}_{\tilde{A}}(x)$; by Lemma 3.8, we have $x \in U(\widetilde{\mu}_{\tilde{A}}, \widetilde{t})$, $x \notin U(\widetilde{\mu}_{\tilde{A}}, \widetilde{s})$ and $x \in L(\widetilde{\lambda}_{\tilde{A}}, \widetilde{s})$, $x \notin L(\widetilde{\lambda}_{\tilde{A}}, \widetilde{t})$. Thus, $f(x) \in f(U(\widetilde{\mu}_{\tilde{A}}, \widetilde{t})) = U(\widetilde{\mu}_{\tilde{A}}, \widetilde{t})$ and $f(x) \in f(L(\widetilde{\lambda}_{\tilde{A}}, \widetilde{s})) = L(\widetilde{\lambda}_{\tilde{A}}, \widetilde{s})$, i.e., $\widetilde{\mu}_{\tilde{A}}(f(x)) \geq \widetilde{t}$ and $\widetilde{\lambda}_{\tilde{A}}(f(x)) \leq \widetilde{s}$. For $\widetilde{\mu}_{\tilde{A}}(f(x)) = \widetilde{t}_1 > \widetilde{t}$, $\widetilde{\lambda}_{\tilde{A}}(f(x)) = \widetilde{s}_1 < \widetilde{s}$, we have $f(x) \in U(\widetilde{\mu}_{\tilde{A}}, \widetilde{t}_1) = f(U(\widetilde{\mu}_{\tilde{A}}, \widetilde{t}_1))$, $f(x) \in L(\widetilde{\lambda}_{\tilde{A}}, \widetilde{s}_1) = f(L(\widetilde{\lambda}_{\tilde{A}}, \widetilde{s}_1))$; whence $x \in U(\widetilde{\mu}_{\tilde{A}}, \widetilde{t}_1)$, $x \in L(\widetilde{\mu}_{\tilde{A}}, \widetilde{s}_1)$. This is a contradiction. Thus, $\widetilde{\mu}_{\tilde{A}}(f(x)) = \widetilde{\mu}_{\tilde{A}}(x)$ and $\widetilde{\lambda}_{\tilde{A}}(f(x)) = \widetilde{\lambda}_{\tilde{A}}(x)$. So, $\tilde{A} = (\widetilde{\mu}_{\tilde{A}}, \widetilde{\lambda}_{\tilde{A}})$ is characteristic.

As a consequence of the above results, we obtain the following theorem.

Theorem 3.24 *If $\tilde{A} = (\widetilde{\mu}_{\tilde{A}}, \widetilde{\lambda}_{\tilde{A}})$ is a fully invariant interval-valued intuitionistic fuzzy Lie ideal, then it is characteristic.*

3.7 Solvable and Nilpotent IIF Lie Ideals

An interval-valued intuitionistic fuzzy subspace of L generated by $\tilde{A} = (\widetilde{\mu}_{\tilde{A}}, \widetilde{\lambda}_{\tilde{A}}) \in J^L$ will be denoted by $[\tilde{A}]$. It is the intersection of all interval-valued intuitionistic fuzzy subspaces of L containing $\tilde{A}$.

For all $x \in L$, we define:

$$[\widetilde{\mu}_{\tilde{A}}](x) = \sup\{\min \widetilde{\mu}_{\tilde{A}}(x_i) | x = \sum \gamma_i x_i, \gamma_i \in \mathbb{F}, x_i \in L\},$$

$$[\widetilde{\lambda}_{\tilde{A}}](x) = \inf\{\max \widetilde{\lambda}_{\tilde{A}}(x_i) | x = \sum \gamma_i x_i, \gamma_i \in \mathbb{F}, x_i \in L\}.$$

Definition 3.22 Let $f : L_1 \to L_2$ be a homomorphism of Lie algebras which has the extension $f : J^{L_1} \to J^{L_2}$ defined by:

$$f(\widetilde{\mu}_{\tilde{A}})(y) = \sup\{\widetilde{\mu}_{\tilde{A}}(x) \,|\, x \in f^{-1}(y)\},$$

$$f(\widetilde{\lambda}_{\tilde{A}})(y) = \inf\{\widetilde{\lambda}_{\tilde{A}}(x) \mid x \in f^{-1}(y)\},$$

for all $\tilde{A} = (\widetilde{\mu}_{\tilde{A}}, \widetilde{\lambda}_{\tilde{A}}) \in J^{L_1}$, $y \in L_2$. Then, $f(A)$ is called the *homomorphic image* of the interval-valued intuitionistic fuzzy set $\tilde{A}$.

Such defined homomorphic image has the following properties:

Proposition 3.3 *Let $f : L_1 \to L_2$ be a homomorphism of Lie algebras, and let $\tilde{A} = (\widetilde{\mu}_{\tilde{A}}, \widetilde{\lambda}_{\tilde{A}})$ be an interval-valued intuitionistic fuzzy Lie ideal of L_1. Then,*

(i) $f(\tilde{A})$ is an interval-valued intuitionistic fuzzy Lie ideal of L_2,
(ii) $f([\tilde{A}]) \supseteq [f(\tilde{A})]$.

Proposition 3.4 *If $\tilde{A} = (\widetilde{\mu}_{\tilde{A}}, \widetilde{\lambda}_{\tilde{A}})$ and $\tilde{B} = (\widetilde{\mu}_{\tilde{B}}, \widetilde{\lambda}_{\tilde{B}})$ are interval-valued intuitionistic fuzzy Lie ideals in L, then also $[\tilde{A}, \tilde{B}]$ is an interval-valued intuitionistic fuzzy Lie ideal of L.*

Theorem 3.25 *Let $\tilde{A}_1$, $\tilde{A}_2$, $\tilde{B}_1$, $\tilde{B}_2$ be interval-valued intuitionistic fuzzy Lie ideals in L such that $\tilde{A}_1 \subseteq \tilde{A}_2$ and $\tilde{B}_1 \subseteq \tilde{B}_2$. Then, $[\tilde{A}_1, \tilde{B}_1] \subseteq [\tilde{A}_2, \tilde{A}_2]$.*

Proof Indeed, for all $x \in L$, we have

$$\begin{aligned}
&\ll \widetilde{\mu}_{\tilde{A}_1}, \widetilde{\mu}_{\tilde{B}_1} \gg (x) \\
&= \sup\{\min\{\widetilde{\mu}_{\tilde{A}_1}(a), \widetilde{\mu}_{\tilde{B}_1}(b)\} | a, b \in L_1, [a, b] = x\} \\
&\geq \sup\{\min\{\widetilde{\mu}_{\tilde{A}_2}(a), \widetilde{\mu}_{\tilde{B}_2}(b)\} | a, b \in L_1, [a, b] = x\} \\
&= \ll \widetilde{\mu}_{\tilde{A}_2}, \widetilde{\mu}_{\tilde{B}_2} \gg (x),
\end{aligned}$$

$$\begin{aligned}
&\ll \widetilde{\lambda}_{\tilde{A}_1}, \widetilde{\lambda}_{\tilde{B}_1} \gg (x) \\
&= \inf\{\max\{\widetilde{\lambda}_{\tilde{A}_1}(a), \widetilde{\lambda}_{\tilde{B}_1}(b)\} \mid a, b \in L_1, [a, b] = x\} \\
&\leq \inf\{\max\{\widetilde{\lambda}_{\tilde{A}_2}(a), \widetilde{\lambda}_{\tilde{B}_2}(b)\} \mid a, b \in L_1, [a, b] = x\} \\
&= \ll \widetilde{\lambda}_{\tilde{A}_2}, \widetilde{\lambda}_{\tilde{B}_2} \gg (x).
\end{aligned}$$

Hence, $[\tilde{A}_1, \tilde{B}_1] \subseteq [\tilde{A}_2, \tilde{B}_2]$.

Definition 3.23 Let $\tilde{A} = (\widetilde{\mu}_{\tilde{A}}, \widetilde{\lambda}_{\tilde{A}})$ be an interval-valued intuitionistic fuzzy Lie ideal of L. Define inductively a sequence of interval-valued intuitionistic fuzzy Lie ideals of L by putting

$$\tilde{A}^{(0)} = A, \quad \text{and} \quad \tilde{A}^{(n)} = [\tilde{A}^{(n-1)}, \tilde{A}^{(n-1)}]$$

for all $n > 0$. $\tilde{A}^{(n)}$ is called the *nth derived interval-valued intuitionistic fuzzy Lie ideal* of L. A series

$$\tilde{A}^{(0)} \subseteq \tilde{A}^{(1)} \subseteq \tilde{A}^{(2)} \subseteq \cdots \subseteq \tilde{A}^{(n)} \subseteq \cdots$$

is called *derived series* of an interval-valued intuitionistic fuzzy Lie ideal $\tilde{A}$ in L. For positive integer n, we define $\tilde{B}^{(n)} = (\widetilde{\mu}^{(n)}_{\tilde{B}}, \widetilde{\lambda}^{(n)}_{\tilde{B}})$ as follows:

$$\widetilde{\mu}^{(n)}_{\tilde{B}} = \sup\{\widetilde{\mu}^{(n)}_{\tilde{A}}(x) : 0 \neq x \in L\},$$

$$\widetilde{\lambda}^{(n)}_{\tilde{B}} = \inf\{\widetilde{\lambda}^{(n)}_{\tilde{A}}(x) : 0 \neq x \in L\}.$$

An interval-valued intuitionistic fuzzy Lie ideal $\tilde{A} = (\widetilde{\mu}_{\tilde{A}}, \widetilde{\lambda}_{\tilde{A}})$ is called *solvable* if there exists a positive integer n such that $\tilde{B}^{(n)} = ([0, 0], [1, 1])$.

Theorem 3.26 *Homomorphic images of solvable interval-valued intuitionistic fuzzy Lie ideals are solvable interval-valued intuitionistic fuzzy Lie ideals.*

Proof Let $f : L_1 \to L_2$ be a Lie algebra homomorphism, and assume that $\tilde{A} = (\widetilde{\mu}_{\tilde{A}}, \widetilde{\lambda}_{\tilde{A}})$ is a solvable interval-valued intuitionistic fuzzy Lie ideal in L_1. Let $f(\tilde{A}) = \tilde{B}$. We prove by induction on n that $f(\tilde{A}^n) \supseteq \tilde{B}^n$, where n is any positive integer. First, we claim that for any Lie ideal $\tilde{A}$ of L that $f([\tilde{A}, \tilde{A}]) \supseteq [f(\tilde{A}), f(\tilde{A})] = [\tilde{B}, \tilde{B}]$.

Let $y \in L_2$, then

$$\begin{aligned}
& f(\ll \widetilde{\mu}_{\tilde{A}}, \widetilde{\mu}_{\tilde{A}} \gg)(y) = \sup\{\ll \widetilde{\mu}_{\tilde{A}}, \widetilde{\mu}_{\tilde{A}} \gg)(x) | f(x) = y\}\\
& = \sup\{\sup\{\min\{\widetilde{\mu}_{\tilde{A}}(a), \widetilde{\mu}_{\tilde{A}}(b) | a, b \in L_1,\\
& [a, b] = x\}, f(x) = y\}\}\\
& = \sup\{\min\{\widetilde{\mu}_{\tilde{A}}(a), \widetilde{\mu}_{\tilde{A}}(b)\} | a, b \in L_1, \ f([a, b]) = y\}\\
& = \sup\{\min\{\widetilde{\mu}_{\tilde{A}}(a), \widetilde{\mu}_{\tilde{A}}(b)\} | a, b \in L_1, \ [f(a), f(b)] = y\}\\
& = \sup\{\min\{\widetilde{\mu}_{\tilde{A}}(a), \widetilde{\mu}_{\tilde{A}}(b)\} | a, b \in L_1,\\
& f(a) = u, f(b) = v, [u, v] = y\}\\
& \geq \sup\{\min\{ \sup_{a \in f^{-1}(u)} \widetilde{\mu}_{\tilde{A}}(a), \ \sup_{b \in f^{-1}(v)} \widetilde{\mu}_{\tilde{A}}(b)\} \mid [u, v] = y\}\\
& = \sup\{\min\{f(\widetilde{\mu}_{\tilde{A}})(u), f(\widetilde{\mu}_{\tilde{A}})(v)\} \mid [u, v] = y\}\\
& = \ll f(\widetilde{\mu}_{\tilde{A}}), f(\widetilde{\mu}_{\tilde{A}}) \gg (y).
\end{aligned}$$

In a similar way, we can prove

$$f(\ll \widetilde{\lambda}_{\tilde{A}}, \widetilde{\lambda}_{\tilde{A}} \gg)(y) \leq \ll f(\widetilde{\lambda}_{\tilde{A}}), f(\widetilde{\lambda}_{\tilde{A}}) \gg (y).$$

Now for $n > 1$, we get

$$\begin{aligned}
& f(\tilde{A}^{(n)}) = f([\tilde{A}^{(n-1)}, \tilde{A}^{(n-1)}])\\
& \supseteq [f(\tilde{A}^{(n-1)}), f(\tilde{A}^{(n-1)})] \supseteq [\tilde{B}^{(n-1)}, \tilde{B}^{(n-1)}]\\
& = \tilde{B}^{(n)}.
\end{aligned}$$

Let m be a positive integer such that $\tilde{A}^{(m)} = \widetilde{0}$, and let $0 \neq y \in L_2$. Then,

$$\begin{aligned}
& \widetilde{\mu}^{(m)}_{\tilde{B}}(y) \leq f(\widetilde{\mu}^{(m)}_{\tilde{A}})(y) = f(0)(y)\\
& = \sup\{0(x) : f(x) = y.\} = [0, 0],\\
& \widetilde{\lambda}^{(m)}_{\tilde{B}}(y) \geq f(\widetilde{\lambda}^{(m)}_{\tilde{A}})(y) = f(1)(y)\\
& = \inf\{1(x) : f(x) = y\} = [1, 1].
\end{aligned}$$

Thus, $\tilde{B}^{(m)} = (\widetilde{\mu}_{\tilde{B}}^{(m)}, \widetilde{\lambda}_{\tilde{B}}^{(m)}) = ([0, 0], [1, 1])$. This completes the proof.

Definition 3.24 Let $\tilde{A} = (\widetilde{\mu}_{\tilde{A}}, \widetilde{\lambda}_{\tilde{A}})$ be an interval-valued intuitionistic fuzzy Lie ideal in L. We define inductively a sequence of interval-valued intuitionistic fuzzy subspace in L by

$$\tilde{A}^0 = \tilde{A}, \quad \text{and} \quad \tilde{A}^n = [\tilde{A}, \tilde{A}^{n-1}]$$

for all natural $n > 0$.

A series

$$\tilde{A}^0 \supseteq \tilde{A}^1 \supseteq \tilde{A}^2 \supseteq \cdots \supseteq \tilde{A}^n \supseteq \cdots$$

is called *descending central series* of an interval-valued intuitionistic fuzzy Lie ideal $\tilde{A}$ in L.

For positive integer n, we define $\tilde{B}^n = (\widetilde{\mu}_{\tilde{B}}^n, \widetilde{\lambda}_{\tilde{B}}^n)$ putting

$$\widetilde{\mu}_{\tilde{B}}^n = \sup\{\widetilde{\mu}_{\tilde{A}}^n(x) : 0 \neq x \in L\},$$

$$\widetilde{\lambda}_{\tilde{B}}^n = \inf\{\widetilde{\lambda}_{\tilde{A}}^n(x) : 0 \neq x \in L\}.$$

An interval-valued intuitionistic fuzzy Lie ideal $\tilde{A} = (\widetilde{\mu}_{\tilde{A}}, \widetilde{\lambda}_{\tilde{A}})$ is called *nilpotent* if there exists a positive integer n such that $\tilde{B}^n = ([0, 0], [1, 1])$.

Using the same method as in Theorem 3.26, we can prove

Theorem 3.27 *Homomorphic images of nilpotent interval-valued intuitionistic fuzzy Lie ideals are nilpotent interval-valued intuitionistic fuzzy Lie ideals.*

Theorem 3.28 *If $\tilde{A}$ is a nilpotent interval-valued intuitionistic fuzzy Lie ideal, then it is solvable.*

Proof It is enough to prove that $\tilde{A}^{(n)} \subseteq \tilde{A}^n$ for all positive integers n. We prove it by induction on n and by the use of Definition 3.23:

$$\tilde{A}^{(1)} = [\tilde{A}, \tilde{A}] = \tilde{A}^1, \tilde{A}^{(2)} = [\tilde{A}^{(1)}, \tilde{A}^{(1)}] \subseteq [\tilde{A}, \tilde{A}^{(1)}] = \tilde{A}^2.$$

$$\tilde{A}^{(n)} = [\tilde{A}^{(n-1)}, \tilde{A}^{(n-1)}] \subseteq [\tilde{A}, \tilde{A}^{(n-1)}] \subseteq [\tilde{A}, \tilde{A}^{(n-1)}] = \tilde{A}^n.$$

This completes the proof.

3.8 Interval-Valued Fuzzy Lie Superalgebras

Definition 3.25 Let $V = V_{\overline{0}} \oplus V_{\overline{1}}$ be a $\mathbb{Z}_2$-graded vector space. Suppose that $M_{\overline{0}} = (x, \widetilde{\mu}_{M_{\overline{0}}}(x))$, $M_{\overline{1}} = (x, \widetilde{\mu}_{M_{\overline{1}}}(x))$ are interval-valued fuzzy subspaces of $V_{\overline{0}}$, $V_{\overline{1}}$, respectively. Define

$$M_{\bar{0}} \oplus M_{\bar{1}} = (x, \widetilde{\mu}_{M_{\bar{0}} \oplus M_{\bar{1}}}(x)), \quad \text{where} \quad \widetilde{\mu}_{M_{\bar{0}} \oplus M_{\bar{1}}}(x) = \min\{\widetilde{\mu}_{M_{\bar{0}}}(x_{\bar{0}}), \widetilde{\mu}_{M_{\bar{1}}}(x_{\bar{1}})\},$$

for $x_{\bar{0}} \in V_{\bar{0}}$, $x_{\bar{1}} \in V_{\bar{1}}$. If $M = (x, \widetilde{\mu}_M(x))$ is an interval-valued fuzzy subspace of V and $M = M_{\bar{0}} \oplus M_{\bar{1}}$, then $M = (x, \widetilde{\mu}_M(x))$ is called a $\mathbb{Z}_2$*-graded interval-valued fuzzy subspace* of V.

The following lemmas are obvious.

Lemma 3.9 *Let $M = (x, \widetilde{\mu}_M(x))$ and $N = (x, \widetilde{\mu}_N(x))$ be interval-valued fuzzy subspaces of V. Then, $M + N$ is also an interval-valued fuzzy subspace of V.*

Lemma 3.10 *Let $M = (x, \widetilde{\mu}_M(x))$ and $N = (x, \widetilde{\mu}_N(x))$ be interval-valued fuzzy subspaces of V. Then, $M \cap N$ is also an interval-valued fuzzy subspace of V.*

Lemma 3.11 *Let $M = (x, \widetilde{\mu}_M(x))$ be an interval-valued fuzzy subspace of V' and ϕ be a mapping from vector space V to V'. Then, the inverse image $\phi^{-1}(M)$ is also an interval-valued fuzzy subspace of V.*

Lemma 3.12 *Let $M = (x, \widetilde{\mu}_M(x))$ be an interval-valued fuzzy subspace of V and f be a mapping from V to V'. Then, the image $\phi(M)$ is also an interval-valued fuzzy subspace of V'.*

Definition 3.26 Let $M = (x, \widetilde{\mu}_M(x))$ be an interval-valued fuzzy set of $\mathscr{L}$. Then, $M = (x, \widetilde{\mu}_M(x))$ is called an *interval-valued fuzzy Lie sub-superalgebra* of $\mathscr{L}$ if for any $x, y \in \mathscr{L}$:

(1) $M = (x, \widetilde{\mu}_M(x))$ is a $\mathbb{Z}_2$-graded interval-valued fuzzy subspace of $\mathscr{L}$,

(2) $\widetilde{\mu}_M([x, y]) \geq \min\{\widetilde{\mu}_M(x), \widetilde{\mu}_M(y)\}$.

Example 3.4 Let $\mathscr{L} = \mathscr{L}_{\bar{0}} \oplus \mathscr{L}_{\bar{1}}$, where $\mathscr{L}_{\bar{0}} =< e >$, $\mathscr{L}_{\bar{1}} =< a_1, a_2, b_1, b_2 >$, and $[a_i, b_i] = e$, $i = 1, 2$, the remaining brackets being zero. Then, $\mathscr{L}$ is a Lie superalgebra and $M_{\bar{0}} = (x, \widetilde{\mu}_{M_{\bar{0}}}(x))$, where $\widetilde{\mu}_{\mathscr{L}_{\bar{0}}} : \mathscr{L}_{\bar{0}} \to D[0, 1]$ is defined by

$$\widetilde{\mu}_{M_{\bar{0}}}(x) = \begin{cases} [0.2, 0.6] & x \in \mathscr{L}_{\bar{0}} \setminus \{0\}, \\ [1, 1] & x = 0, \end{cases}$$

is an interval-valued fuzzy subspace of $\mathscr{L}_{\bar{0}}$.

Also $M_{\bar{1}} = (x, \widetilde{\mu}_{M_{\bar{1}}}(x))$, where $\widetilde{\mu}_{M_{\bar{1}}} : \mathscr{L}_{\bar{1}} \to D[0, 1]$ is defined by

$$\widetilde{\mu}_{M_{\bar{1}}}(x) = \begin{cases} [0.1, 0.5] & x \in \mathscr{L}_{\bar{1}} \setminus \{0\}, \\ [1, 1] & x = 0, \end{cases}$$

is an interval-valued fuzzy subspace of $\mathscr{L}_{\bar{1}}$. Then $M = (x, \widetilde{\mu}_M(x))$, where $\widetilde{\mu}_M : \mathscr{L} \to [0, 1]$ is defined by

$$\widetilde{\mu}_M(x) = \begin{cases} [0.1, 0.5] & x \in \mathscr{L} \setminus \{0\}, \\ [1, 1] & x = 0. \end{cases}$$

is an interval-valued set of $\mathscr{L}$.

For $x \neq 0$, we have $(\widetilde{\mu}_{M_{\bar{0}} \oplus M_{\bar{1}}})(x) = \min\{\widetilde{\mu}_{M_{\bar{0}}}(x_{\bar{0}}), \widetilde{\mu}_{M_{\bar{1}}}(x_{\bar{1}})\} = \min\{[0.2, 0.6], [0.1, 0.5]\} = [0.1, 0.5] = \widetilde{\mu}_M(x)$, so $M = (x, \widetilde{\mu}_M(x))$ is a $\mathbb{Z}_2$-graded interval-valued fuzzy subspace of $\mathscr{L}$.

Let $x, y \in \mathscr{L}$. If $x = a_i$, $y = b_i$, $\widetilde{\mu}_M([x, y]) = \widetilde{\mu}_M(e) = [0.2, 0.6] \geq \min\{\widetilde{\mu}_M(x), \widetilde{\mu}_M(y)\}$, and if not, $\widetilde{\mu}_M([x, y]) = \widetilde{\mu}_M(0) = [1, 1] \geq \min\{\widetilde{\mu}_M(x), \widetilde{\mu}_M(y)\}$. So, $M = (x, \widetilde{\mu}_M(x))$ is an interval-valued fuzzy Lie sub-superalgebra of $\mathscr{L}$.

Definition 3.27 Let $M = (x, \widetilde{\mu}_M(x))$ be an interval-valued fuzzy set of $\mathscr{L}$. Then, $M = (x, \widetilde{\mu}_M(x))$ is called an *interval-valued fuzzy Lie ideal* of $\mathscr{L}$ if for any $x, y \in \mathscr{L}$:

(1) $M = (x, \widetilde{\mu}_M(x))$ is a $\mathbb{Z}_2$-graded interval-valued fuzzy subspace of $\mathscr{L}$,

(2) $\widetilde{\mu}_M([x, y]) \geq \max\{\widetilde{\mu}_M(x), \widetilde{\mu}_M(y)\}$.

Example 3.5 Let $\mathscr{L} = \mathscr{L}_{\bar{0}} \oplus \mathscr{L}_{\bar{1}}$, where $\mathscr{L}_{\bar{0}}$ is the real vector space $\mathbb{R}^3$ and $\mathscr{L}_{\bar{1}} = \{0\}$, define $[x, y] = x \times y$ for $x, y \in \mathscr{L}_{\bar{0}}$, and the other brackets are zero. Then, $\mathscr{L}$ is a Lie superalgebra. Let $M_{\bar{0}} = (x, \widetilde{\mu}_{M_{\bar{0}}}(x))$ be an interval-valued fuzzy set of $\mathscr{L}_{\bar{0}}$. Define $\widetilde{\mu}_{M_{\bar{0}}} : \mathscr{L}_{\bar{0}} \to D[0, 1]$ by

$$\widetilde{\mu}_{M_{\bar{0}}}(x) = \begin{cases} [0.2, 0.5] & \text{if } x = (a, 0, 0),\ a \neq 0, \\ [1, 1] & \text{if } x = 0, \\ [0, 0] & \text{otherwise.} \end{cases}$$

Let $M_{\bar{1}} = (x, \widetilde{\mu}_{M_{\bar{1}}}(x))$ be an interval-valued fuzzy set of $\mathscr{L}_{\bar{1}}$ such that $\widetilde{\mu}_{M_{\bar{1}}}(x) = [1, 1]$ for $x \in \mathscr{L}_{\bar{1}}$. Now, let $M = (x, \widetilde{\mu}_M(x))$, where $\widetilde{\mu}_M(x) = \widetilde{\mu}_{M_{\bar{0}}}(x_{\bar{0}})$ for $x \in \mathscr{L}$. It is easy to get $\widetilde{\mu}_M(x) = \min\{\widetilde{\mu}_{M_{\bar{0}}}(x_{\bar{0}}), \widetilde{\mu}_{M_{\bar{1}}}(x_{\bar{1}})\}$. So, $M = (x, \widetilde{\mu}_M(x))$ is an interval-valued fuzzy Lie sub-superalgebra of $\mathscr{L}$. However, for $\max\{\widetilde{\mu}_M((1, 0, 0)), \widetilde{\mu}_M((1, 1, 1))\} = \max\{[0.2, 0.5], [0, 0]\} = [0.2, 0.5]$ and $\widetilde{\mu}_M([(1, 0, 0), (1, 1, 1)]) = \widetilde{\mu}_M((0, -1, 1)) = [0, 0]$, so it is not an interval-valued fuzzy Lie ideal of $\mathscr{L}$. If we redefine $\widetilde{\mu}_{M_{\bar{0}}}(x) = [1, 1]$ for $x = 0$ and $\widetilde{\mu}_{M_{\bar{0}}}(x) = [0, 0]$ otherwise, then it is obvious that $M = (x, \widetilde{\mu}_M(x))$ is an interval-valued fuzzy Lie ideal of $\mathscr{L}$.

The following theorem explains the relation between interval-valued fuzzy Lie sub-superalgebras (resp. ideals) and fuzzy Lie sub-superalgebras (resp. ideals).

Theorem 3.29 *An interval-valued fuzzy set* $M = (x, \widetilde{\mu}_M(x))$ *of* $\mathscr{L}$ *with* $\widetilde{\mu}_M(x) = [\widetilde{\mu}_M^L(x), \widetilde{\mu}_M^U(x)]$, *is an interval-valued fuzzy Lie sub-superalgebra (resp. ideal) of* $\mathscr{L}$ *if and only if* μ_M^- *and* μ_M^+ *are fuzzy Lie sub-superalgebras (resp. ideals) of* $\mathscr{L}$.

Proof ($\Rightarrow$) Let $M = (x, \widetilde{\mu}_M(x))$ be an interval-valued fuzzy Lie sub-superalgebra of $\mathscr{L}$. Then $M = M_{\bar{0}} \oplus M_{\bar{1}}$, where $M_{\bar{0}} = (x, \widetilde{\mu}_{M_{\bar{0}}}(x))$ is an interval-valued fuzzy subspace of $\mathscr{L}_{\bar{0}}$, $M_{\bar{1}} = (x, \widetilde{\mu}_{M_{\bar{1}}}(x))$ is an interval-valued fuzzy subspace of $\mathscr{L}_{\bar{1}}$.

Let $x \in \mathscr{L}$. Then,

$$\begin{aligned} [\mu_M^-(x), \mu_M^+(x)] &= \widetilde{\mu}_M(x) = \min\{\widetilde{\mu}_{M_{\bar{0}}}(x_{\bar{0}}), \widetilde{\mu}_{M_{\bar{1}}}(x_{\bar{1}})\} \\ &= [\min\{\mu_{M_{\bar{0}}}^-(x_{\bar{0}}), \mu_{M_{\bar{1}}}^-(x_{\bar{1}})\}, \min\{\mu_{M_{\bar{0}}}^+(x_{\bar{0}}), \mu_{M_{\bar{1}}}^+(x_{\bar{1}})\}], \end{aligned}$$

we have $\mu^-_M(x) = \min\{\mu^-_{M_{\bar{0}}}(x_{\bar{0}}), \mu^-_{M_{\bar{1}}}(x_{\bar{1}})\}$ and $\mu^+_M(x) = \min\{\mu^+_{M_{\bar{0}}}(x_{\bar{0}}), \mu^+_{M_{\bar{1}}}(x_{\bar{1}})\}$, so $\mu^-_M = \mu^-_{M_{\bar{0}} \oplus M_{\bar{1}}}$, $\mu^+_M = \mu^+_{M_{\bar{0}} \oplus M_{\bar{1}}}$ are $\mathbb{Z}_2$-graded fuzzy subspaces of $\mathscr{L}$.
Let $x, y \in \mathscr{L}$. Then,

$$\begin{aligned}[]
[\mu^-_M([x, y]), \mu^+_M([x, y])] &= \widetilde{\mu}_M([x, y]) \\
&\geq \min\{\widetilde{\mu}_M(x), \widetilde{\mu}_M(y)\} \\
&= [\min\{\mu^-_M(x), \mu^-_M(y)\}, \min\{\mu^+_M(x), \mu^+_M(y)\}].
\end{aligned}$$

We have $\mu^-_M([x, y]) \geq \min\{\mu^-_M(x), \mu^-_M(y)\}$, and $\mu^+_M([x, y]) \geq \min\{\mu^+_M(x), \mu^+_M(y)\}$. Hence, μ^-_M and μ^+_M are fuzzy Lie sub-superalgebras of $\mathscr{L}$.

($\Leftarrow$) Let μ^-_M and μ^+_M be fuzzy Lie sub-superalgebras of $\mathscr{L}$. Then, $\mu^-_M(x) = \min\{\mu^-_{M_{\bar{0}}}(x_{\bar{0}}), \mu^-_{M_{\bar{1}}}(x_{\bar{1}})\}$ and $\mu^+_M(x) = \min\{\mu^+_{M_{\bar{0}}}(x_{\bar{0}}), \mu^+_{M_{\bar{1}}}(x_{\bar{1}})\}$, where $\mu^-_{M_{\bar{0}}}, \mu^+_{M_{\bar{0}}}$ are fuzzy subspaces of $\mathscr{L}_{\bar{0}}$, $\mu^-_{M_{\bar{1}}}, \mu^+_{M_{\bar{1}}}$ are fuzzy subspaces of $\mathscr{L}_{\bar{1}}$. Then,

$$\begin{aligned}
\widetilde{\mu}_M(x) &= [\mu^-_M(x), \mu^+_M(x)] \\
&= [\min\{\mu^-_{M_{\bar{0}}}(x_{\bar{0}}), \mu^-_{M_{\bar{1}}}(x_{\bar{1}})\}, \min\{\mu^+_{M_{\bar{0}}}(x_{\bar{0}}), \mu^+_{M_{\bar{1}}}(x_{\bar{1}})\}] \\
&= \min\{\widetilde{\mu}_{\bar{0}}(x_{\bar{0}}), \widetilde{\mu}_{\bar{1}}(x_{\bar{1}})\},
\end{aligned}$$

which implies that $M = (x, \widetilde{\mu}_M(x))$ is a $\mathbb{Z}_2$-graded interval-valued fuzzy subspace of $\mathscr{L}$. Let $x, y \in \mathscr{L}$. Then,

$$\begin{aligned}
\widetilde{\mu}_M([x, y]) &= [\mu^-_M([x, y]), \mu^+_M([x, y])] \\
&\geq [\min\{\mu^-_M(x), \mu^-_M(y)\}, \min\{\mu^+_M(x), \mu^+_M(y)\}] \\
&= \min\{\widetilde{\mu}_M(x), \widetilde{\mu}_M(y)\}.
\end{aligned}$$

Hence, $M = (x, \widetilde{\mu}_M(x))$ is an interval-valued fuzzy Lie sub-superalgebra of $\mathscr{L}$.

Theorem 3.30 *Let $M = (x, \widetilde{\mu}_M(x))$ and $N = (x, \widetilde{\mu}_N(x))$ be interval-valued fuzzy Lie sub-superalgebras (resp. ideals) of $\mathscr{L}$. Then, $M + N = (x, \widetilde{\mu}_{M+N}(x))$ is also an interval-valued fuzzy Lie sub-superalgebra (resp. ideal) of $\mathscr{L}$.*

Proof Since $M = (x, \widetilde{\mu}_M(x))$ and $N = (x, \widetilde{\mu}_N(x))$ are interval-valued fuzzy Lie sub-superalgebras, we can define $\widetilde{\mu}_{(M+N)_{\bar{0}}} = \widetilde{\mu}_{M_{\bar{0}}+N_{\bar{0}}}$ and $\widetilde{\mu}_{(M+N)_{\bar{1}}} = \widetilde{\mu}_{M_{\bar{1}}+N_{\bar{1}}}$, where $\widetilde{\mu}_{M_{\bar{0}}}, \widetilde{\mu}_{N_{\bar{0}}}$ are interval-valued fuzzy subspaces of $\mathscr{L}_{\bar{0}}$ and $\widetilde{\mu}_{M_{\bar{1}}}, \widetilde{\mu}_{N_{\bar{1}}}$ are interval-valued fuzzy subspaces of $\mathscr{L}_{\bar{1}}$. By Lemma 3.9, we know that $\widetilde{\mu}_{(M+N)_{\bar{0}}}, \widetilde{\mu}_{(M+N)_{\bar{1}}}$ are interval-valued fuzzy subspaces of $\mathscr{L}_{\bar{0}}, \mathscr{L}_{\bar{1}}$, respectively.

Let $x \in \mathscr{L}$. Then,

$$\begin{aligned}
\widetilde{\mu}_{M+N}(x) &= \sup_{x=a+b} \{\min\{\widetilde{\mu}_M(a), \widetilde{\mu}_N(b)\}\} \\
&= \sup_{x=a+b} \{[\min\{\mu^-_M(a), \mu^-_N(b)\}, \min\{\mu^+_M(a), \mu^+_N(b)\}]\} \\
&= \sup_{x=a+b} \{[\min\{\min\{\mu^-_{M_{\bar{0}}}(a_{\bar{0}}), \mu^-_{M_{\bar{1}}}(a_{\bar{1}})\}, \min\{\mu^-_{N_{\bar{0}}}(b_{\bar{0}}), \mu^-_{N_{\bar{1}}}(b_{\bar{1}})\}\},
\end{aligned}$$

$$
\begin{aligned}
&\min\{\min\{\mu^+_{M_{\bar 0}}(a_{\bar 0}), \mu^+_{M_{\bar 1}}(a_{\bar 1})\}, \min\{\mu^+_{N_{\bar 0}}(b_{\bar 0}), \mu^+_{N_{\bar 1}}(b_{\bar 1})\}\}]\}\\
&= \sup_{x=a+b} \{[\min\{\min\{\mu^-_{M_{\bar 0}}(a_{\bar 0}), \mu^-_{N_{\bar 0}}(b_{\bar 0})\}, \min\{\mu^-_{M_{\bar 1}}(a_{\bar 1}), \mu^-_{N_{\bar 1}}(b_{\bar 1})\}\},\\
&\min\{\min\{\mu^+_{M_{\bar 0}}(a_{\bar 0}), \mu^+_{N_{\bar 0}}(b_{\bar 0})\}, \min\{\mu^+_{M_{\bar 1}}(a_{\bar 1}), \mu^+_{N_{\bar 1}}(b_{\bar 1})\}\}]\}\\
&= \sup_{x=a+b} \{[\min\{\mu^-_{M_{\bar 0}+N_{\bar 0}}(a_{\bar 0}+b_{\bar 0}), \mu^-_{M_{\bar 1}+N_{\bar 1}}(a_{\bar 1}+b_{\bar 1})\},\\
&\min\{\mu^+_{M_{\bar 0}+N_{\bar 0}}(a_{\bar 0}+b_{\bar 0}), \mu^+_{M_{\bar 1}+N_{\bar 1}}(a_{\bar 1}+b_{\bar 1})\}]\}\\
&= \sup_{x=a+b} \{\min\{\widetilde{\mu}_{M_{\bar 0}+N_{\bar 0}}(a_{\bar 0}+b_{\bar 0}), \widetilde{\mu}_{M_{\bar 1}+N_{\bar 1}}(a_{\bar 1}+b_{\bar 1})\}\}\\
&= \min\{\widetilde{\mu}_{M_{\bar 0}+N_{\bar 0}}(x_{\bar 0}), \widetilde{\mu}_{M_{\bar 1}+N_{\bar 1}}(x_{\bar 1})\}\\
&= \min\{\widetilde{\mu}_{(M+N)_{\bar 0}}(x_{\bar 0}), \widetilde{\mu}_{(M+N)_{\bar 1}}(x_{\bar 1})\}.
\end{aligned}
$$

This shows that $\widetilde{\mu}_{M+N}$ is a $\mathbb{Z}_2$-graded interval-valued fuzzy subspace of $\mathscr{L}$.

Let $x, y \in \mathscr{L}$. Then,

$$
\begin{aligned}
\widetilde{\mu}_{M+N}([x, y]) &= [\mu^-_{M+N}([x, y]), \mu^+_{M+N}([x, y])]\\
&\geq [\min\{\mu^-_{M+N}(x), \mu^-_{M+N}(y)\}, \min\{\mu^+_{M+N}(x), \mu^+_{M+N}(y)\}]\\
&= \min\{\widetilde{\mu}_{M+N}(x), \widetilde{\mu}_{M+N}(y)\}.
\end{aligned}
$$

Hence, $M + N = (x, \widetilde{\mu}_{M+N}(x))$ is an interval-valued fuzzy Lie sub-superalgebra of $\mathscr{L}$.

Theorem 3.31 *Let $M = (x, \widetilde{\mu}_M(x))$ and $N = (x, \widetilde{\mu}_N(x))$ be interval-valued fuzzy Lie sub-superalgebras (resp. ideals) of $\mathscr{L}$. Then, $M \cap N = (x, \widetilde{\mu}_{M\cap N}(x))$ is also an interval-valued fuzzy Lie sub-superalgebra (resp. ideal) of $\mathscr{L}$.*

Proof Since $M = (x, \widetilde{\mu}_M(x))$ and $N = (x, \widetilde{\mu}_N(x))$ are interval-valued fuzzy Lie sub-superalgebras, we have $M = M_{\bar 0} \oplus M_{\bar 1}$ and $N = N_{\bar 0} \oplus N_{\bar 1}$, where $M_{\bar 0} = (x, \widetilde{\mu}_{M_{\bar 0}}(x))$, $N_{\bar 0} = (x, \widetilde{\mu}_{N_{\bar 0}}(x))$ are interval-valued fuzzy subspaces of $\mathscr{L}_{\bar 0}$ and $M_{\bar 1} = (x, \widetilde{\mu}_{M_{\bar 1}}(x))$, $N_{\bar 1} = (x, \widetilde{\mu}_{N_{\bar 1}}(x))$ are interval-valued fuzzy subspaces of $\mathscr{L}_{\bar 1}$. Define $\widetilde{\mu}_{(M\cap N)_{\bar 0}} = \widetilde{\mu}_{M_{\bar 0}\cap N_{\bar 0}}$ and $\widetilde{\mu}_{(M\cap N)_{\bar 1}} = \widetilde{\mu}_{M_{\bar 1}\cap N_{\bar 1}}$. By Lemma 3.10, we have that $\widetilde{\mu}_{(M\cap N)_{\bar 0}}$ and $\widetilde{\mu}_{(M\cap N)_{\bar 1}}$ are interval-valued fuzzy subspaces of $\mathscr{L}_{\bar 0}$, $\mathscr{L}_{\bar 1}$, respectively.

Let $x \in \mathscr{L}$. Then,

$$
\begin{aligned}
\widetilde{\mu}_{M\cap N}(x) &= [\mu^-_{M\cap N}(x), \mu^+_{M\cap N}(x)]\\
&= [\min\{\mu^-_M(x), \mu^-_N(x)\}, \min\{\mu^+_M(x), \mu^+_N(x)]\\
&= [\min\{\min\{\mu^-_{M_{\bar 0}}(x_{\bar 0}), \mu^-_{M_{\bar 1}}(x_{\bar 1})\}, \min\{\mu^-_{N_{\bar 0}}(x_{\bar 0}), \mu^-_{N_{\bar 1}}(x_{\bar 1})\}\},\\
&\quad \min\{\min\{\{\mu^+_{M_{\bar 0}}(x_{\bar 0}), \mu^+_{M_{\bar 1}}(x_{\bar 1})\}, \min\{\mu^+_{N_{\bar 0}}(x_{\bar 0}), \mu^+_{N_{\bar 1}}(x_{\bar 1})\}\}]\\
&= [\min\{\min\{\mu^-_{M_{\bar 0}}(x_{\bar 0}), \mu^-_{N_{\bar 0}}(x_{\bar 0})\}, \min\{\mu^-_{M_{\bar 1}}(x_{\bar 1}), \mu^-_{N_{\bar 1}}(x_{\bar 1})\}\},\\
&\quad \min\{\min\{\mu^+_{M_{\bar 0}}(x_{\bar 0}), \mu^+_{N_{\bar 0}}(x_{\bar 0})\}, \min\{\mu^+_{M_{\bar 1}}(x_{\bar 1}), \mu^+_{N_{\bar 1}}(x_{\bar 1})\}\}]\\
&= [\min(\mu^-_{M_{\bar 0}\cap N_{\bar 0}}(x_{\bar 0}), \mu^-_{M_{\bar 1}\cap N_{\bar 1}}(x_{\bar 1})), \min(\mu^+_{M_{\bar 0}\cap N_{\bar 0}}(x_{\bar 0}), \mu^+_{M_{\bar 1}\cap N_{\bar 1}}(x_{\bar 1}))]
\end{aligned}
$$

$$= \min\{\widetilde{\mu}_{M_{\overline{0}}\cap N_{\overline{0}}}(x_{\overline{0}}), \widetilde{\mu}_{M_{\overline{1}}\cap N_{\overline{1}}}(x_{\overline{1}})\}$$
$$= \min\{\widetilde{\mu}_{(A\cap B)_{\overline{0}}}(x_{\overline{0}}), \widetilde{\mu}_{(A\cap B)_{\overline{1}}}(x_{\overline{1}})\}.$$

So, $\widetilde{\mu}_{M\cap N}$ is a $\mathbb{Z}_2$-graded interval-valued fuzzy subspace of $\mathscr{L}$.

Let $x, y \in \mathscr{L}$. Then,

$$\begin{aligned}\widetilde{\mu}_{M\cap N}([x, y]) &= [\mu^-_{M\cap N}([x, y]), \mu^+_{M\cap N}([x, y])] \\ &= [\min\{\mu^-_M([x, y]), \mu^-_N([x, y])\}, \min\{\mu^+_M([x, y]), \mu^+_N([x, y])\}] \\ &\geq [\min\{\min\{\mu^-_M(x), \mu^-_M(y)\}, \min\{\mu^-_N(x), \mu^-_N(y)\}\}, \\ &\quad \min\{\min\{\mu^+_M(x), \mu^+_M(y)\}, \min\{\mu^+_N(x), \mu^+_N(y)\}\}] \\ &= [\min\{\mu^-_{M\cap N}(x), \mu^-_{M\cap N}(y)\}, \min\{\mu^+_{M\cap N}(x), \mu^+_{M\cap N}(y)\}] \\ &= \min\{\widetilde{\mu}_{M\cap N}(x), \widetilde{\mu}_{M\cap N}(y)\}.\end{aligned}$$

Hence, $M \cap N = (x, \widetilde{\mu}_{M\cap N}(x))$ is an interval-valued fuzzy Lie sub-superalgebra of $\mathscr{L}$.

Corollary 3.2 *Let $\{M_i = (x, \widetilde{\mu}_{M_i}(x)) \mid i \in I\}$ be a family of interval-valued fuzzy Lie sub-superalgebras (resp. ideals) of $\mathscr{L}$. Then, $\bigcap_{i\in I} M_i = (x, \widetilde{\mu}_{\bigcap_{i\in I} M_i}(x))$ is also an interval-valued fuzzy Lie sub-superalgebra (resp. ideals) of $\mathscr{L}$.*

We now discuss some fundamental concepts of interval-valued fuzzy Lie sub-superalgebras and ideals of Lie superalgebras under homomorphisms.

Theorem 3.32 *Let $\varphi : \mathscr{L} \to \mathscr{L}'$ be a Lie homomorphism. If $M = (x, \widetilde{\mu}_M(x))$ is an interval-valued fuzzy sub-superalgebra (resp. ideal) of L', then the interval-valued fuzzy set $\varphi^{-1}(M) = (x, \widetilde{\mu}_{\varphi^{-1}(M)}(x))$ of $\mathscr{L}$ is also an interval-valued fuzzy Lie sub-superalgebra (resp. ideal).*

Proof Let $x \in \mathscr{L}$. Then, $x = x_{\overline{0}} + x_{\overline{1}}$ where $x_{\overline{0}} \in \mathscr{L}_{\overline{0}}$ and $x_{\overline{1}} \in \mathscr{L}_{\overline{1}}$. Because φ preserves the grading, we have $\varphi(x) = \varphi(x_{\overline{0}}) + \varphi(x_{\overline{1}})$. Define $\widetilde{\mu}_{\varphi^{-1}(M)_{\overline{0}}} = \widetilde{\mu}_{\varphi^{-1}(M_{\overline{0}})}$ and $\widetilde{\mu}_{\varphi^{-1}(M)_{\overline{1}}} = \widetilde{\mu}_{\varphi^{-1}(M_{\overline{1}})}$, by Lemma 3.11, we have that $\widetilde{\mu}_{\varphi^{-1}(M)_{\overline{0}}}, \widetilde{\mu}_{\varphi^{-1}(M)_{\overline{1}}}$ are interval-valued fuzzy subspaces of $\mathscr{L}_{\overline{0}}, \mathscr{L}_{\overline{1}}$, respectively.

Let $x \in \mathscr{L}$. Then,

$$\begin{aligned}\widetilde{\mu}_{\varphi^{-1}(M)}(x) &= \widetilde{\mu}_M(\varphi(x)) = \widetilde{\mu}_M(\varphi(x_{\overline{0}}) + \varphi(x_{\overline{1}})) \\ &= \min\{\widetilde{\mu}_{M_{\overline{0}}}(\varphi(x_{\overline{0}})), \widetilde{\mu}_{M_{\overline{1}}}(\varphi(x_{\overline{1}}))\} \\ &= \min\{\widetilde{\mu}_{\varphi^{-1}(M_{\overline{0}})}(x_{\overline{0}}), \widetilde{\mu}_{\varphi^{-1}(M_{\overline{1}})}(x_{\overline{1}})\} \\ &= \min\{\widetilde{\mu}_{\varphi^{-1}(M)_{\overline{0}}}(x_{\overline{0}}), \widetilde{\mu}_{\varphi^{-1}(M)_{\overline{1}}}(x_{\overline{1}})\}.\end{aligned}$$

So, we have that $\varphi^{-1}(M) = (x, \widetilde{\mu}_{\varphi^{-1}(M)}(x))$ is a $\mathbb{Z}_2$-graded interval-valued fuzzy subspace of $\mathscr{L}$.

Let $x, y \in \mathscr{L}$. Then,

$$\begin{aligned}\widetilde{\mu}_{\varphi^{-1}(M)}([x, y]) &= \widetilde{\mu}_M(\varphi[x, y]) = \widetilde{\mu}_M([\varphi(x), \varphi(y)]) \\ &\geq \min\{\widetilde{\mu}_M(\varphi(x)), \widetilde{\mu}_M(\varphi(y))\} \\ &= \min\{\widetilde{\mu}_{\varphi^{-1}(M)}(x), \widetilde{\mu}_{\varphi^{-1}(M)}(y)\}.\end{aligned}$$

Hence, $\varphi^{-1}(M) = (x, \widetilde{\mu}_{\varphi^{-1}(M)}(x))$ is an interval-valued fuzzy Lie sub-superalgebra of $\mathscr{L}$. The case of an interval-valued fuzzy ideal can be proved similarly.

Theorem 3.33 *Let $\varphi : \mathscr{L} \to \mathscr{L}'$ be a Lie surjective homomorphism. If $M = (x, \widetilde{\mu}_M(x))$ is an interval-valued fuzzy Lie sub-superalgebra (resp. ideal) of L, then the interval-valued fuzzy set $\varphi(M) = (x, \widetilde{\mu}_{\varphi(M)}(x))$ is also an interval-valued fuzzy Lie sub-superalgebra (resp. ideal) of $\mathscr{L}'$.*

Proof Let $y \in \mathscr{L}'$. Note that $M = (x, \widetilde{\mu}_M(x))$ is a $\mathbb{Z}_2$-graded interval-valued fuzzy vector subspace of L, we define $\widetilde{\mu}_{\varphi(M)_{\overline{0}}} = \widetilde{\mu}_{\varphi(M_{\overline{0}})}, \widetilde{\mu}_{\varphi(M)_{\overline{1}}} = \widetilde{\mu}_{\varphi(M_{\overline{1}})}$, and then they are interval-valued fuzzy subspaces of $\mathscr{L}_{\overline{0}}, \mathscr{L}_{\overline{1}}$, respectively. Then,

$$\begin{aligned}\widetilde{\mu}_{\varphi(M)}(y) &= \sup_{y=\varphi(x)} \{\widetilde{\mu}_M(x)\} = \sup_{y=\varphi(x)} \{\min\{\widetilde{\mu}_{M_{\overline{0}}}(x_{\overline{0}}), \widetilde{\mu}_{M_{\overline{1}}}(x_{\overline{1}})\}\} \\ &= \min\{\sup_{y_{\overline{0}}=\varphi(x_{\overline{0}})} \{\widetilde{\mu}_{M_{\overline{0}}}(x_{\overline{0}})\}, \sup_{y_{\overline{1}}=\varphi(x_{\overline{1}})} \{\widetilde{\mu}_{M_{\overline{1}}}(x_{\overline{1}})\}\} \\ &= \min\{\widetilde{\mu}_{\varphi(M_{\overline{0}})}(y_{\overline{0}}), \widetilde{\mu}_{\varphi(M_{\overline{1}})}(y_{\overline{1}})\} \\ &= \min\{\widetilde{\mu}_{\varphi(M)_{\overline{0}}}(y_{\overline{0}}), \widetilde{\mu}_{\varphi(M)_{\overline{1}}}(y_{\overline{1}})\}.\end{aligned}$$

So, $\varphi(M) = (x, \widetilde{\mu}_{\varphi(M)}(x))$ is a $\mathbb{Z}_2$-graded interval-valued fuzzy vector subspace of L'.

Let $x, y \in \mathscr{L}'$. Then,

$$\begin{aligned}\widetilde{\mu}_{\varphi(M)}([x, y]) &\geq \sup_{[x,y]=\varphi([a,b])} \{\widetilde{\mu}_M([a, b])\} \\ &\geq \sup_{[x,y]=\varphi([a,b])} \{\min\{\widetilde{\mu}_M(a), \widetilde{\mu}_M(b)\}\} \\ &= \min\{\sup_{x=\varphi(a)} \{\widetilde{\mu}_M(a)\}, \sup_{y=\varphi(b)} \{\widetilde{\mu}_M(b)\}\} \\ &= \min\{\widetilde{\mu}_{\varphi(M)}(x), \widetilde{\mu}_{\varphi(M)}(y)\}.\end{aligned}$$

Hence, $\varphi(M) = (x, \widetilde{\mu}_{\varphi(M)}(x))$ is an interval-valued fuzzy Lie sub-superalgebra of $\mathscr{L}'$.

Let J be an ideal of $\mathscr{L}$. Quotient vector space $\mathscr{L}/J$ is a Lie superalgebra, and the bracket operation is defined by $[\overline{x}, \overline{y}] := \overline{[x, y]}$.

Theorem 3.34 *Let $M = (x, \widetilde{\mu}_M(x))$ be an interval-valued fuzzy Lie sub-superalgebra of $\mathscr{L}$ and J be an ideal of $\mathscr{L}$. Define an interval-valued fuzzy set $N = (\overline{x}, \widetilde{\mu}_N(\overline{x}))$ as follows: $\widetilde{\mu}_N(\overline{x}) = \sup_{z\in\overline{x}}\{\widetilde{\mu}_M(z)\}$, where $\overline{x} = x + J \in \mathscr{L}/J$ for*

$x \in \mathscr{L}$. Then, $N = (\overline{x}, \widetilde{\mu}_N(\overline{x}))$ is an interval-valued fuzzy Lie sub-superalgebra of $\mathscr{L}/J$.

Proof Since $M = (x, \widetilde{\mu}_M(x))$ is an interval-valued fuzzy Lie sub-superalgebra of $\mathscr{L}$, we have $M = M_{\overline{0}} \oplus M_{\overline{1}}$ where $M_{\overline{0}} = (x, \widetilde{\mu}_{M_{\overline{0}}}(x))$, $M_{\overline{1}} = (x, \widetilde{\mu}_{M_{\overline{1}}}(x))$ are interval-valued fuzzy subspaces of $\mathscr{L}_{\overline{0}}$, $\mathscr{L}_{\overline{1}}$, respectively.

Define $N_{\overline{0}} = (\overline{x}, \widetilde{\mu}_{N_{\overline{0}}}(\overline{x}))$, where $\widetilde{\mu}_{N_{\overline{0}}}\colon \mathscr{L}_{\overline{0}}/J \to D[0,1]$ by $\widetilde{\mu}_{N_{\overline{0}}}(\overline{x}) = \sup\limits_{z\in\overline{x}}\{\widetilde{\mu}_{M_{\overline{0}}}(z)\}$ for $\overline{x} = x + J \in \mathscr{L}_{\overline{0}}$, and $\widetilde{\mu}_{N_{\overline{1}}}\colon \mathscr{L}_{\overline{1}}/J \to D[0,1]$ by $\widetilde{\mu}_{N_{\overline{1}}}(\overline{x}) = \sup\limits_{z\in\overline{x}}\{\widetilde{\mu}_{M_{\overline{1}}}(z)\}$ for $\overline{x} = x + J \in \mathscr{L}_{\overline{1}}$. Then, $N_{\overline{0}}$, $N_{\overline{1}}$ are interval-valued fuzzy subspaces of $\mathscr{L}_{\overline{0}}/J$, $\mathscr{L}_{\overline{1}}/J$, respectively.

Let $\overline{x} \in \mathscr{L}/J$. Then,

$$\begin{aligned}
\widetilde{\mu}_N(\overline{x}) &= \sup_{z\in\overline{x}}\{\widetilde{\mu}_M(z)\} = \sup_{z\in\overline{x}}\{\min\{\widetilde{\mu}_{M_{\overline{0}}}(z_{\overline{0}}), \widetilde{\mu}_{M_{\overline{1}}}(z_{\overline{1}})\}\} \\
&= \min\{\sup_{z_{\overline{0}}\in\overline{x}_{\overline{0}}}\{\widetilde{\mu}_{M_{\overline{0}}}(z_{\overline{0}}), \sup_{z_{\overline{1}}\in\overline{x}_{\overline{1}}}\{\widetilde{\mu}_{M_{\overline{1}}}(z_{\overline{1}})\}\} \\
&= \min\{\widetilde{\mu}_{N_{\overline{0}}}(\overline{x_{\overline{0}}}), \widetilde{\mu}_{N_{\overline{1}}}(\overline{x_{\overline{1}}})\} = \widetilde{\mu}_{N_{\overline{0}}\oplus N_{\overline{1}}}(x),
\end{aligned}$$

which shows that $N = (\overline{x}, \widetilde{\mu}_N(\overline{x}))$ is a $\mathbb{Z}_2$-graded interval-valued fuzzy subspace of $\mathscr{L}/J$.

Let $\overline{x}, \overline{y} \in \mathscr{L}/J$. Then,

$$\begin{aligned}
\widetilde{\mu}_N([\overline{x}, \overline{y}]) = \widetilde{\mu}_N(\overline{[x, y]}) &= \sup_{z\in\overline{[x,y]}}\{\widetilde{\mu}_M(z)\} \\
&\geq \sup_{[a,b]\in\overline{[x,y]}}\{\widetilde{\mu}_M([a, b])\} \\
&\geq \sup_{[a,b]\in\overline{[x,y]}}\{\min\{\widetilde{\mu}_M(a), \widetilde{\mu}_M(b)\}\} \\
&= \min\{\sup_{a\in\overline{x}}\{\widetilde{\mu}_M(a)\}, \sup_{b\in\overline{y}}\{\widetilde{\mu}_M(b)\}\} \\
&= \min\{\widetilde{\mu}_N(\overline{x}), \widetilde{\mu}_N(\overline{y})\}.
\end{aligned}$$

So, $N = (\overline{x}, \widetilde{\mu}_N(\overline{x}))$ is an interval-valued fuzzy Lie sub-superalgebra of $\mathscr{L}/J$.

Definition 3.28 Let $M = (x, \widetilde{\mu}_M(x))$ be an interval-valued fuzzy Lie sub-superalgebra of $\mathscr{L}$ and $N = (x, \widetilde{\mu}_N(x))$ be an interval-valued fuzzy set of $\mathscr{L}$ with $N \subseteq M$. Then, N is called an *interval-valued fuzzy ideal* of M if it satisfies the following conditions, for $x, y \in \mathscr{L}$ and $\alpha \in \mathbb{F}$:

(1) $N = (x, \widetilde{\mu}_N(x))$ is a $\mathbb{Z}_2$-graded interval-valued fuzzy subspace of $\mathscr{L}$,

(2) $\widetilde{\mu}_N([x, y]) \geq \max\{\min\{\widetilde{\mu}_N(x), \widetilde{\mu}_M(y)\}, \min\{\widetilde{\mu}_N(y), \widetilde{\mu}_M(x)\}\}$.

By Definition 3.28, it is immediately clear that:

1. $M = (x, \widetilde{\mu}_M(x))$ is an interval-valued fuzzy ideal of $\mathscr{L}$ if and only if $M = (x, \widetilde{\mu}_M(x))$ is an interval-valued fuzzy ideal of interval-valued fuzzy Lie sub-superalgebra $[1, 1]_{\mathscr{L}}$.

2. If $M = (x, \widetilde{\mu}_M(x))$ is an interval-valued fuzzy Lie sub-superalgebra of $\mathscr{L}$ and $N = (x, \widetilde{\mu}_N(x))$ is an interval-valued fuzzy Lie ideal of $\mathscr{L}$ with $N \subseteq M$, then N is an interval-valued fuzzy ideal of M.
3. An interval-valued fuzzy ideal of an interval-valued fuzzy Lie sub-superalgebra is an interval-valued fuzzy Lie sub-superalgebra.

Theorem 3.35 *Let $M = (x, \widetilde{\mu}_M(x))$ be an interval-valued fuzzy Lie sub-superalgebra of $\mathscr{L}$ and $N = (x, \widetilde{\mu}_N(x))$ be an interval-valued fuzzy Lie ideal of $\mathscr{L}$. Then, $M \cap N = (x, \widetilde{\mu}_{M\cap N}(x))$ is an interval-valued fuzzy ideal of M.*

Proof Obviously, $M \cap N \subseteq M$ and $M \cap N$ is a $\mathbb{Z}_2$-graded interval-valued fuzzy subspace of $\mathscr{L}$. Then, for $x, y \in \mathscr{L}$, note that $N = (x, \widetilde{\mu}_N(x))$ is an interval-valued fuzzy Lie ideal of $\mathscr{L}$,

$$\begin{aligned}
\widetilde{\mu}_{M\cap N}([x, y]) &= [\mu^-_{M\cap N}([x, y]), \mu^+_{M\cap N}([x, y])] \\
&= [\min\{\mu^-_M([x, y]), \mu^-_N([x, y]), \min\{\mu^+_M([x, y]), \mu^+_N([x, y])] \\
&= \min\{\widetilde{\mu}_M([x, y]), \widetilde{\mu}_N([x, y])\} \\
&\geq \min\{\min\{\widetilde{\mu}_M(x), \widetilde{\mu}_M(y)\}, \max\{\widetilde{\mu}_N(x), \widetilde{\mu}_N(y)\}\} \\
&\geq \min\{\min\{\widetilde{\mu}_M(x), \widetilde{\mu}_M(y)\}, \widetilde{\mu}_N(x)\} \\
&= \min\{\min\{\widetilde{\mu}_N(x), \widetilde{\mu}_M(x)\}, \widetilde{\mu}_M(y)\} \\
&= \min\{\widetilde{\mu}_{M\cap N}(x), \widetilde{\mu}_M(y)\}.
\end{aligned}$$

Similarly,

$$\begin{aligned}
\widetilde{\mu}_{M\cap N}([x, y]) &= [\mu^-_{M\cap N}([x, y]), \mu^+_{M\cap N}([x, y])] \\
&= [\min\{\mu^-_M([x, y]), \mu^-_N([x, y]), \min\{\mu^+_M([x, y]), \mu^+_N([x, y])] \\
&= \min\{\widetilde{\mu}_M([x, y]), \widetilde{\mu}_N([x, y])\} \\
&\geq \min\{\min\{\widetilde{\mu}_M(x), \widetilde{\mu}_M(y)\}, \max\{\widetilde{\mu}_N(x), \widetilde{\mu}_N(y)\}\} \\
&\geq \min\{\min\{\widetilde{\mu}_M(x), \widetilde{\mu}_M(y)\}, \widetilde{\mu}_N(y)\} \\
&= \min\{\min\{\widetilde{\mu}_N(y), \widetilde{\mu}_M(y)\}, \widetilde{\mu}_M(x)\} \\
&= \min\{\widetilde{\mu}_{M\cap N}(y), \widetilde{\mu}_M(x)\}.
\end{aligned}$$

This completes the proof.

Theorem 3.36 *Let $M = (x, \widetilde{\mu}_M(x))$ be an interval-valued fuzzy Lie sub-superalgebra of $\mathscr{L}$ and $N_1 = (x, \widetilde{\mu}_{N_1}(x))$ and $N_2 = (x, \widetilde{\mu}_{N_2}(x))$ be interval-valued fuzzy ideals of M. Then, $N_1 \cap N_2 = (x, \widetilde{\mu}_{N_1\cap N_2}(x))$ is also an interval-valued fuzzy ideal of M.*

Proof Clearly, $N_1 \cap N_2 \subseteq M$ and $N_1 \cap N_2$ is a $\mathbb{Z}_2$-graded interval-valued fuzzy subspace of $\mathscr{L}$. Let $x, y \in \mathscr{L}$. Then,

$$\begin{aligned}\widetilde{\mu}_{N_1\cap N_2}([x,y]) &= \min\{\widetilde{\mu}_{N_1}([x,y]),\widetilde{\mu}_{N_2}([x,y])\}\\ &\geq \min\{\max\{\min\{\widetilde{\mu}_{N_1}(x),\widetilde{\mu}_M(y)\},\min\{\widetilde{\mu}_{N_1}(y),\widetilde{\mu}_M(x)\}\},\\ &\quad \max\ \{\min\{\widetilde{\mu}_{N_2}(x),\widetilde{\mu}_M(y)\},\min\{\widetilde{\mu}_{N_2}(y),\widetilde{\mu}_M(x)\}\}\}\\ &\geq \min\{\min\{\widetilde{\mu}_{N_1}(x),\widetilde{\mu}_M(y)\},\min\{\widetilde{\mu}_{N_2}(x),\widetilde{\mu}_M(y)\}\}\\ &= \min\{\min\{\widetilde{\mu}_{N_1}(x),\widetilde{\mu}_{N_2}(x)\},\widetilde{\mu}_M(y)\}\\ &= \min\{\widetilde{\mu}_{N_1\cap N_2}(x),\widetilde{\mu}_M(y)\},\end{aligned}$$

and

$$\begin{aligned}\widetilde{\mu}_{N_1\cap N_2}([x,y]) &= \min\{\widetilde{\mu}_{N_1}([x,y]),\widetilde{\mu}_{N_2}([x,y])\}\\ &\geq \min\{\max\{\min\{\widetilde{\mu}_{N_1}(x),\widetilde{\mu}_M(y)\},\min\{\widetilde{\mu}_{N_1}(y),\widetilde{\mu}_M(x)\}\},\\ &\quad \max\{\min\{\widetilde{\mu}_{N_2}(x),\widetilde{\mu}_M(y)\},\min\{\widetilde{\mu}_{N_2}(y),\widetilde{\mu}_M(x)\}\}\}\\ &\geq \min\{\min\{\widetilde{\mu}_{N_1}(y),\widetilde{\mu}_M(x)\},\min\{\widetilde{\mu}_{N_2}(y),\widetilde{\mu}_M(x)\}\}\\ &= \min\{\min\{\widetilde{\mu}_{N_1}(y),\widetilde{\mu}_{N_2}(y)\},\widetilde{\mu}_M(x)\}\\ &= \min\{\widetilde{\mu}_{N_1\cap N_2}(y),\widetilde{\mu}_M(x)\}.\end{aligned}$$

So, $\widetilde{\mu}_{N_1\cap N_2}([x,y]) \geq \max\{\min\{\widetilde{\mu}_{N_1\cap N_2}(x),\widetilde{\mu}_M(y)\},\min\{\widetilde{\mu}_{N_1\cap N_2}(y),\widetilde{\mu}_M(x)\}\}$.

Theorem 3.37 *Let $M=(x,\widetilde{\mu}_M(x))$ be an interval-valued fuzzy Lie sub-superalgebra of $\mathscr{L}$ and $N=(x,\widetilde{\mu}_N(x))$ be an interval-valued fuzzy ideal of M. Suppose that $\mathscr{L}'$ is a Lie superalgebra and f is a homomorphism from $\mathscr{L}$ to $\mathscr{L}'$. Then, $f(N)$ is an interval-valued fuzzy ideal of $f(M)$.*

Proof Note first that both $f(M)$ and $f(N)$ are interval-valued fuzzy Lie sub-superalgebra of $\mathscr{L}$ and $f(N)\subseteq f(M)$. Next, note that $N=(x,\widetilde{\mu}_N(x))$ is an interval-valued fuzzy ideal of M, and we have

$$\begin{aligned}\widetilde{\mu}_{f(N)}([x,y]) &= \sup_{z\in f^{-1}[x,y]}\{\widetilde{\mu}_N(z)\} \geq \sup_{z=[a,b]}\{\widetilde{\mu}_N([a,b])\}\\ &\geq \sup_{a\in f^{-1}(x),b\in f^{-1}(y)}\{\max\{\min\{\widetilde{\mu}_M(a),\widetilde{\mu}_N(b)\},\min\{\widetilde{\mu}_M(b),\widetilde{\mu}_N(a)\}\}\\ &\geq \sup_{a\in f^{-1}(x),b\in f^{-1}(y)}\{\min\{\widetilde{\mu}_M(a),\widetilde{\mu}_N(b)\}\}\\ &= \min\{\sup_{a\in f^{-1}(x)}\{\widetilde{\mu}_M(a)\},\sup_{b\in f^{-1}(y)}\{\widetilde{\mu}_M(b)\}\}\\ &= \min\{\widetilde{\mu}_{f(M)}(x),\widetilde{\mu}_{f(N)}(y)\},\end{aligned}$$

and

$$\begin{aligned}\widetilde{\mu}_{f(N)}([x,y]) &= \sup_{z\in f^{-1}[x,y]}\{\widetilde{\mu}_N(z)\} \geq \sup_{z=[a,b]}\{\widetilde{\mu}_N([a,b])\}\\ &\geq \sup_{a\in f^{-1}(x),b\in f^{-1}(y)}\{\max\{\min\{\widetilde{\mu}_M(a),\widetilde{\mu}_N(b)\},\min\{\widetilde{\mu}_M(b),\widetilde{\mu}_N(a)\}\}\end{aligned}$$

$$\geq \sup_{a \in f^{-1}(x), b \in f^{-1}(y)} \{\min\{\widetilde{\mu}_N(a), \widetilde{\mu}_M(b)\}\}$$
$$= \min\{\sup_{a \in f^{-1}(x)} \{\widetilde{\mu}_N(a)\}, \sup_{b \in f^{-1}(y)} \{\widetilde{\mu}_M(b)\}\}$$
$$= \min\{\widetilde{\mu}_{f(N)}(x), \widetilde{\mu}_{f(M)}(y)\}.$$

So, $\widetilde{\mu}_{f(N)}([x, y]) \geq \max\{\min\{\widetilde{\mu}_{f(N)}(x), \widetilde{\mu}_{f(M)}(y)\}, \min\{\widetilde{\mu}_{f(M)}(x), \widetilde{\mu}_{f(N)}(y)\}\}$.

Theorem 3.38 *Let $M = (x, \widetilde{\mu}_M(x))$ be an interval-valued fuzzy Lie sub-superalgebra of $\mathscr{L}'$ and $N = (x, \widetilde{\mu}_N(x))$ be an interval-valued fuzzy ideal of M. Suppose that $\mathscr{L}$ is a Lie superalgebra and f is a homomorphism from $\mathscr{L}$ to $\mathscr{L}'$. Then $f^{-1}(N)$ is an interval-valued fuzzy ideal of $f^{-1}(M)$.*

Definition 3.29 Let $\mathscr{L}'$ be a Lie superalgebra, and let $M = (x, \widetilde{\mu}_M(x))$ be an interval-valued fuzzy Lie sub-superalgebra of $\mathscr{L}$ and $N = (x, \widetilde{\mu}_N(x))$ be an interval-valued fuzzy Lie sub-superalgebra of $\mathscr{L}'$.

1. A *weak homomorphism* from M into N is an epimorphism f of $\mathscr{L}$ onto $\mathscr{L}'$ such that $f(M) \subseteq N$. If f is a weak homomorphism of M onto N, then we say that M is a *weakly homomorphic* to N and we write $M \sim N$.
2. A *weak isomorphism* from M into N is a weak homomorphism f from M into N which is also an isomorphism of $\mathscr{L}$ onto $\mathscr{L}'$. If f is a weak isomorphism from M onto N, then we say that M is *weakly isomorphic* to N and we write $M \simeq N$.
3. A *homomorphism* from M onto N is a weak homomorphism from M onto N such that $f(M) = N$. If f is a homomorphism of M onto N, then we say that M is *homomorphic* to N and we write $M \approx N$.
4. An *isomorphism* from M onto N is a weak isomorphism f from M into N such that $f(M) = N$. If f is an isomorphism from M onto N, then we say that M is *isomorphic* to N and we write $M \cong N$.

For any interval-valued fuzzy set $M = (x, \widetilde{\mu}_M(x))$, we define a new interval-valued fuzzy set $M^* = (x, \overline{\mu}_{M^*}(x))$ as $M^* \colon \mathscr{L}/M \to D[0, 1]$ such that $M^*(x + M) = \widetilde{\mu}_M(x)$ for $x \in \mathscr{L}$. If $M = (x, \widetilde{\mu}_M(x))$ is an interval-valued fuzzy Lie sub-superalgebra of $\mathscr{L}$ and $N = (x, \widetilde{\mu}_N(x))$ is an interval-valued fuzzy ideal of M, then it is clear that N^* is an ideal of M^* and $M|_{M^*} = (x, \widetilde{\mu}_{M|_{M^*}}(x))$ is an interval-valued fuzzy Lie sub-superalgebra of M^*. Thus, by Theorem 3.34, we can talk about the quotient interval-valued fuzzy Lie sub-superalgebra of $M|_{M^*}$ relative to N^*, $M|_{M^*}/N^*$. We call this interval-valued fuzzy Lie sub-superalgebra of M relative to N and denote it by $M/N = (x, \widetilde{\mu}_{M/N}(x))$.

Theorem 3.39 *Let $M = (x, \widetilde{\mu}_M(x))$ be an interval-valued fuzzy Lie sub-superalgebra of $\mathscr{L}$ and $N = (x, \widetilde{\mu}_N(x))$ be an interval-valued fuzzy ideal of M. Then, $M|_{M^*} \approx M/N$.*

Proof Let f be the natural homomorphism from M^* onto M^*/N^*. Then, $\widetilde{\mu}_{f(M|_{M^*})}(\overline{x}) = \sup_{f(z)=\overline{x}} \{\widetilde{\mu}_{M|_{M^*}}(z) \mid z \in M^*\} = \sup_{y \in \overline{x}} \{\widetilde{\mu}_M(y)\} = (\widetilde{\mu}_{M/N})(\overline{x})$. Therefore, $M|_{M^*} \approx M/N$.

Theorem 3.40 *Let* $M = (x, \widetilde{\mu}_M(x))$ *be an interval-valued fuzzy Lie sub-superalgebra of* $\mathscr{L}$ *and* $N_1 = (x, \widetilde{\mu}_{N_1}(x))$ *be an interval-valued fuzzy Lie sub-superalgebra of* L' *such that* $M \approx N_1$. *Then, there exists an interval-valued fuzzy ideal* $N_2 = (x, \widetilde{\mu}_{N_2}(x))$ *of* M *such that* $M/N_2 \cong N_1|_{N_1^*}$.

Proof Since $M \approx N_1$, there exists an epimorphism f of $\mathscr{L}$ onto $\mathscr{L}'$ such that $f(M) = N_1$. Define $N_2 = (x, \widetilde{\mu}_{N_2}(x))$ as follows:

$$\forall\, x \in \mathscr{L}, \quad \widetilde{\mu}_{N_2}(x) = \begin{cases} \widetilde{\mu}_M(x) & x \in \ker f, \\ [0, 0] & \text{otherwise.} \end{cases}$$

Clearly, $N_2 = (x, \widetilde{\mu}_{N_2}(x))$ is an interval-valued fuzzy Lie sub-superalgebra and $N_2 \subseteq M$. We say that N_2 is an interval-valued fuzzy ideal of M. If $x \in \ker f$, then $[x, y] \in \ker f$, for any $y \in \mathscr{L}$, so $\widetilde{\mu}_{N_2}([x, y]) = \widetilde{\mu}_M([x, y]) \geq \min\{\widetilde{\mu}_M(x), \widetilde{\mu}_M(y)\} = \min\{\widetilde{\mu}_{N_2}(x), \widetilde{\mu}_M(y)\}$. Similarly, if $y \in \ker f$, then $[x, y] \in \ker f$, for any $x \in \mathscr{L}$, so $\widetilde{\mu}_{N_2}([x, y]) = \widetilde{\mu}_M([x, y]) \geq \min\{\widetilde{\mu}_M(x), \widetilde{\mu}_M(y)\} = \min\{\widetilde{\mu}_M(x), \widetilde{\mu}_{N_2}(y)\}$. Hence, if $x, y \in \ker f$, then $\widetilde{\mu}_{N_2}([x, y]) \geq \max\{\min\{\widetilde{\mu}_{N_2}(x), \widetilde{\mu}_M(y)\}, \min\{\widetilde{\mu}_M(x), \widetilde{\mu}_{N_2}(y)\}\}$.

If $x \in \ker f$ and $y \in \mathscr{L} \setminus \ker f$, then $\widetilde{\mu}_{N_2}(y) = 0$, so $\widetilde{\mu}_{N_2}([x, y]) \geq \min\{\widetilde{\mu}_M(x), \widetilde{\mu}_{N_2}(y)\}$ and $\widetilde{\mu}_{N_2}([x, y]) = \widetilde{\mu}_M([x, y]) \geq \min\{\widetilde{\mu}_M(x), \widetilde{\mu}_M(y)\} = \min\{\widetilde{\mu}_{N_2}(x), \widetilde{\mu}_M(y)\}$, we have $\widetilde{\mu}_{N_2}([x, y]) \geq \max\{\min\{\widetilde{\mu}_{N_2}(x), \widetilde{\mu}_M(y)\}, \min\{\widetilde{\mu}_M(x), \widetilde{\mu}_{N_2}(y)\}\}$.

If $y \in \ker f$ and $x \in \mathscr{L} \setminus \ker f$, the case can be proved similarly.

If $x, y \in \mathscr{L} \setminus \ker f$, then $\widetilde{\mu}_{N_2}(x) = \widetilde{\mu}_{N_2}(y) = 0$, so in this case we have $\widetilde{\mu}_{N_2}([x, y]) \geq \max\{\min\{\widetilde{\mu}_{N_2}(x), \widetilde{\mu}_M(y)\}, \min\{\widetilde{\mu}_M(x), \widetilde{\mu}_{N_2}(y)\}\}$. Hence, N_2 is an interval-valued fuzzy ideal of M.

Next, since $M \approx N_1$, $f(M^*) = {N_1}^*$. Let $g = f|_{M^*}$. Then, g is a homomorphism of M^* onto ${N_1}^*$ and $\ker g = {N_2}^*$. Thus, there exists an isomorphism h of $M^*/{N_2}^*$ onto ${N_1}^*$ such that $h(\overline{x}) = g(x) = f(x)$ for all $x \in M^*$. We have

$$\begin{aligned} h(\widetilde{\mu}_{M/N_2})(z) &= \sup_{h(\overline{x})=z} \{\widetilde{\mu}_{M/N_2}(\overline{x}) \mid x \in M^*\} \\ &= \sup\{\sup_{y\in\overline{x}}\{\widetilde{\mu}_M(y)\} \mid x \in M^*,\ g(x) = z\} \\ &= \sup\{\widetilde{\mu}_M(y) \mid y \in M^*,\ g(y) = z\} \\ &= \sup\{\widetilde{\mu}_M(y) \mid y \in \mathscr{L},\ f(y) = z\} = \widetilde{\mu}_{N_1}(z). \end{aligned}$$

Thus, $M/N_2 \cong N_1|_{{N_1}^*}$.

Chapter 4
Generalized Fuzzy Lie Subalgebras

In this chapter, we present certain generalized fuzzy Lie subalgebras, including (α, β)-fuzzy Lie subalgebras, implication-based fuzzy Lie subalgebras, $(\alpha, \beta)^*$-fuzzy Lie subalgebras, interval-valued $(\in, \in \vee q)$-fuzzy Lie ideals, and (γ, δ)-intuitionistic fuzzy Lie subalgebras. We describe some of their related properties.

4.1 Introduction

The concept of fuzzy set theory was first initiated by Zadeh in 1965. Fuzzy set theory has become an important tool in studying different disciplines, including computer science, medical science, management science, social science, and engineering. The notion of fuzzy subgroup was made by Rosenfeld [115] in 1971. Das [47] characterized fuzzy subgroups by their level subgroups. Liu [96] introduced and developed basic results concerning the notion of fuzzy subrings and fuzzy ideals of a ring. The concept of quasicoincidence of a fuzzy point with a fuzzy subset was introduced by Pu and Liu [113]. Using the idea of quasicoincidence of a fuzzy point with a fuzzy subset, Bhakat and Das defined in [35] different types of fuzzy subgroups called, (α, β)-fuzzy subgroups. In particular, they introduced the concept of $(\in, \in \vee q)$-fuzzy subgroups which was an important and useful generalization of Rosenfeld's fuzzy subgroups.

Definition 4.1 A fuzzy set μ in a set X of the form

$$\mu(y) = \begin{cases} t \in (0, 1], & \text{if } y = x, \\ 0, & \text{if } y \neq x, \end{cases}$$

is said to be a *fuzzy point* with support x and value t and is denoted by x_t. For a fuzzy point x_t and a fuzzy set μ in a set X, Pu and Liu gave meaning to the symbol $x_t \alpha \mu$, where $\alpha \in \{\in, q, \in \vee q, \in \wedge q\}$. A fuzzy point x_t is called *belong to* a fuzzy set μ, written as $x_t \in \mu$, if $\mu(x) \geqslant t$. A fuzzy point x_t is said to be *quasicoincident with*

M. Akram, *Fuzzy Lie Algebras*, Infosys Science Foundation Series,
https://doi.org/10.1007/978-981-13-3221-0_4

a fuzzy set μ, written as $x_t q\mu$, if $\mu(x) + t > 1$. To say that $x_t \in \vee q\mu$ (respectively, $x_t \in \wedge q\mu$) means that $x_t \in \mu$ or $x_t q\mu$ (respectively, $x_t \in \mu$ and $x_t q\mu$). $x_t\overline{\alpha}\mu$ means that $x_t\alpha\mu$ does not hold, where $\alpha \in \{\in, q, \in\vee q, \in\wedge q\}$.

Definition 4.2 Let V be a vector space. A fuzzy set μ on V is called an (α, β)*-fuzzy subspace* of V if it satisfies the following conditions:

(1) $x_s\alpha\mu, y_t\alpha\mu \Rightarrow (x+y)_{\min(s,t)}\beta\mu$,
(2) $x_s\alpha\mu \Rightarrow (mx)_s\beta\mu$

for all $x, y \in V, m \in \mathbb{F}, s, t \in (0, 1]$.

From (2), it follows that:

- $x_s\alpha\mu \Rightarrow (-x)_s\beta\mu$,
- $x_s\alpha\mu \Rightarrow (0)_s\beta\mu$.

Let α and β denote one of the symbols $\in, q, \in\vee q$ or $\in\wedge q$ unless otherwise specified.

4.2 (α, β)-Fuzzy Lie Subalgebras

Definition 4.3 Let L be a Lie algebra. A fuzzy set μ in L is called an (α, β)*-fuzzy Lie subalgebra* of L if it satisfies the following conditions:

(1) $x_s\alpha\mu, y_t\alpha\mu \Rightarrow (x+y)_{\min(s,t)}\beta\mu$,
(2) $x_s\alpha\mu \Rightarrow (mx)_s\beta\mu$,
(3) $x_s\alpha\mu, y_t\alpha\mu \Rightarrow [x, y]_{\min(s,t)}\beta\mu$

for all $x, y \in L, m \in \mathbb{F}, s, t \in (0, 1]$.

The proofs of the following propositions are obvious.

Proposition 4.1 *Every* $(\in, \in)$*-fuzzy Lie subalgebra is an* $(\in, \in\vee q)$*-fuzzy Lie subalgebra.*

Proposition 4.2 *Every* $(\in\vee q, \in\vee q)$*-fuzzy Lie subalgebra is an* $(\in, \in\vee q)$*-fuzzy Lie subalgebra.*

Converse of Propositions 4.1 and 4.2 may not be true as seen in the following example.

Example 4.1 Let V be a vector space over a field $\mathbb{F}$ such that $dim(V) = 5$. Let $\{e_1, e_2, \ldots, e_5\}$ be a basis of the vector space V over the field $\mathbb{F}$ with Lie brackets as follows:

$$[e_1, e_2] = e_3, \quad [e_1, e_3] = e_5, \quad [e_1, e_4] = e_5, \quad [e_1, e_5] = 0,$$

$$[e_2, e_3] = e_5, \quad [e_2, e_4] = 0, \quad [e_2, e_5] = 0, \quad [e_3, e_4] = 0,$$

$$[e_3, e_5] = 0, \quad [e_4, e_5] = 0, \quad [e_i, e_j] = -[e_j, e_i]$$

and $[e_i, e_j] = 0$ for all $i = j$. Then, V is a Lie algebra over $\mathbb{F}$. We define a fuzzy set $\mu : V \rightarrow [0, 1]$ by

$$\mu(x) := \begin{cases} 0.5 \text{ if } x = 0, \\ 0.6 \text{ if } x \in \{e_3, e_5\}, \\ 1.0 \text{ if } x \in \{e_1, e_2, e_4\}. \end{cases}$$

By routine computations, it is easy to see that μ is an $(\in, \in \vee q)$-fuzzy Lie subalgebra of L. But, it is easy to see that μ is not $(\in, \in)$- and $(\in \vee q, \in \vee q)$-fuzzy Lie subalgebra of L.

For a fuzzy set μ in L, we denote $L_0 = \{x \in L : \mu(x) > 0\}$.

We now establish a series of lemmas:

Lemma 4.1 *If μ is a nonzero $(\in, \in)$-fuzzy Lie subalgebra of L, then the set L_0 is a fuzzy Lie subalgebra of L.*

Proof Let $x, y \in L_0$. Then $\mu(x) > 0$ and $\mu(y) > 0$.
(1) If $\mu(x + y) = 0$, then we can see that $x_{\mu(x)} \in \mu$ and $y_{\mu(y)} \in \mu$, but $(x + y)_{\min(\mu(x),\mu(y))}\overline{\in}\mu$ since $\mu(x + y) = 0 < \min(\mu(x), \mu(y))$. This is clearly a contradiction, and hence $\mu(x + y) > 0$, which shows that $x + y \in L_0$.
(2) If $\mu(mx) = 0$, then we can see that $x_{\mu(x)} \in \mu$, but $(mx)_{\mu(x)}\overline{\in}\mu$ since $\mu(mx) = 0 < \mu(x)$. This is clearly a contradiction, and hence $\mu(mx) > 0$, which shows that $mx \in L_0$.
(3) If $\mu([x, y]) = 0$, then we can see that $x_{\mu(x)} \in \mu$ and $y_{\mu(y)} \in \mu$, but $([x, y])_{\min(\mu(x),\mu(y))}\overline{\in}\mu$ since $\mu([x, y]) = 0 < \min(\mu(x), \mu(y))$, a contradiction, and hence $\mu([x, y]) > 0$, which shows that $[x, y] \in L_0$. Consequently, L_0 is a Lie subalgebra of L.

Lemma 4.2 *If μ is a nonzero $(\in, q)$-fuzzy Lie subalgebra of L, then the set L_0 is a fuzzy Lie subalgebra of L.*

Proof Let $x, y \in L_0$. Then $\mu(x) > 0$ and $\mu(y) > 0$.
(1) Suppose that $\mu(x + y) = 0$, then

$$\mu(x + y) + \min(\mu(x), \mu(y)) = \min(\mu(x), \mu(y)) \leq 1.$$

Hence $(x + y)_{\min(\mu(x),\mu(y))}\overline{q}\mu$, which is a contradiction since $x_{\mu(x)} \in \mu$ and $y_{\mu(y)} \in \mu$. Thus $\mu(x + y) > 0$, so $x + y \in L_0$.
(2) Suppose that $\mu(mx) = 0$, then

$$\mu(mx) + \mu(x) = \mu(x) \leq 1.$$

Hence $mx_{\mu(x)}\overline{q}\mu$, a contradiction since $x_{\mu(x)} \in \mu$. Thus $\mu(mx) > 0$, so $mx \in L_0$.
(3) Suppose that $\mu([x, y]) = 0$, then

$$\mu([x, y]) + \min(\mu(x), \mu(y)) = \min(\mu(x), \mu(y)) \leq 1.$$

Hence $[x, y]_{\min(\mu(x),\mu(y))}\overline{q}\mu$, which is a contradiction since $x_{\mu(x)} \in \mu$ and $y_{\mu(y)} \in \mu$. Thus $\mu([x, y]) > 0$, so $[x, y] \in L_0$. Hence, L_0 is a fuzzy Lie subalgebra of L.

Lemma 4.3 *If μ is a nonzero $(q, \in)$-fuzzy Lie subalgebra of L, then the set L_0 is a fuzzy Lie subalgebra of L.*

Proof Let $x, y \in L_0$. Then $\mu(x) > 0$ and $\mu(y) > 0$. Thus, $\mu(x) + 1 > 1$ and $\mu(y) + 1 > 1$, which imply that $x_1 q\mu$ and $y_1 q\mu$.
(1) If $\mu(x + y) = 0$, then $\mu(x + y) < 1 = \min(1, 1)$. Therefore, $(x + y)_{\min(1,1)}\overline{\in}\mu$, which is a contradiction. It follows that $\mu(x + y) > 0$ so that $x + y \in L_0$.
(2) If $\mu(mx) = 0$, then $\mu(mx) < 1 = 1$. Therefore, $mx\overline{\in}\mu$, a contradiction. It follows that $\mu(mx) > 0$ so that $mx \in L_0$.
(3) If $\mu([x, y]) = 0$, then $\mu([x, y]) < 1 = \min(1, 1)$. Therefore, $[x, y]_{\min(1,1)}\overline{\in}\mu$, which is a contradiction. It follows that $\mu([x, y]) > 0$ so that $[x, y] \in L_0$.
This ends the proof.

Lemma 4.4 *If μ is a nonzero (q, q)-fuzzy Lie subalgebra of L, then the set L_0 is a fuzzy Lie subalgebra of L.*

Proof Let $x, y \in L_0$. Then $\mu(x) > 0$ and $\mu(y) > 0$. Thus $\mu(x) + 1 > 1$ and $\mu(y) + 1 > 1$. This implies that $xq\mu$ and $yq\mu$.
(1) If $\mu(x + y) = 0$, then $\mu(x + y) + \min(1, 1) = 0 + 1 = 1$, and so $(x + y)_{\min(1,1)}\overline{q}\mu$. This is impossible, and hence $\mu(x + y) > 0$, i.e., $x + y \in L_0$.
(2) If $\mu(mx) = 0$, then $\mu(mx) + 1 = 0 + 1 = 1$, and so $(mx)\overline{q}\mu$. This is impossible, and hence $\mu(mx) > 0$, i.e., $mx \in L_0$.
(3) If $\mu([x, y]) = 0$, then $\mu([x, y]) + \min(1, 1) = 0 + 1 = 1$, and so $[x, y]_{\min(1,1)}\overline{q}\mu$. This is impossible, and hence $\mu([x, y]) > 0$, i.e., $[x, y] \in L_0$.
This completes the proof.

By using similar method as given in the above lemmas, we can also prove the following lemma.

Lemma 4.5 *If μ is a nonzero $(\in, \in \vee q)$-, $(\in, \in \wedge q)$-, $(\in \vee q, q)$-, $(\in \vee q, \in)$-, $(\in \vee q, \in \wedge q)$-, $(q, \in \wedge q)$-, $(q, \in \vee q)$-, or $(\in \vee q, \in \vee q)$-fuzzy Lie subalgebra of L, then the set L_0 is a fuzzy Lie subalgebra of L.*

In summarizing the above lemmas, we obtain the following theorem.

Theorem 4.1 *If μ is a nonzero (α, β)-fuzzy Lie subalgebra of L, then the set L_0 is a fuzzy Lie subalgebra of L.*

Theorem 4.2 *Let $L_0 \subset L_1 \subset \cdots \subset L_n = L$ be a strictly increasing chain of an $(\in, \in)$-fuzzy Lie subalgebras of a Lie algebra L, then there exists $(\in, \in)$-fuzzy Lie subalgebra μ of L whose level subalgebras are precisely the members of the chain with $\mu_{0.5} = L_0$.*

Proof Let $\{t_i : t_i \in (0, 0.5], i = 1, 2, \ldots, n\}$ be such that $t_1 > t_2 > t_3 > \cdots > t_n$. Let $\mu : L \to [0, 1]$ defined by

$$\mu(x) = \begin{cases} t, & \text{if x} = 0, \\ n, & \text{if x} = 0, x \in L_0 \\ t_1, & \text{if x} \in L_1 \setminus L_0, \\ t_2, & \text{if x} \in L_2 \setminus L_1, \\ \vdots \\ t_n, & \text{if x} \in L_n \setminus L_{n-1}. \end{cases}$$

Let $x, y \in L$. If $x + y \in L_0$, then

$$\mu(x + y) \geq 0.5 \geq \min(\mu(x), \mu(y), 0.5).$$

On the other hand, if $x + y \notin L_0$, then there exists i, $1 \leq i \leq n$ such that $x + y \in L_i \setminus L_{i-1}$ so that $\mu(x + y) = t_i$. Now there exists $j (\geq i)$ such that $x \in L_j$ or $y \in L_j$. If $x, y \in L_k (k < i)$, then L_k is a Lie subalgebra of L, $x + y \in L_k$ which contradicts $x + y \notin L_{i-1}$. Thus

$$\mu(x + y) \geq t_i \geq t_j \geq \min(\mu(x), \mu(y), 0.5).$$

The verification is analogous (2–3), and we omit the details. Hence, μ is $(\in, \in)$-fuzzy Lie subalgebra of L. It follows from the contradiction of μ that $\mu_{0.5} = L_0$, $\mu_{t_i} = L_i$ for $i = 1, 2, \ldots, n$. This completes the proof.

4.3 Implication-Based Fuzzy Lie Subalgebras

Fuzzy logic is an extension of set-theoretical multivalued logic in which the truth values are linguistic variables or terms of the linguistic variable truth. Some operators, for example, $\vee$; $\wedge$; $\neg$; $\to$ in fuzzy logic are also defined by using truth tables, and the extension principle can be applied to derive definitions of the operators. In fuzzy logic, the truth value of fuzzy proposition p is denoted by $[p]$. For a universe of discourse U, we write here the fuzzy logical and corresponding set-theoretical notations:

1. $[x \in \mu] = \mu(x)$,
2. $[p \wedge q] = \min([p], [q])$,
3. $[p \to q] = \min(1, 1 - [p] + [q])$,
4. $[\forall x p(x)] = \inf_{x \in U}[p(x)]$,
5. $\models p$ if and only if $[p] = 1$ for all valuations.

The truth valuation rules given in (4) are those in the Lukasiewicz system of continuous-valued logic. Of course, various implication operators have been defined. We show only a selection of them in the following:

A. Gaines–Rescher implication operator ($\mathscr{I}_{GR}$):

$$\mathscr{I}_{GR}(x, y) := \begin{cases} 1 \text{ if } x \leq y, \\ 0 \text{ otherwise} . \end{cases}$$

B. Gödel implication operator ($\mathscr{I}_G$):

$$\mathscr{I}_G(x, y) := \begin{cases} 1 \text{ if } x \leq y, \\ y \text{ otherwise} . \end{cases}$$

C. The contraposition of Gödel implication operator ($\overline{\mathscr{I}}_G$):

$$\overline{\mathscr{I}}_G(x, y) := \begin{cases} 1 & \text{if } x \leq y, \\ 1 - x & \text{otherwise} . \end{cases}$$

Ying [135, 136] introduced the concept of fuzzifying topology. We can expand this concept to Lie algebras, and we define a fuzzifying Lie subalgebra as follows:

Definition 4.4 A fuzzy set μ in L is called a *fuzzfying Lie subalgebra* of L if it satisfies the following conditions:

(a) for any $x, y \in L$,

$$\models \min([x \in \mu], [y \in \mu]) \to [x + y \in u],$$

(b) for any $x \in L$ and $m \in \mathbb{F}$,

$$\models [x \in \mu] \to [mx \in u],$$

(c) for any $x, y \in L$,

$$\models \min([x \in \mu], [y \in \mu]) \to [[x, y] \in u].$$

Definition 4.5 Let μ be a fuzzy set of L and $t \in (0, 1]$. Then, μ is called a *t-implication-based Lie subalgebra* of L if it satisfies the following conditions:

(d) For any $x, y \in L \models_t \min([x \in \mu], [y \in \mu]) \to [x + y \in \mu]$,
(e) For any $x \in L, m \in \mathbb{F} \models_t [x \in \mu] \to [mx \in \mu]$,
(f) For any $x, y \in L \models_t \min([x \in \mu], [y \in \mu]) \to [[x, y] \in \mu]$.

Proposition 4.3 *Let $\mathscr{I}$ be an implication operator. A fuzzy set μ of L is a t-implication-based fuzzy Lie subalgebra of L if and only if it satisfies the following:*

(g) $\mathcal{I}(\min(\mu(x), \mu(y)), \mu(x+y)) \geq t$,
(h) $\mathcal{I}(\mu(x), \mu(mx)) \geq t$,
(i) $\mathcal{I}(\min(\mu(x), \mu(y)), \mu([x, y])) \geq t$

for all $x, y \in L$, $m \in \mathbb{F}$.

Proof Straightforward.

Definition 4.6 Let $\lambda_1, \lambda_2 \in [0, 1]$ and $\lambda_1 < \lambda_2$. If μ is a fuzzy set of a Lie algebra L, then μ is called a *fuzzy Lie subalgebra with thresholds* (λ_1, λ_2) if

(j) $\max(\mu(x+y), \lambda_1) \geq \min(\mu(x), \mu(y), \lambda_2)$,
(k) $\max(\mu(mx), \lambda_1) \geq \min(\mu(x), \lambda_2)$,
(l) $\max(\mu([x, y]), \lambda_1) \geq \min(\mu(x), \mu(y), \lambda_2)$

for all $x, y \in L$, $m \in \mathbb{F}$.

We now give characterization theorems.

Theorem 4.3 *Let* μ *be a fuzzy set in* L. *If* $\mathcal{I} = \mathcal{I}_G$, *then* μ *is a 0.5-implication-based fuzzy Lie subalgebra of* L *if and only if* μ *is a fuzzy Lie subalgebra with thresholds* $\lambda_1 = 0$ *and* $\lambda_2 = 0.5$ *of* L.

Proof Suppose that μ is a 0.5-implication-based Lie subalgebra of L. Then (i) $I_G(\min(\mu(x), \mu(y)), \mu(x+y)) \geq 0.5$, and hence $\mu(x+y) \geq \min(\mu(x), \mu(y))$ or $\min(\mu(x), \mu(y)) \geq \mu(x+y) \geq 0.5$. It follows that

$$\mu(x+y) \geq \min(\mu(x), \mu(y), 0.5).$$

(ii) $I_G(\min(\mu(x), \mu(mx)) \geq 0.5$, and hence $\mu(mx) \geq \mu(x)$ or $\mu(x) \geq \mu(mx) \geq 0.5$. It follows that

$$\mu(mx) \geq \min(\mu(x), 0.5).$$

(iii) $I_G(\min(\mu(x), \mu(y)), \mu([x, y])) \geq 0.5$, and hence $\mu([x, y]) \geq \min(\mu(x), \mu(y))$ or $\min(\mu(x), \mu(y)) \geq \mu([x, y]) \geq 0.5$. It follows that

$$\mu([x, y]) \geq \min(\mu(x), \mu(y), 0.5).$$

so that μ is a fuzzy Lie subalgebra with thresholds $\lambda_1 = 0$ and $\lambda_2 = 0.5$ of L.

Conversely, if μ is a fuzzy Lie subalgebra with thresholds $\lambda_1 = 0$ and $\lambda_2 = 0.5$ of L, then
(i) $\mu(x+y) = \max(\mu(x+y), 0) \geq \min(\mu(x), \mu(y), 0.5)$. If $\min(\mu(x), \mu(y), 0.5) = \min(\mu(x), \mu(y))$, then

$$I_G(\min(\mu(x), \mu(y)), \mu(x+y)) = 1 \geq 0.5.$$

Otherwise, $I_G(\min(\mu(x), \mu(y)), \mu(x+y)) \geq 0.5$.
(ii) $\mu(mx) = \max(\mu(mx), 0) \geq \min(\mu(x), 0.5)$. If $\min(\mu(x), 0.5) = \mu(x)$, then

$$I_G(\mu(x), \mu(mx)) = 1 \geq 0.5.$$

Otherwise, $I_G(\mu(x), \mu(mx)) \geq 0.5$.
(iii) $\mu([x, y]) = \max(\mu([x, y]), 0) \geq \min(\mu(x), \mu(y), 0.5)$. If $\min(\mu(x), \mu(y), 0.5) = \min(\mu(x), \mu(y))$, then

$$I_G(\min(\mu(x), \mu(y)), \mu([x, y])) = 1 \geq 0.5.$$

Otherwise, $I_G(\min(\mu(x), \mu(y)), \mu([x, y])) \geq 0.5$. Hence, μ is a 0.5-implication-based Lie subalgebra of L.

Theorem 4.4 *Let μ be a fuzzy set in L. If $\mathscr{I} = \overline{\mathscr{I}}_G$, then μ is a 0.5-implication-based fuzzy Lie subalgebra of L if and only if μ is a fuzzy Lie subalgebra with thresholds $\lambda_1 = 0.5$ and $\lambda_2 = 1$ of L.*

Proof Suppose that μ is a 0.5-implication-based Lie subalgebra of L. Then
(i) $\overline{\mathscr{I}}_G(\min(\mu(x), \mu(y)), \mu(x + y)) \geq 0.5$ which implies that $\mu(x + y) \geq \min(\mu(x), \mu(y))$ or $1 - \min(\mu(x), \mu(y)) \geq 0.5$, i.e, $\min(\mu(x), \mu(y)) \leq 0.5$.
Thus

$$\max(\mu(x + y), 0.5) \geq \min(\mu(x), \mu(y), 1).$$

(ii) $\overline{\mathscr{I}}_G(\mu(mx) \geq 0.5$ which implies that $\mu(mx) \geq \mu(x)$ or $1 - \mu(x) \geq 0.5$, i.e, $\mu(x) \leq 0.5$.
Thus

$$\max(\mu(mx), 0.5) \geq \min(\mu(x), 1).$$

(iii) $\overline{\mathscr{I}}_G(\min(\mu(x), \mu(y)), \mu([x, y])) \geq 0.5$ which implies that $\mu([x, y]) \geq \min(\mu(x), \mu(y))$ or $1 - \min(\mu(x), \mu(y)) \geq 0.5$, i.e, $\min(\mu(x), \mu(y)) \leq 0.5$.
Thus

$$\max(\mu([x, y]), 0.5) \geq \min(\mu(x), \mu(y), 1).$$

Hence, μ is a fuzzy Lie subalgebra with thresholds $\lambda_1 = 0.5$ and $\lambda_2 = 1$ of L

The proof of converse part is obvious.

Theorem 4.5 *Let μ be a fuzzy set in L. If $\mathscr{I} = \mathscr{I}_{GR}$, then μ is a 0.5-implication-based fuzzy Lie subalgebra of L if and only if μ is a fuzzy Lie subalgebra with thresholds $\lambda_1 = 0$ and $\lambda_2 = 1$ of L.*

Proof Obvious.

As a consequence of the above theorems, we obtain the following corollary.

Corollary 4.1 *(1) Let $\mathscr{I} = \mathscr{I}_{GR}$. Then, μ is an implication-based fuzzy Lie subalgebra of L if and only if μ is a Yehia's fuzzy Lie subalgebra of L.*
(2) Let $\mathscr{I} = \mathscr{I}_G$. Then μ is an implication-based fuzzy Lie subalgebra of L if and only if μ is an $(\in, \in \vee q)$- fuzzy Lie subalgebra of L.

4.4 $(\alpha, \beta)^*$-Fuzzy Lie Subalgebras

Definition 4.7 A fuzzy set ν on X of the form

$$\nu(y) = \begin{cases} t \in [0, 1) & \text{if } y = x, \\ 1, & \text{if } y \neq x \end{cases}$$

is called an *anti-fuzzy point* with support x and value t and is denoted by x_t. A fuzzy set ν in X is said to be *nonunit* if there exists $x \in X$ such that $\nu(x) < 1$. An anti-fuzzy point x_t is said to *"besides to"* a fuzzy set ν, written as $x_t \prec \nu$ if $\nu(x) \leqslant t$. An anti-fuzzy point x_t is said to be *"non-quasicoincident with"* a fuzzy set ν, denoted by $x_t \vdash \nu$ if $\nu(x) + t \leqslant 1$.

Definition 4.8 Let V be a vector space. A fuzzy set ν in V is called an $(\alpha, \beta)^*$*-fuzzy subspace* of V if it satisfies the following conditions:

(1) $x_s\alpha\nu,\ y_t\alpha\nu \Rightarrow (x + y)_{\max(s,t)}\beta\nu$,
(2) $x_s\alpha\nu \Rightarrow (mx)_s\beta\nu$,

for all $x, y \in V$, $m \in \mathbb{F}$, $s, t \in [0, 1)$.

Let α and β denote one of the symbols $\prec$, $\vdash$, $\prec\vee\vdash$ or $\prec\wedge\vdash$ unless otherwise specified.

Notations: The following notations will be used:

- "$x_t \prec \nu$" and "$x_t \vdash \nu$" will be denoted by $x_t \prec\wedge\vdash \nu$.
- "$x_t \prec \nu$" or "$x_t \vdash \nu$" will be denoted by $x_t \prec\vee\vdash \nu$.
- The symbol $\overline{\prec\wedge\vdash}$ means neither $\prec$ nor $\vdash$ hold.

Definition 4.9 Let L be a Lie algebra. A fuzzy set ν in L is called an $(\alpha, \beta)^*$*-fuzzy Lie subalgebra* of L if it satisfies the following conditions:

(1) $x_s\alpha\nu,\ y_t\alpha\nu \Rightarrow (x + y)_{\max(s,t)}\beta\nu$,
(2) $x_s\alpha\nu \Rightarrow (mx)_s\beta\nu$,
(3) $x_s\alpha\nu,\ y_t\alpha\nu \Rightarrow ([x, y])_{\min(s,t)}\beta\nu$

for all $x, y \in L$, $m \in \mathbb{F}$, $s, t \in [0, 1)$.

Remark 4.1 If ν is a fuzzy set in L such that $\nu(x) \geqslant 0.5$ for all $x \in L$, then $\{x_t \mid x_t \prec\wedge\vdash \mu\} = \emptyset$.

The proof of the following proposition is trivial.

Proposition 4.4 *For any fuzzy set ν in L, Definition 4.9 is equivalent to the following conditions:*

(4) $x_s, y_t \prec \nu \Rightarrow (x + y)_{\max(s,t)} \prec \nu$,
(5) $x_s \prec \nu \Rightarrow (mx)_s \prec \nu$,
(6) $x_s, y_t \prec \nu \Rightarrow ([x, y])_{\min(s,t)} \prec \nu$

for all $x, y \in L$, $m \in \mathbb{F}$, $s, t \in [0, 1)$. □

For a fuzzy set ν in a Lie algebra L, we denote $L^* = \{x \in L : \nu(x) < 1\}$.

Proposition 4.5 *If ν is a nonunit $(\prec, \prec)^*$-fuzzy Lie subalgebra of L, then L^* is a Lie subalgebra of L.*

Proof Let $x, y \in L^*$. Then $\nu(x) < 1$ and $\nu(y) < 1$.

(1) Assume $\nu(x + y) = 1$. Then we can see that $x_{\nu(x)} \prec \nu$ and $y_{\nu(y)} \prec \nu$, but $(x + y)_{\max(\nu(x),\nu(y))} \overline{\prec} \nu$ since $\nu(x + y) = 1 > \max(\nu(x), \nu(y))$. This is clearly a contradiction, and hence $\nu(x + y) < 1$, which shows that $x + y \in L^*$.

(2) Assume $\nu(mx) = 1$. Then we can see that $x_{\nu(x)} \prec \nu$, but $(mx)_{\nu(x)} \overline{\prec} \nu$ since $\nu(mx) = 1 > \nu(x)$. This is clearly a contradiction, and hence $\nu(mx) < 1$, which shows that $mx \in L^*$.

(3) Assume $\nu([x, y]) = 1$. Then we can see that $x_{\nu(x)} \prec \nu$ and $y_{\nu(y)} \prec \nu$, but $([x, y])_{\min(\nu(x),\nu(y))} \overline{\prec} \nu$ since $\nu([x, y]) = 1 > \min(\nu(x), \nu(y))$. This is clearly a contradiction, and hence $\nu([x, y]) < 1$, which shows that $[x, y] \in L^*$. Hence, L^* is a Lie subalgebra of L.

Proposition 4.6 *If ν is a nonunit $(\prec, \vdash)^*$-fuzzy Lie subalgebra of L, then the set L^* is a Lie subalgebra of L.*

Proof Let $x, y \in L^*$. Then $\nu(x) < 1$ and $\nu(y) < 1$.

(1) Suppose that $\nu(x + y) = 1$, then

$$\nu(x + y) + \max(\nu(x), \nu(y)) \geqslant 1.$$

Hence $(x + y)_{\max(\nu(x),\nu(y))} \overline{\vdash} \nu$, which is a contradiction since $x_{\nu(x)} \prec \nu$ and $y_{\nu(y)} \prec \nu$. Thus $\nu(x + y) < 1$, so $x + y \in L^*$.

(2) Suppose that $\nu(mx) = 1$, then

$$\nu(mx) + \nu(x) \geqslant 1.$$

Hence $mx_{\nu(x)} \overline{\vdash} \nu$, a contradiction since $x_{\nu(x)} \prec \nu$. Thus $\nu(mx) < 1$, so $mx \in L^*$.

(3) Suppose that $\nu([x, y]) = 1$, then

$$\nu([x, y]) + \min(\nu(x), \nu(y)) \geqslant 1.$$

Hence $[x, y]_{\min(\nu(x),\nu(y))} \overline{\vdash} \nu$, which is a contradiction since $x_{\nu(x)} \prec \nu$ and $y_{\nu(y)} \prec \nu$. Thus $\nu([x, y]) < 1$, so $[x, y] \in L^*$. Hence, L^* is a Lie subalgebra of L.

Proposition 4.7 *If ν is a nonunit $(\vdash, \prec)^*$-fuzzy Lie subalgebra of L, then L^* is a Lie subalgebra of L.*

Proof Let $x, y \in L^*$. Then $\nu(x) < 1$ and $\nu(y) < 1$. Thus $x \vdash \nu$ and $y \vdash \nu$.

(1) If $\nu(x + y) = 1$, then $\nu(x + y) = 1 > 0 = \max(0, 0)$. Therefore, $(x + y)_{\max(0,0)} \overline{\prec} \nu$, which is a contradiction. It follows that $\nu(x + y) < 1$ so that $x + y \in L^*$.

(2) If $\nu(mx) = 1$, then $\nu(mx) = 1 > 0$. Therefore, $mx\overline{\prec}\nu$, a contradiction. It follows that $\nu(mx) < 1$ so that $mx \in L^*$.

(3) If $\nu([x, y]) = 1$, then $\nu([x, y]) = 1 > 0 = \min(0, 0)$. Therefore, $[x, y]_{\min(0,0)}\overline{\prec}\nu$, which is a contradiction. It follows that $\nu([x, y]) < 1$ so that $[x, y] \in L^*$. Hence, L^* is a Lie subalgebra of L.

Proposition 4.8 *If ν is a nonunit $(\vdash, \vdash)^*$-fuzzy Lie subalgebra of L, then L^* is a Lie subalgebra of L.*

Proof Let $x, y \in L^*$. Then $\nu(x) < 1$ and $\nu(y) < 1$.

(1) If $\nu(x + y) = 1$, then $\nu(x + y) + \max(0, 0) = 1$, and so $(x + y)_{\max(0,0)}\overline{\vdash}\nu$. This is impossible, and hence $\nu(x + y) < 1$, i.e., $x + y \in L^*$.

(2) If $\nu(mx) = 1$, then $\nu(mx) + 0 = 1$, and so $(mx)\overline{\vdash}\nu$. This is impossible, and hence $\nu(mx) < 1$, i.e., $mx \in L^*$.

(3) If $\nu([x, y]) = 1$, then $\nu([x, y]) + \min(0, 0) = 1$, and so $[x, y]_{\min(0,0)}\overline{\vdash}\nu$. This is impossible, and hence $\nu([x, y]) < 1$, i.e., $[x, y] \in L^*$. Hence L^* is a Lie subalgebra of L.

Proposition 4.9 *If ν is a nonunit $(\prec, \prec\vee\vdash)^*$-fuzzy Lie subalgebra of L, then L^* is a Lie subalgebra of L.*

Proof Let $x,\ y \in L^*$. Then $\nu(x) < 1$ and $\nu(y) < 1$. Thus $\nu(x) = s_1$ and $\nu(y) = s_1$ for some $s_1, s_2 \in [0, 1)$. It follow that $x_{s_1} \prec \nu$ and $y_{s_2} \prec \nu$ so that $(x + y)_{\max(s_1,s_2)} \prec\vee\vdash \nu$, i.e., $(x + y)_{\max(s_1,s_2)} \prec \nu$ or $(x + y)_{\max(s_1,s_2)} \vdash \nu$. If $(x + y)_{\max(s_1,s_2)} \prec \nu$, then $\nu(x + y) \leqslant \max(s_1, s_2) < 1$ and hence $x + y \in L^*$. On the other hand, If $(x + y)_{\max(s_1,s_2)} \vdash \nu$, then $\nu(x + y) \leqslant \nu(x + y) + \max(s_1, s_2) < 1$, and hence $x + y \in L^*$. Verification of conditions (2) and (3) in Definition 4.9 is similar, and we omit the details.

By using similar argumentations, we can also prove the following two propositions.

Proposition 4.10 *If ν is a nonunit $(\vdash, \prec \vee \vdash)^*$-fuzzy Lie subalgebra of L, then L^* is a Lie subalgebra of L.* □

Proposition 4.11 *If ν is a nonunit $(\prec, \prec\wedge\vdash)^*$-, $(\prec\vee\vdash, \vdash)^*$-, $(\prec\vee\vdash, \prec)^*$-, $(\prec\vee\vdash, \prec\wedge\vdash)^*$-, $(\vdash, \prec\wedge\vdash)^*$-, or $(\prec\vee\vdash, \prec \vee \vdash)^*$-fuzzy Lie subalgebra of L, then L^* is a Lie subalgebra of L.* □

Definition 4.10 A fuzzy set ν in L is called an $(\prec, \prec\vee\vdash)^*$-*fuzzy Lie subalgebra* of L if the following conditions are satisfied:

(a) $x_s, y_t \prec \nu \Rightarrow (x + y)_{\max(s,t)} \prec\vee\vdash \nu$,
(b) $x_s \prec \nu \Rightarrow (mx)_s \prec\vee\vdash \nu$,
(c) $x_s, y_t \prec \nu \Rightarrow ([x, y])_{\min(s,t)} \prec\vee\vdash \nu$

for all $x, y \in L$, $m \in \mathbb{F}$, $s, t \in [0, 1)$.

Example 4.2 Let V be a vector space over a field $\mathbb{F}$ such that $dim(V) = 5$. Let $\{e_1, e_2, \ldots, e_5\}$ be a basis of the vector space V over $\mathbb{F}$ with Lie brackets as follows:

$$[e_1, e_2] = e_3, \quad [e_1, e_3] = e_5, \quad [e_1, e_4] = e_5, \quad [e_1, e_5] = 0,$$

$$[e_2, e_3] = e_5, \quad [e_2, e_4] = 0, \quad [e_2, e_5] = 0, \quad [e_3, e_4] = 0,$$

$$[e_3, e_5] = 0, \quad [e_4, e_5] = 0, \quad [e_i, e_j] = -[e_j, e_i]$$

and $[e_i, e_j] = 0$ for all $i = j$. Then, V is a Lie algebra over $\mathbb{F}$. We define a fuzzy set $\nu : V \to [0, 1]$ by

$$\nu(x) := \begin{cases} 0.25 \text{ if } x = 0, \\ 0.46 \text{ if } x \in \{e_3, e_5\}, \\ \quad 0 \quad \text{if } x \in \{e_1, e_2, e_4\}. \end{cases}$$

By routine computations, it is easy to see that ν is an $(\prec, \prec\vee\vdash)^*$-fuzzy Lie subalgebra of L.

Theorem 4.6 *Let ν be a fuzzy set in a Lie algebra L. Then, ν is an $(\prec, \prec\vee\vdash)^*$-fuzzy Lie subalgebra of L if and only if*

(d) $\nu(x + y) \leqslant \max(\nu(x), \nu(y), 0.5)$,
(e) $\nu(mx) \leqslant \max(\nu(x), 0.5)$,
(f) $\nu([x, y]) \leqslant \min(\nu(x), \nu(y), 0.5)$

hold for all $x, y \in L$, $m \in \mathbb{F}$.

Proof $(a) \Rightarrow (d)$: Let $x, y \in L$. We consider the following two cases:

(1) $\max(\nu(x), \nu(y)) > 0.5$,
(2) $\max(\nu(x), \nu(y)) \leqslant 0.5$.

Case (1): Assume that $\nu(x + y) > \max(\nu(x), \nu(y), 0.5)$ Then $\nu(x + y) > \max(\nu(x), \nu(y))$. Take s such that $\nu(x + y) > s > \max(\nu(x), \nu(y))$. Then $x_s \prec \nu$, $y_s \prec \nu$, but $(x + y)_s \overline{\prec\vee\vdash} \nu$, which is contradiction with (a).

Case (2): Assume that $\nu(x + y) > 0.5$. Then $x_{0.5}, y_{0.5} \prec \nu$ but $(x + y)_{0.5} \overline{\prec\vee\vdash} \nu$, a contradiction. Hence (d) holds.

$(d) \Rightarrow (a)$: Let $x_s, y_t \prec \nu$, then $\nu(x) \leqslant s$, $\nu(y) \leqslant t$. Now, we have

$$\nu(x + y) \leqslant \max(\nu(x), \nu(y), 0.5) \leqslant \max(s, t, 0.5).$$

If $\max(s, t) < 0.5$, then $\nu(x + y) \leqslant 0.5 \Rightarrow \nu(x + y) + \max(s, t) < 1$. On the other hand, if $\max(s, t) \geqslant 0.5$, then $\nu(x + y) \leqslant \max(s, t)$. Hence $(x + y)_{\max(s,t)} \prec\vee\vdash \nu$.

The verifications of $(b) \Leftrightarrow (e)$ and $(c) \Leftrightarrow (f)$ are analogous, and we omit the details. This completes the proof.

Theorem 4.7 *Let ν be an $(\prec, \prec\vee\vdash)^*$-fuzzy Lie subalgebra of L.*

(i) If there exists $x \in L$ such that $\nu(x) \leqslant 0.5$, then $\nu(0) \leqslant 0.5$.
(ii) If $\nu(0) > 0.5$, then ν is an $(\prec, \prec)^*$-fuzzy Lie subalgebra of L.

Proof (i) Let $x \in L$ such that $\nu(x) \leqslant 0.5$. Then $\nu(-x) = \max(\nu(x), 0.5) = 0.5$. Hence, $\nu(0) = \nu(x - x) \leqslant \max(\nu(x), \nu(-x), 0.5) = 0.5$.

(ii) If $\nu(0) > 0.5$ then $\nu(x) > 0.5$ for all $x \in L$. Thus, we conclude that $\nu(x + y) \leqslant \max(\nu(x), \nu(y))$, $\nu(mx) \leqslant \nu(x)$, $\nu([x, y]) \leqslant \min(\nu(x), \nu(y))$ for all $x, y \in L$, $m \in \mathbb{F}$. Hence, ν is an $(\prec, \prec)^*$-fuzzy Lie subalgebra of L.

Theorem 4.8 *Let ν be a fuzzy set of fuzzy Lie subalgebra of L. Then, ν is an $(\prec, \prec\vee\vdash)^*$-fuzzy Li subalgebra of L if and only if each nonempty $L(\nu; t)$, $t \in [0.5, 1)$ is a Lie subalgebra of L.*

Proof Assume that ν is an $(\prec, \prec\vee\vdash)^*$ fuzzy Lie subalgebra of L and let $t \in [0.5, 1)$. If $x, y \in L(\nu; t)$ and $m \in \mathbb{F}$, then $\nu(x) \leq t$ and $\nu(y) \leq t$. Thus,

$$\nu(x + y) \leqslant \max(\nu(x), \nu(y), 0.5) \leqslant \max(t, 0.5) = t,$$

$$\nu(mx) \leqslant \max(\nu(x), 0.5) \leqslant \max(t, 0.5) = t,$$

$$\nu([x, y]) \leqslant \max(\nu(x), \nu(y), 0.5) \leqslant \max(t, 0.5) = t,$$

and so $x + y, mx, [x, y] \in L(\nu; t)$. This shows that $L(\nu; t)$ is a Lie subalgebra of L.

Conversely, let ν be a fuzzy set such that $L(\nu; t)$ is a Lie subalgebra of L, for all $t \in [0.5, 1)$. If there exist $x, y \in L$ such that $\nu(x + y) > \max(\nu(x), \nu(y), 0.5)$, then we can take $t \in (0, 1)$ such that

$$\nu(x + y) > t > \max(\nu(x), \nu(y), 0.5).$$

Thus $x, y \in L(\nu; t)$ and $t > 0.5$, and so $x + y \notin L(\nu; t)$, which contradicts to the assumption that all $L(\nu; t)$ are Lie ideals. Therefore,

$$\nu(x + y) \leqslant \max(\nu(x), \nu(y), 0.5).$$

The verification is analogous for other conditions, and we omit the details. Hence, ν is an $(\prec, \prec\vee\vdash)^*$ fuzzy Lie subalgebra of L.

Theorem 4.9 *Let ν be a fuzzy set in a Lie algebra L. Then, $L(\nu; t)$ is a Lie subalgebra of L if and only if*

(g) $\min(\nu(x + y), 0.5) \leqslant \max(\nu(x), \nu(y))$,
(h) $\min(\nu(mx), 0.5) \leqslant \nu(x)$,
(i) $\min(\nu([x, y]), 0.5)) \leqslant \max(\nu(x), \nu(y))$

for all $x, y \in L$, $m \in \mathbb{F}$.

Proof Suppose that $L(\nu; t)$ is a Lie subalgebra of L. Let $\min(\nu(x+y), 0.5) > \max(\nu(x), \nu(y)) = t$ for some $x, y \in L$, then $t \in [0.5, 1)$, $\nu(x+y) > t$, $x \prec L(\nu; t)$ and $y \prec L(\nu; t)$. Since $x, y \prec L(\nu; t)$ and $L(\nu; t)$ is a Lie subalgebra of L, so $x + y \prec L(\nu; t)$ or $\nu(x+y) \leqslant t$, which is contradiction with $\nu(x+y) > t$. Hence (g) holds. For (h), (i) the verification is analogous.

Conversely, suppose that (g), (h) and (i) hold. Assume that $t \in [0.5, 1)$, $x, y \prec L(\nu; t)$. Then

$$0.5 > t \geqslant \max(\nu(x), \nu(y)) \geqslant \min(\nu(x+y), 0.5) \Rightarrow \nu(x+y) \leqslant t,$$

$$0.5 > t \geqslant \nu(x) \geqslant \min(\nu(mx), 0.5) \Rightarrow \nu(mx) \leqslant t,$$

$$0.5 > t \geqslant \max(\nu(x), \nu(y)) \geqslant \min(\nu([x, y], 0.5) \Rightarrow \nu([x, y]) \leqslant t,$$

and so $x + y \prec L(\nu; t)$, $mx \prec L(\nu; t)$,$[x, y] \prec L(\nu; t)$. This shows that $L(\nu; t)$ is a Lie subalgebra of L.

Definition 4.11 An $(\in, \in \vee q)$-fuzzy Lie subalgebra of L is called *proper* if $\mathrm{Im}(\nu)$ has at least two elements. Two $(\in, \in \vee q)$- fuzzy Lie subalgebras ν_1 and ν_2 are said to be *equivalent* if they have the same family of level Lie subalgebras.

Theorem 4.10 *Any proper $(\in, \in\vee q)$-fuzzy Lie subalgebra of L for which the cardinality of $\{\nu(x) : \nu(x) > 0.5\} \leqslant 2$ can be expressed as the union of two proper nonequivalent $(\in, \in \vee q)$-fuzzy Lie subalgebras of L.*

Proof Let ν be a proper $(\in, \in \vee q)$-fuzzy Lie subalgebra of L such that $\{\nu(x) : \nu(x) > 0.5\} = \{t_1, t_2, \ldots, t_n\}$ where $t_1 < t_2 < \cdots < t_n$ and $n \geqslant 2$. Then

$$\nu_{0.5} \subseteq \nu_{t_1} \subseteq \cdots \subseteq \nu_{t_n} = L$$

is the chain of $(\in, \in\vee q)$-fuzzy Lie subalgebras of ν. Define μ_1 and μ_2 by

$$\mu_1(x) = \begin{cases} t_1, & \text{if } x \in \nu_{t_1}, \\ t_2, & \text{if } x \in \nu_{t_2} \setminus \nu_{t_1}, \\ \vdots & \\ t_n, & \text{if } x \in \nu_{t_n} \setminus \nu_{t_{n-1}}, \end{cases}$$

$$\mu_2(x) = \begin{cases} \nu(x), & \text{if } x \in \nu_{0.5}, \\ n, & \text{if } x \in \nu_{t_2} \setminus \nu_{0.5}, \\ t_3, & \text{if } x \in \nu_{t_3} \setminus \nu_{t_2}, \\ \vdots & \\ t_n, & \text{if } x \in \nu_{t_n} \setminus \nu_{t_{n-1}}, \end{cases}$$

respectively, where $t_3 > n > t_2$. Then, μ_1 and μ_2 are $(\in, \in \vee q)$-fuzzy Lie subalgebras of L with

$$\nu_{t_1} \subseteq \nu_{t_2} \subseteq \cdots \subseteq \nu_{t_n}$$

and

$$\nu_{t_{0.5}} \subseteq \nu_{t_2} \subseteq \cdots \subseteq \nu_{t_n}$$

being, respectively, chains of $(\in, \in \vee q)$-fuzzy Lie subalgebras of μ_1 and μ_2.

Hence, ν can be expressed as the union of two proper nonequivalent $(\in, \in \vee q)$-fuzzy Lie subalgebras of L.

Theorem 4.11 *Let $\{\nu_i : i \in I\}$ be a family of $(\prec, \prec)^*$-fuzzy Lie subalgebras of L. Then, $\nu = \bigcup_{i\in I} \nu_i$ is an $(\prec, \prec)^*$-fuzzy Lie subalgebra of L.*

Proof Let $x_s \prec \nu$ and $y_t \prec \nu$, where $s, t \in [0, 1)$. Then $\nu(x) \leqslant s$ and $\nu(y) \leqslant t$. Thus, we have $\nu_i(x) \leqslant s$ and $\nu_i(y) \leqslant t$ for all $i \in I$. Hence $\nu_i(x + y) \leqslant \max(s, t)$. Therefore, $\nu(x + y) \leqslant \max(s, t)$, which implies that $(x + y)_{\max\{s,t\}} \prec \nu$. For other conditions, the verification is analogous.

Theorem 4.12 *Let $\{\nu_i : i \in I\}$ be a family of $(\prec, \prec \vee \vdash)^*$-fuzzy Lie subalgebras of L. Then, $\nu := \bigcap_{i\in I} \nu_i$ is an $(\prec, \prec \vee \vdash)^*$-fuzzy Lie subalgebra of L.*

Proof By Theorem 4.6, we have $\nu(x + y) \leqslant \max(\nu(x), \nu(y), 0.5)$, and hence

$$\begin{aligned}
\nu(x + y) &= \inf_{i\in I} \nu_i(x + y) \\
&\leqslant \inf_{i\in I} \max(\nu_i(x), \nu_i(y), 0.5) \\
&= \max(\inf_{i\in I} \nu_i(x), \inf_{i\in I} \nu_i(y), 0.5) \\
&= \max(\textstyle\bigcap_{i\in I} \nu_i(x), \bigcap_{i\in I} \nu_i(y), 0.5) \\
&= \max(\nu(x), \nu(y), 0.5).
\end{aligned}$$

For other conditions, the verification is analogous. By Theorem 4.6, it follows that ν is an $(\prec, \prec \vee \vdash)^*$-fuzzy Lie subalgebra of L.

Remark. Let $\{\nu_i : i \in I\}$ be a family of $(\prec, \prec \vee \vdash)^*$-fuzzy Lie subalgebras of L. Is $\nu = \bigcup_{i\in I} \nu_i$ an $(\prec, \prec \vee \vdash)^*$-fuzzy Lie subalgebra of L? When? The following example shows that it is not an $(\prec, \prec \vee \vdash)^*$-fuzzy Lie subalgebra in general.

Example 4.3 Let V be a vector space over a field $\mathbb{F}$ such that $dim(V) = 5$. Let $\{e_1, e_2, e_3, e_4, e_5\}$ be its basis.If we define fuzzy sets $\mu_1, \mu_2 : V \to [0, 1]$ by putting

$$\mu_1(x) := \begin{cases} 0.6 \text{ if } x = 0, \\ 1 \text{ if } x \in \{e_3, e_5\}, \\ 0 \text{ if } x \in \{e_1, e_2, e_4\}, \end{cases}$$

$$\mu_2(x) := \begin{cases} 0.3 \text{ if } x = 0, \\ 1 \text{ if } x \in \{e_3, e_5\}, \\ 0 \text{ if } x \in \{e_1, e_2, e_4\}, \end{cases}$$

then both μ_1 and μ_2 will be $(\prec, \prec \vee \vdash)^*$-fuzzy Lie subalgebras of L, but $\mu_1 \cup \mu_2$ is not an $(\prec, \prec \vee \vdash)^*$-fuzzy Lie subalgebra of L since

$$\begin{aligned} 1 = \max(\mu_1(e_3), \mu_2(e_3)) &= (\mu_1 \cup \mu_2)(e_3) = (\mu_1 \cup \mu_2)([e_1, e_2]) \\ &\leqslant \min((\mu_1 \cup \mu_2)(e_1), (\mu_1 \cup \mu_2)(e_2), 0.5) = \min(0, 0, 0.5) = 0. \end{aligned}$$

Theorem 4.13 *Let $\{\nu_i : i \in I\}$ be a family of $(\prec, \prec \vee \vdash)^*$-fuzzy Lie subalgebras of L such that $\nu_i \subseteq \nu_j$ or $\nu_j \subseteq \nu_i$ for all $i, j \in I$. Then, the fuzzy set $\nu := \bigcup_{i \in I} \nu_i$ is an $(\prec, \prec \vee \vdash)^*$-fuzzy Lie subalgebra of L.*

Proof By Theorem 4.6, we have $\nu(x + y) \leqslant \max(\nu(x), \nu(y), 0.5)$, and hence

$$\begin{aligned} \nu(x + y) &= \sup_{i \in I} \nu_i(x + y) \\ &\leqslant \sup_{i \in I} \max(\nu_i(x), \nu_i(y), 0.5) \\ &= \max(\sup_{i \in I} \nu_i(x), \sup_{i \in I} \nu_i(y), 0.5) \\ &= \max(\textstyle\bigcup_{i \in I} \nu_i(x), \bigcup_{i \in I} \nu_i(y), 0.5) \\ &= \max(\nu(x), \nu(y), 0.5). \end{aligned}$$

It is easy to see that

$$\sup_{i \in I} \max(\nu_i(x), \nu_i(y), 0.5) \geqslant \bigcup_{i \in I} \max(\nu_i(x), \nu_i(y), 0.5).$$

Suppose that

$$\sup_{i \in I} \max(\nu_i(x), \nu_i(y), 0.5) \neq \bigcup_{i \in I} \max(\nu_i(x), \nu_i(y), 0.5),$$

then there exists s such that

$$\sup_{i \in I} \max(\nu_i(x), \nu_i(y), 0.5) > s > \bigcup_{i \in I} \max(\nu_i(x), \nu_i(y), 0.5).$$

Since $\nu_i \subseteq \nu_j$ or $\nu_j \subseteq \nu_i$ for all $i, j \in I$, there exists $k \in I$ such that $s > \max(\nu_k(x), \nu_k(y), 0.5)$. On the other hand, $\max(\nu_i(x), \nu_i(y), 0.5) > s$ for all $i \in I$, a contradiction. Hence

$$\begin{aligned} \sup_{i \in I} \max\{\nu_i(x), \nu_i(y), 0.5\} &= \max(\textstyle\bigcup_{i \in I} \nu_i(x), \bigcup_{i \in I} \nu_i(y), 0.5) \\ &= \max\{\nu(x), \nu(y), 0.5\}. \end{aligned}$$

The verification of other conditions is analogous. By Theorem 4.6, it follows that ν is an $(\prec, \prec\vee\vdash)^*$-fuzzy Lie subalgebra of L.

Finally, we study anti-fuzzy Lie subalgebras with thresholds.

Definition 4.12 Let $m_1, m_2 \in [0, 1]$ and $m_1 < m_2$. If ν is a fuzzy set of a Lie algebra L, then ν is called an *anti-fuzzy Lie subalgebra with thresholds* (m_1, m_2) if

(1) $\min(\nu(x + y), m_1) \leqslant \max(\nu(x), \nu(y), m_2)$,
(2) $\min(\nu(mx), m_1) \leqslant \max(\nu(x), m_2)$,
(3) $\min(\nu([x, y]), m_1) \leqslant \max(\nu(x), \nu(y), m_2)$

for all $x, y \in L, m \in \mathbb{F}$.

Theorem 4.14 *A fuzzy set ν of Lie algebra L is an anti-fuzzy Lie subalgebra with thresholds (m_1, m_2) of L if and only if $L(\nu; t)(\neq \emptyset)$, for $t \in (m_1, m_2]$, is a Lie subalgebra of L.*

Proof Assume that ν is an anti-fuzzy Lie subalgebra with thresholds (m_1, m_2) of L. Let $x, y \in L(\nu; t)$. Then $\nu(x) \leqslant t$ and $\nu(y) \leqslant t$, $t \in (m_1, m_2]$. Then, it follows that

$$\min(\nu(x + y), m_1) \leqslant \max(\nu(x), \nu(y), m_2) - t \longrightarrow \nu(x + y) \leqslant t,$$

$$\min(\nu(mx), m_1) \leqslant \max(\nu(x), m_2) = t \Longrightarrow \nu(mx) \leqslant t,$$

$$\min(\nu([x, y]), m_1) \leqslant \min(\nu(x), \nu(y), m_2) = t \Longrightarrow \nu([x, y]) \leqslant t,$$

and hence $x + y, mx, [x, y] \in L(\nu; t)$. This shows that $L(\nu; t)$ is a Lie subalgebra of L.

Conversely, assume that ν is a fuzzy set such that $L(\nu; t) \neq \emptyset$ is a Lie subalgebra of L for $m_1, m_2 \in [0, 1]$ and $m_1 < m_2$. Suppose that $\min(\nu(x + y), m_1) > \max(\nu(x), \nu(y), m_2) = t$, then $\nu(x + y) > t, x \in L(\nu; t), y \in L(\nu; t), t \in (m_1, m_2]$. Since $x, y \in L(\nu; t)$ and $L(\nu; t)$ are Lie subalgebras, $x + y \in L(\nu; t)$, i.e., $\nu(x + y) \leqslant t$. This is a contradiction. Therefore condition (1) holds. The verification of (2) and (3) is analogous.

Remark. By Definition 4.12, we have the following result: If ν is an anti-fuzzy subalgebra with thresholds m_1, m_2, then we can conclude that: ν is an anti-fuzzy subalgebra when $m_1 = 0$ and $m_2 = 1$; ν is an $(\prec, \prec\vee\vdash)^*$-fuzzy Lie subalgebra when $m_1 = 0.5$ and $m_2 = 1$.

By Definition 4.12, one can define other anti-fuzzy subalgebra of L, such as $[0.2, 0.6)$-fuzzy subalgebra of L, $[0.3, 0.8)$-fuzzy subalgebra of L.

4.5 Interval-Valued $(\in, \in \vee q)$-Fuzzy Lie Ideals

We now assume that any two interval numbers of $D[0, 1]$ are comparable. Based on Bhakat and Das [35], we can extend the concept of quasicoincidence of fuzzy point within a fuzzy set to the concept of quasicoincidence of a fuzzy interval value with an interval-valued fuzzy set as follows:

Definition 4.13 An *interval-valued fuzzy set* $\widetilde{\mu}$ of a Lie algebra L of the form

$$\widetilde{\mu}(y) = \begin{cases} \widetilde{t} \in (\widetilde{0}, \widetilde{1}], & \text{if } y = x \\ \widetilde{0}, & y \neq x \end{cases}$$

is called *fuzzy interval value with support* x and interval value $\widetilde{t}$ and is denoted by $\gamma(x; \widetilde{t})$. A fuzzy interval value $\gamma(x; \widetilde{t})$ is said to be *belong to an interval-valued fuzzy set* $\widetilde{\mu}$ written as $\gamma(x; \widetilde{t}) \in \widetilde{\mu}$ if $\widetilde{\mu}(x) \geq \widetilde{t}$. A fuzzy interval value $\gamma(x; \widetilde{t})$ is said to be *quasi-coincident with an interval-valued fuzzy set* $\widetilde{\mu}$ written as $\gamma(x; \widetilde{t}) q \widetilde{\mu}$ if $\widetilde{\mu}(x) + \widetilde{t} > \widetilde{1}$.

Definition 4.14 Let V be a vector space. An interval-valued fuzzy set $\widetilde{\mu}$ in V is called an *interval-valued* $(\in, \in \vee q)$*-fuzzy subspace* of V if it satisfies the following conditions:

(1) $\gamma(x; \widetilde{s}), \gamma(y; \widetilde{t}) \in \widetilde{\mu} \Rightarrow \gamma(x + y; \min(\widetilde{s}, \widetilde{t})) \in \vee q \widetilde{\mu}$,
(2) $\gamma(x; \widetilde{s}) \in \widetilde{\mu} \Rightarrow \gamma(\alpha x; \widetilde{s}) \in \vee q \widetilde{\mu}$

for all $x, y \in V, \alpha \in \mathbb{F}, \widetilde{s}, \widetilde{t} \in D(0, 1]$.

Definition 4.15 An interval-valued fuzzy set $\widetilde{\mu}$ in the Lie algebra L is called an *interval-valued* $(\in, \in \vee q)$*-fuzzy Lie ideal* of L if it satisfies the following conditions:

(1) $\gamma(x; \widetilde{s}), \gamma(y; \widetilde{t}) \in \widetilde{\mu} \Rightarrow \gamma(x + y; \min(\widetilde{s}, \widetilde{t})) \in \vee q \widetilde{\mu}$,
(2) $\gamma(x; \widetilde{s}) \in \widetilde{\mu} \Rightarrow \gamma(\alpha x; \widetilde{s}) \in \vee q \widetilde{\mu}$,
(3) $\gamma(x; \widetilde{s}), \gamma(y; \widetilde{t}) \in \widetilde{\mu} \Rightarrow \gamma([x, y]; \widetilde{s}) \in \vee q \widetilde{\mu}$

for all $x, y \in L, \alpha \in \mathbb{F}, \widetilde{s}, \widetilde{t} \in D(0, 1]$.

The following lemma is a direct consequence of Definition 4.14.

Lemma 4.6 *Every interval-valued* $(\in, \in)$*-fuzzy Lie ideal is an* $(\in, \in \vee q)$*-fuzzy Lie ideal.*

We remark here that the converse of Lemma 4.6 does not hold in general. This can be illustrated in the following example.

Example 4.4 Let V be a vector space over a field $\mathbb{F}$ such that $dim(V) = 5$. Let $\{e_1, e_2, \ldots, e_5\}$ be a basis of a vector space V over field $\mathbb{F}$ with the Lie brackets given below :

$$[e_1, e_2] = e_3, \quad [e_1, e_3] = e_5, \quad [e_1, e_4] = e_5, \quad [e_1, e_5] = 0,$$

$$[e_2, e_3] = e_5, \quad [e_2, e_4] = 0, \quad [e_2, e_5] = 0, \quad [e_3, e_4] = 0,$$

$$[e_3, e_5] = 0, \quad [e_4, e_5] = 0, \quad [e_i, e_j] = -[e_j, e_i]$$

and $[e_i, e_j] = 0$ for all $i = j$.

Then, V under the Lie product clearly forms a Lie algebra over $\mathbb{F}$.

We now define an interval-valued set $\widetilde{\mu} : V \to D[0, 1]$ by

$$\widetilde{\mu}(x) := \begin{cases} [0.5, 0.6] \text{ if } x = 0, \\ [0.6, 0.7] \text{ if } x \in \{e_3, e_5\}, \\ [1.0, 1.0] \text{ if } x \in \{e_1, e_2, e_4\}. \end{cases}$$

Then, by routine computation, we can see immediately that $\widetilde{\mu}$ is an interval-valued $(\in, \in \vee q)$-fuzzy Lie ideal of L, but $\widetilde{\mu}$ is not an interval-valued $(\in, \in)$-fuzzy Lie ideal of L.

Lemma 4.7 *The conditions (1)–(3) in Definition 4.14 are equivalent to the following inequalities:*

(l) $\widetilde{\mu}(x + y) \geq \min(\widetilde{\mu}(x), \widetilde{\mu}(y), [0.5, 0.5])$,
(m) $\widetilde{\mu}(\alpha x) \geq \min(\widetilde{\mu}(x), [0.5, 0.5])$,
(n) $\widetilde{\mu}([x, y]) \geq \min(\widetilde{\mu}(x), [0.5, 0.5])$

for all $x, y \in L$, $\alpha \in \mathbb{F}$.

Proof (1) $\Rightarrow$ (l) : Let $x, y \in L$. We only need to consider the following two cases:

(i) $\min(\widetilde{\mu}(x), \widetilde{\mu}(y)) < [0.5, 0.5]$.
(ii) $\min(\widetilde{\mu}(x(, \widetilde{\mu}(y)) \geq [0.5, 0.5]$.

Case (i): Assume that $\widetilde{\mu}(x + y) < \min(\widetilde{\mu}(x), \widetilde{\mu}(y), [0.5, 0.5])$. Then, $\widetilde{\mu}(x + y) < \min(\widetilde{\mu}(x), \widetilde{\mu}(y))$. Take $\widetilde{s}$ such that $\widetilde{\mu}(x + y) < \widetilde{s} < \min(\widetilde{\mu}(x), \widetilde{\mu}(y))$. Then $\gamma(x; \widetilde{s}) \in \widetilde{\mu}$, $\gamma(y; \widetilde{s}) \in \widetilde{\mu}$, but $\gamma(x + y; \widetilde{s})\overline{\in \vee q}\widetilde{\mu}$. This contradicts condition (1) in Definition 4.15.
Case (ii): Assume that $\widetilde{\mu}(x + y) < [0.5, 0.5]$. Then $\gamma(x; [0.5, 0.5])$, $\gamma(y; [0.5, 0.5]) \in \widetilde{\mu}$ but $\gamma(x + y; [0.5, 0.5])\overline{\in \vee q}\widetilde{\mu}$, which is a contradiction. Hence condition (l) holds.
$(l) \Rightarrow (1)$: Let $\gamma(x; \widetilde{s}), \gamma(y; \widetilde{t}) \in \widetilde{\mu}$. Then, $\widetilde{\mu}(x) \geq \widetilde{s}$ and $\widetilde{\mu}(y) \geq \widetilde{t}$. Thus, it follows that

$$\widetilde{\mu}(x + y) \geq \min(\widetilde{\mu}(x), \widetilde{\mu}(y), [0.5, 0.5]) \geq \min(\widetilde{s}, \widetilde{t}, [0.5, 0.5]).$$

If $\min(\widetilde{s}, \widetilde{t}) > [0.5, 0.5]$, then $\widetilde{\mu}(x + y) \geq [0.5, 0.5] \Rightarrow \widetilde{\mu}(x + y) + \min(\widetilde{s}, \widetilde{t}) \geq [1, 1]$. On the other hand, if $\min(\widetilde{s}, \widetilde{t}) \leq [0.5, 0.5]$, then $\widetilde{\mu}(x + y) \geq \min(\widetilde{s}, \widetilde{t})$. Hence $\gamma(x + y; \min(\widetilde{s}, \widetilde{t})) \in \vee q\widetilde{\mu}$.

By using the same arguments, we can prove (2) $\Leftrightarrow$ (m) and (3) $\Leftrightarrow$ (n). We hence omit the details.

The following propositions are easy to prove, and we hence omit the details.

Proposition 4.12 *Let $\widetilde{\mu}$ be an interval-valued $(\in, \in \vee q)$-fuzzy Lie ideal of L. Then, the following properties hold:*

(i) *If there exists $x \in L$ such that $\widetilde{\mu}(x) \geq [0.5, 0.5]$, then $\widetilde{\mu}(0) \geq [0.5, 0.5]$.*
(ii) *If $\widetilde{\mu}(0) < [0.5, 0.5]$, then $\widetilde{\mu}$ is an interval-valued $(\in, \in)$-fuzzy Lie ideal of L.*

Proposition 4.13 *Let $\{\widetilde{\mu}_i : i \in I\}$ be a family of interval-valued $(\in, \in \vee q)$-fuzzy Lie ideals of L. Then, $\widetilde{\mu} = \bigcap_{i \in I} \widetilde{\mu}_i$ is an interval-valued $(\in, \in \vee q)$-fuzzy Lie ideal of L.*

Remark 1. If $\{\widetilde{\mu}_i : i \in I\}$ is a family of interval-valued $(\in, \in \vee q)$-fuzzy Lie ideals of L. We naturally ask: Is $\widetilde{\mu} = \bigcup_{i \in I} \widetilde{\mu}_i$ an interval-valued $(\in, \in \vee q)$-fuzzy Lie ideal of L? The answer is negative as can be seen in the following example:

Example 4.5 Let V be a vector space over a field $\mathbb{F}$ such that $dim(V) = 5$. Let $\{e_1, e_2, e_3, e_4, e_5\}$ be a basis of the vector space V over the field $\mathbb{F}$ with Lie brackets as defined in Example 4.4. If we define fuzzy sets $\widetilde{\mu}_1, \widetilde{\mu}_2 : V \to D[0, 1]$ such that

$$\widetilde{\mu}_1(x) := \begin{cases} [0.6, 0.7] & \text{if } x = 0, \\ [0, 0] & \text{if } x \in \{e_3, e_5\}, \\ [1, 1] & \text{if } x \in \{e_1, e_2, e_4\}. \end{cases}$$

$$\widetilde{\mu}_2(x) := \begin{cases} [0.3, 0.4] & \text{if } x = 0, \\ [0, 0] & \text{if } x \in \{e_3, e_5\}, \\ [1, 1] & \text{if } x \in \{e_1, e_2, e_4\}. \end{cases}$$

then both $\widetilde{\mu}_1$ and $\widetilde{\mu}_2$ are interval-valued $(\in, \in \vee q)$-fuzzy Lie ideals of L, but $\widetilde{\mu}_1 \bigcup \widetilde{\mu}_2$ is not an interval-valued $(\in, \in \vee q)$-fuzzy Lie ideal of L because

$$\begin{aligned} [0, 0] &= \min(\widetilde{\mu}_1(e_3), \widetilde{\mu}_2(e_3)) = (\widetilde{\mu}_1 \cup \widetilde{\mu}_2)(e_3) = (\widetilde{\mu}_1 \cup \widetilde{\mu}_2)([e_1, e_2]) \\ &\geq \min((\widetilde{\mu}_1 \cup \widetilde{\mu}_2)(e_1), (\widetilde{\mu}_1 \cup \widetilde{\mu}_2)(e_2), [0.5, 0.5]) = \min([1, 1], [1, 1], [0.5, 0.5]) = [0.5, 0.5]. \end{aligned}$$

Theorem 4.15 *Let $\{\widetilde{\mu}_i : i \in I\}$ be a family of interval-valued $(\in, \in \vee q)$-fuzzy Lie ideals of L such that $\widetilde{\mu}_i \subseteq \widetilde{\mu}_j$ or $\widetilde{\mu}_j \subseteq \widetilde{\mu}_i$ for all $i, j \in I$. Then, $\widetilde{\mu} := \bigcup_{i \in I} \widetilde{\mu}_i$ is an interval-valued $(\in, \in \vee q)$-fuzzy Lie ideal of L.*

Proof By Definition, for any fixed $x, y \in L$ and given any small $\varepsilon > \widetilde{0}$, there are $i_0, j_0 \in I$ such that

$$\bigcup_{i \in I} \widetilde{\mu}_i(x) < \widetilde{\mu}_{i_0}(x) + \varepsilon,$$

$$\bigcup_{i \in I} \widetilde{\mu}_i(y) < \widetilde{\mu}_{j_0}(y) + \varepsilon.$$

Since $\widetilde{\mu}_{i_0} \subseteq \widetilde{\mu}_{j_0}$ or $\widetilde{\mu}_{j_0} \subseteq \widetilde{\mu}_{i_0}$, we suppose, without loss of generality, that $\widetilde{\mu}_{j_0} \subseteq \widetilde{\mu}_{i_0}$. Now, we have

$$\bigcup_{i \in I} \widetilde{\mu}_i(x) < \widetilde{\mu}_{i_0}(x) + \varepsilon,$$

$$\bigcup_{i \in I} \widetilde{\mu}_i(y) < \widetilde{\mu}_{i_0}(y) + \varepsilon.$$

It follows that $\widetilde{\mu}_i(x + y) \geq \min(\widetilde{\mu}_i(x), \widetilde{\mu}_i(y), [0.5, 0.5])$, by Proposition 4.12. Now for all $i \in I$, we can deduce that

$$\begin{aligned}
\widetilde{\mu}(x + y) + \varepsilon \geq &= \sup_{i \in I} \widetilde{\mu}_i(x + y) + \varepsilon \\
&\geq \widetilde{\mu}_{i_0}(x + y) + \varepsilon \\
&\geq \min\{\widetilde{\mu}_{i_0}(x),\ \widetilde{\mu}_{i_0}(y),\ [0.5, 0.5]\} + \varepsilon \\
&= \min\{\widetilde{\mu}(x),\ \widetilde{\mu}(y),\ [0.5, 0.5]\}.
\end{aligned}$$

Therefore $\widetilde{\mu}(x + y) = \min\{\widetilde{\mu}(x),\ \widetilde{\mu}(y),\ [0.5, 0.5]\}$.

By Lemma 4.7, $\widetilde{\mu}$ is an interval-valued ($\in, \in \vee q$)-fuzzy Lie ideal of L. The verifications for other conditions are analogous and are hence omitted.

Theorem 4.16 *For any chain $G_0 \subset G_1 \subset G_2 \subset \cdots G_n = L$ of Lie ideals of a Lie algebra L, there exists an interval-valued ($\in, \in \vee q$)-fuzzy Lie ideal of L for which level sets coincide with this chain.*

Proof Let $\{\widetilde{t}_k \mid k = 0, 1, \ldots, n\}$ be finite decreasing in $D[0, 1]$. Consider interval-valued fuzzy set $\widetilde{\mu}$ in L defined by $\widetilde{\mu}(G_0) = \widetilde{t}_0$, $\widetilde{\mu}(G_k \setminus G_{k-1}) = \widetilde{t}_k$ for $0 < k \leq n$. Let $x, y \in L$. If $x, y \in G_k \setminus G_{k-1}$, then $x + y \in G_k$ and

$$\widetilde{\mu}(x + y) \geq \widetilde{t}_k = \min\{\widetilde{\mu}(x), \widetilde{\mu}(y)\}.$$

For $i > j$, if $x \in G_i \setminus G_{i-1}$ and $y \in G_j \setminus G_{j-1}$, then $\widetilde{\mu}(x) = t_i = \widetilde{\mu}(y)$ and $x + y \in G_i$. Thus

$$\widetilde{\mu}(x + y) \geq \widetilde{t}_i = \min\{\widetilde{\mu}(x), \widetilde{\mu}(y)\}.$$

The verifications of conditions (2–3) are similar, and we omit the details. So $\widetilde{\mu}$ is an interval-valued fuzzy Lie ideal of a Lie algebra L. Hence, it is easy to see that $\widetilde{\mu}$ is an interval-valued ($\in, \in \vee q$)-fuzzy ideal of L. Such defined $\widetilde{\mu}$ has only the values $\widetilde{t}_0, \widetilde{t}_1, \ldots, \widetilde{t}_n$. Their level subsets are Lie ideals and form the chain

$$U(\widetilde{\mu}, \widetilde{t}_0) \subset U(\widetilde{\mu}, \widetilde{t}_1) \subset \cdots \subset U(\widetilde{\mu}, \widetilde{t}_n) = L$$

We now prove that

$$U(\widetilde{\mu}, \widetilde{t}_k) = G_k \text{ for } 0 < k \leq n.$$

Indeed,

$$U(\widetilde{\mu}, \widetilde{t_0}) = \{x \in L \mid \widetilde{\mu}(x) \geq \widetilde{t_0}\} = G_0,$$

Clearly, $G_k \subseteq U(\widetilde{\mu}, \widetilde{t_k})$. If $x \in U(\widetilde{\mu}, \widetilde{t_k})$, then $\widetilde{\mu}(x) \geq \widetilde{t_k}$ and so $x \notin G_i$ for $i > k$. Hence

$$\widetilde{\mu}(x) \in \{\widetilde{t_0}, \widetilde{t_1}, \ldots, \widetilde{t_k}\},$$

which implies $x \in G_i$ for some $i \leq k$. Since $G_i \subseteq G_k$, it follows that $x \in G_k$. Consequently, $U(\widetilde{\mu}, \widetilde{t_k}) = G_k$ for some $0 < k \leq n$, which implies $x \in G_j$ for some $j \leq k$. Since $G_j \subseteq G_k$, it follows that $y \in G_k$. This completes the proof

We now characterize the Lie ideals of Lie algebras by their level subsets.

Theorem 4.17 *Let $\widetilde{\mu}$ be an interval-valued fuzzy set of fuzzy Lie ideal of L. Then, $\widetilde{\mu}$ is an interval-valued $(\in, \in \vee q)$-fuzzy ideal of L if and only if $U(\widetilde{\mu}; \widetilde{t})$ is a nonempty ideal of L, when $t \in D(0, 0.5]$.*

Proof Assume that $\widetilde{\mu}$ is an interval-valued $(\in, \in \vee q)$-fuzzy Lie ideal of L. Let $\widetilde{t} \in D(0, 0.5]$ be such that $U(\widetilde{\mu}; \widetilde{t}) \neq \emptyset$. If $x, y \in U(\widetilde{\mu}; \widetilde{t})$ and $\alpha \in \mathbb{F}$, then $\widetilde{\mu}(x) \geq \widetilde{t}$ and $\widetilde{\mu}(y) \geq \widetilde{t}$. Thus, it follows that

$$\widetilde{\mu}(x + y) \geq \min(\widetilde{\mu}(x), \widetilde{\mu}(y), [0.5, 0.5]) \geq \min(\widetilde{t}, [0.5, 0.5]) = \widetilde{t},$$

$$\widetilde{\mu}(\alpha x) \geq \min(\widetilde{\mu}(x), [0.5, 0.5]) \geq \min(\widetilde{t}, [0.5, 0.5]) = \widetilde{t},$$

and

$$\widetilde{\mu}([x, y]) \geq \min(\widetilde{\mu}(x), [0.5, 0.5]) \geq \min(\widetilde{t}, [0.5, 0.5]) = \widetilde{t},$$

so that $x + y, \alpha x, [x, y] \in U(\widetilde{\mu}; \widetilde{t})$. This shows that $U(\widetilde{\mu}; \widetilde{t})$ is an interval-valued $(\in, \in \vee q)$-fuzzy Lie ideal of L.

The proof of the converse part is straightforward and is hence omitted.

Naturally, we can establish a result when each nonempty $U(\widetilde{\mu}; \widetilde{t})$ is a Lie ideal of L for $\widetilde{t} \in D(0.5, 1]$.

Theorem 4.18 *Let $\widetilde{\mu}$ be an interval-valued fuzzy set of a Lie algebra L. Then, $U(\widetilde{\mu}; \widetilde{t})(\neq \emptyset)$ is a Lie ideal of L for all $\widetilde{t} \in D(0.5, 1]$ if and only if the following conditions hold:*

(a) $\max(\widetilde{\mu}(x + y), [0.5, 0.5]) \geq \min(\widetilde{\mu}(x), \widetilde{\mu}(y))$,
(b) $\max(\widetilde{\mu}(\alpha x), [0.5, 0.5]) \geq \widetilde{\mu}(x)$,
(c) $\max(\widetilde{\mu}([x, y]), [0.5, 0.5]) \geq \widetilde{\mu}(x)$

for all $x, y \in L$, $\alpha \in \mathbb{F}$.

Proof Suppose that $U(\widetilde{\mu}; \widetilde{t})(\neq \emptyset)$ is a Lie ideal of L. We first prove the necessity of the theorem; that is, we need to verify conditions (a)–(c) hold.

(a) Assume that $\max(\widetilde{\mu}(x+y), [0.5, 0.5]) < \min(\widetilde{\mu}(x), \widetilde{\mu}(y)) = \widetilde{t}$ holds for some $x, y \in L$. Then we have $\widetilde{t} \in D(0.5, 1]$, $\widetilde{\mu}(x+y) < \widetilde{t}$, $x \in U(\widetilde{\mu}; \widetilde{t})$ and $y \in U(\widetilde{\mu}; \widetilde{t})$. Since $x, y \in U(\widetilde{\mu}; \widetilde{t})$, $U(\widetilde{\mu}; \widetilde{t})$ is a Lie ideal of L. This leads to $x + y \in U(\widetilde{\mu}; \widetilde{t})$, a contradiction. Hence, condition (a) holds.

(b) Suppose that $\max(\widetilde{\mu}(\alpha x), [0.5, 0.5]) < \widetilde{\mu}(x) = \widetilde{t}$ for some $x \in L, \alpha \in \mathbb{F}$. Then $\widetilde{t} \in D(0.5, 1], \widetilde{\mu}(\alpha x) < \widetilde{t}, x \in U(\widetilde{\mu}; \widetilde{t})$. Since $x \in U(\widetilde{\mu}; \widetilde{t}), U(\widetilde{\mu}; \widetilde{t})$ is a Lie ideal of L. This leads to $\alpha x \in U(\widetilde{\mu}; \widetilde{t})$, which is also a contradiction and so (b) holds.

(c) Assume that $\max(\widetilde{\mu}([x, y]), [0.5, 0.5]) < \widetilde{\mu}(x) = \widetilde{t}$, for some $x, y \in L$. Then $\widetilde{t} \in D(0.5, 1]$, $\widetilde{\mu}([x, y]) < \widetilde{t}, x \in U(\widetilde{\mu}; \widetilde{t})$ and $y \in U(\widetilde{\mu}; \widetilde{t})$. Since $x \in U(\widetilde{\mu}; \widetilde{t})$, $U(\widetilde{\mu}; \widetilde{t})$ is a Lie ideal of L. This leads to $[x, y] \in U(\widetilde{\mu}; \widetilde{t})$, which is a contradiction. This proves that (c) also holds.

The proof of the sufficiency part is straightforward and is hence omitted.

Definition 4.16 Let $\widetilde{\mu}$ be an interval-valued fuzzy set of L and $\widetilde{t} \in D(0, 1]$. The set $O(\widetilde{\mu}; \widetilde{t}) := \{x \in L \mid \gamma(x; \widetilde{t}) \in \vee q\, \widetilde{\mu}\} = \{x \in L \mid \widetilde{\mu}(x) \geq \widetilde{t} \text{ or } \widetilde{t} + \widetilde{\mu}(x) > [1, 1]\}$ is called $\in \vee q$-*level subset* of L determined by $\widetilde{\mu}$ and $\widetilde{t}$.

Theorem 4.19 *An interval-valued fuzzy set $\widetilde{\mu}$ of L is an interval-valued ($\in, \in \vee q$)-fuzzy Lie ideal of L if and only if $O(\widetilde{\mu}; \widetilde{t})$ is a Lie ideal of L when $O(\widetilde{\mu}; t) \neq \emptyset$ for all $\widetilde{t} \in D(0, 1]$.*

Proof Let $\widetilde{\mu}$ be an interval-valued ($\in, \in \vee q$)-fuzzy Lie ideal of L and $O(\widetilde{\mu}; \widetilde{t}) \neq \emptyset$. Then, there exists $a \in O(\widetilde{\mu}; \widetilde{t})$ such that $\widetilde{\mu}(a) \geq \widetilde{t}$ or $\widetilde{\mu}(a) + \widetilde{t} > [1, 1]$. Let $x \in O(\widetilde{\mu}; \widetilde{t})$ and $y \in O(\widetilde{\mu}; \widetilde{t})$. Then $\widetilde{\mu}(x) \geq t$ or $\widetilde{\mu}(x) + \widetilde{t} > [1, 1]$ and $\widetilde{\mu}(y) \geq \widetilde{t}$ or $\widetilde{\mu}(y) + \widetilde{t} > [1, 1]$. From (1), it follows that

$$\begin{aligned}\widetilde{\mu}(x+y) &\geq \min\{\widetilde{\mu}(x),\ \widetilde{\mu}(y),\ [0.5, 0.5]\}\\ &= \min\{\min\{\widetilde{\mu}(x),\ \widetilde{\mu}(y)\},\ [0.5, 0.5]\}.\end{aligned}$$

We consider the following two cases:
(1) $\min\{\widetilde{\mu}(x), \widetilde{\mu}(y)\} \leq [0.5, 0.5]$. Then

$$\widetilde{\mu}(x+y) \geq \min\{\widetilde{\mu}(x),\ \widetilde{\mu}(y)\}.$$

If $\min\{\widetilde{\mu}(x), \widetilde{\mu}(y)\} \geq \widetilde{t}$, then $\widetilde{\mu}(x+y) \geq \widetilde{t}$, so $x + y \in O(\widetilde{\mu}; \widetilde{t})$. If $\min\{\widetilde{\mu}(x), \widetilde{\mu}(y)\} + \widetilde{t} > [1, 1]$, then $\widetilde{\mu}(x+y) + \widetilde{t} > [1, 1]$, and so $x + y \in O(\widetilde{\mu}; \widetilde{t})$.
(2) $\min\{\widetilde{\mu}(x), \widetilde{\mu}(y)\} > [0.5, 0.5]$. Then $\widetilde{\mu}(x+y) \geq [0.5, 0.5]$. If $\widetilde{t} \leq [0.5, 0.5]$, then $\widetilde{\mu}(x+y) \geq \widetilde{t}$. Thus $x + y \in O(\widetilde{\mu}; \widetilde{t})$. If $\widetilde{t} > [0.5, 0.5]$, then $\widetilde{\mu}(x+y) + \widetilde{t} > [1, 1]$, and so $x + y \in O(\widetilde{\mu}; \widetilde{t})$. By summarizing the above results, we see that $O(\widetilde{\mu}; \widetilde{t})$ satisfies (1). The verification is analogous for other conditions. Therefore $O(\widetilde{\mu}; \widetilde{t})$ is a Lie ideal of L. The proof of the converse part is routine, and we omit the details.

Remark 4.2 From the above discussion, we know that: $U(\widetilde{\mu}; \widetilde{t})$ is a Lie ideal of the Lie algebra L when (i) $\widetilde{t} \in D(0, 1]$ (ii) $\widetilde{t} \in D(0, 0.5]$ (iii) $\widetilde{t} \in D(0.5, 1]$. An obvious question is whether $\widetilde{\mu}$ is a kind of an interval-valued fuzzy Lie ideal or not when

$U(\widetilde{\mu}; \widetilde{t})(\neq \emptyset)$ (e.g., $\widetilde{s}_1, \widetilde{s}_2 \in D[0, 1]$ and $\widetilde{s}_1 < \widetilde{s}_2$)? Based on Yuan et al. [137], we can extend the concept of fuzzy subgroup with thresholds to fuzzy Lie ideals with thresholds in the following way:

Definition 4.17 Let $\widetilde{\lambda_1}, \widetilde{\lambda_2} \in D[0, 1]$ and $\widetilde{\lambda_1} < \widetilde{\lambda_2}$. If $\widetilde{\mu}$ is an interval-valued fuzzy set of a Lie algebra L, then $\widetilde{\mu}$ is said to be an *interval-valued fuzzy Lie ideal with thresholds* if the following conditions are satisfied:

(i) $\max(\widetilde{\mu}(x + y), \widetilde{\lambda_1}) \geq \min(\widetilde{\mu}(x), \widetilde{\mu}(y), \widetilde{\lambda_2})$,
(ii) $\max(\widetilde{\mu}(\alpha x), \widetilde{\lambda_1}) \geq \min(\widetilde{\mu}(x), \widetilde{\lambda_2})$,
(iii) $\max(\widetilde{\mu}([x, y]), \widetilde{\lambda_1}) \geq \min(\widetilde{\mu}(x), \widetilde{\lambda_2})$

for all $x, y \in L, \alpha \in \mathbb{F}$.

Example 4.6 Let $\Re^2 = \{(x, y) : x, y \in \mathbb{R}\}$ be the set of all two-dimensional real vectors. Define the bracket $[\cdot, \cdot]$ as usual cross product in $\Re^2$; that is, $[x, y] = x \times y$ is a real Lie operation on $\Re^2$. Consider an IF set $\widetilde{\mu} : \Re^2 \to D[0, 1]$ defined by

$$\widetilde{\mu}(x, y) = \begin{cases} [0.21, 0.32] \text{ if } x = y = 0, \\ [0.71, 0.82] \text{ otherwise.} \end{cases}$$

Then, it can be easily verified that $\widetilde{\mu}$ is an interval-valued fuzzy Lie ideal with thresholds $\widetilde{\rho_1} = [0.21, 0.25]$ and $\widetilde{\rho_2} = [0.52, 0.57]$ of L, but $\widetilde{\mu}$ is not an interval-valued $(\in, \in)$-fuzzy Lie ideal because

$$\widetilde{\mu}([3, 3]) = \widetilde{\mu}(0) = [0.21, 0.32] < [0.71, 0.82] = \widetilde{\mu}(3).$$

Theorem 4.20 *An interval-valued fuzzy set $\widetilde{\mu}$ of Lie algebra L is an interval-valued fuzzy Lie ideal with thresholds of L if and only if $U(\widetilde{\mu}; \widetilde{t})(\neq \emptyset)$, for $\widetilde{t} \in (\widetilde{\lambda_1}, \widetilde{\lambda_2}]$, is a Lie ideal of L.*

Proof Assume that $\widetilde{\mu}$ is an interval-valued fuzzy Lie ideal with thresholds of L. Let $x, y \in U(\widetilde{\mu}; \widetilde{t})$. Then, $\widetilde{\mu}(x) \geq \widetilde{t}$ and $\widetilde{\mu}(y) \geq \widetilde{t}, \widetilde{t} \in D(\widetilde{\rho_1}, \widetilde{\rho_2}]$. It hence follows that

$$\max(\widetilde{\mu}(x + y), \widetilde{\rho_1}) \geq \min(\widetilde{\mu}(x), \widetilde{\mu}(y), \widetilde{\rho_2}) = \widetilde{t} \Longrightarrow \widetilde{\mu}(x + y) \geq \widetilde{t},$$

$$\max(\widetilde{\mu}(\alpha x), \widetilde{\rho_1}) \geq \min(\widetilde{\mu}(x), \widetilde{\rho_2}) = \widetilde{t} \Longrightarrow \widetilde{\mu}(\alpha x) \geq \widetilde{t},$$

$$\max(\widetilde{\mu}([x, y]), \widetilde{\rho_1}) \geq \min(\widetilde{\mu}(x), \widetilde{\rho_2}) = \widetilde{t} \Longrightarrow \widetilde{\mu}([x, y]) \geq \widetilde{t},$$

and so $x + y, \alpha x, [x, y] \in U(\widetilde{\mu}; \widetilde{t})$. This shows that $U(\widetilde{\mu}; \widetilde{t})$ is a Lie ideal of L.

Conversely, we can assume that $\widetilde{\mu}$ is an interval-valued fuzzy set such that $U(\widetilde{\mu}; \widetilde{t}) \neq \emptyset$ is a Lie ideal of L for $\widetilde{\rho_1}, \widetilde{\rho_2} \in D[0, 1]$ and $\widetilde{\rho_1} < \widetilde{\rho_2}$. Suppose that $\max(\widetilde{\mu}(x + y), \widetilde{\rho_1}) < \min(\widetilde{\mu}(x), \widetilde{\mu}(y), \widetilde{\rho_2}) = \widetilde{t}$. Then $\widetilde{\mu}(x + y) < \widetilde{t}$, $x \in U(\widetilde{\mu}; \widetilde{t})$, $y \in U(\widetilde{\mu}; \widetilde{t})$, $\widetilde{t} \in D(\rho_1, \rho_2]$. Since $x, y \in U(\widetilde{\mu}; \widetilde{t})$ and $U(\widetilde{\mu}; \widetilde{t})$ are Lie ideals, we have $x + y \in U(\widetilde{\mu}; \widetilde{t})$, that is, $\widetilde{\mu}(x + y) \geq \widetilde{t}$. This is clearly a contradiction, and consequently, condition (1) holds. The verifications conditions of (2–3) are similar, and we omit the details. Thus, the proof is completed.

The proof of the following theorem is straightforward, and we hence omit the proof.

Theorem 4.21 (i) *An interval-valued fuzzy set $\widetilde{\mu}$ of Lie algebra L is an interval-valued fuzzy Lie ideal with thresholds of L if and only if $U(\widetilde{\mu}; \widetilde{t})(\neq \emptyset)$, for $\widetilde{t} \in (\widetilde{\lambda}_1, \widetilde{\lambda}_2]$, is a Lie ideal of L.*

(ii) *Let $f : L_1 \to L_2$ be an epimorphism of Lie algebras. If $\widetilde{\nu}$ is an interval-valued fuzzy Lie ideal with thresholds ($\widetilde{\lambda}_1$ and $\widetilde{\lambda}_2$) in L_2 then $f^{-1}(\widetilde{\nu})$ is an interval-valued fuzzy Lie ideal with thresholds ($\widetilde{\lambda}_1, \widetilde{\lambda}_2$) in L_1, where*

$$f^{-1}(\widetilde{\nu})(x) = \widetilde{\nu}(f(x)) \quad for\ all \ \ x \in L_1.$$

Corollary 4.2 *Let $f : L_1 \to L_2$ be an onto homomorphism of Lie algebras. If μ is an interval-valued ($\in, \in \vee q$)-fuzzy Lie ideal of L_2, then $f^{-1}(\mu)$ is an interval-valued ($\in, \in \vee q$)-fuzzy Lie ideal of L_1.*

4.6 Interval-Valued ($\in, \in \vee q_{\widetilde{m}}$)-Fuzzy Lie Algebras

Let m be an element of [0,1) and let $\widetilde{m}$ be an element of D[0,1) unless otherwise specified. By $x_{\widetilde{t}}\, q_{\widetilde{m}} \widetilde{\mu}$, we mean $\widetilde{\mu}(x) + \widetilde{t} + \widetilde{m} > \widetilde{1}$, $\widetilde{t} \in D(0, \frac{1-m}{2}]$. For brevity, we write the following notions:

- $x_{\widetilde{t}} \in \widetilde{\mu}$ or $x_{\widetilde{t}} q_{\widetilde{m}} \widetilde{\mu}$ will be denoted by $x_{\widetilde{t}} \in \vee q_{\widetilde{m}} \widetilde{\mu}$.
- $x_{\widetilde{t}} \in \widetilde{\mu}$ and $x_{\widetilde{t}} q_{\widetilde{m}} \widetilde{\mu}$ will be denoted by $x_{\widetilde{t}} \in \wedge q_{\widetilde{m}} \widetilde{\mu}$.
- The symbol $\overline{\in \wedge q_{\widetilde{m}}}$ means neither $\in$ nor $q_{\widetilde{m}}$ hold.

We formulate a technical lemma.

Lemma 4.8 *Let $\widetilde{\mu}$ be an interval-valued fuzzy set of a Lie algebra L. Then, the nonempty level set $\cup(\widetilde{\mu}; \widetilde{t})$ is a Lie algebra of L for all $\widetilde{t} \in D(\frac{1-m}{2}, 1]$ if and only if*

(a) *rmax* $(\widetilde{\mu}(x + y), [\frac{1-m}{2}, \frac{1-m}{2}]) \geq \text{rmin}(\widetilde{\mu}(x), \widetilde{\mu}(y))$,
(b) *rmax* $(\widetilde{\mu}(\alpha x), [\frac{1-m}{2}, \frac{1-m}{2}]) \geq \widetilde{\mu}(x)$,
(c) *rmax* $(\widetilde{\mu}([x + y]), [\frac{1-m}{2}, \frac{1-m}{2}]) \geq \text{rmin}(\widetilde{\mu}(x), \widetilde{\mu}(y))$
for all $x, y \in L, \alpha \in \mathbb{F}$.

Proof Let $\widetilde{t} \in D(\frac{1-m}{2}, 1])$ be such that $\cup(\widetilde{\mu}; \widetilde{t})(\neq \emptyset)$ and $\cup(\widetilde{\mu}, \widetilde{t})$ is a Lie algebra of L.

(a) Assume that *rmax* $(\widetilde{\mu}(x + y), [\frac{1-m}{2}, \frac{1-m}{2}]) < \text{rmin}(\widetilde{\mu}(x), \widetilde{\mu}(y)) = \widetilde{t}$ for some $x, y \in L$, then $\widetilde{t} \in D(\frac{1-m}{2}, 1]$, $\widetilde{\mu}(x + y) < \widetilde{t}, x \in \cup(\widetilde{\mu}; \widetilde{t})$ and $y \in \cup(\widetilde{\mu}; \widetilde{t})$. Since $x, y \in \cup(\widetilde{\mu}; \widetilde{t})$, $\cup(\widetilde{\mu}; \widetilde{t})$ is a Lie algebra of L, so $x + y \in \cup(\widetilde{\mu}; \widetilde{t})$, a contradiction. Hence (a) holds.

(b) Assume that *rmax* $(\widetilde{\mu}(\alpha x), [\frac{1-m}{2}, \frac{1-m}{2}]) < \widetilde{\mu}(x) = \widetilde{t}$ for some $\alpha \in X, x \in L$, then $\widetilde{t} \in D(\frac{1-m}{2}, 1], \widetilde{\mu}(\alpha x) < \widetilde{t}, x \in \mathsf{U}(\widetilde{\mu}; \widetilde{t})$ and $\alpha \in \mathsf{U}(\widetilde{\mu}; \widetilde{t})$. Since $\alpha, x \in \mathsf{U}(\widetilde{\mu}; \widetilde{t})$, $\mathsf{U}(\widetilde{\mu}; \widetilde{t})$ is a Lie algebra of L, so $\alpha x \in \mathsf{U}(\widetilde{\mu}; \widetilde{t})$, a contradiction. Hence (b) holds.

(c) Assume that *rmax* $(\widetilde{\mu}([x + y]), [\frac{1-m}{2}, \frac{1-m}{2}]) < \mathrm{rmin}(\widetilde{\mu}(x), \widetilde{\mu}(y)) = \widetilde{t}$ for some $x, y \in L$, then $\widetilde{t} \in D(\frac{1-m}{2}, 1], \widetilde{\mu}([x, \widetilde{\mu} y]) < \widetilde{t}, x \in \mathsf{U}(\widetilde{\mu}; \widetilde{t})$ and $y \in \mathsf{U}(\widetilde{\mu}; \widetilde{t})$. Since $x, y \in \mathsf{U}(\widetilde{\mu}; \widetilde{t}), \mathsf{U}(\widetilde{\mu}; \widetilde{t})$ is a Lie algebra of L, so $[x, y] \in \mathsf{U}(\widetilde{\mu}; \widetilde{t})$, a contradiction. Hence (c) holds.

The proof of the sufficiency part is straightforward and is hence omitted. This completes the proof.

Corollary 4.3 *Let $\widetilde{\mu}$ be an interval-valued fuzzy set of a Lie algebra L. Then, the nonempty level set $\mathsf{U}(\widetilde{\mu}; \widetilde{t})$ is a Lie algebra of L for all $\widetilde{t} \in D(0.5, 1]$ if and only if*

(a) *rmax* $(\widetilde{\mu}(x + y), [0.5, 0.5]) \geq \mathrm{rmin}(\widetilde{\mu}(\mathrm{x}), \widetilde{\mu}(\mathrm{y}))$,
(b) *rmax* $(\widetilde{\mu}(\alpha x), [0.5, 0.5]) \geq \widetilde{\mu}(x)$,
(c) *rmax* $(\widetilde{\mu}([x + y]), [0.5, 0.5]) \geq \mathrm{rmin}(\widetilde{\mu}(\mathrm{x}), \widetilde{\mu}(\mathrm{y}))$.
for all $x, y \in L, \alpha \in \mathbb{F}$.

Definition 4.18 An interval-valued fuzzy set $\widetilde{\mu} \in L$ is called an $(\in, \in \vee q_{\widetilde{m}})$*-fuzzy Lie* algebra over field $\mathbb{F}$ of L, if it satisfies the following conditions.

1. $x_{\widetilde{t_1}}, y_{\widetilde{t_2}} \in \widetilde{\mu} \Longrightarrow (x + y)_{\mathrm{rmin}(\widetilde{t}_1, \widetilde{t}_2)} \in \vee q_{\widetilde{m}} \widetilde{\mu}$,
2. $x_{\widetilde{t_2}} \in \widetilde{\mu} \Longrightarrow (\alpha x)_{\widetilde{t_2}} \in \vee q_{\widetilde{m}} \widetilde{\mu}$,
3. $x_{\widetilde{t_1}}, y_{\widetilde{t_2}} \in \widetilde{\mu} \Longrightarrow ([x + y])_{\mathrm{rmin}(\widetilde{t}_1, \widetilde{t}_2)} \in \vee q_{\widetilde{m}} \widetilde{\mu}$

for all $x, y \in L, \alpha \in \mathbb{F}, \widetilde{t_1}, \widetilde{t_2} \in D(0, 1]$.

From Definition 4.18, it follows that we can develop different types of fuzzy Lie algebras for different values of $\widetilde{m} \in D[0, 1)$. Hence, an interval-valued $(\in, \in \vee q_{\widetilde{m}})$-*fuzzy lie algebra* with $\widetilde{m} = [0, 0]$ is called an interval-valued $(\in, \in \vee q)$-fuzzy Lie algebra.

Example 4.7 Let $\mathfrak{R}^3 = \{(x, y, z) : x, y, z \in \mathbb{R}\}$ be the set of all three-dimensional real vectors. Then $\mathfrak{R}^3$ with $[x, y] = x \times y$ form a real Lie algebra. Define an interval-valued fuzzy set $\widetilde{\mu} : \mathfrak{R}^3 \to D[0, 1]$ by

$$\widetilde{\mu}(x, y, z) = \begin{cases} [0.4, 0.4], & \text{if } x = y = z = 0, \\ [0.5, 0.5], & \text{otherwise.} \end{cases}$$

By routine computations, we can easily check that $\widetilde{\mu}$ forms $(\in, \in \vee q)$-, $(\in, \in \vee q_{[0.2, 0.2]})$-, $(\in, \in \vee q_{[0.4, 0.4]})$-fuzzy Lie algebras of L for $\widetilde{m} = [0, 0], [0.2, 0.2], [0.4, 0.4]$, respectively.

Theorem 4.22 *An interval-valued fuzzy set $\widetilde{\mu}$ in L is an interval-valued $(\in, \in \vee q_{\widetilde{m}})$-fuzzy Lie algebra of L if and only if it satisfies*

(1) $\widetilde{\mu}(x + y) \geq \mathrm{rmin}\{\widetilde{\mu}(x), \widetilde{\mu}(y), [\frac{1-m}{2}, \frac{1-m}{2}]\}$,
(2) $\widetilde{\mu}(\alpha x) \geq \mathrm{rmin}\{\widetilde{\mu}(x), [\frac{1-m}{2}, \frac{1-m}{2}]\}$,

(3) $\widetilde{\mu}([x+y]) \geq \text{rmin}\{\widetilde{\mu}(x), \widetilde{\mu}(y), [\frac{1-m}{2}, \frac{1-m}{2}]\}$.

for all $x, y \in L, \alpha \in \mathbb{F}$.

Proof Let $\widetilde{\mu}$ be an interval-valued ($\in, \in \vee q_{\widetilde{m}}$)-fuzzy Lie algebra of L. Assume that (1) is not valid. Then, there exist $x', y' \in L$ such that $\widetilde{\mu}(x'+y') < \text{rmin}\{\widetilde{\mu}(x'), \widetilde{\mu}(y'), [\frac{1-m}{2}, \frac{1-m}{2}]\}$.

If $\text{rmin}(\widetilde{\mu}(x'), \widetilde{\mu}(y')) < [\frac{1-m}{2}, \frac{1-m}{2}]$, then $\widetilde{\mu}(x'+y') < \text{rmin}(\widetilde{\mu}(x'), \widetilde{\mu}(y'))$. Thus $\widetilde{\mu}(x'+y') < \widetilde{t} \leq rmin\{\widetilde{\mu}(x'), \widetilde{\mu}(y')\}$ for some $\widetilde{t} \in D(0, 1]$.

It follows that $x'_{\widetilde{t}} \in \widetilde{\mu}$ and $y'_{\widetilde{t}} \in \widetilde{\mu}$, but $(x'+y')_{\widetilde{t}} \overline{\in} \widetilde{\mu}$, a contradiction. Moreover, $\widetilde{\mu}(x'+y') + \widetilde{t} < 2\widetilde{t} < [1-m, 1-m]$, and so $(x'+y')_{\widetilde{t}} \overline{q_{\widetilde{m}}} \widetilde{\mu}$. Hence, consequently $(x'+y')_{\widetilde{t}} \overline{\in \vee q_{\widetilde{m}}} \widetilde{\mu}$, a contradiction. On the other hand, if $\text{rmin}\{\widetilde{\mu}(x'), \widetilde{\mu}(y')\} \geq [\frac{1-m}{2}, \frac{1-m}{2}]$, then $\widetilde{\mu}(x') \geq [\frac{1-m}{2}, \frac{1-m}{2}]$, $\widetilde{\mu}(y') \geq [\frac{1-m}{2}, \frac{1-m}{2}]$ and $\widetilde{\mu}(x'+y') < [\frac{1-m}{2}, \frac{1-m}{2}]$. Thus $x'_{[\frac{1-m}{2}, \frac{1-m}{2}]} \in \widetilde{\mu}$ and $y'_{[\frac{1-m}{2}, \frac{1-m}{2}]} \in \widetilde{\mu}$, but $(x'+y')_{[\frac{1-m}{2}, \frac{1-m}{2}]} \overline{\in} \widetilde{\mu}$. Also

$\widetilde{\mu}(x'+y') + [\frac{1-m}{2}, \frac{1-m}{2}] < [\frac{1-m}{2}, \frac{1-m}{2}] + [\frac{1-m}{2}, \frac{1-m}{2}] = [1-m, 1-m]$, i.e., $(x'+y')_{[\frac{1-m}{2}, \frac{1-m}{2}]} \overline{q_{\widetilde{m}}} \widetilde{\mu}$. Hence $(x'+y')_{[\frac{1-m}{2}, \frac{1-m}{2}]} \overline{\in \vee q_{\widetilde{m}}} \widetilde{\mu}$, a contradiction. So (1) is valid. By using a very similar argumentations as in the proof of the (1), we can easily prove (2) and (3) are also valid.

Conversely, assume that $\widetilde{\mu}$ satisfies (1). Let $x, y \in L$ and $\widetilde{t_1}, \widetilde{t_2} \in D(0, 1]$ be such that $x_{\widetilde{t_1}} \widetilde{\mu}$ and $y_{\widetilde{t_2}} \widetilde{\mu}$. Then

$\widetilde{\mu}(x+y) \geq \text{rmin}\{\widetilde{\mu}(x), \widetilde{\mu}(y), [\frac{1-m}{2}, \frac{1-m}{2}]\} \geq \text{rmin}\{\widetilde{t_1}, \widetilde{t_2}, [\frac{1-m}{2}, \frac{1-m}{2}]\}$.

Assume that $\widetilde{t_1} \leq [\frac{1-m}{2}, \frac{1-m}{2}]$ or $\widetilde{t_2} \leq [\frac{1-m}{2}, \frac{1-m}{2}]$. Then $\widetilde{\mu}(x+y) \geq rmin\{\widetilde{t_1}, \widetilde{t_2}\}$, which implies that $(x+y)_{rmin}\{\widetilde{t_1}, \widetilde{t_2}\} \in \widetilde{\mu}$. Now suppose that $\widetilde{t_1} > [\frac{1-m}{2}, \frac{1-m}{2}]$ and $\widetilde{t_2} > [\frac{1-m}{2}, \frac{1-m}{2}]$. Then $\widetilde{\mu}(x+y) \geq [\frac{1-m}{2}, \frac{1-m}{2}]$, and thus

$\widetilde{\mu}(x+y) + \text{rmin}\{\widetilde{t_1}, \widetilde{t_2}\} > [\frac{1-m}{2}, \frac{1-m}{2}] + [\frac{1-m}{2}, \frac{1-m}{2}] = [1-m, 1-m]$,
i.e., $(x+y)_{rmin\{\widetilde{t_1}, \widetilde{t_2}\}} q_{\widetilde{m}} \widetilde{\mu}$. Hence $(x+y)_{rmin\{\widetilde{t_1}, \widetilde{t_2}\}} \in \vee q_{\widetilde{m}} \widetilde{\mu}$.

By using a very similar argumentations as in the proof of the (1), we can easily prove (2) and (3) are also valid. Consequently, $\widetilde{\mu}$ is an interval-valued ($\in, \in \vee q_{\widetilde{m}}$)-fuzzy Lie algebra of L.

Corollary 4.4 *An interval-valued fuzzy set $\widetilde{\mu}$ in L is an interval-valued ($\in, \in \vee q$)-fuzzy Lie algebra of L if and only if it satisfies*

(1) $\widetilde{\mu}(x+y) \geq \text{rmin}\{\widetilde{\mu}(x), \widetilde{\mu}(y), [0.5, 0.5]\}$,
(2) $\widetilde{\mu}(\alpha x) \geq \text{rmin}\{\widetilde{\mu}(x), [0.5, 0.5]\}$,
(3) $\widetilde{\mu}([x+y]) \geq \text{rmin}\{\widetilde{\mu}(x), \widetilde{\mu}(y), [0.5, 0.5]\}$.

Theorem 4.23 *Let $\widetilde{\mu}$ be an in interval-valued fuzzy set of fuzzy Lie algebra of L. Then, $\widetilde{\mu}$ in L is an interval-valued ($\in, \in \vee q_{\widetilde{m}}$)-fuzzy Lie algebra of L if and only if the level set $\text{U}(\widetilde{\mu}; \widetilde{t})$, $\widetilde{t} \in D(0, \frac{1-m}{2}]$, is a Lie algebra of L.*

Proof Assume that $\widetilde{\mu}$ is an interval-valued ($\in, \in \vee q$)-fuzzy Lie algebra of L. Let $\widetilde{t} \in (0, \frac{1-m}{2}]$ and $x, y, \alpha \in \text{U}(\widetilde{\mu}; \widetilde{t})$. Then $\widetilde{\mu}(x) \geq \widetilde{t}$, $\widetilde{\mu}(y) \geq \widetilde{t}$. It follows from Theorem 4.22 that:

$$\begin{aligned}\widetilde{\mu}(x+y) &\geq \text{rmin}(\widetilde{\mu}(x), \widetilde{\mu}(y), [\frac{1-m}{2}, \frac{1-m}{2}])\\ &\geq \text{rmin}(\widetilde{t}, [\frac{1-m}{2}, \frac{1-m}{2}]) = \widetilde{t},\\ \widetilde{\mu}(\alpha x) &\geq \text{rmin}(\widetilde{\mu}(x), [\frac{1-m}{2}, \frac{1-m}{2}])\\ &\geq \text{rmin}(\widetilde{t}, [\frac{1-m}{2}, \frac{1-m}{2}]) = \widetilde{t},\\ \widetilde{\mu}([x+y]) &\geq \text{rmin}(\widetilde{\mu}(x), \widetilde{\mu}(y), [\frac{1-m}{2}, \frac{1-m}{2}])\\ &\geq \text{rmin}(\widetilde{t}, [\frac{1-m}{2}, \frac{1-m}{2}]) = \widetilde{t},\end{aligned}$$

so that $x+y, \alpha x, [x, y] \in \cup(\widetilde{\mu}; \widetilde{t})$. Hence $\cup(\widetilde{\mu}; \widetilde{t})$ is a Lie algebra of L.

Conversely, suppose that the nonempty set $\cup(\widetilde{\mu}; \widetilde{t})$ is a Lie algebra of L for all $\widetilde{t} \in D(0, \frac{1-m}{2}]$. If condition (1) is not true, then there exist $a, b \in L$ such that $\widetilde{\mu}(a+b) < \text{rmin}(\widetilde{\mu}(a), \widetilde{\mu}(b), [\frac{1-m}{2}, \frac{1-m}{2}])$. Hence, we can take $\widetilde{t} \in D(0, 1]$ such that $\widetilde{\mu}(a+b) < \widetilde{t_1} < \text{rmin}(\widetilde{\mu}(a), \widetilde{\mu}(b), [\frac{1-m}{2}, \frac{1-m}{2}])$. Then $\widetilde{t} \in [\frac{1-m}{2}, \frac{1-m}{2}]$ and $a, b \in \cup(\widetilde{\mu}; \widetilde{t})$. Since $\cup(\widetilde{\mu}; \widetilde{t})$ is a Lie algebra of L, it follows that $a+b \in \cup(\widetilde{\mu}; \widetilde{t})$ so that $\widetilde{\mu}(a+b) \geq \widetilde{t}$. This is a contradiction. Therefore the condition (1) is valid. By using a very similar argumentation as in the proof of (1), we can easily prove (2) and (3) are also valid. Hence, $\widetilde{\mu}$ is an $(\in, \in \vee q_{\widetilde{m}})$-fuzzy Lie algebra of L.

Corollary 4.5 *Let $\widetilde{\mu}$ be an interval-valued fuzzy set of fuzzy Lie algebra of L. Then, $\widetilde{\mu}$ in L is an interval-valued $(\in, \in \vee q)$-fuzzy Lie algebra of L if and only if the level set $\cup(\widetilde{\mu}; \widetilde{t})$, $\widetilde{t} \in D(0, 0.5]$, is a Lie algebra of L.*

Theorem 4.24 *Let $\widetilde{\mu}$ be an interval-valued $(\in, \in \vee q_{\widetilde{m}})$-fuzzy Lie algebra of L such that $\widetilde{\mu}(x) < [\frac{1-m}{2}, \frac{1-m}{2}]$ for all $x \in L$, then $\widetilde{\mu}$ is an interval-valued fuzzy Lie algebra of L.*

Proof Let $x, y \in L$, $\alpha \in \mathbb{F}$ and $\widetilde{t_1}, \widetilde{t_1} \in D(0, 1]$ be such that $x_{\widetilde{t_1}} \in \widetilde{\mu}$, $y_{\widetilde{t_2}} \in \widetilde{\mu}$. Then $\widetilde{\mu}(x) \geq \widetilde{t_1}$, $\widetilde{\mu}(y) \geq \widetilde{t_2}$. It follows that

$$\begin{aligned}\widetilde{\mu}(x+y) &\geq \text{rmin}(\widetilde{\mu}(x), \widetilde{\mu}(y), [\frac{1-m}{2}, \frac{1-m}{2}])\\ &= \text{rmin}(\widetilde{\mu}(x), \widetilde{\mu}(y) = \text{rmin}(\widetilde{t_1}, \widetilde{t_2}),\\ \widetilde{\mu}(\alpha x) &\geq \text{rmin}(\widetilde{\mu}(x), [\tfrac{1-m}{2}, \tfrac{1-m}{2}]) = \widetilde{t_1},\\ \widetilde{\mu}([x+y]) &\geq \text{rmin}(\widetilde{\mu}(x), \widetilde{\mu}(y), [\frac{1-m}{2}, \frac{1-m}{2}])\\ &= \text{rmin}(\widetilde{\mu}(x), \widetilde{\mu}(y) = \text{rmin}(\widetilde{t_1}, \widetilde{t_2}),\end{aligned}$$

so that $(x+y)_{\text{rmin}(\widetilde{t_1}, \widetilde{t_2})} \in \widetilde{\mu}$, $(\alpha x)_{\widetilde{t_1}} \in \widetilde{\mu}$, $([x, y])_{\text{rmin}\widetilde{t_1}, \widetilde{t_2}} \in \widetilde{\mu}$. Hence $\widetilde{\mu}$ is an interval-valued fuzzy Lie algebra of L.

Corollary 4.6 *Let $\widetilde{\mu}$ be an $(\in, \in \vee q)$-fuzzy Lie algebra of L such that $\widetilde{\mu}(x) < [0.5, 0.5]$ for all $x \in L$, then $\widetilde{\mu}$ is an interval-valued fuzzy Lie algebra of L.*

Lemma 4.9 *Let $\widetilde{\mu}$ be an interval-valued $(\in, \in \vee q_{\tilde{m}})$-fuzzy Lie algebra of L.*
1. If there exist $x \in L$ such that $\widetilde{\mu}(x) \geq [\frac{1-m}{2}, \frac{1-m}{2}]$, then $\widetilde{\mu}(0) \geq [\frac{1-m}{2}, \frac{1-m}{2}]$.
2. If $\widetilde{\mu}(0) < [\frac{1-m}{2}, \frac{1-m}{2}]$, then $\widetilde{\mu}$ is an interval-valued fuzzy Lie algebra of L.

For an interval-valued fuzzy lie algebra $\widetilde{\mu}$ of L, we state the following theorems without their proofs.

Theorem 4.25 *Let $\{\widetilde{\mu}_i : i \in I\}$ be a family of interval-valued $(\in, \in \vee q_{\tilde{m}})$-fuzzy Lie algebras of L over a field. Then, $\widetilde{\mu} = \cap_i \widetilde{\mu}_i$ is an $(\in, \in \vee q_{\tilde{m}})$-fuzzy Lie algebra of L.*

Theorem 4.26 *An interval-valued fuzzy set $\widetilde{\mu}$ of L is an interval-valued $(\in, \in \vee q_{\tilde{m}})$-fuzzy Lie algebra of L if and only if for every $\widetilde{t} \in D(\frac{1-m}{2}, 1]$ each nonempty level $\cup(\widetilde{\mu}; \widetilde{t})$ is a Lie algebra of L.*

Theorem 4.27 *An interval-valued fuzzy set $\widetilde{\mu}$ of L is an interval-valued $(\in, \in \vee q_{\tilde{m}})$-fuzzy Lie algebra of L if and only if for every $\widetilde{t} \in D(\frac{1-m}{2}, 1]$ each nonempty level $\cup(\widetilde{\mu}; \widetilde{t})$ is a Lie algebra of L.*

4.7 (γ, δ)-Intuitionistic Fuzzy Lie Algebras

Definition 4.19 Let c be a point in a nonempty set X. If $\gamma \in (0, 1]$ and $\delta \in [0, 1)$ are two real numbers such that $0 \leq \gamma + \delta \leq 1$, then the intuitionistic fuzzy set $c(\gamma, \delta) =< x, c_\gamma, 1 - c_{1-\delta} >$ is called an *intuitionistic fuzzy point* in X, where γ(respectively, δ) is the degree of membership (respectively, nonmembership) of $c(\gamma, \delta)$ and $c \in X$ is the support of $c(\gamma, \delta)$. Let $c(\gamma, \delta)$ be an intuitionistic fuzzy point in X, and let $A =< x, \mu_A, \nu_A >$ be an intuitionistic fuzzy set in X. Then, $c(\gamma, \delta)$ is said to belong to A, written $c(\gamma, \delta) \in A$ if $\mu_A(c) \geq \gamma$ and $\nu_A(c) \leq \delta$. We say that $c(\gamma, \delta)$ is quasicoincident with A, written $c(\gamma, \delta)qA$, if $\mu_A(c) + \gamma > 1$ and $\nu_A(c) + \delta < 1$. To say that $c(\gamma, \delta) \in \vee qA$ (respectively, $c(\gamma, \delta) \in \wedge qA$) means that $c(\gamma, \delta) \in A$ or $c(\gamma, \delta)qA$ (respectively, $c(\gamma, \delta) \in A$ and $c(\gamma, \delta)qA$) and $c(\gamma, \delta)\overline{\in \vee q}A$ means that $c(\gamma, \delta) \in \vee qA$ does not hold.

Definition 4.20 Let f and g be any two intuitionistic fuzzy subsets of L. Then, the sum $f + g$ is an intuitionistic fuzzy subset of L defined by

$$(\mu_f + \mu_g)(z) = \begin{cases} \bigvee_{z=[x,y]}(\mu_f(x) \wedge \mu_f(y) \wedge \mu_g(x) \wedge \mu_g(y)) \text{ for } z = [x, y], \\ 0 & \textit{otherwise.} \end{cases}$$

$$(\nu_f + \nu_g)(z) = \begin{cases} \bigwedge_{z=[x,y]}(\nu_f(x) \vee \nu_f(y) \vee \nu_g(x) \vee \nu_g(y)) \text{ for } z = [x, y], \\ 1 & \textit{otherwise.} \end{cases}$$

Definition 4.21 An intuitionistic fuzzy set $A = (\mu_A, \nu_A)$ in L is called an (γ, δ)-*intuitionistic fuzzy Lie algebra* of L if it satisfies the following conditions:

(a) $x(s_1, t_1)\gamma A, y(s_2, t_2)\gamma A \Rightarrow (x + y)(\min(s_1, s_2), \max(t_1, t_2))\delta A,$

(b) $x(s,t)\gamma A \Rightarrow (mx)(s,t)\delta A$,
(c) $x(s_1,t_1)\gamma A, y(s_2,t_2)\gamma A \Rightarrow [x,y](\max(s_1,s_2),\min(t_1,t_2))\delta A$

for all $x, y \in L$, $m \in \mathbb{F}$, $s, s_1, s_2 \in (0,1]$, $t, t_1, t_2 \in [0,1)$.

From (b), it follows that:

(d) $x(s,t)\gamma A \Rightarrow (-x)(s,t)\delta A$.
(e) $x(s,t)\gamma A \Rightarrow (0)(s,t)\delta A$.

Example 4.8 Let V be a vector space over a field $\mathbb{F}$ such that $dim(V)=5$. Let $\{e_1,e_2,\ldots,e_5\}$ be a basis of the vector space V over $\mathbb{F}$ with Lie brackets as follows:

$$[e_1,e_2]=e_3,\ \ [e_1,e_3]=e_5,\ \ [e_1,e_4]=e_5,\ \ [e_1,e_5]=0,$$

$$[e_2,e_3]=e_5,\ \ [e_2,e_4]=0,\ \ [e_2,e_5]=0,\ \ [e_3,e_4]=0,$$

$$[e_3,e_5]=0,\ \ [e_4,e_5]=0,\ \ [e_i,e_j]=-[e_j,e_i]$$

and $[e_i,e_j]=0$ for all $i=j$. Then V is a Lie algebra over $\mathbb{F}$.

We define an intuitionistic fuzzy set $A=(\mu_A,\nu_A): V \rightarrow [0,1]\times[0,1]$ by

$$\mu_A(x) := \begin{cases} 1 & \text{if } x=0, \\ 0.5 & \text{otherwise} \end{cases}$$

$$\nu_A(x) := \begin{cases} 0 & \text{if } x=0, \\ 0.3 & \text{otherwise} \end{cases}$$

Take $s=0.4 \in (0,1]$ and $t=0.5 \in [1,0)$. By routine computations, it is easy to see that A is not an (γ,δ)-intuitionistic fuzzy Lie subalgebra of L.

For an intuitionistic fuzzy set A in L, we denote $L(0,1)=\{x \in L : \mu(x)>0$ and $\nu(x)<1\}$.

Theorem 4.28 *Let $A=(\mu_A,\nu)$ be an (γ,δ)-intuitionistic fuzzy Lie subalgebra of L, then the nonzero set $L(0,1)$ is a Lie subalgebra of L.*

Proof Let $x, y \in L(0,1)$. Then $\mu_A(x)>0$ and $\nu_A(x)<1$, $\mu_A(y)>0$ and $\nu_A(y)<1$. Assume that $\mu_A(x+y)=0$ and $\nu_A(x+y)=1$. If $\gamma \in \{\in, \in\vee q\}$, then we can see that $x(\mu_A(x),\nu_A(x))\gamma A$ and $y(\mu_A(y),\nu_A(y))\gamma A$, but $(x+y)(\min\{\mu_A(x),\nu_A(x)\},\max\{\mu_A(y),\nu_A(y)\})\overline{\delta}A$ for all $\delta \in \{\in, \in\vee q, \in\wedge q\}$, a contradiction. Also, $x(1,0)qA$ and $y(1,0)qA$, but $(x+y)(1,0)\overline{\delta}A$ for all $\delta \in \{\in, \in\vee q, \in\wedge q\}$, a contradiction. Thus $\mu_A(x+y)>0$ and $\nu_A(x+y)<1$. Thus $x+y \in L(0,1)$. For other conditions, the verification is analogous. Consequently, $L(0,1)$ is a Lie subalgebra of L.

Theorem 4.29 *If H is a Lie subalgebra of L, then an intuitionistic fuzzy set $A=(\mu_A,\nu_A)$ of L such that*

(i) $\mu_A(x) \geq 0.5$ *and* $\nu_A(x) \leq 0.5$ *for all* $x \in H$,
(ii) $\mu_A(0) = 0$ *and* $\nu_A(0) = 1$ *otherwise.*

Then, $A = (\mu_A, \nu_A)$ *is an* $(\gamma, \in \vee q)$*-intuitionistic fuzzy Lie subalgebra of* L.

Proof (a) Let $x, y \in L, s_1, s_2 \in (0, 1], t_1, t_2 \in [0, 1)$ be such that $x(s_1, t_1), y(s_2, t_2) \in A$. Then $\mu_A(x) \geq s_1$ and $\nu_A(x) \leq t_1$, $\mu_A(y) \geq s_1$ and $\nu_A(y) \leq t_1$. Thus $x, y \in H$ and so $x + y \in H$, that is, $\mu_A(x + y) \geq 0.5$ and $\nu_A(x + y) \leq 0.5$. If $\min(s_1, s_2) \leq 0.5$ and $\max(t_1, t_2) \geq 0.5$, then $\mu_A(x + y) \geq 0.5 \geq \min(s_1, s_2)$ and $\nu_A(x + y) \leq 0.5 \leq \max(t_1, t_2)$. Hence $(x + y)(\min(s_1, s_2), \max(t_1, t_2)) \in A$. If $\min(s_1, s_2) > 0.5$ and $\max(t_1, t_2) < 0.5$, then $\mu_A(x + y) + \min(s_1, s_2) > 0.5 + 0.5 = 1$, $\nu_A(x + y) + \max(t_1, t_2) < 0.5 + 0.5 = 1$. Thus $(x + y)(\min(s_1, s_2), \max(t_1, t_2))qA$. Hence $(x + y)(\min(s_1, s_2), \max(t_1, t_2)) \in \vee qA$. The verification for other conditions is analogous. $A = (\mu_A, \nu_A)$ is an $(\in, \in \vee q)$-intuitionistic fuzzy Lie subalgebra of L.
(b) Let $x, y \in L$, $s_1, s_2 \in (0, 1]$, $t_1, t_2 \in [0, 1)$ be such that $x(s_1, t_1), y(s_2, t_2)qA$. Then $\mu_A(x) + s_1 \geq 1$ and $\nu_A(x) + t_1 \leq 1$, $\mu_A(y) + s_2 \geq 1$ and $\nu_A(y) + t_2 \leq 1$. Thus $x, y \in H$ and so $x + y \in H$, that is, $\mu_A(x + y) \geq 0.5$ and $\nu_A(x + y) \leq 0.5$. If $\min(s_1, s_2) \leq 0.5$ and $\max(t_1, t_2) \geq 0.5$, then $\mu_A(x + y) \geq 0.5 \geq \min(s_1, s_2)$ and $\nu_A(x + y) \leq 0.5 \leq \max(t_1, t_2)$. Hence $(x + y)(\min(s_1, s_2), \max(t_1, t_2)) \in A$. If $\min(s_1, s_2) > 0.5$ and $\max(t_1, t_2) < 0.5$, then $\mu_A(x + y) + \min(s_1, s_2) > 0.5 + 0.5 = 1$, $\nu_A(x + y) + \max(t_1, t_2) < 0.5 + 0.5 = 1$. Thus $(x + y)(\min(s_1, s_2), \max(t_1, t_2))qA$. Hence $(x + y)(\min(s_1, s_2), \max(t_1, t_2)) \in \vee qA$. The verification for other conditions is analogous. This shows that $A = (\mu_A, \nu_A)$ is an $(q, \in \vee q)$-intuitionistic fuzzy Lie subalgebra of L. The remaining is a consequence of (a) and (b). This completes the proof.

Definition 4.22 An intuitionistic fuzzy set $A = (\mu_A, \nu_A)$ in L is called an $(\in, \in \vee q)$-*intuitionistic fuzzy Lie algebra* of L if it satisfies the following conditions:

(f) $x(s_1, t_1) \in A, y(s_2, t_2) \in A \Rightarrow (x + y)(\min(s_1, s_2), \max(t_1, t_2)) \in \vee qA$,
(g) $x(s, t) \in A \Rightarrow (mx)(s, t) \in \vee qA$,
(h) $x(s_1, t_1) \in A, y(s_2, t_2) \in A \Rightarrow [x, y](\max(s_1, s_2), \min(t_1, t_2)) \in \vee qA$

for all $x, y \in L, m \in \mathbb{F}, s, s_1, s_2 \in (0, 1], t, t_1, t_2 \in [0, 1)$.

Theorem 4.30 *Let* $A = (\mu_A, \nu_A)$ *be an intuitionistic fuzzy set in a Lie algebra* L. *Then,* A *is an* $(\in, \in \vee q)$*-intuitionistic fuzzy Lie subalgebra of* L *if and only if*
(i) $\mu_A(x + y) \geqslant \min(\mu_A(x), \mu_A(y), 0.5)$, $\nu_A(x + y) \leqslant \max(\nu_A(x), \nu_A(y), 0.5)$,
(j) $\mu_a(mx) \geqslant \min(\mu_A(x), 0.5)$, $\nu_a(mx) \leqslant \max(\nu_A(x), 0.5)$,
(k) $\mu_A([x, y]) \geqslant \max(\mu_A(x), \mu_A(y), 0.5)$, $\nu_A([x, y]) \leqslant \min(\nu_A(x), \nu_A(y), 0.5)$
hold for all $x, y \in L$, $m \in \mathbb{F}$.

Proof $(f) \Rightarrow (i)$: Let $x, y \in L$. We consider the following two cases:
(1) $\min(\mu_A(x), \mu_A(y)) < 0.5$, $\max(\nu_A(x), \nu_A(y)) > 0.5$,
(2) $\min(\mu_A(x, \mu_A(y)) \geqslant 0.5$, $\max(\nu_A(x, \nu_A(y)) \leqslant 0.5$.
Case (1): Assume that $\mu_A(x + y) < \min(\mu_A(x), \mu_A(y), 0.5)$, $\nu_A(x + y) > \max(\nu_A(x), \nu_A(y), 0.5)$. Then $\mu_A(x + y) < \min(\mu_A(x), \mu_A(y))$, $\nu_A(x + y) >$

$\max(\nu_A(x), \nu_A(y))$. Take s, t such that $\mu_A(x+y) < s < \min(\mu_A(x), \mu_A(y))$, $\nu_A(x+y) > t > \max(\nu_A(x), \nu_A(y))$. Then $x_s, y_s \in \mu_A$ and $x_t, y_t \in \nu_A$, but $(x+y)(\min(s_1, s_2), \max(t_1, t_2))\overline{\in \vee q}A$, which is contradiction with (f). Case (2): Assume that $\mu_A(x+y) < 0.5$, $\nu_A(x+y) > 0.5$. Then $x(0.5, 0.5), y(0.5, 0.5) \in A$ but $(x+y)(0.5, 0.5)\overline{\in \vee q}A$, a contradiction. Hence (i) holds. $(i) \Rightarrow (f)$: Let $x(s_1, t_1), y(s_2, t_2) \in A$, then $\mu_A(x) \geqslant s_1, \mu_A(y) \geqslant s_2, \nu_A(x) \leqslant t_1, \nu_A(y) \leqslant t_2$. Now, we have

$$\mu_A(x+y) \geqslant \min(\mu_A(x), \mu_A(y), 0.5) \geqslant \min(s_1, s_2, 0.5),$$

$$\nu_A(x+y) \leqslant \max(\nu_A(x), \nu_A(y), 0.5) \leqslant \max(t_1, t_2, 0.5).$$

If $\min(s_1, s_2) > 0.5$, $\max(t_1, t_2 < 0.5$, then $\mu_A(x+y) \geqslant 0.5 \Rightarrow \mu_A(x+y) + \min(s_1, s_2) > 1$, $\nu_A(x+y) \leqslant 0.5 \Rightarrow \nu_A(x+y) + \max(t_1, t_2) < 1$. On the other hand, if $\min(s_1, s_2) \leqslant 0.5$, $\max(t_1, t_2) \geqslant 0.5$, then $\mu_A(x+y) \geqslant \min(s_1, s_2)$, $\nu_A(x+y) \leqslant \max(t_1, t_2)$. Hence $(x+y)(\min(s_1, s_2), \max(t_1, t_2)) \in \vee q A$. The verification of $(g) \Leftrightarrow (j)$ and $(h) \Leftrightarrow (k)$ is analogous, and we omit the details. This completes the proof.

Theorem 4.31 *Let $A = (\mu_A, \nu_A)$ be an intuitionistic fuzzy set of Lie algebra L. Then, A is an $(\in, \in \vee q)$-intuitionistic fuzzy Lie subalgebra of L if and only if each nonempty $A_{(s,t)}$, $s \in (0.5, 1]$, $t \in [0.5, 1)$ is a Lie subalgebra of L.*

Proof Assume that $A = (\mu_A, \nu_A)$ is an $(\in, \in \vee q)$-intuitionistic fuzzy Lie subalgebra of L and let $s \in (0.5, 1]$, $t \in [0.5, 1)$. If $x, y \in A_{(s,t)}$ and $m \in \mathbb{F}$, then $\mu_A(x) \geq s$ and $\mu_A(y) \geq s$, $\nu_A(x) \leq t$ and $\nu_A(y) \leq t$. Thus,

$$\mu_A(x+y) \geqslant \min(\mu_A(x), \mu_A(y), 0.5) \geqslant \min(s, 0.5) = s,$$

$$\nu_A(x+y) \leqslant \max(\nu_A(x), \nu_A(y), 0.5) \leqslant \max(t, 0.5) = t,$$

$$\mu_A(mx) \geqslant \min(\mu_A(x), 0.5) \geqslant \min(s, 0.5) = s,$$

$$\nu_A(mx) \leqslant \max(\nu_A(x), 0.5) \leqslant \max(t, 0.5) = t,$$

$$\mu_A([x, y]) \geqslant \min(\mu_A(x), \mu_A(y), 0.5) \geqslant \min(t, 0.5) = t,$$

$$\nu_A([x, y]) \leqslant \max(\nu_A(x), \nu_A(y), 0.5) \leqslant \max(t, 0.5) = t,$$

and so $x + y, mx, [x, y] \in A_{(s,t)}$. This shows that $A_{(s,t)}$ are Lie subalgebras of L. The proof of converse part is obvious. This ends the proof.

Theorem 4.32 *Let A be an intuitionistic fuzzy set in a Lie algebra L. Then, $A_{(s,t)}$ is a Lie subalgebra of L if and only if*

(1) $\max(\mu_A(x+y), 0.5) \geqslant \min(\mu_A(x), \mu_A(y))$, $\min(\nu_A(x+y), 0.5) \leqslant \max(\nu_A(x), \nu_A(y))$,

(2) $\max(\mu_A(mx), 0.5) \geqslant \mu_A(x)$, $\min(\nu_A(mx), 0.5) \leqslant \nu_A(x)$,

(3) $\max(\mu_A([x, y]), 0.5)) \geqslant \min(\mu_A(x), \mu_A(y))$, $\min(\nu_A([x, y]), 0.5)) \leqslant \max(\nu_A(x), \nu_A(y))$
for all $x, y \in L$, $m \in \mathbb{F}$.

Proof Suppose that $A_{(s,t)}$ is a Lie subalgebra of L. Let $\max(\mu_A(x + y), 0.5) < \min(\mu_A(x), \mu_A(y)) = s$, $\min(\nu_A(x + y), 0.5) > \max(\nu_A(x), \nu_A(y)) = t$ for some $x, y \in L$, then $s \in (0.5, 1], t \in [0.5, 1)$, $\mu_A(x + y) < s$, $\nu_A(x + y) > t$, $x, y \in A_{(s,t)}$. Since $x, y \in A_{(s,t)}$ and $A_{(s,t)}$ is a Lie subalgebra of L, so $x + y \in A_{(s,t)}$ or $\mu_A(x + y) \geqslant s$, $\nu_A(x + y) \leqslant t$, which is contradiction with $\mu_A(x + y) < s$, $\nu_A(x + y) > t$. Hence (1) holds. For (2), (3) the verification is analogous.

Conversely, suppose that (1–3) holds. Assume that $s \in (0.5, 1]$, $t \in [0.5, 1)$, $x, y \in A_{(s,t)}$. Then

$$0.5 < s \leqslant \min(\mu_A(x), \mu_A(y)) \leqslant \max(\mu_A(x + y), 0.5) \Rightarrow \mu_A(x + y) \geqslant s,$$

$$0.5 > t \geqslant \max(\nu_A(x), \nu_A(y)) \geqslant \min(\nu_A(x + y), 0.5) \Rightarrow \nu_A(x + y) \leqslant t,$$

$$0.5 < s \leqslant \mu_A(x) \leqslant \max(\mu_A(mx), 0.5) \Rightarrow \mu_A(mx) \geqslant s,$$

$$0.5 > t \geqslant \nu_A(x) \geqslant \min(\nu_A(mx), 0.5) \Rightarrow \nu_A(mx) \leqslant t,$$

$$0.5 < s \leqslant \min(\mu_A(x), \mu_A(y)) \leqslant \max(\mu_A([x, y], 0.5) \Rightarrow \mu_A([x, y]) \geqslant s,$$

$$0.5 > t \geqslant \max(\nu_A(x), \nu_A(y)) \geqslant \min(\nu_A([x, y], 0.5) \Rightarrow \nu_A([x, y]) \leqslant t,$$

and so $x + y, mx, [x, y] \in A_{(s,t)}$. This shows that $A_{(s,t)}$ is a Lie subalgebra of L.

Theorem 4.33 *The intersection of any family of* $(\in, \in \vee q)$*-intuitionistic fuzzy Lie subalgebras of L is an* $(\in, \in \vee q)$*-intuitionistic fuzzy Lie subalgebra.*

Proof Let $\{A_i : i \in I\}$ be a family of $(\in, \in \vee q)$-intuitionistic fuzzy Lie subalgebras of L, and let $A := \bigcap_{i \in I} A_i = (\sup_{i \in I} \mu_i, \inf_{i \in I} \nu_i)$. Let x, $y \in L$, then by Theorem 4.30, we have $\mu_A(x + y) \geqslant \min(\mu_A(x), \mu_A(y), 0.5)$, $\nu_A(x + y) \leqslant \max(\nu_A(x), \nu_A(y), 0.5)$, and hence

$$\begin{aligned}
\mu_A(x + y) &= \sup_{i \in I} \mu_i(x + y) \\
&\geqslant \sup_{i \in I} \min(\mu_i(x), \mu_i(y), 0.5) \\
&= \min(\sup_{i \in I} \mu_i(x), \sup_{i \in I} \mu_i(y), 0.5) \\
&= \min(\bigcap_{i \in I} \mu_i(x), \bigcap_{i \in I} \mu_i(y), 0.5) \\
&= \min(\mu_A(x), \mu_A(y), 0.5),
\end{aligned}$$

$$\begin{aligned}\nu_A(x+y) &= \inf_{i\in I} \nu_i(x+y)\\ &\leqslant \inf_{i\in I} \max(\nu_i(x), \nu_i(y), 0.5)\\ &= \max(\inf_{i\in I} \nu_i(x), \inf_{i\in I} \nu_i(y), 0.5)\\ &= \max(\bigcap_{i\in I} \nu_i(x), \bigcap_{i\in I} \nu_i(y), 0.5)\\ &= \max(\nu_A(x), \nu_A(y), 0.5).\end{aligned}$$

For other conditions, the verification is analogous. By Theorem 4.30, it follows that A is an $(\in, \in \vee q)$-intuitionistic fuzzy Lie subalgebra of L.

Theorem 4.34 *Let $L_0 \subset L_1 \subset \cdots \subset L_n = L$ be a strictly increasing chain of an $(\in, \in)$-intuitionistic fuzzy Lie subalgebras of a Lie algebra L, then there exists $(\in, \in)$-intuitionistic fuzzy Lie subalgebra $A = (\mu_A, \nu_A)$ of L whose level subalgebras are precisely the members of the chain with $A_{0.5} = (\mu_{0.5}, \nu_{0.5}) = L(0, 1)$.*

Proof Let $\{s_i : s_i \in (0, 0.5], i = 1, 2, \ldots, n\}$ be such that $s_1 > s_2 > s_3 > \cdots > s_n$ and $\{t_i : t_i \in [0, 0.5), i = 1, 2, \ldots, n\}$ be such that $t_1 < t_2 < t_3 < \cdots < t_n$. Define $A = (\mu_A, \nu_A) : L \to [0, 1] \times [0, 1]$ by

$$\mu_A(x) = \begin{cases} s, & \text{if x } = 0,\\ n, & \text{if x } = 0, x \in L_0\\ s_1, & \text{if x } \in L_1 \setminus L_0,\\ s_2, & \text{if x } \in L_2 \setminus L_1,\\ \vdots & \\ s_n, & \text{if x } \in L_n \setminus L_{n-1},\end{cases}$$

$$\nu_A(x) = \begin{cases} t, & \text{if x } = 0,\\ n, & \text{if x } = 0, x \in L_0\\ t_1, & \text{if x } \in L_1 \setminus L_0,\\ t_2, & \text{if x } \in L_2 \setminus L_1,\\ \vdots & \\ t_n, & \text{if x } \in L_n \setminus L_{n-1}.\end{cases}$$

Let $x, y \in L$. If $x + y \in L_0$, then

$$\mu_A(x+y) \geq 0.5 \geq \min(\mu_A(x), \mu_A(y), 0.5),$$

$$\nu_A(x+y) \leq 0.5 \leq \max(\nu_A(x), \nu_A(y), 0.5).$$

On the other hand, If $x + y \notin L_0$, then there exists i, $1 \leq i \leq n$ such that $x + y \in L_i \setminus L_{i-1}$ so that $\mu_A(x+y) = s_i$ and $\nu_A(x+y) = t_i$. Now there exists $j (\geq i)$ such that $x \in L_j$ or $y \in L_j$. If $x, y \in L_k (k < i)$, then L_k is a Lie subalgebra of L, $x + y \in L_k$ which contradicts $x + y \notin L_{i-1}$. Thus

$$\mu_A(x + y) \geq s_i \geq s_j \geq \min(\mu_A(x), \mu_A(y), 0.5),$$

$$\nu_A(x + y) \leq t_i \leq t_j \leq \max(\nu_A(x), \nu_A(y), 0.5).$$

The verification is analogous (2–3), and we omit the details. Hence, A is an $(\in, \in)$-intuitionistic fuzzy Lie subalgebra of L. It follows from the contradiction of A that $A_{0.5} = (\mu_{0.5}, \nu_{0.5}) = L(0, 1)$, $A_{(}s_i, t_i) = L_i$ for $i = 1, 2, \ldots, n$. This completes the proof.

Chapter 5
Fuzzy Lie Structures Over a Fuzzy Field

In this chapter, we present the concept of fuzzy Lie ideals of a Lie algebra over a fuzzy field. We characterize the Artinian and Noetherian Lie algebras by considering their fuzzy Lie ideals over a fuzzy field. We describe $(\in, \in \vee q_m)$-fuzzy Lie subalgebras over a fuzzy field. We discuss vague Lie subalgebras over a vague field. We also present anti-fuzzy Lie sub-superalgebras over anti-fuzzy field.

5.1 Introduction

In handling information regarding various aspects of uncertainty, nonclassical logic (a great extension and development of classical logic) is considered to be a more powerful technique than the classical logic. The nonclassical logic has nowadays become a useful tool in computer science. Moreover, nonclassical logic deals with fuzzy information and uncertainty. The fuzzy set theory states that there are propositions with an infinite number of truth values, assuming two extreme values, 1 (totally true), 0 (totally false) and a continuum in between, that justify the term "fuzzy". Present day Science and Technology is featured with complex processes and phenomena for which complete information is not always available. For such cases, mathematical models are developed to handle various types of system containing uncertainty. The fuzzy sets provide us a meaningful and powerful representation of measurement of such uncertainties and vague concepts. Applications of this theory can be found in different domains, including artificial intelligence, engineering, decision theory, operation research, and robotics. Rosenfeld [115] applied this concept to abstract algebra and formulated the notion of fuzzy subgroups and showed how some basic notions of group theory can be extended in an elementary manner to fuzzy subgroups. This object of fuzzy set theory was successively redefined and generalized by Malik et al. [99]. Liu [96] introduced and developed basic results concerning the notion of fuzzy subrings and fuzzy ideals of a ring. Katsaras and Liu,

M. Akram, *Fuzzy Lie Algebras*, Infosys Science Foundation Series,
https://doi.org/10.1007/978-981-13-3221-0_5

in their pioneering paper [86] introduced the notion of a fuzzy subspace of a vector space. Malik and Mordeson [100, 101] introduced the concepts of fuzzy subfields, and fuzzy subspaces of a vector space over fuzzy field. Nanda [107] dealt with fuzzy algebra over a fuzzy field. Yehia [133] considered the notions of fuzzy ideals and fuzzy subalgebras of Lie algebras over a field. Antony and Lilly [26] defined fuzzy Lie algebra over fuzzy field. Later, Akram and Shum [23] discussed properties of fuzzy Lie ideals over fuzzy field.

Definition 5.1 Let $\mathbb{F}$ be a field. A fuzzy set λ of $\mathbb{F}$ is called a *fuzzy field* if the following conditions are satisfied:

- $(\forall\, \alpha, \beta \in \mathbb{F})(\lambda(\alpha - \beta) \geq \min\{\lambda(\alpha), \lambda(\beta)\})$,
- $(\forall\, \alpha,\ \beta \in \mathbb{F}, \beta \neq 0)(\lambda(\alpha\beta^{-1}) \geq \min\{\lambda(\alpha), \lambda(\beta)\})$.

Lemma 5.1 *If λ is a fuzzy subfield of $\mathbb{F}$, then $\lambda(0) \geq \lambda(1) \geq \lambda(\alpha) = \lambda(-\alpha)$ for all $\alpha \in \mathbb{F}$ and $\lambda(-\alpha) = \lambda(\alpha^{-1})$ for all $\alpha \in \mathbb{F} - \{0\}$.*

Lemma 5.2 *Let λ be a fuzzy subfield of $\mathbb{F}$. Then for $t \in [0, 1]$, the fuzzy-cut $U(\lambda, t)$ is a crisp subfield of $\mathbb{F}$.*

Definition 5.2 Let L be a Lie algebra. Let μ be a fuzzy set of L and λ a fuzzy field of $\mathbb{F}$. Then, μ is called a *fuzzy Lie subalgebra over a fuzzy field* λ if the following conditions:

(1) $\mu(x + y) \geq \min\{\mu(x), \mu(y)\}$,
(2) $\mu(\alpha x) \geq \max\{\lambda(\alpha), \mu(x)\}$,
(3) $\mu([x, y]) \geq \min\{\mu(x), \mu(x)\}$

hold for all $x, y \in L$ and $\alpha \in \mathbb{F}$.

Definition 5.3 Let L be a Lie algebra. Let μ be a fuzzy set of L and λ a fuzzy field of $\mathbb{F}$. Then, μ is called a *fuzzy Lie ideal over a fuzzy field* λ if the following conditions:

(a) $\mu(x + y) \geq \min\{\mu(x), \mu(y)\}$,
(b) $\mu(\alpha x) \geq \max\{\lambda(\alpha), \mu(x)\}$,
(c) $\mu([x, y]) \geq \mu(x)$

hold for all $x, y \in L$ and $\alpha \in \mathbb{F}$.

From condition (b), it follows that $\mu(0) \geq \lambda(0)$.

Example 5.1 Let $\Re^2 = \{(x, y) : x, y \in \mathbb{R}\}$ be the set of all *two*-dimensional real vectors. Then, it is clear that $\Re^2$ endowed with the operation defined by $[x, y] = x \times y$ form a real Lie algebra. Define a fuzzy set $\mu : \Re^2 \to [0, 1]$ by

$$\mu(x, y) = \begin{cases} 0 & \text{if } x = y = 0, \\ 1 & \text{otherwise.} \end{cases}$$

and define $\lambda : \mathbb{R} \to [0, 1]$ for all $\alpha \in \mathbb{R}$ by

$$\lambda(\alpha) = \begin{cases} 0 & \text{if } \alpha \in \mathbb{Q}, \\ 1 & \text{if } \alpha \in \mathbb{R} - \mathbb{Q}(\sqrt{3}). \end{cases}$$

By routine computations, one can easily check that μ is both fuzzy Lie subalgebra and fuzzy Lie ideal over fuzzy field.

We now formulate the following theorem of fuzzy Lie ideals over fuzzy field of L.

Theorem 5.1 *Let μ be a fuzzy Lie ideal over fuzzy field of L and ν the closure of the image of μ. Then, the following conditions are equivalent:*

1. *μ is a fuzzy Lie ideal over fuzzy field of L,*
2. *the nonempty strong level subset $^{>}U(\mu, t)$ of μ is a Lie ideal of L, for all $t \in [0, 1]$,*
3. *the nonempty strong level subset $^{>}U(\mu, t)$ of μ is a Lie ideal of L, for all $t \in Im(\mu) \setminus \nu$,*
4. *the nonempty level subset $U(\mu, t)$ of μ is a Lie ideal of L, for all $t \in Im(\mu)$,*
5. *the nonempty level subset $U(\mu, t)$ of μ is a Lie ideal of L, for all $t \in [0, 1]$.*

Definition 5.4 Let μ be a fuzzy Lie ideal over fuzzy field in L and $\mu_n = [\mu, \mu_{n-1}]$ for $n > 0$, where $\mu_0 = \mu$. If there exists a positive integer n such that $\mu_n = 0$, then a fuzzy Lie ideal over fuzzy field is called *nilpotent*.

Definition 5.5 Let μ be a fuzzy Lie ideal over fuzzy field in L. Define a sequence of fuzzy Lie ideals over fuzzy fields in L by $\mu^0 = \mu$, $\mu^n = [\mu^{n-1}, \mu^{n-1}]$ for $n > 0$. If there exists a positive integer n such that $\mu^n = 0$, then a fuzzy Lie ideal over fuzzy field is called *solvable*.

Proposition 5.1
1. *The homomorphic image of a solvable fuzzy Lie ideal over fuzzy field is a solvable fuzzy Lie ideal over fuzzy field.*
2. *The homomorphic image of a nilpotent fuzzy Lie ideal over fuzzy field is a nilpotent fuzzy Lie ideal over fuzzy field.*
3. *If μ is a nilpotent fuzzy Lie ideal over fuzzy field, then it is solvable.*

5.2 $(\in, \in \vee q_m)$-Fuzzy Lie Subalgebras Over a Fuzzy Field

Definition 5.6 A fuzzy point $F(x, t)$ is called *belong to* a fuzzy set μ, written as $F(x, t) \in \mu$, if $\mu(x) \geq t$. A fuzzy point $F(x, t)$ is said to be *quasicoincident with* a fuzzy set μ, written as $F(x, t)q\mu$, if $\mu(x) + t > 1$.
For brevity, we use the following notations:

- $F(x, t) \in \mu$ or $F(x, t)q\mu$ is written as $F(x, t) \in \vee q\mu$.
- $F(x, t) \in \mu$ and $F(x, t)q\mu$ is written as $F(x, t) \in \wedge q\mu$.

Let m be an element of [0, 1) unless otherwise specified. By $F(x,t)\ q_m\ \mu$, we mean $\mu(x)+t+m>1$. The notation $F(x,t) \in \vee\ q_m\ \mu$ means $F(x,t) \in \mu$ or $F(x,t)\ q_m\ \mu$.

We formulate a technical lemma.

Lemma 5.3 *Let μ be a fuzzy set of a Lie algebra L. Then, the nonempty level set $U(\mu, t)$ is a Lie subalgebra of L for all $t \in (\frac{1-m}{2}, 1]$ if and only if*

(a) $\max(\mu(x+y), \frac{1-m}{2}) \geq \min(\mu(x), \mu(y))$,
(b) $\max(\mu(\alpha x), \frac{1-m}{2}) \geq \min(\lambda(\alpha), \mu(x))$,
(c) $\max(\mu([x, y]), \frac{1-m}{2}) \geq \min(\mu(x), \mu(y))$

for all $x, y \in L$, $\alpha \in \mathbb{F}$.

Proof Let $t \in (\frac{1-m}{2}, 1]$ be such that $U(\mu,t)(\neq \emptyset)$ and $U(\mu,t)$ is a Lie subalgebra of L.

(a) Assume that $\max(\mu(x+y), \frac{1-m}{2}) < \min(\mu(x), \mu(y)) = t$ for some $x, y \in L$, then $t \in (\frac{1-m}{2}, 1]$, $\mu(x+y) < t$, $x \in U(\mu,t)$ and $y \in U(\mu,t)$. Since $x, y \in U(\mu,t)$, $U(\mu,t)$ is a Lie subalgebra of L, so $x+y \in U(\mu,t)$, a contradiction. Hence, (a) holds.
(b) Assume that $\max(\mu(\alpha x), \frac{1-m}{2}) < \min(\lambda(\alpha), \mu(x)) = t$ for some $\alpha \in \mathbb{F}$, $x \in L$, then $t \in (\frac{1-m}{2}, 1]$, $\mu(\alpha x) < t$, $x \in U(\mu,t)$ and $\alpha \in U(\mu,t)$. Since $\alpha, x \in U(\mu,t)$, $U(\mu,t)$ is a Lie subalgebra of L, so $\alpha x \in U(\mu,t)$, a contradiction. Hence, (b) holds.
(c) Assume that $\max(\mu([x,y]), \frac{1-m}{2}) < \min(\mu(x), \mu(y)) = t$ for some $x, y \in L$, then $t \in (\frac{1-m}{2}, 1]$, $\mu([x,y]) < t$, $x \in U(\mu,t)$ and $y \in U(\mu,t)$. Since $x, y \in U(\mu,t)$, $U(\mu,t)$ is a Lie subalgebra of L, so $[x,y] \in U(\mu,t)$, a contradiction. Hence, (c) holds.

The proof of the sufficiency part is straightforward and is hence omitted. This completes the proof.

Corollary 5.1 *Let μ be a fuzzy set of a Lie algebra L. Then, the nonempty level set $U(\mu,t)$ is a Lie subalgebra of L for all $t \in (0.5, 1]$ if and only if*

(a) $\max(\mu(x+y), 0.5) \geq \min(\mu(x), \mu(y))$,
(b) $\max(\mu(\alpha x), 0.5) \geq \min(\lambda(\alpha), \mu(x))$,
(c) $\max(\mu([x,y]), 0.5) \geq \min(\mu(x), \mu(y))$

for all $x, y \in L$, $\alpha \in \mathbb{F}$.

Definition 5.7 A fuzzy set μ in L is called an $(\in, \in \vee q_m) - fuzzy\ Lie\ subalgebra$ over a fuzzy field λ of L, if it satisfies the following conditions:

(1) $F(x,t_1), F(y,t_2) \in \mu \Longrightarrow F(x+y, \min(t_1,t_2)) \in \vee\ q_m\ \mu$,
(2) $F(\alpha,t_1) \in \lambda, F(x,t_2) \in \mu \Longrightarrow F(\alpha x, \min(t_1,t_2)) \in \vee\ q_m \mu$,
(3) $F(x,t_1), F(y,t_2) \in \mu \Longrightarrow F([x,y], \min(t_1,t_2)) \in \vee\ q_m\ \mu$

for all $x, y \in L, \alpha \in \mathbb{F}, t_1, t_2 \in (0, 1]$.

From (2), it follows that:

(4) $F(-1, t_1) \in \lambda,\ F(x, t_2) \in \mu \Longrightarrow F(-x, \min(t_1, t_2)) \in \vee\, q_m\ \mu$.
(5) $F(0, t_1) \in \lambda,\ F(x, t_2) \in \mu \Longrightarrow F(0, \min(t_1, t_2)) \in \vee\, q_m\ \mu$.

From Definition 5.7, it follows that we can develop different types of fuzzy Lie subalgebras for different values of $m \in [0, 1)$. Hence, an $(\in, \in \vee\, q_m)$-*fuzzyLie subalgebra* over a fuzzy field λ with $m = 0$ is called an $(\in, \in \vee\, q)$-fuzzy Lie subalgebra over a fuzzy field.

Example 5.2 Let $\Re^3 = \{(x, y, z) : x, y, z \in \mathbb{R}\}$ be the set of all *three*-dimensional real vectors. Then, $\Re^3$ with $[x, y] = x \times y$ form a real Lie algebra. Define a fuzzy set $\mu : \Re^3 \to [0, 1]$ by

$$\mu(x, y, z) = \begin{cases} 0.4 & \text{if } x = y = z = 0, \\ 0.5 & \text{otherwise.} \end{cases}$$

and define fuzzy set $\lambda : \mathbb{R} \to [0, 1]$ for all $\alpha \in \mathbb{R}$ by

$$\lambda(\alpha) = \begin{cases} 0.4 & \text{if } \alpha \in \mathbb{Q}, \\ 0.5 & \text{if } \alpha \in \mathbb{R} - \mathbb{Q}(\sqrt{3}). \end{cases}$$

By routine computations, we can easily check that μ forms $(\in, \in \vee\, q)$-, $(\in, \in \vee\, q_{0.2})$-, $(\in, \in \vee q_{0.4})$-fuzzy Lie subalgebras of L over the fuzzy field for $m = 0, 0.2, 0.4$, respectively.

The proof of the following proposition is obvious.

Proposition 5.2 *Every $(\in, \in)$-fuzzy Lie subalgebra is an $(\in, \in \vee q_m)$-fuzzy Lie subalgebra.*

Corollary 5.2 *Every $(\in, \in)$-fuzzy Lie subalgebra is an $(\in, \in \vee q)$-fuzzy Lie subalgebra.*

Theorem 5.2 *A fuzzy set μ in L is an $(\in, \in \vee q_m)$-fuzzy Lie subalgebra of L over fuzzy field λ if and only if it satisfies*

(I) $\mu(x + y) \geq \min\{\mu(x), \mu(y), \frac{1-m}{2}\}$,
(II) $\mu(\alpha x) \geq \min\{\lambda(\alpha), \mu(x), \frac{1-m}{2}\}$,
(III) $\mu([x, y]) \geq \min\{\mu(x), \mu(y), \frac{1-m}{2}\}$

for all $x, y \in L, \alpha \in \mathbb{F}$.

Proof Let μ be an $(\in, \in \vee q_m)$-fuzzy Lie subalgebra of L over fuzzy field. Assume that (I) is not valid. Then there exist $x_0, y_0 \in L$ such that

$$\mu(x_0 + y_0) < \min\left\{\mu(x_0), \mu(y_0), \frac{1 - m}{2}\right\}.$$

If $\min(\mu(x_0), \mu(y_0)) < \frac{1-m}{2}$, then $\mu(x_0 + y_0) < \min(\mu(x_0), \mu(y_0))$. Thus,

$$\mu(x_0 + y_0) < t \leq \min\{\mu(x_0), \mu(y_0)\} \text{ for some } t \in (0, 1].$$

It follows that $F(x_0, t) \in \mu$ and $F(y_0, t) \in \mu$, but $F(x_0 + y_0, t)\overline{\in}\mu$, a contradiction. Moreover, $\mu(x_0 + y_0) + t < 2t < 1 - m$, and so $F(x_0 + y_0, t)\overline{q_m}\mu$. Consequently $F(x_0 + y_0, t)\overline{\in \vee q_m}\ \mu$, a contradiction. On the other hand, if $\min(\mu(x_0), \mu(y_0) \geq \frac{1-m}{2}$, then $\mu(x_0) \geq \frac{1-m}{2}$, $\mu(y_0) \geq \frac{1-m}{2}$ and $\mu(x_0 + y_0) < \frac{1-m}{2}$. Thus, $F(x_0, \frac{1-m}{2}) \in \mu$, $F(y_0, \frac{1-m}{2}) \in \mu$, but $F(x_0 + y_0, \frac{1-m}{2})\overline{\in}\mu$. Also

$$\mu(x_0 + y_0) + \frac{1-m}{2} < \frac{1-m}{2} + \frac{1-m}{2} = 1 - m,$$

i.e., $F(x_0 + y_0, \frac{1-m}{2})\overline{q_m}\mu$. Hence, $F(x_0 + y_0, \frac{1-m}{2})\overline{\in \vee q_m}\ \mu$, a contradiction. So (I) is valid. By using a very similar argumentation as in the proof of the (I), we can easily prove that (II) and (III) also valid.

Conversely, assume that μ satisfies (I). Let $x, y \in L$ and $t_1, t_2 \in (0, 1]$ be such that $F(x, t_1) \in \mu$ and $F(y, t_2) \in \mu$. Then,

$$\begin{aligned} \mu(x + y) &\geq \min\left(\mu(x), \mu(y), \frac{1-m}{2}\right) \\ &\geq \min\left(t_1, t_2, \frac{1-m}{2}\right). \end{aligned}$$

Assume that $t_1 \leq \frac{1-m}{2}$ or $t_2 \leq \frac{1-m}{2}$. Then, $\mu(x + y) \geq \min(t_1, t_2)$, which implies that $F(x + y, \min(t_1, t_2)) \in \mu$. Now suppose that $t_1 > \frac{1-m}{2}$ and $t_2 > \frac{1-m}{2}$. Then, $\mu(x + y) \geq \frac{1-m}{2}$, and thus

$$\mu(x + y) + \min(t_1, t_2) > \frac{1-m}{2} + \frac{1-m}{2} = 1 - m,$$

i.e., $F(x + y, \min(t_1, t_2))q_m\mu$. Hence, $F(x + y, \min(t_1, t_2)) \in \vee q_m\mu$. By using a very similar argumentation, it is easy to see $F(\alpha x, \min(t_1, t_2)) \in \vee q_m\mu$ and $F([x, y], \min(t_1, t_2)) \in \vee q_m\mu$. Consequently, μ is an $(\in, \in \vee q_m)$-fuzzy Lie subalgebra of L over a fuzzy field. This ends the proof.

Corollary 5.3 *A fuzzy set μ in L is an $(\in, \in \vee q)$-fuzzy Lie subalgebra of L over fuzzy field λ if and only if it satisfies*

(I) $\mu(x + y) \geq \min\{\mu(x), \mu(y), 0.5\}$,
(II) $\mu(\alpha x) \geq \min\{\lambda(\alpha), \mu(x), 0.5\}$,
(III) $\mu([x, y]) \geq \min\{\mu(x), \mu(y), 0.5\}$.

Theorem 5.3 *Let μ be a fuzzy set of fuzzy Lie subalgebra of L. Then, μ is an $(\in, \in \vee q_m)$-fuzzy Lie subalgebra of L over a fuzzy field if and only if the level set $U(\mu, t)$, $t \in (0, \frac{1-m}{2}]$, is a Lie subalgebra of L.*

Proof Assume that μ is an $(\in, \in \vee q)$-fuzzy Lie subalgebra of L over a fuzzy field λ. Let $t \in (0, \frac{1-m}{2}]$ and $x, y, \alpha \in U(\mu, t)$. Then, $\mu(x) \geq t$, $\mu(y) \geq t$ and $\lambda(\alpha) \geq t$. It follows from Theorem 5.2 that:

$$\begin{aligned}\mu(x+y) &\geq \min\left(\mu(x), \mu(y), \frac{1-m}{2}\right) \\ &\geq \min\left(t, \frac{1-m}{2}\right) = t, \\ \mu(\alpha x) &\geq \min\left(\lambda(\alpha), \mu(x), \frac{1-m}{2}\right) \\ &\geq \min\left(t, \frac{1-m}{2}\right) = t, \\ \mu([x, y]) &\geq \min\left(\mu(x), \mu(y), \frac{1-m}{2}\right) \\ &\geq \min\left(t, \frac{1-m}{2}\right) = t,\end{aligned}$$

so that $x + y, \alpha x, [x, y] \in U(\mu, t)$. Hence, $U(\mu, t)$ is a Lie subalgebra of L.

Conversely, suppose that the nonempty set $U(\mu, t)$ is a Lie subalgebra of L for all $t \in (0, \frac{1-m}{2}]$. If the condition (I) is not true, then there exist $a, b \in L$ such that $\mu(a+b) < \min(\mu(a), \mu(b), \frac{1-m}{2})$. Hence, we can take $t \in (0, 1]$ such that $\mu(a+b) < t_1 < \min(\mu(a), \mu(b), \frac{1-m}{2})$. Then $t \in \frac{1-m}{2}$ and $a, b \in U(\mu, t)$. Since $U(\mu, t)$ is a Lie subalgebra of L, it follows that $a + b \in U(\mu, t)$ so that $\mu(a + b) \geq t$. This is a contradiction. Therefore, the condition (I) is valid. By using a very similar argumentation as in the proof of the (I), we can easily prove (II) and (III) are also valid. Hence, μ is an $(\in, \in \vee q_m)$-fuzzy Lie subalgebra of L over a fuzzy field.

Corollary 5.4 *Let μ be a fuzzy set of fuzzy Lie subalgebra of L. Then, μ is an $(\in, \in \vee q)$-fuzzy Lie subalgebra of L over a fuzzy field if and only if the level set $U(\mu, t)$, $t \in (0, 0.5]$, is a Lie subalgebra of L.*

Theorem 5.4 *Let μ be an $(\in, \in \vee q_m)$-fuzzy Lie subalgebra of L over a fuzzy field such that $\mu(x) < \frac{1-m}{2}$ for all $x \in L$, then μ is a fuzzy Lie subalgebra of L over a fuzzy field.*

Proof Let $x, y \in L, \alpha \in \mathbb{F}$ and $t_1, t_2 \in (0, 1]$ be such that $F(x, t_1) \in \mu$, $F(y, t_2) \in \mu$, $F(\alpha, t_3) \in \lambda$. Then, $\mu(x) \geq t_1$, $\mu(y) \geq t_2$ and $\lambda(\alpha) \geq t_3$. It follows from Theorem 5.2 that

$$\begin{aligned}\mu(x+y) &\geq \min\left(\mu(x), \mu(y), \frac{1-m}{2}\right) \\ &= \min(\mu(x), \mu(y)) = \min(t_1, t_2), \\ \mu(\alpha x) &\geq \min\left(\lambda(\alpha), \mu(x), \frac{1-m}{2}\right)\end{aligned}$$

$$= \min(\mu(x), \mu(y)) = \min(t_3, t_1),$$
$$\mu([x, y]) \geq \min\left(\mu(x), \mu(y), \frac{1-m}{2}\right)$$
$$= \min(\mu(x), \mu(y)) = \min(t_1, t_2)$$

so that $F(x + y, \min(t_1, t_2)) \in \mu$, $F(\alpha x, \min(t_1, t_3)) \in \mu$, $F([x, y], \min(t_1, t_2)) \in \mu$. Hence, μ is a fuzzy Lie subalgebra of L over a fuzzy field.

Corollary 5.5 *Let μ be an $(\in, \in \vee q)$-fuzzy Lie subalgebra of L over a fuzzy field such that $\mu(x) < 0.5$ for all $x \in L$, then μ is a fuzzy Lie subalgebra of L over a fuzzy field.*

Theorem 5.5 *Let μ be an $(\in, \in \vee q_m)$-fuzzy Lie subalgebra of L over a fuzzy field.*

(i) If there exists $x \in L$ such that $\mu(x) \geq \frac{1-m}{2}$, then $\mu(0) \geq \frac{1-m}{2}$.
(ii) If $\mu(0) < \frac{1-m}{2}$, then μ is a fuzzy Lie subalgebra of L.

Corollary 5.6 *Let μ be an $(\in, \in \vee q)$-fuzzy Lie subalgebra of L over a fuzzy field.*

(i) If there exists $x \in L$ such that $\mu(x) \geq 0.5$, then $\mu(0) \geq 0.5$.
(ii) If $\mu(0) < 0.5$, then μ is a fuzzy Lie subalgebra of L.

Definition 5.8 An $(\in, \in \vee q_m) - fuzzyLiesubalgebra$ of L over a fuzzy field is said to be *proper* if $Im(\mu)$ has at least two elements. Two $(\in, \in \vee q_m)$- fuzzy Lie subalgebras μ and λ are said to be *equivalent* if they have same family of level Lie subalgebras. Otherwise, they are said to be nonequivalent.

Theorem 5.6 *A proper $(\in, \in \vee q_m)$-fuzzy Lie subalgebra of L over a fuzzy field such that cardinality of $\{\mu(x) : \mu(x) < \frac{1-m}{2}\} \geq 2$. Then, there exist two proper nonequivalent $(\in, \in \vee q_m)$-fuzzy Lie subalgebras of L over a fuzzy field such that μ can be expressed as the union of them.*

Proof Let μ be a proper $(\in, \in \vee q_m)$-fuzzy Lie subalgebra of L over a fuzzy field with $\{\mu(x) : \mu(x) < \frac{1-m}{2}\} = \{t_1, t_2, \ldots, t_n\}$ where $t_1 > t_2 > \cdots > t_n$ and $n \geq 2$. Then,

$$[\mu]_{\frac{1-m}{2}} \subseteq [\mu]_{t_1} \subseteq \cdots \subseteq [\mu]_{t_n} = L$$

is the chain of $(\in, \in \vee q_m)$-Lie subalgebras of μ. Define two fuzzy sets $\lambda_1, \lambda_2 \leq \mu$ defined by

$$\lambda_1(x) = \begin{cases} t_1, & \text{if x} \in [\mu]_{t_1}, \\ t_2, & \text{if x} \in [\mu]_{t_2} \setminus [\mu]_{t_1}, \\ \vdots & \\ t_n, & \text{if x} \in [\mu]_{t_n} \setminus [\mu]_{t_{n-1}}, \end{cases}$$

$$\lambda_2(x) = \begin{cases} \mu(x), & \text{if } x \in [\mu]_{\frac{1-m}{2}}, \\ n, & \text{if } x \in [\mu]_{t_2} \setminus [\mu]_{\frac{1-m}{2}}, \\ t_3, & \text{if } x \in [\mu]_{t_3} \setminus [\mu]_{t_2}, \\ \vdots & \\ t_n, & \text{if } x \in [\mu]_{t_n} \setminus [\mu]_{t_{n-1}}, \end{cases}$$

respectively, where $t_3 < n < t_2$. Then, λ_1 and λ_2 are $(\in, \in \vee q_m)$-fuzzy Lie subalgebra of L over a fuzzy field with

$$[\mu]_{t_1} \subseteq [\mu]_{t_2} \subseteq \cdots \subseteq [\mu]_{t_n}$$

and

$$[\mu]_{\frac{1-m}{2}} \subseteq [\mu]_{t_2} \subseteq \cdots \subseteq [\mu]_{t_n}$$

being, respectively, chains of $(\in, \in \vee q_m)$-fuzzy Lie subalgebras over a fuzzy field. Hence, λ_1 and λ_2 are nonequivalent and $\mu = \lambda_1 \cup \lambda_2$.

Corollary 5.7 *A proper $(\in, \in \vee q)$-fuzzy Lie subalgebra of L over a fuzzy field such that cardinality of $\{\mu(x) : \mu(x) < \frac{1-m}{2}\} \geq 2$. Then, there exist two proper nonequivalent $(\in, \in \vee q)$-fuzzy Lie subalgebras of L over a fuzzy field such that μ can be expressed as the union of them.*

Theorem 5.7 *Let $\{\mu_i : i \in I\}$ be a family of $(\in, \in \vee q_m)$-fuzzy Lie subalgebras of L over a fuzzy field. Then, $\mu = \cap_i \mu_i$ is an $(\in, \in \vee q_m)$-fuzzy Lie subalgebra of L over a fuzzy field.*

Proof By Theorem 5.2, we have $\mu(x+y) \geq \min(\mu(x), \mu(y), \frac{1-m}{2})$, and hence

$$\begin{aligned} \mu(x+y) &= \sup_{i \in I} \mu_i(x+y) \\ &\geq \sup_{i \in I} \min\left\{\mu_i(x), \mu_i(y), \frac{1-m}{2}\right\} \\ &= \min\left\{\sup_{i \in I} \mu_i(x), \sup_{i \in I} \mu_i(y), \frac{1-m}{2}\right\} \\ &= \min\left\{\bigcap_{i \in I} \mu_i(x), \bigcap_{i \in I} \mu_i(y), \frac{1-m}{2}\right\} \\ &= \min\left\{\mu(x), \mu(y), \frac{1-m}{2}\right\}. \end{aligned}$$

For other conditions, the verification is analogous. By Theorem 5.2, it follows that μ is an $(\in, \in \vee q_m)$-fuzzy Lie subalgebra of L over a fuzzy field.

Taking $m = 0$ in Theorem 5.7, we obtain the following corollary.

Corollary 5.8 *Let $\{\mu_i : i \in I\}$ be a family of $(\in, \in \vee q)$-fuzzy Lie subalgebras of L over a fuzzy field. Then, $\mu = \cap_i \mu_i$ is an $(\in, \in \vee q)$-fuzzy Lie subalgebra of L over a fuzzy field.*

Theorem 5.8 *Let $\{\mu_i : i \in I\}$ be a family of $(\in, \in \vee q_m)$-fuzzy Lie subalgebras of L over a fuzzy field. such that $\mu_i \subseteq \mu_j$ or $\mu_j \subseteq \mu_i$ for all $i, j \in I$. Then, $\nu := \bigcup_{i\in I} \nu_i$ is an $(\in, \in \vee q_m)$-fuzzy Lie subalgebra of L over a fuzzy field.*

Proof By Theorem 5.2, we have $\mu(x+y) \geq \min(\mu(x), \mu(y), \frac{1-m}{2})$, and hence

$$\begin{aligned}\mu(x+y) &= \inf_{i\in I} \mu_i(x+y)\\ &\geq \inf_{i\in I} \min\left\{\mu_i(x)), \mu_i(y), \tfrac{1-m}{2}\right\}\\ &= \min\left\{\inf_{i\in I} \mu_i(x), \inf_{i\in I} \mu_i(y), \tfrac{1-m}{2}\right\}\\ &= \min\left\{\bigcup_{i\in I} \mu_i(x), \bigcup_{i\in I} \mu_i(y), \tfrac{1-m}{2}\right\}\\ &= \min\left\{\mu(x), \mu(y), \tfrac{1-m}{2}\right\}.\end{aligned}$$

It is easy to see that

$$\inf_{i\in I} \min\left\{\mu_i(x), \mu_i(y), \frac{1-m}{2}\right\} \leq \bigcup_{i\in I} \min\left(\mu_i(x), \mu_i(y), \frac{1-m}{2}\right).$$

Suppose that

$$\inf_{i\in I} \min\left\{\mu_i(x), \mu_i(y), \frac{1-m}{2}\right\} \neq \bigcup_{i\in I} \min\left(\mu_i(x), \mu_i(y), \frac{1-m}{2}\right),$$

then there exists s such that

$$\inf_{i\in I} \min\left\{\mu_i(x), \mu_i(y), \frac{1-m}{2}\right\} < s < \bigcup_{i\in I} \min\left(\mu_i(x), \mu_i(y), \frac{1-m}{2}\right).$$

Since $\mu_i \subseteq \mu_j$ or $\mu_j \subseteq \mu_i$ for all $i, j \in I$, there exists $k \in I$ such that $s < \min(\mu_k(x), \mu_k(y), \frac{1-m}{2})$. On the other hand, $\min(\mu_i(x), \mu_i(y), \frac{1-m}{2}) > s$ for all $i \in I$, a contradiction. Hence,

$$\begin{aligned}\inf_{i\in I} \min\left\{\mu_i(x), \mu_i(y), \tfrac{1-m}{2}\right\} &= \min\left(\bigcup_{i\in I} \mu_i(x), \bigcup_{i\in I} \mu_i(y), \tfrac{1-m}{2}\right)\\ &= \min\left\{\mu(x), \mu(y), \tfrac{1-m}{2}\right\}.\end{aligned}$$

For other conditions, the verification is analogous. By Theorem 5.2, it follows that μ is an $(\in, \in \vee q_m)$-fuzzy Lie subalgebra of L over a fuzzy field.

Taking $m = 0$ in Theorem 5.8, we obtain the following Corollary.

Corollary 5.9 *Let $\{\mu_i : i \in I\}$ be a family of $(\in, \in \vee q)$-fuzzy Lie subalgebras of L over a fuzzy field. such that $\mu_i \subseteq \mu_j$ or $\mu_j \subseteq \mu_i$ for all $i, j \in I$. Then, $\nu := \bigcup_{i\in I} \nu_i$ is an $(\in, \in \vee q)$-fuzzy Lie subalgebra of L over a fuzzy field.*

5.3 Vague Lie Subalgebras Over a Vague Field

Different authors from time to time have made a number of generalizations of Zadeh's fuzzy set theory. The notion of vague set was introduced by Gau and Buehrer [76]. This is because in most cases of judgments, the evaluation is done by human beings and so the certainty is a limitation of knowledge or intellectual functionaries. Naturally, every decision maker hesitates more or less on every evaluation activity. For example, in order to judge whether a patient has cancer or not, a medical doctor (the decision maker) will hesitate because of the fact that a fraction of evaluation he thinks in favor of the truthness, another fraction in favor of the falseness and the rest part remains undecided to him. This is the breaking philosophy in the notion of vague set theory introduced by Gau and Buehrer.

Definition 5.9 A *vague set* A in the universe of discourse X is a pair (t_A, f_A), where $t_A : X \rightarrow [0, 1]$, $f_A : X \rightarrow [0, 1]$ are true and false memberships, respectively, such that $t_A(x) + f_A(x) \leq 1$ for all $x \in X$. The interval $[t_A(x), 1 - f_A(x)]$ is called the *vague value* of x in A and is denoted by $V_A(x)$.

A vague set operation is an operation on vague sets. These operations are generalization of fuzzy set operations.

Definition 5.10 Let $A = (t_A, f_A)$ and $B = (t_B, f_B)$ be two vague sets, then the following operations are defined as:

1. $\overline{A} = (f_A, 1 - t_A)$,
2. $A \subset B \Leftrightarrow V_A(x) \leq V_B(x)$, that is, $t_A(x) \leq t_B(x)$ and $1 - f_A(x) \leq 1 - f_B(x)$,
3. $A = B \Leftrightarrow V_A(x) = V_B(x)$,
4. $C = A \cap B \Leftrightarrow V_C(x) = \min(V_A(x), V_B(x))$,
5. $C = A \cup B \Leftrightarrow V_C(x) = \max(V_A(x), V_B(x))$

for all $x \in X$.

Definition 5.11 A vague set $A = (t_A, f_A)$ of a set X is called

1. the *zero vague set* if $t_A(x) = 0$ and $f_A(x) = 1$ for all $x \in X$.
2. the *unit vague set* if $t_A(x) = 1$ and $f_A(x) = 0$ for all $x \in X$.
3. the *s-vague set* if $t_A(x) = s$ and $f_A(x) = 1 - s$ for all $x \in X$, $s \in [0, 1]$.

We denote zero vague and unit vague value by $\mathbf{0} = [0, 0]$ and $\mathbf{1} = [1, 1]$, respectively.

For $s, t \in [0, 1]$, we define (s, t)-cut and s-cut of a vague set.

Definition 5.12 Let $A = (t_A, f_A)$ be vague set of a universal set X. The (s, t)- cut of a vague set A is a crisp set $A_{(s,t)}$ of X given by

$$A_{(s,t)} = \{x \in X : V_A(x) \geq [s, t]\}.$$

Obviously, $A_{(0,0)} = X$. The (s, t)-cuts are also vague-cuts of the vague set A.

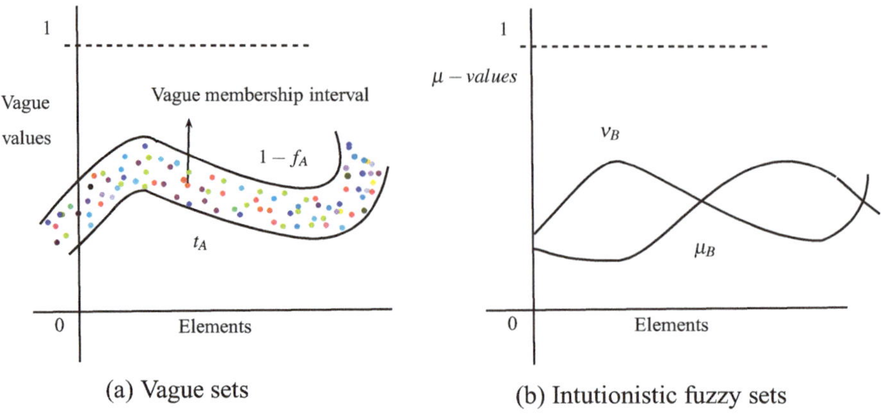

Fig. 5.1 Comparison between vague sets and intutionistic fuzzy sets

Definition 5.13 The s-cut of the vague set $A = (t_A, f_A)$ is a crisp set A_s of the set X given by $A_s = A_{(s,s)}$. Note that $A_0 = X$. Equivalently, we can define the s-cut as

$$A_s = \{x \in X : t_A(x) \geq s\}.$$

Remark 5.1 The intuitionistic fuzzy sets and vague sets look similar,and analytically vague sets are more appropriate when representing vague data. The difference between them is discussed below. The membership interval of element x for vague set A is $[t_A(x).1 - f_A(x)]$. But, the membership value for element x in an intuitionistic fuzzy set B is $< x, \mu_B(x), \nu_B(x) >$. Here the semantics of t_A is the same as with A and μ_B is the same as with B. However, the boundary is able to indicate the possible existence of a data value. This difference gives rise to a simpler but meaningful graphical view of data sets (see Fig. 5.1).

We now present the concept of vague subfield.

Definition 5.14 A *vague set* $F = (t_F, f_F)$ of X is said to be a *vague subfield* of the field $\mathbb{F}$ if the following conditions are satisfied:

(1) $(\forall\, \alpha, \beta \in \mathbb{F})(V_F(\alpha - \beta) \geq \min\{V_F(\alpha), V_F(\beta)\})$,
(2) $(\forall\, \alpha, \beta \in \mathbb{F}, \beta \neq 0)(V_F(\alpha\beta^{-1}) \geq \min\{V_F(\alpha), V_F(\beta)\})$,

that is,

$$(3)\quad \begin{cases} t_F(\alpha - \beta) \geq \min\{t_F(\alpha), t_F(\beta)\}, \\ 1 - f_F(\alpha - \beta) \geq \min\{1 - f_F(\alpha), 1 - f_F(\beta)\}, \end{cases}$$

$$(4)\quad \begin{cases} t_F(\alpha\beta^{-1}) \geq \min\{t_F(\alpha), t_F(\beta)\}, \\ 1 - f_F(\alpha\beta^{-1}) \geq \min\{1 - f_F(\alpha), 1 - f_F(\beta)\}. \end{cases}$$

Example 5.3 Consider a field $\mathbb{F} = \{0, 1, w, w^2\}$, where $w = \frac{-1+\sqrt{-3}}{2}$, with the following Cayley Table 5.1: It can be easily seen that the vague set

Table 5.1 Vague subfield

+	0	1	w	w^2
0	0	1	w	w^2
1	1	0	w^2	w
w	w	w^2	0	1
w^2	w^2	w	1	0
.	0	1	w	w^2
0	0	0	0	0
1	0	1	w	w^2
w	0	w	w^2	1
w^2	0	w^2	1	w

$$\{(0, [0.3, 0.2]), (1, [0.4, 0.5]), (w, [0.3, 0.6]), (w^2, [0.5, 0.4])\}$$

forms a vague subfield of the field $\mathbb{F}$.

The following Lemmas can be easily proved and hence, we omit their proofs.

Lemma 5.4 *If $F = (t_F, f_F)$ is a vague subfield of $\mathbb{F}$, then*

$$V_F(0) \geq V_F(1) \geq V_F(\alpha) = V_F(-\alpha) \ \ for\ \alpha \in \mathbb{F}, \ \ and$$

$$V_F(-\alpha) = V_F(\alpha^{-1}) \ \ for\ \alpha \in \mathbb{F} - \{0\}.$$

Lemma 5.5 *A vague set $A = (t_A, f_A)$ of $\mathbb{F}$ is a vague subfield of $\mathbb{F}$ if and only if t_A and $1 - f_A$ are fuzzy subfields.*

Proposition 5.3 *If A and B are vague subfields of $\mathbb{F}$, then $A \cap B$ is a vague subfield of $\mathbb{F}$.*

Proof Let $\alpha, \beta \in \mathbb{F}$. Then, we have

$$\begin{aligned} t_{A\cap B}(\alpha - \beta) &= \min\{t_A(\alpha - \beta), t_B(\alpha - \beta)\} \\ &\geq \min\{\min\{t_A(\alpha), t_A(\beta)\}, \min\{t_B(\alpha), t_B(\beta)\}\} \\ &= \min\{\min\{t_A(\alpha), t_B(\alpha)\}, \min\{t_A(\beta), t_B(\beta)\}\} \\ &= \min\{t_{A\cap B}(\alpha), t_{A\cap B}(\beta)\}, \end{aligned}$$

and hence, we derive that

$$\begin{aligned} t_{A\cap B}(\alpha\beta^{-1}) &= \min\{t_A(\alpha\beta^{-1}), t_B(\alpha\beta^{-1})\} \\ &\geq \min\{\min\{t_A(\alpha), t_A(\beta)\}, \min\{t_B(\alpha), t_B(\beta)\}\} \\ &= \min\{\min\{t_A(\alpha), t_B(\alpha)\}, \min\{t_A(\beta), t_B(\beta)\}\} \end{aligned}$$

$$= \min\{t_{A\cap B}(\alpha), t_{A\cap B}(\beta)\},$$

$$\begin{aligned}1-f_{A\cap B}(\alpha-\beta) &= \min\{1-f_A(\alpha-\beta), 1-f_B(\alpha-\beta)\}\\ &\geq \min\{\min\{1-f_A(\alpha), 1-f_A(\beta)\}, \min\{1-f_B(\alpha), 1-f_B(\beta)\}\}\\ &= \min\{\min\{1-f_A(\alpha), 1-f_B(\alpha)\}, \min\{1-f_A(\beta), 1-f_B(\beta)\}\}\\ &= \min\{1-f_{A\cap B}(\alpha), 1-f_{A\cap B}(\beta)\},\end{aligned}$$

$$\begin{aligned}1-f_{A\cap B}(\alpha\beta^{-1}) &= \min\{1-f_A(\alpha\beta^{-1}), 1-f_B(\alpha\beta^{-1})\}\\ &\geq \min\{\min\{1-f_A(\alpha), 1-f_A(\beta)\}, \min\{1-f_B(\alpha), 1-f_B(\beta)\}\}\\ &= \min\{\min\{1-f_A(\alpha), 1-f_B(\alpha)\}, \min\{1-f_A(\beta), 1-f_B(\beta)\}\}\\ &= \min\{1-f_{A\cap B}(\alpha), 1-f_{A\cap B}(\beta)\}.\end{aligned}$$

Therefore, we have proved that $A \cap B$ is indeed a vague subfield of $\mathbb{F}$.

Proposition 5.4 *The zero vague set, unit vague set, and s-vague set are all vague subfields of X.*

Proof Let $A = (t_A, f_A)$ be a vague subfield of $\mathbb{F}$. For $\alpha,\ \beta \in \mathbb{F}$, we have

$$t_A(\alpha-\beta) \geq \min\{t_A(\alpha), t_A(\beta)\} = \min\{s, s\} = s,$$

$$1-f_A(\alpha-\beta) \geq \min\{1-f_A(\alpha), 1-f_A(\beta)\} = \min\{s, s\} = s,$$

$$t_A(\alpha\beta^{-1}) \geq \min\{t_A(\alpha), t_A(\beta)\} = \min\{s, s\} = s,$$

$$1-f_A(\alpha\beta^{-1}) \geq \min\{1-t_A(\alpha), 1-t_A(\beta)\} = \min\{s, s\} = s.$$

This shows that s-vague set of $\mathbb{F}$ is a vague subfield of $\mathbb{F}$. The proofs for the other cases are similar.

Proposition 5.5 *Let A be a vague subfield of $\mathbb{F}$. Then, for $s \in [0, 1]$, the vague-cut A_s is a crisp subfield of $\mathbb{F}$.*

Proof Suppose that $A = (t_A, f_A)$ is a vague subfield of $\mathbb{F}$. For $\alpha, \beta \in A_s$ we can deduce that

$$t_A(\alpha) \geq s,\quad 1-f_A(\alpha) \geq s,\quad t_A(\beta) \geq s,\quad 1-f_A(\beta) \geq s,$$

so that

$$t_A(\alpha-\beta) \geq \min\{t_A(\alpha), t_A(\beta)\} \geq \min\{s, s\} = s,$$

$$1-f_A(\alpha-\beta) \geq \min\{1-f_A(\alpha), 1-f_A(\beta)\} \geq \min\{s, s\} = s,$$

$$t_A(\alpha\beta^{-1}) \geq \min\{t_A(\alpha), t_A(\beta)\} \geq \min\{s, s\} = s,$$

$$1 - f_A(\alpha\beta^{-1}) \geq \min\{1 - t_A(\alpha), 1 - t_A(\beta)\} \geq \min\{s, s\} = s.$$

This implies that $\alpha - \beta, \alpha\beta^{-1} \in A_s$. Hence, A_s is a crisp subfield of $\mathbb{F}$.

Proposition 5.6 *Let $\mathbb{K}$ be a vague set of $\mathbb{F}$ which is defined by*

$$V_{\mathbb{K}}(\alpha) = \begin{cases} [s, s] & \text{if } \alpha \in \mathbb{K}, \\ [t, t] & \text{otherwise} \end{cases}$$

for all $s, t \in [0, 1]$ with $s \geq t$. Then, $\mathbb{K}$ is a vague subfield of $\mathbb{F}$ if and only if $\mathbb{K}$ is a (crisp) subfield of $\mathbb{F}$.

We now present the concept of vague Lie subalgebras over vague field.

Definition 5.15 A *vague set* $A = (t_A, f_A)$ of L is called a *vague Lie subalgebra over a vague field* $F = (t_F, f_F)$ (briefly, *vague Lie* $\mathbb{F}$*-subalgebra*) of L if the following conditions are satisfied

(a) $V_A(x + y) \geq \min\{V_A(x), V_A(y)\}$,
(b) $V_A(\alpha x) \geq \min\{V_F(\alpha), V_A(x)\}$,
(c) $V_A([x, y]) \geq \min\{V_A(x), V_A(y)\}$

for all $x, y \in L$ and $\alpha \in \mathbb{F}$.

In other words,

(d) $\begin{cases} t_A(x + y) \geq \min\{t_A(x), t_A(y)\}, \\ 1 - f_A(x + y) \geq \min\{1 - f_A(x), 1 - f_A(y)\}, \end{cases}$

(e) $\begin{cases} t_A(\alpha x) \geq \min\{t_F(\alpha), t_A(x)\}, \\ 1 - f_A(\alpha x) \geq \min\{1 - f_F(\alpha), 1 - f_A(x)\}, \end{cases}$

(f) $\begin{cases} t_A([x, y]) \geq \min\{t_A(x), t_A(y)\}, \\ 1 - f_A([x, y]) \geq \min\{1 - f_A(x), 1 - f_A(y)\}. \end{cases}$

From (b), it follows that $V_A(0) \geq V_F(0)$.

Example 5.4 Let $\Re^2 = \{(x, y) : x, y \in \mathbb{R}\}$ be the set of all *two*-dimensional real vectors. Then, $\Re^2$ with $[x, y] = x \times y$ form a real Lie algebra. Define a vague set $A = (t_A, f_A) : \Re^2 \to [0, 1]$ by

$$t_A(x, y) = \begin{cases} 0.4 & \text{if } x = y = 0, \\ 0.3 & \text{otherwise}, \end{cases} \qquad f_A(x, y) = \begin{cases} 0.3 & \text{if } x = y = 0, \\ 0.4 & \text{otherwise}, \end{cases}$$

and define $F = (t_F, f_F) : \mathbb{R} \to [0, 1]$ for all $\alpha \in \mathbb{R}$ by

$$t_F(\alpha) = \begin{cases} 0.3 & \text{if } \alpha \in \mathbb{Q}, \\ 0.2 & \text{if } 0\alpha \in \mathbb{R} - \mathbb{Q}(\sqrt{3}), \end{cases}$$

$$f_F(\alpha) = \begin{cases} 0.2 & \text{if} \alpha \in \mathbb{Q}, \\ 0.4 & \text{if} 0 \alpha \in \mathbb{R} - \mathbb{Q}(\sqrt{3}). \end{cases}$$

By routine verification, we can easily check that A is a vague Lie $\mathbb{F}$-subalgebra.

The proofs of the following propositions are obvious.

Proposition 5.7 *A vague set $A = (t_A, f_A)$ of L is a vague Lie $\mathbb{F}$-subalgebra of L if and only if t_A and $1 - f_A$ are fuzzy Lie $\mathbb{F}$-subalgebras over a fuzzy field.*

Proposition 5.8 *Let $\{A_i : i \in I\}$ be a family of vague Lie $\mathbb{F}$-subalgebras of L. Then $\bigcap_{i \in I} A_i$ is a vague Lie $\mathbb{F}$-subalgebra of L.*

Proposition 5.9 *The zero vague set, unit vague set, and s-vague set are vague Lie $\mathbb{F}$-subalgebras of L.*

Theorem 5.9 *Let A be a vague Lie $\mathbb{F}$-subalgebra of L. Then, for any $s, t \in [0, 1]$, the vague-cut $A_{(s,t)}$ is a crisp Lie subalgebra of L.*

Proof Suppose that $A = (t_A, f_A)$ is a vague Lie subalgebra of L over a vague field $F = (t_F, f_F)$. Let $x, y, \alpha \in A_{(s,t)}$, $x, y \in L$, $\alpha \in \mathbb{F}$. Then,

$$t_A(x) \geq s, \quad 1 - f_A(x) \geq t, \quad t_A(y) \geq s, \quad 1 - f_A(y) \geq t, \quad t_F(\alpha) \geq s$$

and $1 - f_F(\alpha) \geq t$.

From Definition 4.1, it follows that

$$\begin{aligned} t_A(x+y) &\geq \min\{t_A(x), t_A(y)\} \geq \min\{s, s\} = s, \\ 1 - f_A(x+y) &\geq \min\{1 - f_A(x), 1 - f_A(y)\} \geq \min\{t, t\} = t, \\ t_A(\alpha x) &\geq \min\{t_F(\alpha), t_A(x)\} \geq \min\{s, s\} = s, \\ 1 - f_A(\alpha x) &\geq \min\{1 - t_F(\alpha), 1 - t_A(x)\} \geq \min\{t, t\} = t, \\ t_A([x, y]) &\geq \min\{t_A(x), t_A(y)\} \geq \min\{s, s\} = s, \\ 1 - f_A([x, y]) &\geq \min\{1 - f_A(x), 1 - f_A(y)\} \geq \min\{t, t\} = t. \end{aligned}$$

This implies that $x + y, \alpha x, [x, y] \in A_{(s,t)}$. Hence, $A_{(s,t)}$ is a crisp Lie subalgebra of L.

Corollary 5.10 *Let A be a vague Lie $\mathbb{F}$-subalgebra of L. Then, for $s \in [0, 1]$, the vague-cut A_s is a crisp Lie subalgebra of L.*

The proofs of the following propositions are obvious.

Proposition 5.10 *(i) Let $f : L_1 \to L_2$ be an onto homomorphism of Lie algebras. If $B = (t_B, f_B)$ is a vague Lie $\mathbb{F}$-subalgebra of L_2, then the pre-image $f^{-1}(B)$ of B under f is a vague Lie $\mathbb{F}$-subalgebra of L_1.*

(ii) Let $f : L_1 \to L_2$ be an epimorphism of Lie algebras. If $A = (t_A, f_A)$ is a vague Lie $\mathbb{F}$-subalgebra of L_2, then $f^{-1}(A^c) = (f^{-1}(A))^c$.

(iii) *Let $f : L_1 \to L_2$ be an epimorphism of Lie algebras. If $A = (t_A, f_A)$ is a vague Lie $\mathbb{F}$-subalgebra of L_2 and $B = (t_B, f_B)$ is the pre-image of $A = (\mu_A, \lambda_A)$ under f. Then, $B = (t_B, f_B)$ is a vague Lie $\mathbb{F}$-subalgebra of L_1.*

Definition 5.16 Let $g : L_1 \to L_2$ be a homomorphism of Lie algebras. For any vague fuzzy set $A = (t_A, f_A)$ in a Lie algebra L_2, we define a vague fuzzy set $A^g = (t_A^g, f_A^g)$ in L_1 by

$$t_A^g(x) = t_A(g(x)), \quad f_A^g(x) = f_A(g(x))$$

for all $x \in L_1$. Clearly, $A^g(x_1) = A^g(x_2) = A(x)$ for all $x_1, x_2 \in g^{-1}(x)$.

Lemma 5.6 *Let $g : L_1 \to L_2$ be a homomorphism of Lie algebras. If $A = (t_A, f_A)$ is a vague Lie $\mathbb{F}$-subalgebra of L_2, then A^g is a vague Lie $\mathbb{F}$-subalgebra of L_1.*

Proof Let $x, y \in L_1$ and $\alpha \in \mathbb{F}$. Then,

$$\begin{aligned} t_A^g(x + y) &= t_A(g(x + y)) = t_A(g(x) + g(y)) \\ &\geq \min\{t_A(g(x)), t_A(g(y))\} = \min\{t_A^g(x), t_A^g(y)\}, \end{aligned}$$

$$\begin{aligned} 1 - f_A^g(x + y) &= 1 - f_A((g(x + y)) = 1 - f_A(g(x) + g(y)) \\ &\geq \min\{1 - f_A(g(x)), 1 - f_A(g(y))\} \\ &= \min\{1 - f_A^g(x), 1 - f_A^g(y)\}. \end{aligned}$$

The verification of the other conditions is similar. Hence, A^g is a vague Lie $\mathbb{F}$-subalgebra of L_1.

Theorem 5.10 *Let $g : L_1 \to L_2$ be an epimorphism of Lie algebras. Then, A^g is a vague Lie $\mathbb{F}$-subalgebra of L_1 if and only if A is a vague Lie $\mathbb{F}$-subalgebra of L_2.*

Proof The sufficiency follows from Lemma 5.6. In proving the necessity, we first recall that g is a surjective mapping. Hence for any $x, y \in L_2$,, there exist $x_1, y_1 \in L_1$ such that $x = g(x_1), y = g(y_1)$. Thus, $t_A(x) = t_A^g(x_1), t_A(y) = t_A^g(y_1), 1 - f_A(x) = 1 - f_A^g(x_1), 1 - f_A(y) = 1 - f_A^g(y_1)$, whence

$$\begin{aligned} t_A(x + y) &= t_A(g(x_1) + g(y_1)) = t_A(g(x_1 + y_1)) \\ &= t_A^g(x_1 + y_1) \geq \min\{t_A^g(x_1), t_A^g(y_1)\} = \min\{t_A(x), t_A(y)\}, \\ 1 - f_A(x + y) &= 1 - f_A(g(x_1) + g(y_1)) = 1 - f_A(g(x_1 + y_1)) \\ &= 1 - f_A^g(x_1 + y_1) \geq \min\{1 - f_A^g(x_1), 1 - f_A^g(y_1)\} \\ &= \min\{1 - f_A(x), 1 - f_A(y)\}. \end{aligned}$$

The verification of the other conditions is similar. This proves that $A = (t_A, f_A)$ is a vague Lie $\mathbb{F}$-subalgebra of L_2.

5.4 Special Types of Vague Lie Subalgebras

Definition 5.17 Let $A = (t_A, f_A)$ be a vague Lie $\mathbb{F}$-subalgebra in L. Define inductively a sequence of vague Lie $\mathbb{F}$-subalgebras in L by Lie brackets

$$A^0 = A, \quad A^1 = [A^0, A^0], \quad A^2 = [A^1, A^1], \quad \ldots, \quad A^n = [A^{n-1}, A^{n-1}].$$

Then, A^n is said to be the *nth derived vague Lie* $\mathbb{F}$ *-subalgebra* of L. Moreover, a series

$$A^0 \supseteq A^1 \supseteq A^2 \supseteq \cdots \supseteq A^n \supseteq \cdots$$

is said to be a *derived series* of a vague Lie $\mathbb{F}$-subalgebra A in L. A vague Lie $\mathbb{F}$-subalgebra A in L is called a *solvable vague Lie* $\mathbb{F}$*-subalgebra* if there exists a positive integer n such that $A^n = \mathbf{0}$.

Definition 5.18 Let $A = (t_A, f_A)$ be a vague Lie $\mathbb{F}$-subalgebra in L. We define inductively a sequence of vague Lie $\mathbb{F}$-subalgebras in L by Lie brackets

$$A_0 = A, \quad A_1 = [A, A_0], \quad A_2 = [A, A_1], \quad \ldots, \quad A_n = [A, A_{n-1}].$$

Then we call the series

$$A_0 \supseteq A_1 \supseteq A_2 \supseteq \cdots \supseteq A_n \supseteq \cdots$$

the *descending central series* of a vague Lie $\mathbb{F}$-subalgebra A in L. A vague Lie $\mathbb{F}$-subalgebra A in L is called a *nilpotent vague Lie* $\mathbb{F}$ *-subalgebra* if there exists a positive integer n such that $A_n = \mathbf{0}$.

By using similar arguments as in the proof of Theorem 4.7 in [4], we obtain the following theorem.

Theorem 5.11 *(I) The homomorphic image of a solvable vague Lie* $\mathbb{F}$*-subalgebra is a solvable vague Lie* $\mathbb{F}$*-subalgebra.*

(II) The homomorphic image of a nilpotent vague Lie $\mathbb{F}$*-subalgebra is a nilpotent vague Lie* $\mathbb{F}$*-subalgebra.*

(III) If A is a nilpotent vague Lie $\mathbb{F}$*-subalgebra, then it is solvable.*

Definition 5.19 A vague Lie $\mathbb{F}$-subalgebra $A = (t_A, f_A)$ of a Lie algebra L is said to be *normal* if there exists an element $x_0 \in L$ such that $V_A(x_0) = \mathbf{1}$, i.e., $t_A(x_0) = 1$ and $f_A(x_0) = 0$.

The following Lemma is easy to prove and we hence omit the proof.

Lemma 5.7 *Let* $A = (t_A, f_A)$ *be a vague Lie* $\mathbb{F}$*-subalgebra of* L *such that* $t_A(x) + f_A(x) \leq t_A(0) + f_A(0)$ *for all* $x \in L$*. Define* $A^+ = (t_A^+, f_A^+)$*, where* $t_A^+(x) = t_A(x) + 1 - t_A(0)$*,* $f_A^+(x) = f_A(x) - f_A(0)$ *for all* $x \in L$*. Then,* A^+ *is normal vague set.*

By using the above lemma, we deduce the following theorem.

Theorem 5.12 *Let $A = (t_A, f_A)$ be a vague Lie $\mathbb{F}$-subalgebra of a Lie algebra L. Then, the vague set A^+ is a normal vague Lie $\mathbb{F}$-subalgebra of L containing A.*

Proof Let $x, y \in L$ and $\alpha \in \mathbb{F}$. Then,

$$\begin{aligned}
\min\{V_A^+(x), V_A^+(y)\} &= \min\{V_A(x) + 1 - V_A(0), V_A(y) + 1 - V_A(0)\} \\
&= \min\{V_A(x), V_A(y)\} + 1 - V_A(0)\} \\
&\leq V_A(x + y) + 1 - V_A(0) = V_{A^+}(x + y), \\
\min\{V_F^+(\alpha), V_A^+(x)\} &= \min\{V_F(\alpha) + 1 - V_F(0), V_A(x) + 1 - V_A(0)\} \\
&= \min\{V_F(\alpha), V_A(x)\} + 1 - (V_F(0) + V_A(0))\} \\
&\leq V_A(\alpha x) + 1 - (V_F(0) + V_A(0)) = V_{A^+}(\alpha x), \\
\min\{V_A^+(x), V_A^+(y)\} &= \min\{V_A(x) + 1 - V_A(0), V_A(y) + 1 - V_A(0)\} \\
&= \min\{V_A(x), V_A(y)\} + 1 - V_A(0)\} \\
&\leq V_A([x, y]) + 1 - V_A(0) = V_{A^+}([x, y]).
\end{aligned}$$

Thus, A^+ is a normal vague Lie $\mathbb{F}$-subalgebra of L. Clearly $A \subseteq A^+$.

The following theorems are obvious.

Theorem 5.13 *A vague Lie $\mathbb{F}$-subalgebra A of a Lie algebra L is normal if and only if $A^+ = A$.*

Theorem 5.14 *If $A = (t_A, f_A)$ is a vague Lie $\mathbb{F}$-subalgebra of a Lie algebra L, then $(A^+)^+ = A^+$.*

Corollary 5.11 *If A is normal vague Lie $\mathbb{F}$-subalgebra of a Lie algebra L, then $(A^+)^+ = A$.*

Theorem 5.15 *Let A and B be vague Lie $\mathbb{F}$-subalgebras of a Lie algebra L. Then, $(A \cup B)^+ = A^+ \cup B^+$.*

Proof Let $A = (t_A, f_A)$ and $B = (t_B, f_B)$ be two vague Lie $\mathbb{F}$-subalgebras of a Lie algebra L. Then, $A \cup B = (t_{A\cup B}, f_{A\cup B})$, where

$$t_{A\cup B}(x) = \max\{t_A(x), t_B(x)\}, \quad f_{A\cup B}(x) = \min\{f_A(x), f_B(x)\}, \quad \forall\, x \in L.$$

Thus, $(A \cup B)^+ = (t_{(A\cup B)^+}(x), f_{(A\cup B)^+}(x))$, where

$$\begin{aligned}
t_{(A\cup B)^+}(x) &= t_{(A\cup B)}(x) + 1 - t_{(A\cup B)}(0) \\
&= \max\{t_A(x), t_B(x)\} + 1 - \max\{t_A(0), t_B(0)\} \\
&= \max\{t_A(x) + 1 - t_A(0), t_B(x) + 1 - t_B(0)\} \\
&= \max\{t_{A^+}(x), t_{B^+}(x)\} = t_{A^+\cup B^+}(x).
\end{aligned}$$

Similarly, we can prove that $f_{(A\cup B)^+}(x) = f_{A^+\cup B^+}(x)$ for $x \in L$. Hence, $(A \cup B)^+ = A^+ \cup B^+$.

The proof of the following theorem is obvious.

Theorem 5.16 *Let A be a vague Lie $\mathbb{F}$-subalgebra of a Lie algebra L. If there exist a vague Lie $\mathbb{F}$-subalgebra B of L satisfying $B \subset A^+$, then A is normal.*

Corollary 5.12 *Let A be a vague Lie $\mathbb{F}$-subalgebra of a Lie algebra L. If there exists a vague Lie $\mathbb{F}$-subalgebra B of L satisfying $B^+ \subset A$, then $A^+ = A$.*

Denote the family of all vague Lie $\mathbb{F}$-subalgebras of a Lie algebra L by $VLS(L)$, and the set of all normal vague Lie $\mathbb{F}$-subalgebra of L by $\mathcal{N}(L)$. It is clear that $\mathcal{N}(L)$ is a poset under set inclusion.

Theorem 5.17 *A nonconstant maximal element of $(\mathcal{N}(L), \subseteq)$ takes only the values $\mathbf{0}$ and $\mathbf{1}$.*

Proof Let $A \in \mathcal{N}(L)$ be a nonconstant maximal element of $(\mathcal{N}(L), \subseteq)$. Then, $t_A(x_0) = 1$ and $f_A(x_0) = 0$ for some $x_0 \in L$. Let $x \in L$ be such that $V_A(x) \neq \mathbf{1}$. We claim that $V_A(x) = \mathbf{0}$. If not, then there exists $a \in L$ such that $\mathbf{0} < V_A(a) < \mathbf{1}$. Let B be a vague set in L over vague field $\mathbb{K}$ defined by $V_B(x) := \frac{1}{2}\{V_A(x) + V_A(a)\}$, $V_{\mathbb{K}}(x) := \frac{1}{2}\{V_F(x) + V_F(a)\}$ for all $x \in L$. For $x, y \in L$ and $\alpha \in \mathbb{F}$, we have

$$\begin{aligned}
V_B(x+y) &= \frac{1}{2}\{V_A(x+y) + V_A(a)\} \geq \frac{1}{2}\{\min\{V_A(x), V_A(y)\} + V_A(a)\} \\
&= \min\left\{\frac{1}{2}(V_A(x) + V_A(a)), \frac{1}{2}(V_A(y) + V_A(a))\right\} \\
&= \min\{V_B(x), V_B(y)\}, \\
V_B(\alpha x) &= \frac{1}{2}\{V_A(\alpha x) + V_A(a)\} \geq \frac{1}{2}\{\min\{V_F(\alpha), V_A(x)\} + V_A(a)\} \\
&= \min\left\{\frac{1}{2}(V_F(\alpha) + V_F(a)), \frac{1}{2}(V_A(x) + V_A(a))\right\} \\
&= \min\{V_{\mathbb{K}}(\alpha), V_B(x)\}, \\
V_B([x,y]) &= \frac{1}{2}\{V_A([x,y]) + V_A(a)\} \geq \frac{1}{2}\{\min\{V_A(x), V_A(y)\} + V_A(a)\} \\
&= \min\left\{\frac{1}{2}(V_A(x) + V_A(a)), \frac{1}{2}(V_A(y) + V_A(a))\right\} \\
&= \min\{V_B(x), V_B(y)\}.
\end{aligned}$$

This proves that B is a vague Lie $\mathbb{F}$-subalgebra of L. Now we have

$$\begin{aligned}
V_{B^+}(x) &= V_B(x) + 1 - V_B(0) \\
&= \frac{1}{2}\{\min\{V_A(x), V_A(a)\} + 1 - \frac{1}{2}\{\min\{V_A(0), V_A(a)\} \\
&= V_A(x) + 1,
\end{aligned}$$

which implies that $V_{B^+}(0) = \frac{1}{2}\{V_A(0) + 1\} = 1$. Thus, B^+ forms a normal vague Lie $\mathbb{F}$-subalgebra of L. But $V_{B^+}(0) = \mathbf{1} > V_{B^+}(a) = \frac{1}{2}\{V_A(a) + 1\} > V_A(a)$, so B^+ is a nonconstant normal vague Lie $\mathbb{F}$-subalgebra of L and $V_{B^+}(a) > V_A(a)$, which is a contradiction. Hence, a nonconstant maximal element of $(\mathscr{N}(L), \subseteq)$ takes only two values: $\mathbf{0}$ and $\mathbf{1}$.

Definition 5.20 A nonconstant vague Lie $\mathbb{F}$-subalgebra $A \in VLS(L)$ is called *maximal* if A^+ is a maximal element of the poset $(\mathscr{N}(L), \subseteq)$.

Theorem 5.18 *A maximal vague Lie $\mathbb{F}$-subalgebra $A \in VLS(L)$ is normal and takes only two values:* $\mathbf{0}$ *and* $\mathbf{1}$.

Proof Let $A \in VLS(L)$ be maximal. Then, A^+ is a nonconstant maximal element of the poset $(\mathscr{N}(L), \subseteq)$ and, by Theorem 5.17, the possible values of $V_A^+(x)$ are $\mathbf{0}$ and $\mathbf{1}$, that is, t_A^+ takes only two values 0 and 1. Clearly, $t_A^+(x) = 1$ if and only if $t_A(x) = t_A(0) = 0, t_A^+(x) = 0$ if and only if $t_A(x) = t_A(0) = 1$. But $A \subseteq A^+$ implies $t_A(x) \leq t_A^+(x)$ for all $x \in L$. Hence, $t_A^+(x) = 0$ implies $t_A(x) = 0$. Consequently, $V_A(0) = 1$.

Theorem 5.19 *A level subset of a maximal $A \in VLS(L)$ is a maximal Lie subalgebra of L.*

Proof Let S be a level subset of a maximal $A \in VLS(L)$, i.e., $S = L = \{x \in L \mid V_A(x) - 1\}$. It is not difficult to verify that S is a Lie subalgebra of L. Obviously $S \neq L$ because V_A takes only two values. Let M be a Lie subalgebra of L containing S. Then, $V_S \subseteq V_M$. Since $V_A = V_S$ and V_A takes only two values, V_M also takes only these two values. But, by our assumption, $A \in VLS(L)$ is maximal so that $V_S = V_A = V_M$ or $V_M(x) = 1$, for all $x \in L$. In the last case, we have $S = L$ which is impossible. So, we must have $V_A = V_S = V_M$ which implies that $S = M$. This means that S is a maximal Lie subalgebra of L.

Definition 5.21 A normal vague Lie $\mathbb{F}$-subalgebra $A \in VLS(L)$ is called *completely normal* if there exists $x \in L$ such that $A(x) = \mathbf{0}$. The set of all completely normal $A \in VLS(L)$ is denoted by $\mathscr{C}(L)$. Clearly, $\mathscr{C}(L) \subseteq \mathscr{N}(L)$.

Theorem 5.20 *A nonconstant maximal element of $(\mathscr{N}(L), \subseteq)$ is also a maximal element of $(\mathscr{C}(L), \subseteq)$.*

Proof Let A be a nonconstant maximal element of $(\mathscr{N}(L), \subseteq)$. Then, by Theorem 5.17, A takes only two values $\mathbf{0}$ and $\mathbf{1}$ and so $V_A(x_0) = \mathbf{1}$ and $V_A(x_1) = \mathbf{0}$, for some $x_0, x_1 \in L$. Hence $A \in \mathscr{C}(L)$. Assume that there exists $B \in \mathscr{C}(L)$ such that $A \subseteq B$. Then, it follows that $A \subseteq B$ in $\mathscr{N}(L)$. Since A is maximal in $(\mathscr{N}(L), \subseteq)$ and B is nonconstant, we have $A = B$. Thus, A is maximal element of $(\mathscr{C}(L), \subseteq)$. This completes the proof.

Theorem 5.21 *Every maximal $A \in VLS(L)$ is completely normal.*

Proof Let $A \in VLS(L)$ be maximal. Then, by Theorem 5.18, A is normal and $A = A^+$ takes only two values $\mathbf{0}$ and $\mathbf{1}$. Since A is nonconstant, it follows that $V_A(x_0) = \mathbf{1}$ and $V_A(x_1) = \mathbf{0}$ for some $x_0, x_1 \in L$. Hence, A is completely normal, ending the proof.

We state a method of construction for a new normal vague Lie $\mathbb{F}$-subalgebra from an old one.

Theorem 5.22 *Let $f : [0, 1] \to [0, 1]$ be an increasing function and $A = (t_A, f_A)$ a vague set on a Lie algebra L. Then $A_f = (t_{A_f}, f_{A_f})$ defined by $t_{A_f}(x) = f(t_A(x))$ and $f_{A_f}(x) = f(f_A(x))$ is a vague Lie $\mathbb{F}$-subalgebra if and only if $A = (t_A, f_A)$ is a vague Lie $\mathbb{F}$-subalgebra. Moreover, if $f(t_A(0)) = 1$ and $f(f_A(0)) = 0$, then A_f is normal.*

5.5 Anti-fuzzy Lie Sub-superalgebras Over Anti-fuzzy Field

We first discuss the concept of an anti-fuzzy field.

Definition 5.22 A fuzzy set λ of a field $\mathbb{F}$ is called an *anti-fuzzy subfield* if the following conditions are satisfied:

(1) $(\forall\, x, y \in \mathbb{F})(\lambda(x - y) \leq \lambda(x) \vee \lambda(y))$,
(2) $(\forall\, x, y \in \mathbb{F},\ x \neq 0)(\lambda(xy^{-1}) \leq \lambda(x) \vee \lambda(y))$.

Example 5.5 Consider a field $Z_5 = \{0, 1, 2, 3, 4\}$ with the following Cayley tables:

+	0	1	2	3	4
0	0	1	2	3	4
1	1	2	3	4	0
2	2	3	4	0	1
3	3	4	0	1	2
4	4	0	1	2	3

·	0	1	2	3	4
0	0	0	0	0	0
1	0	1	2	3	4
2	0	2	4	1	3
3	0	3	1	4	2
4	0	4	3	2	1

Let $\lambda : Z_5 \to [0, 1]$ be a fuzzy set defined by

$$\lambda(x) = \begin{cases} 0.1 & \text{if } x = 0, \\ 0.8 & \text{otherwise.} \end{cases}$$

By routine computations, it is easy to check that λ is an anti-fuzzy subfield of Z_5.

The following Lemma is trivial.

Lemma 5.8 *If λ is an anti-fuzzy field of $\mathbb{F}$, then*

- $(\forall\, x \in \mathbb{F})\ (\lambda(0) \leq \lambda(1) \leq \lambda(x) = \lambda(-x))$ *and*
- $(\forall\, x \in \mathbb{F} - \{0\})\ (\lambda(x) = \lambda(x^{-1}))$.

Theorem 5.23 *Let λ be an anti-fuzzy field in a field $\mathbb{F}$. Then, λ is an anti-fuzzy field of $\mathbb{F}$ if and only if the set $L(\lambda, s) = \{x \in \mathbb{F} \mid \lambda(x) \leq s\}$, $s \in [0, 1]$, is a field of $\mathbb{F}$ when it is nonempty.*

Definition 5.23 An anti-fuzzy field λ of a field $\mathbb{F}$ is said to be an *anti-fuzzy characteristic* if $\lambda^f(x) = \lambda(x)$ for all $x \in \mathbb{F}$ and $f \in \mathrm{Aut}(\mathbb{F})$. An anti-fuzzy field λ of field $\mathbb{F}$ is said to be *fully invariant anti-fuzzy field* if $\lambda(f(x)) \geq \lambda(x)$ for all $x \in \mathbb{F}$ and $f \in \mathrm{End}(\mathbb{F})$.

Theorem 5.24 *An anti-fuzzy field is characteristic if and only if each its level set is a characteristic field.*

As a consequence of the above Theorem, we obtain the following theorem.

Theorem 5.25 *If λ is a fully invariant anti-fuzzy field of $\mathbb{F}$, then it is characteristic.*

Theorem 5.26 *A field $\mathbb{F}$ is Northerian if and only if for any anti-fuzzy field λ, $(Im(\lambda), \leq)$ is well ordered.*

Proof ($\Rightarrow$) Let $\mathbb{F}$ be Northerian. If for some anti-fuzzy field λ of $\mathbb{F}$, $(Im(\lambda), \leq)$ is not well ordered, then there is a strictly increasing number sequence in $Im(\lambda)$: $t_1 < t_2 < \cdots$. Denote $p = \sup\{t_i \mid i = 1, 2, \ldots\}$. It is easy to verify that $U = \{x \in \mathbb{F} \mid \lambda(x) < p\}$ is a subfield of $\mathbb{F}$. Thus, there are $a_1, \ldots, a_n \in U$ such that $U = (a_1, \ldots, a_n]$, and so $\lambda(a_1) \vee \cdots \vee \lambda(a_n)$ is the greatest element of $(\{\lambda(x) \mid x \in U\}; \leq)$. We observe that $(\{t_i \mid i = 1, 2, \ldots\}; <)$ is a subset of $(\{\lambda(x) \mid x \in U\}; \leq)$, a contradiction. Hence, $(Im(\lambda), \leq)$ is well ordered.

($\Leftarrow$) Suppose that for any anti-fuzzy field λ of $\mathbb{F}$, $(Im(\lambda), \leq)$ are well-ordered subsets of $[0, 1]$. If $\mathbb{F}$ is not Noetherian, then there is a strictly ascending chain of fields of $\mathbb{F}$:

$$U_1 \subsetneqq U_2 \subsetneqq \cdots \subsetneqq U_n \subsetneqq \cdots,$$

where $U_i \subsetneqq U_j$ expresses $U_i \subseteq U_j$ but $U_i \neq U_j$. We construct the fuzzy set λ of $\mathbb{F}$ by

$$\lambda(x) := \begin{cases} \frac{n}{n+1} & \text{if } x \in U_n \setminus U_{n-1},\ n = 1, 2, \ldots, \\ 1 & \text{if } x \notin \bigcup\limits_{n=1}^{\infty} U_n, \end{cases}$$

where $U_0 = \emptyset$. We now prove that λ is an anti-fuzzy field of $\mathbb{F}$. We consider the following cases.

Case I. If at least one of x and y belong to $\mathbb{F} \setminus \bigcup\limits_{n=1}^{\infty} U_n$, then at least one of $\lambda(x)$ and $\lambda(y)$ is equal to 1, thereby $\lambda(x - y) \leq \lambda(x) \vee \lambda(y)$.

Case II. If $x \in U_i \setminus U_{i-1}$ and $y \in U_j \setminus U_{j-1}$ with $i \leq j$, then $x, y \in U_j$. Hence, $x - y \in U_j$, and so

$$\lambda(x - y) \leq \frac{j}{j+1} = \frac{i}{i+1} \vee \frac{j}{j+1} = \lambda(x) \vee \lambda(y).$$

Case III. If $x \in U_i \setminus U_{i-1}$ and $y \in U_j \setminus U_{j-1}$ with $j \leq i$, then by the way similar to Case II we can prove $\lambda(x - y) \leq \lambda(x) \vee \lambda(y)$.

Therefore, λ is an anti-fuzzy field of $\mathbb{F}$. But

$$(Im(\lambda), \leq) = \left(\left\{ \frac{1}{2}, \frac{2}{3}, \ldots, \frac{n}{n+1}, \ldots, 1 \right\}, \leq \right)$$

is not well ordered, a contradiction. The verification for other condition is obvious. Hence, $\mathbb{F}$ is Noetherian.

Let $\mathbb{F}/\mathbb{K}$ be a field extension and let $\mathbb{I}(\mathbb{F}/\mathbb{K}) = \{S \mid S\ subfield\ of\ \mathbb{F}, \mathbb{K} \subseteq S\}$ be the lattice of its intermediate fields.

Definition 5.24 Let $\mathbb{F}/\mathbb{K}$ be an extension of fields. A fuzzy set $\lambda : \mathbb{F} \to [0, 1]$ is called an *anti-fuzzy intermediate field* of $\mathbb{F}/\mathbb{K}$ if the following conditions are satisfied:

(a) $(\forall\, x, y \in \mathbb{F})(\lambda(x - y) \leq \lambda(x) \vee \lambda(y))$,
(b) $(\forall\, x, y \in \mathbb{F}, x \neq 0)(\lambda(xy^{-1}) \leq \lambda(x) \vee \lambda(y))$,
(c) $(\forall\, x \in \mathbb{F})(\forall\, k \in \mathbb{K})(\lambda(x) \geq \lambda(k))$.

Let $\mathbb{CFI}(\mathbb{F}/\mathbb{K})$ denote the set of all anti-fuzzy intermediate fields of $\mathbb{F}/\mathbb{K}$.

Lemma 5.9 *λ is an anti-fuzzy intermediate field if and only if for all $Im(\lambda)$, the level set $L(\lambda, s)$ is an intermediate field of $\mathbb{F}/\mathbb{K}$.*

A fuzzy subset λ of $\mathbb{F}$ is said to have the inf property if, for every nonempty subset A of $Im(\lambda)$, there exists $x \in \{y \in \mathbb{F} \mid \lambda(y) \in A\}$ such that $\lambda(x) = \inf(A)$.

Definition 5.25 Let $\mathbb{F}/\mathbb{K}$ be an extension of fields and $\lambda \in \mathbb{CFI}(\mathbb{F}/\mathbb{K})$. Then, λ is called an *anti-fuzzy chain subfield* of $\mathbb{F}/\mathbb{K}$ if for all $x, y \in \mathbb{F}$, $\lambda(x) = \lambda(y) \Longleftrightarrow \mathbb{K}(x) = \mathbb{K}(y)$.

We now give characterizations without their proofs.

Theorem 5.27 *Let $\mathbb{F}/\mathbb{K}$ be a field extension. Then, every anti-fuzzy intermediate field of $\mathbb{F}/\mathbb{K}$ has the* inf *property if and only if there are no infinite strictly increasing sequences of intermediate fields of $\mathbb{F}/\mathbb{K}$.*

Theorem 5.28 *The intermediate fields of $\mathbb{F}/\mathbb{K}$ are chained if and only if $\mathbb{F}/\mathbb{K}$ has an anti-fuzzy chain subfield.*

Theorem 5.29 *Let $\mathbb{F}/\mathbb{K}$ be an extension such that the intermediate fields of $\mathbb{F}/\mathbb{K}$ are chained. If:*

(a) $\mathbb{F}/\mathbb{K}$ is algebraic,
(b) any intermediate field S of $\mathbb{F}/\mathbb{K}$ with $S \neq \mathbb{F}$ is a finite simple extension of $\mathbb{K}$,
(c) $(\mathbb{CFI}(\mathbb{F}/\mathbb{K}), \supseteq)$ satisfies the ascending chain condition.

Then $\mathbb{CFI}(\mathbb{F}/\mathbb{K})$ is well ordered.

We now present the notion of anti-fuzzy Lie sub-superalgebras of Lie superalgebras over an anti-fuzzy field.

Definition 5.26 Let V be a vector space. A fuzzy subset μ of V is called an *anti-fuzzy subspace* of V over an anti-fuzzy field λ if the following axioms are satisfied:

(i) $\mu(x+y) \leq \mu(x) \vee \mu(y)$,
(ii) $\mu(\alpha x) \leq \lambda(\alpha) \vee \mu(x)$

for all $x, y \in V$ and $\alpha \in \mathbb{F}$.

Definition 5.27 Let $V = V_{\bar{0}} \oplus V_{\bar{1}}$ be a $\mathbb{Z}_2$-graded vector space and let $\mu_{\bar{0}}$ and $\mu_{\bar{1}}$ be anti-fuzzy subspaces of subspaces $V_{\bar{0}}$ and $V_{\bar{1}}$ over an anti-fuzzy field λ, respectively. Then, $\mu(x) = (\mu_{\bar{0}} \oplus \mu_{\bar{1}})(x) = \mu_{\bar{0}}(x_{\bar{0}}) \vee \mu_{\bar{1}}(x_{\bar{1}})$, where $x_{\bar{0}} \in V_{\bar{0}}$, $x_{\bar{1}} \in V_{\bar{1}}$ and $x = x_{\bar{0}} + x_{\bar{1}} \in V$, is called a *$\mathbb{Z}_2$-graded anti-fuzzy subspace over an anti-fuzzy field.*

Definition 5.28 A fuzzy set $\mu : \mathscr{L} \to [0, 1]$ is called an *anti-fuzzy Lie sub-superalgebra* of $\mathscr{L}$ over an anti-fuzzy field λ if

(iii) μ is an anti-fuzzy subspace over an anti-fuzzy field,
(iv) μ is a $\mathbb{Z}_2$-graded anti-fuzzy subspace over an anti-fuzzy field,
(v) $\mu([x, y]) \leq \mu(x) \vee \mu(y)$

hold for all $x, y \in \mathscr{L}$.

Example 5.6 Let $\mathscr{L} = \mathscr{L}_0 \oplus \mathscr{L}_1$ be a Lie superalgebra, where $\mathscr{L}_0 = \Re^2$ (the real vector space), $\mathscr{L}_1 = 0$ (null (sub)space) and $[x, y] = x \times y$ ($\times$ is cross product of vectors) for $x, y \in \mathscr{L}_0$ and the remaining Lie brackets are zero. In fact, $\mathscr{L}$ is a Lie algebra over a field $\mathbb{R}$.
Let $\mu_{\bar{0}} : \mathscr{L}_0 \to [0, 1]$ be a fuzzy set defined by

$$\mu_{\bar{0}}(x, y) = \begin{cases} 0 & \text{if } x = y = 0, \\ 1 & \text{otherwise,} \end{cases}$$

Let $\mu_{\bar{1}} : \mathscr{L}_1 \to [0, 1]$ be a fuzzy set defined by $\mu_{\bar{1}}(x) = 0$ for all $x \in \mathscr{L}_1$. Let $\lambda : \mathbb{R} \to [0, 1]$ be fuzzy set defined by

$$\lambda(x) = \begin{cases} 0 & \text{if } x \in \mathbb{Q}, \\ 1 & x \in \mathbb{R} - \mathbb{Q}(\sqrt{3}). \end{cases}$$

Then it is easy to see that $\mu_{\bar{0}}$ and $\mu_{\bar{1}}$ are anti-fuzzy subspaces of $\mathscr{L}_0$ and $\mathscr{L}_1$ over an anti-fuzzy field, respectively. Let $\mu : \mathscr{L} \to [0, 1]$ be a fuzzy set defined by $\mu(x) = \mu_{\bar{0}}(x)$ for all $x \in \mathscr{L}$. By routine computations, it is easy to check that μ is an anti-fuzzy Lie sub-superalgebra of $\mathscr{L}$ over an anti-fuzzy field.

Example 5.7 Let $\mathscr{L} = \mathscr{L}_0 \oplus \mathscr{L}_1$ be a Lie superalgebra, where $\mathscr{L}_0 =< e >$, $\mathscr{L}_1 = < x_1, x_2, y_1, y_2 >$, $[x_i, y_i] = e$ for $i = 1, 2$, and remaining Lie brackets are zero.
Let $\mu_{\bar{0}} : \mathscr{L}_0 \to [0, 1]$ be a fuzzy set defined by

$$\mu_{\bar{0}}(x) = \begin{cases} 0.6 & \text{if } x \in \mathscr{L}_0 - \{0\}, \\ 0 & x = 0, \end{cases}$$

Let $\mu_{\bar{1}} : \mathscr{L}_1 \to [0, 1]$ be a fuzzy set defined by

$$\mu_{\bar{1}}(x) = \begin{cases} 0.5 & \text{if } x \in \mathcal{L}_1 - \{0\}, \\ 0 & x = 0. \end{cases}$$

Let $\lambda : \mathbb{R} \to [0, 1]$ be fuzzy set defined by

$$\lambda(x) = \begin{cases} 0 & \text{if } x \in \mathbb{Q}, \\ 1 & x \in \mathbb{R} - \mathbb{Q}(\sqrt{3}). \end{cases}$$

Then, clearly $\mu_{\bar{0}}$ and $\mu_{\bar{1}}$ are anti-fuzzy subspaces of $\mathcal{L}_0$ and $\mathcal{L}_1$ over an anti-fuzzy field, respectively. Let $\mu : \mathcal{L} \to [0, 1]$ be a fuzzy set defined by

$$\mu(x) = \begin{cases} 0.6 & \text{if } x \in \mathcal{L} - \{0\}, \\ 0 & x = 0. \end{cases}$$

It is easy to see that:

(i) For $x \neq 0$, $(\mu_{\bar{0}} + \mu_{\bar{1}})(x) = \mu_{\bar{0}}(x) \vee \mu_{\bar{1}}(x) = 0.6 = \mu(x)$. So, μ is $\mathbb{Z}_2$ anti-fuzzy subspace of $\mathcal{L}$ over an anti-fuzzy field.
(ii) Let $x = a_i$, $y = b_i \in \mathcal{L}$, then $0.6 = \mu(e) = \mu([a_i, b_i]) = \mu([x, y]) \leq \mu(x) \vee \mu(y) = 0.6$.

If not, $0 = \mu(0) = \mu([x, y]) \leq \mu(x) \vee \mu(y) = 0$. Hence, μ is an anti-fuzzy Lie sub-superalgebra of $\mathcal{L}$ over an anti-fuzzy field.

The proofs of the following propositions are obvious.

Proposition 5.11 *Let μ be a fuzzy set of $\mathcal{L}$. Then, μ is an anti-fuzzy Lie sub-superalgebra of $\mathcal{L}$ over an anti-fuzzy field if and only if μ^c is a fuzzy Lie sub-superalgebra of $\mathcal{L}$ over a fuzzy field.*

Proposition 5.12 *If μ and ν are anti-fuzzy Lie sub-superalgebras over an anti-fuzzy field, then $\mu + \nu$ and $\mu \cap \nu$ are anti-fuzzy Lie sub-superalgebras in $\mathcal{L}$ over an anti-fuzzy field.*

Definition 5.29 For a family of fuzzy sets $\{\mu_i \mid i \in I\}$ in a Lie superalgebra $\mathcal{L}$, the *union* $\bigvee \mu_i$ of $\{\mu_i \mid i \in I\}$ is defined by

$$\left(\bigvee \mu_i\right)(x) = \sup\{\mu_i(x) \mid i \in I\}$$

for each $x \in \mathcal{L}$.

Proposition 5.13 *If $\{\mu_i \mid i \in I\}$ is a family of anti-fuzzy Lie sub-superalgebras of $\mathcal{L}$ over an anti-fuzzy field λ then so is $\bigvee \mu_i$.*

Proof For $x, y \in \mathcal{L}$ and $\alpha \in \mathbb{F}$, we have

$$\begin{aligned}
&\left(\bigvee \mu_i\right)(x+y) = \sup\{\mu_i(x+y) \mid i \in I\} \\
&\leq \sup\{\mu_i(x) \vee \mu_i(y) \mid i \in I\} \\
&= (\sup\{\mu_i(x) \mid i \in I\}) \vee (\sup\{\mu_i(y) \mid i \in I\}) \\
&= \left(\left(\bigvee \mu_i\right)(x)\right) \vee \left(\left(\bigvee \mu_i\right)(y)\right), \\
&\left(\bigvee \mu_i\right)(\alpha x) = \sup\{\mu_i(\alpha x) \mid i \in I\} \\
&\leq \sup\{\lambda_i(\alpha) \vee \mu_i(x) \mid i \in I\} \\
&= (\sup\{\lambda_i(\alpha) \mid i \in I\}) \vee (\sup\{\mu_i(x) \mid i \in I\}) \\
&= \left(\left(\bigvee \lambda_i\right)(\alpha)\right) \vee \left(\left(\bigvee \mu_i\right)(x)\right).
\end{aligned}$$

So $\bigvee \mu_i$ is an anti-fuzzy subspace of $\mathscr{L}$ over an anti-fuzzy field λ.
For $x \in \mathscr{L}$, $x = x_{\bar{0}} + x_{\bar{1}}$, where $x_{\bar{0}} \in \mathscr{L}_0$ and $x_{\bar{1}} \in \mathscr{L}_1$, we have

$$\begin{aligned}
&\left(\bigvee \mu_i\right)(x) = \left(\bigvee \mu_i\right)(x_{\bar{0}} + x_{\bar{1}}) \\
&= \sup\{\mu_i(x_{\bar{0}} + x_{\bar{1}}) \mid i \in I\} \\
&= \sup\{\mu_i(x_{\bar{0}}) \vee \mu_i(x_{\bar{1}}) \mid i \in I\} \\
&= (\sup\{\mu_i(x_{\bar{0}}) \mid i \in I\}) \vee (\sup\{\mu_i(x_{\bar{1}}) \mid i \in I\}) \\
&= \left(\left(\bigvee \mu_i\right)(x_{\bar{0}})\right) \vee \left(\left(\bigvee \mu_i\right)(x_{\bar{1}})\right).
\end{aligned}$$

This shows that $\bigvee \mu_i$ is a $\mathbb{Z}_2$-graded anti-fuzzy subspace of $\mathscr{L}$. For $x, y \in \mathscr{L}$, we have

$$\begin{aligned}
&\left(\bigvee \mu_i\right)([x, y]) = \sup\{\mu_i([x, y]) \mid i \in I\} \\
&\leq \sup\{\mu_i(x) \vee \mu_i(y) \mid i \in I\} \\
&= (\sup\{\mu_i(x) \mid i \in I\}) \vee (\sup\{\mu_i(y) \mid i \in I\}) \\
&= \left(\left(\bigvee \mu_i\right)(x)\right) \vee \left(\left(\bigvee \mu_i\right)(y)\right).
\end{aligned}$$

Hence, $\bigvee \mu_i$ is an anti-fuzzy Lie sub-superalgebra of $\mathscr{L}$ over an anti-fuzzy field.

Definition 5.30 Let μ be a fuzzy set and $s \in [0, 1]$. Then, lower-level subset $L(\mu, s)$ and weak level subset ${}^{>}L(\mu, t)$ of $\mathscr{L}$ are defined by

- $L(\mu, s) = \{x \in \mathscr{L} \mid \mu(x) \leq s\}$ and ${}^{>}L(\mu, s) = \{x \in \mathscr{L} \mid \mu(x) < s\}$, respectively.

Theorem 5.30 *Let μ be an anti-fuzzy Lie sub-superalgebra of $\mathscr{L}$ over an anti-fuzzy field λ and let ν be the closure of the co-image of μ. Then, the following conditions are equivalent:*

(a) μ is an anti-fuzzy Lie sub-superalgebra of $\mathscr{L}$ over an anti-fuzzy field λ,
(b) the nonempty weak level subset ${}^{>}L(\mu, s)$ of μ is a Lie sub-superalgebra of $\mathscr{L}$ for all $s \in [0, 1]$,
(c) the nonempty weak level subset ${}^{>}L(\mu, s)$ of μ is a Lie sub-superalgebra of $\mathscr{L}$ for all $s \in Im(\mu) \setminus \nu$,
(d) the nonempty level subset $L(\mu, s)$ of μ is a Lie sub-superalgebra of $\mathscr{L}$ for all $s \in Im(\mu)$,
(e) the nonempty level subset $L(\mu, s)$ of μ is a Lie sub-superalgebra of $\mathscr{L}$ for all $s \in [0, 1]$.

Proof (a)⇒(b): Let $x \in {}^{>}L(\mu, s) \subseteq \mathscr{L} = \mathscr{L}_0 \oplus \mathscr{L}_1$. Then $x = x_{\bar{0}} + x_{\bar{1}}$, where $x_{\bar{0}} \in \mathscr{L}_0$ and $x_{\bar{1}} \in \mathscr{L}_1$. Since $\mu(x) = \max\{\mu_0(x_{\bar{0}}), \mu_1(x_{\bar{1}})\} < s$, $\mu_0(x_{\bar{0}}) < s$ and $\mu_1(x_{\bar{1}}) < s$ which imply that $x_{\bar{0}}, x_{\bar{1}} \in {}^{>}L(\mu, s)$. Let $x, y \in {}^{>}L(\mu, s)$. Then $\mu(x) < s$ and $\mu(y) < s$. $\mu([x, y]) \leq \mu(x) \vee \mu(y) \leq s$ which imply that $[x, y] \in L(\mu, s)$.
(b)⇒(c) and (c)⇒(d) are obvious.
(d)⇒(e): Let $x, y \in L(\mu, s)$ for $s \in [0, 1]$. Then $\mu(x) \leq s$ and $\mu(y) \leq s$. Assume that $t = \mu(x) \vee \mu(y)$. Then $t \leq s$, $\mu(x) \leq t$ and $\mu(y) \leq t$. Since $\mu(x) \vee \mu(y) = t$, $\mu(x) = t$ and $\mu(y) = t$, i.e., $t \in Im(\mu)$. Thus, $x, y \in L(\mu, t)$. Since $L(\mu, t)$ is a Lie sub-superalgebra of $\mathscr{L}$, $[x, y] \in L(\mu, t)$. So $\mu(x) \leq t \leq s$. Thus, $[x, y] \in L(\mu, s)$. Hence, $L(\mu, s)$ is Lie sub-superalgebra of $\mathscr{L}$.
(e)⇒(a): Let $x, y \in \mathscr{L}$ and $\alpha \in \mathbb{F}$. Suppose that $\mu(x) \vee \mu(y) = s$, then $\mu(x) \leq s$ and $\mu(y) \leq s$, so $x, y \in L(\mu, s)$. Since $L(\mu, s)$ is Lie sub-superalgebra of $\mathscr{L}$, $x + y$, $\alpha x \in L(\mu, s)$. Thus, $\widetilde{m}(x + y) \leq \mu(x) \vee \mu(y) \leq s$ and $\mu(\alpha x) \leq F(\alpha) \vee \mu(x) \leq s$. Hence, μ is anti-fuzzy subspace of $\mathscr{L}$ over an anti-fuzzy field.
Let $x \in \mathscr{L}$. Assume that $\mu(x) = s$. Then $x \in L(\mu, s)$. Since $L(\mu, s)$ is $\mathbb{Z}_2$-graded subspace of $\mathscr{L}$, we can express $x = x_{\bar{0}} + x_{\bar{1}}$, where $x_{\bar{0}} \in \mathscr{L}_0 \cap L(\mu, s)$ and $x_{\bar{1}} \in \mathscr{L}_1 \cap L(\mu, s)$. Thus, $\mu(x) \leq \max(\mu(x_{\bar{0}}), \mu(x_{\bar{1}}))$. Define a mapping $\widetilde{m} : \mathscr{L}_0 \to [0, 1]$ by $\widetilde{m}(x) = \mu(x)$ and $\mu_{\bar{1}} : \mathscr{L}_1 \to [0, 1]$ by $\mu_{\bar{1}}(x) = \mu(x)$. Then $s = \mu(s) \leq \mu(x_{\bar{0}}) \vee \mu(x_{\bar{1}}) = \widetilde{m}(x_{\bar{1}}) \vee \widetilde{m}(x_{\bar{1}})$ and $\widetilde{m}(x_{\bar{1}}) \vee \widetilde{m}(x_{\bar{1}}) \leq s$. Thus, $\widetilde{m}(x_{\bar{0}}) \vee \widetilde{m}(x_{\bar{1}}) = s = \mu(x)$. Hence, μ is $\mathbb{Z}_2$-graded anti-fuzzy subspace.
Let $x, y \in \mathscr{L}$ be such that $x, y \in L(\mu, s)$. Then $\mu(x) \leq s$, $\mu(y) \leq s$. It follows that $\mu([x, y]) \leq \mu(x) \vee \mu(y) \leq s$ so that $[x, y] \in L(\mu, s)$. Hence, μ is an anti-fuzzy Lie sub-superalgebra of $\mathscr{L}$ over an anti-fuzzy field λ.

Theorem 5.31 *If μ is an anti-fuzzy Lie sub-superalgebra of $\mathscr{L}$ over an anti-fuzzy field, then for all $x \in \mathscr{L}$*

$$\mu(x) = \inf\{s \in [0, 1] \mid x \in L(\mu, s)\}.$$

Proof Let $t := \inf\{s \in [0, 1] \mid x \in L(\mu, s)\}$, and let $\epsilon > 0$. Then, $t - \epsilon > s$ for some $s \in [0, 1]$ such that $x \in L(\mu, s)$, and so $t - \epsilon > \mu(x)$. Since ϵ is an arbitrary, it follows that $t \geq \mu(x)$. Now let $\mu(x) = v$, then $x \in L(\mu, v)$ and so $v \in \{s \in [0, 1] \mid x \in L(\mu, s)\}$. Thus, $\mu(x) = v \geq \inf\{s \in [0, 1] \mid x \in L(\mu, s)\} = t$. Hence, $\mu(x) = t$. This completes the proof.

An anti-fuzzy Lie sub-superalgebra μ of a Lie superalgebra $\mathscr{L}$ is said to be *abnormal* if $\mu(0) = 0$.

Theorem 5.32 *Let μ be an anti-fuzzy Lie sub-superalgebra of $\mathscr{L}$ over an anti-fuzzy field and μ^* be a fuzzy set in $\mathscr{L}$ defined by $\mu^*(x) = \mu(x) - \mu(0)$ for all $x \in \mathscr{L}$. Then, μ^* is an abnormal Lie sub-superalgebra of $\mathscr{L}$ containing μ.*

Proof For $x, y \in \mathscr{L}$ and $\alpha \in \mathbb{F}$, we have

$$\begin{aligned}\mu^*(x+y) &= \mu(x+y) - \mu(0) \leq \mu(x) \vee \mu(y) - \mu(0) \\ &= (\mu(x) - \mu(0)) \vee (\mu(y) - \mu(0)) = \mu^*(x) \vee \mu^*(y), \\ \mu^*(\alpha x) &= \mu(\alpha x) - \mu(0) \leq \lambda(\alpha) \vee \mu(x) - \mu(0) \\ &= (\lambda(\alpha) - \lambda(0)) \vee (\mu(x) - \mu(0)) = \lambda^*(\alpha) \vee \mu^*(x).\end{aligned}$$

This shows that μ^* is an anti-fuzzy subspace over an anti-fuzzy field. For $x \in \mathscr{L}$, $x = x_{\bar{0}} + x_{\bar{1}}$, where $x_{\bar{0}} \in \mathscr{L}_0$ and $x_{\bar{1}} \in \mathscr{L}_1$, we have

$$\begin{aligned}\mu^*(x) = \mu^*(x_{\bar{0}} + x_{\bar{1}}) &= \mu(x_{\bar{0}} + x_{\bar{1}}) - \mu(0) \\ &= \mu(x_{\bar{0}}) \vee \mu(x_{\bar{1}}) - \mu(0) \\ &= (\mu(x_{\bar{0}}) - \mu(0)) \vee (\mu(x_{\bar{1}}) - \mu(0)) \\ &= \mu^*(x_{\bar{0}}) \vee \mu^*(x_{\bar{1}}).\end{aligned}$$

This shows that μ^* is a $\mathbb{Z}_2$-graded anti-fuzzy subspace of $\mathscr{L}$.

For any $x, y \in \mathscr{L}$,

$$\begin{aligned}\mu^*([x, y]) &= \mu([x, y]) - \mu(0) \leq \mu(x) \vee \mu(y) - \mu(0) \\ &= (\mu(x) - \mu(0)) \vee (\mu(y) - \mu(0)) = \mu^*(x) \vee \mu^*(y).\end{aligned}$$

Hence, μ^* is an abnormal Lie sub-superalgebra of $\mathscr{L}$. Clearly, $\mu^*(0) = 1$ and $\mu \subset \mu^*$. This ends the proof.

Corollary 5.13 *If μ is an anti-fuzzy Lie sub-superalgebra of $\mathscr{L}$ over an anti-fuzzy field satisfying $\mu^+(x) = 1$ for some $x \in \mathscr{L}$, then $\mu(x) = 1$.*

We now present the concept of anti-fuzzy Lie sub-superalgebras of Lie superalgebras over an anti-fuzzy field under homomorphisms.

Theorem 5.33 *Let $f : \mathscr{L} \to \acute{\mathscr{L}}$ be an epimorphism of Lie superalgebras. If ν is an anti-fuzzy Lie sub-superalgebra of $\acute{\mathscr{L}}$ over an anti-fuzzy field and μ is the pre-image of ν under f. Then μ is an anti-fuzzy Lie sub-superalgebra of $\mathscr{L}$ over an anti-fuzzy field.*

Proof For any $x, y \in \mathscr{L}$ and $\alpha \in \mathbb{F}$,

$$\begin{aligned}\mu(x+y) &= \nu(f(x+y)) = \nu(f(x) + f(y)) \\ &\leq \nu(f(x)) \vee \nu(f(y)) = \mu(x) \vee \mu(y), \\ \mu(\alpha x) &= \nu(f(\alpha x)) = \nu(f(\alpha) f(x)) \\ &\leq \lambda(f(\alpha)) \vee \nu(f(x)) = \lambda(\alpha) \vee \mu(x).\end{aligned}$$

This show that μ is an anti-fuzzy subspace of $\mathscr{L}$ over an anti-fuzzy field.
Let $x \in \mathscr{L}$. Then, $x = x_{\bar{0}} + x_{\bar{1}}$, where $x_{\bar{0}} \in \mathscr{L}_0$ and $x_{\bar{1}} \in \mathscr{L}_1$. Since f preserves the grading, $f(x) = f(x_{\bar{0}} + x_{\bar{1}}) = f(x_{\bar{0}}) + f(x_{\bar{1}})$. So

$$\mu(x) = \nu(f(x)) = \nu(f(x_{\bar{0}}) + f(x_{\bar{1}}))$$
$$= \nu(f(x_{\bar{0}})) \vee \nu(f(x_{\bar{1}}))$$
$$= \mu(x_{\bar{0}}) \vee \mu(x_{\bar{1}}).$$

This shows that μ is a $\mathbb{Z}_2$-graded anti-fuzzy subspace of $\mathscr{L}$. For any $x, y \in \mathscr{L}$,

$$\mu([x, y]) = \nu(f([x, y])) = \nu([f(x), f(y)])$$
$$\leq \nu(f(x)) \vee \nu(f(y)) = \mu(x) \vee \mu(y).$$

Hence, μ is an anti-fuzzy Lie sub-superalgebra of $\mathscr{L}$ over an anti-fuzzy field.

Definition 5.31 Let $\mathscr{L}$ and $\acute{\mathscr{L}}$ be two Lie superalgebras and let f be a function of $\mathscr{L}$ into $\acute{\mathscr{L}}$. If μ is a fuzzy set in $\acute{\mathscr{L}}$, then the *pre-image* of μ under f is the fuzzy set in $\mathscr{L}$ defined by $f^{-1}(\mu)(x) = \mu(f(x))$ for all $x \in \mathscr{L}$.

Theorem 5.34 *Let $f : \mathscr{L} \to \acute{\mathscr{L}}$ be an onto homomorphism of Lie superalgebras. If μ is an anti-fuzzy Lie sub-superalgebra of $\acute{\mathscr{L}}$ over an anti-fuzzy field λ, then $f^{-1}(\mu)$ is an anti-fuzzy Lie sub-superalgebra of $\mathscr{L}$ over an anti-fuzzy field λ.*

Proof Let $x_1, x_2 \in \mathscr{L}$ and $\alpha \in \mathbb{F}$, then

$$\begin{aligned} f^{-1}(\mu)(x_1 + x_2) &= \mu(f(x_1) + f(x_2)) \\ &\leq \mu(f(x_1) \vee \mu(f(x_2)) \\ &= f^{-1}(\mu)(x_1) \vee f^{-1}(\mu)(x_2), \\ f^{-1}(\mu)(\alpha x_1) &= \mu(f(\alpha) + f(x_1)) \\ &\leq \lambda(f(\alpha)) \vee \mu(f(x_1)) \\ &= f^{-1}(\lambda)(\alpha) \vee f^{-1}(\mu)(x_1). \end{aligned}$$

This shows that $f^{-1}(\mu)$ is an anti-fuzzy subspace of $\mathscr{L}$ over an anti-fuzzy field. Let $x \in \mathscr{L}$. Then, $x = x_{\bar{0}} + x_{\bar{1}}$, where $x_{\bar{0}} \in \mathscr{L}_0$ and $x_{\bar{1}} \in \mathscr{L}_1$. Since f preserves the grading, $f(x) = f(x_{\bar{0}} + x_{\bar{1}}) = f(x_{\bar{0}}) + f(x_{\bar{1}})$.

$$f^{-1}(\mu)(x) = \mu(f(x)) = \mu(f(x_{\bar{0}}) + f(x_{\bar{1}}))$$
$$= \mu(f(x_{\bar{0}})) \vee \mu(f(x_{\bar{1}}))$$
$$= f^{-1}(\mu)(x_{\bar{0}}) \vee f^{-1}(\mu)(x_{\bar{1}}).$$

This shows that μ is a $\mathbb{Z}_2$-graded anti-fuzzy subspace of $\mathscr{L}$.

$$\begin{aligned} f^{-1}(\mu)([x_1, x_2]) &= \mu([f(x_1), f(x_2)]) \\ &\leq \mu(f(x_1) \vee \mu(f(x_2)) \\ &= f^{-1}(\mu)(x_1) \vee f^{-1}(\mu)(x_2). \end{aligned}$$

Hence, $f^{-1}(\mu)$ is an anti-fuzzy Lie sub-superalgebra of $\mathscr{L}$ over an anti-fuzzy field.

Corollary 5.14 *Let $f : \mathscr{L} \to \acute{\mathscr{L}}$ be an onto homomorphism of Lie superalgebras. If μ is an anti-fuzzy Lie sub-superalgebra of $\acute{\mathscr{L}}$, then $f^{-1}(\mu^c) = (f^{-1}(\mu))^c$.*

Theorem 5.35 *Let μ be an anti-fuzzy Lie sub-superalgebra of $\mathcal{L}$ over an anti-fuzzy field λ and Let $f : \mathcal{L} \to \acute{\mathcal{L}}$. Then, the image $f(\mu)$ is an anti-fuzzy Lie sub-superalgebra of $\acute{\mathcal{L}}$ over an anti-fuzzy field λ.*

Proof Consider $f(x)$, $f(y) \in f(\mathcal{L})$. Let x_0, $y_0 \in f^{-1}(f(x))$ be such that $\mu(x_0) = \inf_{t\in f^{-1}(f(x))} \mu(t)$ and $\mu(y_0) = \inf_{t\in f^{-1}(f(y))} \mu(t)$, respectively. Then we can deduce that

$$\begin{aligned}
&\mu(f(x+y)) = \mu(f(x) + f(y)) = \inf_{t\in f^{-1}(f(x)+f(y))} \mu(t)\\
&\leq \mu(x_0 + y_0) \leq \mu(x_0) \vee \mu(y_0)\\
&= \inf_{t\in f^{-1}(f(x))} \mu(t) \vee \inf_{t\in f^{-1}(f(y))} \mu(t)\\
&= \mu(f(x)) \vee \mu(f(y)),\\
&\mu(f(\alpha x)) = \mu(f(\alpha) f(x)) = \inf_{t\in f^{-1}(f(\alpha) f(x))} \mu(t)\\
&\leq \mu(\alpha_0 x_0) \leq \lambda(\alpha_0) \vee \mu(x_0)\\
&= \inf_{t\in f^{-1}(f(\alpha))} \mu(t) \vee \inf_{t\in f^{-1}(f(x))} \mu(t)\\
&= \lambda(f(\alpha)) \vee \mu(f(x)).
\end{aligned}$$

So μ is an anti-fuzzy subspace of $\acute{\mathcal{L}}$ over an anti-fuzzy field.

Let $f(x) \in f(\mathcal{L})$. Then, $f(x) = f(x_{\bar{0}}) + f(x_{\bar{1}})$, where $f(x_{\bar{0}}) \in f(\mathcal{L}_0)$ and $f(x_{\bar{1}}) \in f(\mathcal{L}_1)$.

$$\begin{aligned}
\mu(f(x)) &= \mu(f(x_{\bar{0}}) + f(x_{\bar{1}}))\\
&= \inf_{t\in f^{-1}(f(x_{\bar{0}})+f(x_{\bar{1}}))} \mu(t)\\
&= \mu(x_0 + y_0) = \mu(x_0) \vee \mu(y_0)\\
&= \inf_{t\in f^{-1}(f(x_{\bar{0}}))} \mu(t) \vee \inf_{t\in f^{-1}(f(x_{\bar{1}}))} \mu(t)\\
&= \mu(f(x_{\bar{0}})) \vee \mu(f(x_{\bar{1}})).
\end{aligned}$$

This shows that μ is a $\mathbb{Z}_2$-graded anti-fuzzy subspace of $\acute{\mathcal{L}}$.

$$\begin{aligned}
&\mu(f([x, y])) = \mu([f(x), f(y)]) = \inf_{t\in f^{-1}([f(x), f(y)])} \mu(t)\\
&\leq \mu([x_0, y_0]) \leq \mu(x_0) \vee \mu(y_0)\\
&= \inf_{t\in f^{-1}(f(x))} \mu(t) \vee \inf_{t\in f^{-1}(f(y))} \mu(t)\\
&= \mu(f(x)) \vee \mu(f(y)).
\end{aligned}$$

Hence, μ is an anti-fuzzy Lie sub-superalgebra of $\acute{\mathcal{L}}$ over an anti-fuzzy field.

Definition 5.32 Let $\mathcal{L}$ and $\acute{\mathcal{L}}$ be two Lie superalgebras and let f be a function of μ is a fuzzy set in $\mathcal{L}$, then the *co-image* of μ under f is the fuzzy set defined by
$f(\mu)(y) = \begin{cases} \inf\{\mu(t) \mid t \in \mathcal{L}, f(t) = y\}, & \text{if } f^{-1}(y) \neq \emptyset,\\ 1, & \text{otherwise.}\end{cases}$

Definition 5.33 Let $\mathscr{L}$ and $\acute{\mathscr{L}}$ be any sets and let $f : \mathscr{L} \to \acute{\mathscr{L}}$ be any function. A fuzzy set μ is called *f-invariant* if and only if for all $x, y \in \mathscr{L}$, $f(x) = f(y)$ implies $\mu(x) = \mu(y)$.

Theorem 5.36 *Let $f : \mathscr{L} \to \acute{\mathscr{L}}$ be an epimorphism of Lie superalgebras. Then, μ is an f-invariant anti-fuzzy Lie sub-superalgebra of $\mathscr{L}$ over an anti-fuzzy field λ if and only if $f(\mu)$ is an anti-fuzzy Lie sub-superalgebra of $\acute{\mathscr{L}}$ over an anti-fuzzy field λ.*

Proof Let $x, y \in \acute{\mathscr{L}}$ and $\alpha \in \mathbb{F}$. Then, there exist $a, b \in \mathscr{L}$ and $\beta \in \mathbb{F}$ such that $f(a) = x$, $f(b) = y$, $f(\beta) = \alpha$, $x + y = f(a + b)$ and $\alpha x = \beta f(a)$. Since μ is f-invariant, by assumption, we have

$$\begin{aligned} f(\mu)(x + y) &= \mu(a + b) \leq \mu(a) \vee \mu(b) = f(\mu)(x) \vee f(\mu)(y), \\ f(\mu)(\alpha x) &= \mu(\beta a) \leq \lambda(\beta) \vee \mu(b) = f(\lambda)(\alpha) \vee f(\mu)(x). \end{aligned}$$

Thus, $f(\mu)$ is an anti-fuzzy subspace of $\acute{\mathscr{L}}$ over an anti-fuzzy field.

Let $x \in \acute{\mathscr{L}}$ and $x = x_{\bar{0}} + x_{\bar{1}}$, where $x_{\bar{0}} \in \acute{\mathscr{L}}_{\bar{0}}$ and $x_{\bar{1}} \in \acute{\mathscr{L}}_{\bar{0}}$. Then, there exist $a \in \mathscr{L}$ and $a = a_{\bar{0}} + a_{\bar{1}}$, where $a_{\bar{0}} \in \mathscr{L}_0$, $a_{\bar{1}} \in \mathscr{L}_1$ such that $f(a) = x$, $f(a_{\bar{0}}) = x_{\bar{0}}$, $f(a_{\bar{1}}) = x_{\bar{1}}$.

$$\begin{aligned} f^{-1}(\mu)(x) = f(\mu)(x) = \mu(a) &= \mu(a_{\bar{0}} + a_{\bar{1}}) \\ &= \mu(a_{\bar{0}}) \vee \mu(a_{\bar{1}}) \\ &= f^{-1}(\mu)(x_{\bar{0}}) \vee f^{-1}(\mu)(x_{\bar{1}}). \end{aligned}$$

This shows that $f(\mu)$ is a $\mathbb{Z}_2$-graded anti-fuzzy subspace of $\acute{\mathscr{L}}$.

$$f(\mu)([x, y]) = \mu([a, b]) \leq \mu(a) \vee \mu(b) = f(\mu)(x) \vee f(\mu)(y).$$

Hence, $f(\mu)$ is an anti-fuzzy Lie sub-superalgebra of $\acute{\mathscr{L}}$ over an anti-fuzzy field. Conversely, if $f(\mu)$ is an anti-fuzzy Lie sub-superalgebra of $\acute{\mathscr{L}}$, then for any $x \in \mathscr{L}$

$$\begin{aligned} f^{-1}(f(\mu))(x) &= f(\mu)(f(x)) \\ &= \inf\{\mu(t) \mid t \in \mathscr{L}, f(t) = f(x)\} \\ &= \inf\{\mu(t) \mid t \in \mathscr{L}, \mu(t) = \mu(x)\} \\ &= \mu(x). \end{aligned}$$

Hence, $f^{-1}(f(\mu)) = \mu$ is an anti-fuzzy Lie sub-superalgebra by Theorem 5.34. This completes the proof.

Theorem 5.37 *Let $f : \mathscr{L} \to \acute{\mathscr{L}}$ be an epimorphism of Lie superalgebras. If μ and ν are anti-fuzzy Lie sub-superalgebras of $\mathscr{L}$ over an anti-fuzzy field, then $f(\mu + \nu) = f(\mu) + f(\nu)$.*

Chapter 6
Bipolar Fuzzy Lie Structures

In this chapter, we present properties of bipolar fuzzy Lie ideals, bipolar fuzzy Lie sub-superalgebras, bipolar fuzzy bracket product, solvable bipolar fuzzy Lie ideals, and nilpotent bipolar fuzzy Lie ideals. We describe some properties of nilpotency of bipolar fuzzy Lie ideals. We deal the concept of bipolar fuzzy adjoint representation of Lie algebras and discuss the relationship between this representation and nilpotent bipolar fuzzy Lie ideals. We also define Killing form in the bipolar fuzzy case and study some of its properties.

6.1 Introduction

A wide variety of human decision making is based on double-sided or bipolar judgmental thinking on a positive side and a negative side, for instances, cooperation and competition, friendship and hostility, common interests and conflict interests, effect and side effect, likelihood and unlikelihood, feedforward and feedback. In Chinese medicine, *Yin* and *Yang* are the two sides. *Yin* is the negative side of a system, and *Yang* is the positive side of a system. The notion of bipolar fuzzy sets (YinYang bipolar fuzzy sets) was introduced by Zhang [146, 147] in the space $\{\forall\ (x, y) \mid (x, y) \in [-1, 0] \times [0, 1]\}$. Although bipolar fuzzy sets and intuitionistic fuzzy sets look similar to each other, they are essentially different sets [93, 94].

Definition 6.1 Let X be a nonempty set. A *bipolar fuzzy set* B in X is an object having the form

$$B = \left(\mu_B^P, \mu_B^N\right) = \left\{\left(x, \mu_B^P(x), \mu_B^N(x)\right) \mid x \in X\right\}$$

where $\mu_B^P : X \to [0, 1]$ and $\mu_B^N : X \to [-1, 0]$ are mappings.

Positive membership degree $\mu_B^P(x)$ denotes the satisfaction degree of an element x to the property corresponding to a bipolar fuzzy set B, and negative membership

M. Akram, *Fuzzy Lie Algebras*, Infosys Science Foundation Series,
https://doi.org/10.1007/978-981-13-3221-0_6

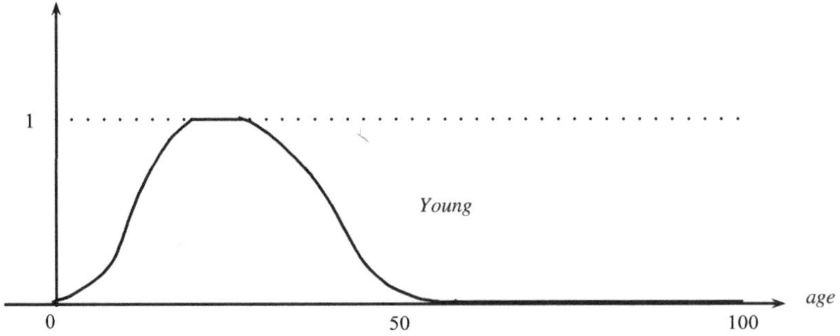

Fig. 6.1 A fuzzy set "young"

degree $\mu_B^N(x)$ denotes the satisfaction degree of x to some implicit counter-property corresponding to B. If $\mu_B^P(x) \neq 0$ and $\mu_B^N(x) = 0$, it is the state when x has only positive satisfaction for B. If $\mu_B^P(x) = 0$ and $\mu_B^N(x) \neq 0$, it is the state when x does not satisfy the property of B but somewhat satisfies the counter-property of B. It is possible for an element x to be such that $\mu_B^P(x) \neq 0$ and $\mu_B^N(x) \neq 0$ when the membership function of the property coincides with its counter-property over $x \in X$.

Example 6.1 Suppose that there is a fuzzy set "young" defined on the age domain [0, 100] like Fig. 6.1. In that fuzzy set, consider two ages 50 and 95 with membership degree 0. Although both of them do not satisfy the property "young," we may say that age 95 is more apart from the property rather than age 50. Only with the membership degrees ranged on the interval [0, 1], it is difficult to express this kind of meaning.

We define a bipolar fuzzy set as in Fig. 6.2 for the same fuzzy set "young" of Fig. 6.1. The negative membership degrees indicate the satisfaction range of elements to an implicit counter-property (e.g., old against the property young). This kind of bipolar fuzzy set representation enables the elements with 0 degree of membership in traditional fuzzy sets, to be expressed into the elements with zero degree of membership (when irrelevant elements) and negative degree of membership (when contrary elements). The age elements 50 and 95, with membership degree 0 in the fuzzy set of Fig. 6.1, have 0 and a negative membership degree in the bipolar fuzzy set of Fig. 6.2, respectively. Now it is manifested that 50 is an irrelevant age to the property young and 95 is more apart from the property young than 50(i.e., 95 is a contrary age to the property young).

Example 6.2 Let $X = \{P_1, P_2, P_3, P_4, P_5, P_6\}$ be a set of products manufactured in a company. The products can be categorized according to their profit and loss. The profit and loss of every product vary from time to time. The possibilities of profit and loss of all the products are given in Table 6.1. Table 6.1 shows that the product P_1 has 60% profit and 40% loss on the average. Profit is the positive and loss is the negative behavior of the product, that is, two-sided behavior. It can be written in the form of a bipolar fuzzy set as, $A = \{(P_1, 0.6, -0.4), (P_2, 0.8, -0.5), (P_3, 0.9, -0.1), (P_4, 0.7, -0.2), (P_5, 0.5, -0.6), (P_6, 0.6, -0.4)\}$.

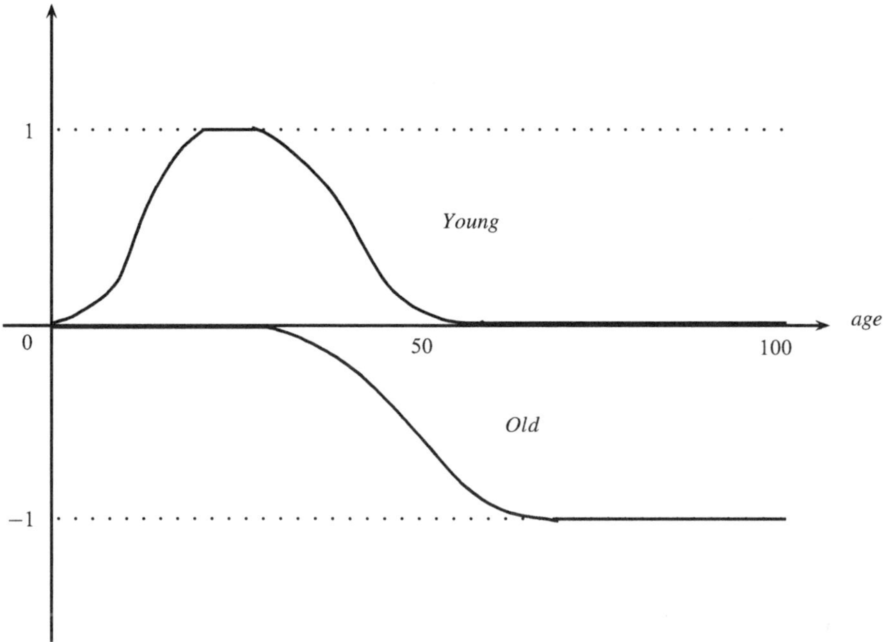

Fig. 6.2 A bipolar fuzzy set "young"

Table 6.1 Profit and loss of products

Product	Profit	Loss
P_1	0.6	0.4
P_2	0.8	0.5
P_3	0.9	0.1
P_4	0.7	0.2
P_5	0.5	0.6
P_6	0.6	0.4

Example 6.3 Consider a fuzzy set

$$\text{frog's prey} = \{(\text{mosquito},1.0),(\text{dragon fly},0.4),(\text{turtle},0.0),(\text{snake},0.0)\}.$$

In this fuzzy set, both turtle and snake have the membership degree 0. It is known that frog and turtle are indifferent from each other concerning the prey-hunting relationship, but snake is a predator of frog. Turtle is an irrelevant animal and snake is related to frog by a counter implicit counter-property, but they both seem irrelevant in fuzzy set. As we can see from this example, it is difficult to express the difference of the irrelevant elements in fuzzy sets.

The same fuzzy set "frog's prey" can be redefined in the form of a bipolar fuzzy set, as follows:

$$\text{frog's prey} = \{(\text{mosquito},1,0),(\text{dragon fly},0.4,0),(\text{turtle},0,0),(\text{snake},0,-1)\}.$$

We can see that membership degree 0 and nonmembership degree 0 of turtle mean that frog never hunts turtle and turtle never hunts frog. While membership degree 0 and nonmembership degree –1 of snake mean that frog never hunts snake but snake always hunts frog. Here, the counter implicit counter-property is "predator of frog," which created the difference between fuzzy set and bipolar fuzzy set of frog's prey.

Definition 6.2 For every two bipolar fuzzy sets $A = (\mu_A^P, \mu_A^N)$ and $B = (\mu_B^P, \mu_B^N)$ in X the following operations hold:

- $(A \bigcap B)(x) = (\min(\mu_A^P(x), \mu_B^P(x)), \max(\mu_A^N(x), \mu_B^N(x))),$
- $(A \bigcup B)(x) = (\max(\mu_A^P(x), \mu_B^P(x)), \min(\mu_A^N(x), \mu_B^N(x))),$
- $\overline{A}(x) = (1 - \mu_A^P(x), -1 - \mu_A^N(x)),$
- $A \subseteq B$ if and only if $\mu_A^P(x) \leq \mu_B^P(x)$ *and* $\mu_A^N(x) \geq \mu_B^N(x) \quad \forall x \in X.$

Definition 6.3 Let A be a bipolar fuzzy set on a universe of discourse X, then cardinality of A, denoted as $Card(A)$, is defined as

$$Card(A) = \sum_{x \in X} (\mu_A^P(x)) + \mid \mu_A^N(x) \mid).$$

Definition 6.4 Let A be a bipolar fuzzy set on X. Then, for $\alpha \in [0, 1]$, α-cut of A, denoted by A_α, is defined as

$$A^\alpha = U(\mu_A^P, \alpha) \cup U(\mu_A^N, \alpha),$$

such that

$$U(\mu_A^P, \alpha) = \{x \mid \mu_A^P(x) \geq \alpha\},$$
$$U(\mu_A^N, \alpha) = \{x \mid \mu_A^N(x) \leq -\alpha\}.$$

$U(\mu_A^P, \alpha)$ is positive $\alpha - cut$ and $U(\mu_A^N, \alpha)$ is negative $\alpha - cut$.

Definition 6.5 Let A be a bipolar fuzzy set on a universe of discourse X, then the support of A denoted by $Supp(A)$ is defined as

$$Supp(A) = Supp^P(A) \cup Supp^N(A),$$

such that

$$Supp^P(A) = \{x \mid \mu_A^P(x) > 0\},$$
$$Supp^N(A) = \{x \mid \mu_A^N(x) < 0\}.$$

$Supp^P(A)$ is positive support and $Supp^N(A)$ is negative support.

Definition 6.6 Let X be a nonempty set. We call a mapping $A = (\mu_A^P, \mu_A^N) : X \times X \to [0, 1] \times [-1, 0]$ a *bipolar fuzzy relation* on X such that $\mu_A^P(x, y) \in [0, 1]$ and $\mu_A^N(x, y) \in [-1, 0]$.

Definition 6.7 Let A and B be any two bipolar fuzzy sets defined on universes of discourse X and Y, respectively. We call a mapping $A \times B = (\mu_{A\times B}^P, \mu_{A\times B}^N) = X \times Y \to [0, 1] \times [-1, 0]$, a bipolar fuzzy relation from A to B such that

$$\mu_{A\times B}^P(x, y) = \min\{\mu_A^P(x), \mu_B^P(y)\},$$
$$\mu_{A\times B}^N(x, y) = \max\{\mu_A^N(x), \mu_B^N(y)\}.$$

Definition 6.8 (Extension principle for bipolar fuzzy sets) Let X be a Cartesian product of universes $X = X_1 \times X_2 \times ...X_n$ and $A_1, A_2, ...A_n$ be n bipolar fuzzy sets in $X_1, X_2, ..., X_n$, respectively. f is a mapping from X to a universe Y, i.e., $y = f(x_1, ..., x_n)$. Then, a bipolar fuzzy set B in Y is defined as follows:

$$B = \{(y, (\mu_B^P(y), \mu_B^N(y))) \mid y = f(x_1, x_2, ..., x_n), (x_1, x_2, ..., x_n) \in X_1 \times X_2 \times X_n\}$$

$$\mu_B^P(y) = \begin{cases} \max\limits_{(x_1,x_2,...,x_n)\in f^{-1}(y)} \min\limits_i \mu_{A_i}^P(x_i) & \text{if } f^{-1}(y) \neq \emptyset, \\ 0, & \text{otherwise.} \end{cases}$$

$$\mu_B^N(y) = \begin{cases} \min\limits_{(x_1,x_2,...,x_n)\in f^{-1}(y)} \max\limits_i \mu_{A_i}^N(x_i) & \text{if } f^{-1}(y) \neq \emptyset, \\ 0, & \text{otherwise.} \end{cases}$$

Remark 6.1 In the fuzzy literature, several extensions for fuzzy sets have been proposed. Some of them are concerned with fuzzy sets whose membership degrees are expressed with a pair of membership values. The intuitionistic fuzzy sets and the bipolar fuzzy logic belong to each extension. This remark briefly introduces the intuitionistic fuzzy sets and bipolar fuzzy logic and compares them with bipolar fuzzy set.

When we match a bipolar fuzzy set $A = \{(x, (\mu_A^P(x), \mu_A^N(x))) \mid x \in X\}$ to an intuitionistic fuzzy set $A = \{(x, \mu_A(x), \nu_A(x)) | x \in X\}$ under the condition $\mu_A^P(x) = \mu_A(x)$ and $\mu_A^N(x) = -\nu_A(x)$, bipolar fuzzy sets and intuitionistic fuzzy sets look similar to each other. However, they are different from each other in the following senses: In bipolar fuzzy sets, the positive membership degree $\mu_A^P(x)$ characterizes the extent that the element x satisfies the property A, and the negative membership degree $\nu_A^N(x)$ characterizes the extent that the element x satisfies an implicit counter-property of A. On the other hand, in intuitionistic fuzzy sets, the membership degree $\mu_A(x)$ denotes the degree that the element x satisfies the property A and the membership degree $\nu_A(x)$ indicates the degree that x does not satisfy property of A. Since a counter-property is not usually equivalent to not-property, both bipolar fuzzy sets and intuitionistic fuzzy sets are the different extensions of fuzzy sets. Their difference can be manifested in the interpretation of an element x with membership degree $(0, 0)$. In the perspective of bipolar fuzzy set A, it is interpreted that the element x does not satisfy both the property A and its implicit counter-property. It means that it is indifferent (i.e., neutral) from the property and its implicit counter-property. In the perspective of intuitionistic fuzzy set A, it is interpreted that the element x does not satisfy the property and its not-property. When we regard an intuitionistic fuzzy set as an interval-valued fuzzy set, the element with the membership degree (0, 0) in an intuitionistic fuzzy set has the membership degree [0, 1] in interval-valued fuzzy set. It means that we have no knowledge about the element. On the other hand, their set operations union, intersection, and negation are also different from each other. These things differentiate bipolar fuzzy sets from intuitionistic fuzzy sets. The intuitionistic fuzzy set representation is useful when there are some uncertainties in assigning membership degrees. The bipolar fuzzy set representation is useful when irrelevant elements and contrary elements are needed to be discriminated.

Definition 6.9 Let V be a vector space. A bipolar fuzzy set $A = (\mu_A^P, \mu_A^N)$ on V is called a *bipolar fuzzy subspace* if the following conditions are satisfied:

1. $\mu_A^P(x+y) \geq \min(\mu_A^P(x), \mu_A^P(y))$ and $\mu_A^N(x+y) \leq \max(\mu_A^N(x), \mu_A^N(y))$,
2. $\mu_A^P(\alpha x) \geq \mu_A^P(x)$ and $\mu_A^N(\alpha x) \leq \mu_A^N(x)$,

for all $x, y \in V$ and $\alpha \in \mathbb{F}$.

The following lemmas are obvious.

Lemma 6.1 *$A = (\mu_A^P, \mu_A^N)$ is a bipolar fuzzy subspace of V if and only if μ_A^P and μ_A^N are fuzzy subspaces of V.*

Lemma 6.2 *Let $A = (\mu_A^P, \mu_A^N)$ and $B = (\mu_B^P, \mu_B^N)$ be bipolar fuzzy subspaces of V. Then, $A + B$ is also a bipolar fuzzy subspace of V.*

Lemma 6.3 *Let $A = (\mu_A^P, \mu_A^N)$ and $B = (\mu_B^P, \mu_B^N)$ be bipolar fuzzy subspaces of V. Then, $A \cap B$ is also a bipolar fuzzy subspace of V.*

Lemma 6.4 *Let $A = (\mu_A^P, \mu_A^N)$ be a bipolar fuzzy subspace of V' and ϕ be a mapping from vector space V to V'. Then, the inverse image $\phi^{-1}(A)$ is also a bipolar fuzzy subspace of V.*

Lemma 6.5 *Let $A = (\mu_A^P, \mu_A^N)$ be a bipolar fuzzy subspace of V and f be a mapping from V to V'. Then, the image $\phi(A)$ is also a bipolar fuzzy subspace of V'.*

6.2 Bipolar Fuzzy Lie Ideals

Definition 6.10 Let L be a Lie algebra. A bipolar fuzzy set $A = (\mu_A^P, \mu_A^N)$ on L is called a *bipolar fuzzy Lie ideal* if the following conditions are satisfied:

1. $\mu_A^P(x+y) \geq \min(\mu_A^P(x), \mu_A^P(y))$ and $\mu_A^N(x+y) \leq \max(\mu_A^N(x), \mu_A^N(y))$,
2. $\mu_A^P(\alpha x) \geq \mu_A^P(x)$ and $\mu_A^N(\alpha x) \leq \mu_A^N(x)$,
3. $\mu_A^P([x, y]) \geq \mu_A^P(x)$ and $\mu_A^N([x, y]) \leq \mu_A^N(x)$

for all $x, y \in L$ and $\alpha \in \mathbb{F}$.

Example 6.4 Let $\Re^2 = \{(x, y) | x, y \in R\}$ be the set of all two-dimensional real vectors. Then, $\Re^2$ with the bracket $[\cdot, \cdot]$ defined as usual cross product, i.e., $[x, y] = x \times y$, is a real Lie algebra. We define a bipolar fuzzy set $A = (\mu_A^P, \mu_A^N)$

$$\mu_A^P(x, y) = \begin{cases} 1 \text{ if } x = y = 0, \\ 0 \text{ otherwise,} \end{cases} \qquad \mu_A^N(x, y) = \begin{cases} 0 & \text{if } x = y = 0, \\ -1 & \text{otherwise.} \end{cases}$$

By routine computations, we can check that it is bipolar fuzzy Lie ideal of a Lie algebra L.

Definition 6.11 Let $A = (\mu_A^P, \mu_A^N) \in J^L$, a bipolar fuzzy subspace of L generated by A will be denoted by $[A]$. It is the intersection of all bipolar fuzzy subspaces of L containing A. For all $x \in L$, we define:

$$[\mu_A^P](x) = \sup\{\min \mu^P(x_i) \mid x = \sum \alpha_i x_i, \alpha_i \in \mathbb{F}, x_i \in L\},$$

$$[\mu_A^N](x) = \inf\{\max \mu^N(x_i) \mid x = \sum \alpha_i x_i, \alpha_i \in \mathbb{F}, x_i \in L\}.$$

Definition 6.12 Let $f : L_1 \to L_2$ be a homomorphism of Lie algebras which has an extension $f : J^{L_1} \to J^{L_2}$ defined by:

$$f(\mu_A^P)(y) = \sup\{\mu^P(x), x \in f^{-1}(y)\},$$

$$f(\mu_A^N)(y) = \inf\{\mu^N(x), x \in f^{-1}(y)\},$$

for all $A = (\mu_A^P, \mu_A^N) \in J^{L_1}$, $y \in L_2$. Then, $f(A)$ is called the *homomorphic image* of A.

The following two propositions are obvious.

Proposition 6.1 *Let $f : L_1 \to L_2$ be a homomorphism of Lie algebras, and let $A = (\mu_A^P, \mu_A^N)$ be a bipolar fuzzy Lie ideal of L_1. Then,*
(i) *$f(A)$ is a bipolar fuzzy Lie ideal of L_2,*
(ii) *$f([A]) \supseteq [f(A)]$.*

Proposition 6.2 *If A and B are bipolar fuzzy Lie ideals in L, then $[A, B]$ is a bipolar fuzzy Lie ideal of L.*

Theorem 6.1 *Let A_1, A_2, B_1, B_2 be bipolar fuzzy Lie ideals in L such that $A_1 \subseteq A_2$ and $B_1 \subseteq B_2$, then $[A_1, B_1] \subseteq [A_2, B_2]$.*

Proof Indeed,

$$\begin{aligned}\ll \mu^P_{A_1}, \mu^P_{B_1} \gg (x) &= \sup\{\min(\mu^P_{A_1}(a), \mu^P_{B_1}(b)) \mid a, b \in L_1, [a, b] = x\}\\ &\geqslant \sup\{\min(\mu^P_{A_2}(a), \mu^P_{B_2}(b)) \mid a, b \in L_1, [a, b] = x\}\\ &= \ll \mu^P_{A_2}, \mu^P_{B_2} \gg (x),\\ \ll \mu^N_{A_1}, \mu^N_{B_1} \gg (x) &= \inf\{\max(\mu^N_{A_1}(a), \mu^N_{B_1}(b)) \mid a, b \in L_1, [a, b] = x\}\\ &\leqslant \inf\{\max(\mu^N_{A_2}(a), \mu^N_{B_2}(b)) \mid a, b \in L_1, [a, b] = x\}\\ &= \ll \mu^N_{A_2}, \mu^N_{B_2} \gg (x).\end{aligned}$$

Hence, $[A_1, B_1] \subseteq [A_2, B_2]$.

Let $A = (\mu^P_A, \mu^N_A)$ be a bipolar fuzzy Lie ideal in L. Putting

$$A^0 = A,\ A^1 = [A, A_0],\ A^2 = [A, A_1],\ \ldots,\ A^n = [A, A^{n-1}]$$

we obtain a descending series of a bipolar fuzzy Lie ideals

$$A^0 \supseteq A^1 \supseteq A^2 \supseteq \cdots \supseteq A^n \supseteq \cdots$$

and a series of bipolar fuzzy sets $B^n = (\mu^P_{B^n}, \mu^N_{B^n})$ such that

$$\mu^P_{B^n} = \sup\{\mu^P_{A^n}(x) \mid 0 \neq x \in L\},\quad \mu^N_{B^n} = \inf\{\mu^N_{A^n}(x) \mid 0 \neq x \in L\}.$$

Definition 6.13 A bipolar fuzzy Lie ideal $A = (\mu^P_A, \mu^N_A)$ is called *nilpotent* if there exists a positive integer n such that $B^n = (0, 1)$.

Theorem 6.2 *A homomorphic image of a nilpotent bipolar fuzzy Lie ideal is a nilpotent bipolar fuzzy Lie ideal.*

Proof Let $f : L_1 \to L_2$ be a homomorphism of Lie algebras, and let $A = (\mu^P_A, \mu^N_A)$ be a nilpotent bipolar fuzzy Lie ideal in L_1. Assume that $f(A) = B$. We prove by induction that $f(A^n) \supseteq B^n$ for every natural n. First, we claim that $f([A, A]) \supseteq [f(A), f(A)] = [B, B]$. Let $y \in L_2$, then

$$\begin{aligned}&f(\ll \mu^P_A, \mu^P_A \gg)(y) = \sup\{\ll \mu^P_A, \mu^P_A \gg (x) \mid f(x) = y\}\\ &= \sup\{\sup\{\min(\mu^P_A(a), \mu^P_A(b)) \mid a, b \in L_1, [a, b] = x, f(x) = y\}\}\\ &= \sup\{\min(\mu^P_A(a), \mu^P_A(b)) \mid a, b \in L_1, [a, b] = x, f(x) = y\}\\ &= \sup\{\min(\mu^P_A(a), \mu^P_A(b)) \mid a, b \in L_1, [f(a), f(b)] = y\}\\ &= \sup\{\min(\mu^P_A(a), \mu^P_A(b)) \mid a, b \in L_1, f(a) = u, f(b) = v, [u, v] = y\}\\ &\geqslant \sup\{\min(\sup_{a \in f^{-1}(u)} \mu^P_A(a), \sup_{b \in f^{-1}(v)} \mu^P_A(b)) \mid [u, v] = y\}\\ &= \sup\{\min(f(\mu^P_A)(u), f(\mu^P_A)(v)) \mid [u, v] = y\} = \ll f(\mu^P_A), f(\mu^P_A) \gg (y),\end{aligned}$$

$$\begin{aligned}
f(\ll \mu_A^N, \mu_A^N \gg)(y) &= \inf\{\ll \mu_A^N, \mu_A^N \gg (x) \mid f(x) = y\} \\
&= \inf\{\inf\{\max(\mu_A^N(a), \mu_A^N(b)) \mid a, b \in L_1, [a, b] = x, f(x) = y\}\} \\
&= \inf\{\max(\mu_A^N(a), \mu_A^N(b)) \mid a, b \in L_1, [a, b] = x, f(x) = y\} \\
&= \inf\{\max(\mu_A^N(a), \mu_A^N(b)) \mid a, b \in L_1, [f(a), f(b)] = y\} \\
&= \inf\{\max(\mu_A^N(a), \mu_A^N(b)) \mid a, b \in L_1, f(a) = u, f(b) = v, [u, v] = y\} \\
&\leqslant \inf\{\max(\inf_{a \in f^{-1}(u)} \mu_A^N(a), \inf_{b \in f^{-1}(v)} \mu_A^N(b)) \mid [u, v] = y\} \\
&= \inf\{\max(f(\mu_A^N)(u), f(\mu_A^N)(v)) \mid [u, v] = y\} = \ll f(\mu_A^N), f(\mu_A^N) \gg (y).
\end{aligned}$$

Thus

$$f([A, A]) \supseteq f(\ll A, A \gg) \supseteq \ll f(A), f(A) \gg = [f(A), f(A)].$$

For $n > 1$, we get

$$f(A^n) = f([A, A^{n-1}]) \supseteq [f(A), f(A^{n-1})] \supseteq [B, B^{n-1}] = B^n.$$

Let m be a positive integer such that $A^m = (0, 1)$. Then, for $0 \neq y \in L_2$ we have

$$\mu_{B^m}^P(y) \leqslant f(\mu_{A^n}^P)(y) = f(0)(y) = \sup\{0(a) \mid f(a) = y\} = 0,$$

$$\mu_{B^m}^N(y) \geqslant f(\mu_{A^n}^N)(y) = f(1)(y) = \inf\{1(a) \mid f(a) = y\} = 1.$$

Thus, $B^m = (0, 1)$. This completes the proof.

Let $A = (\mu_A^P, \mu_A^N)$ be a bipolar fuzzy Lie ideal in L. Putting

$$A^{(0)} = A,\ A^{(1)} = [A^{(0)}, A^{(0)}],\ A^{(2)} = [A^{(1)}, A^{(1)}], \ldots, A^{(n)} = [A^{(n-1)}, A^{(n-1)}]$$

we obtain series

$$A^{(0)} \subseteq A^{(1)} \subseteq A^{(2)} \subseteq \cdots \subseteq A^{(n)} \subseteq \cdots$$

of bipolar fuzzy Lie ideals and a series of bipolar fuzzy sets $B^{(n)} = (\mu_{B^n}^P, \mu_{B^n}^N)$ such that

$$\mu_{B^n}^P = \sup\{\mu_{A^n}^P(x) \mid 0 \neq x \in L\}, \quad \mu_{B^n}^N = \inf\{\mu_{A^n}^N(x) \mid 0 \neq x \in L\}.$$

Definition 6.14 A bipolar fuzzy Lie ideal $A = (\mu_A^P, \mu_A^N)$ is called *solvable* if there exists a positive integer n such that $B^{(n)} = (0, 1)$.

Theorem 6.3 *A nilpotent bipolar fuzzy Lie ideal is solvable.*

Proof It is enough to prove that $A^{(n)} \subseteq A^n$ for all positive integers n. We prove it by induction on n and by the use of Theorem 6.1:

$$A^{(1)} = [A, A] = A^1, \qquad A^{(2)} = [A^{(1)}, A^{(1)}] \subseteq [A, A^{(1)}] = A^2.$$

$$A^{(n)} = [A^{(n-1)}, A^{(n-1)}] \subseteq [A, A^{(n-1)}] \subseteq [A, A^{(n-1)}] = A^n.$$

This completes the proof.

Definition 6.15 Let $A = (\mu_A^P, \mu_A^N)$ and $B = (\mu_B^P, \mu_B^N)$ be two bipolar fuzzy Lie ideals of a Lie algebra L. The sum $A \oplus B$ is called a *direct sum* if $A \cap B = (0, 1)$.

Theorem 6.4 *The direct sum of two nilpotent bipolar fuzzy Lie ideals is also a nilpotent bipolar fuzzy Lie ideal.*

Proof Suppose that $A = (\mu_A^P, \mu_A^N)$ and $B = (\mu_B^P, \mu_B^N)$ are two bipolar fuzzy Lie ideals such that $A \cap B = (0, 1)$. We claim that $[A, B] = (0, 1)$. Let $x(\neq 0) \in L$, then

$$\ll \mu_A^P, \mu_B^P \gg (x) = \sup\{\min(\mu_A^P(a), \mu_B^P(b)) \mid [a, b] = x\} \leqslant \min(\mu_A^P(x), \mu_B^P(x)) = 0$$

and

$$\ll \mu_A^N, \mu_B^N \gg (x) = \inf\{\max(\mu_A^N(a), \mu_B^N(b)) \mid [a, b] = x\} \geqslant \max(\mu_A^N(x), \mu_B^N(x)) = 1.$$

This proves our claim. Thus, we obtain $[A^m, B^n] = (0, 1)$ for all positive integers m, n. Now, we again claim that $(A \oplus B)^n \subseteq A^n \oplus B^n$ for positive integer n. We prove this claim by induction on n. For $n = 1$,

$$(A \oplus B)^1 = [A \oplus B, A \oplus B] \subseteq [A, A] \oplus [A, B] \oplus [B, A] \oplus [B, B] = A^1 \oplus B^1.$$

Now for $n > 1$,

$$\begin{aligned}(A \oplus B)^n &= [A \oplus B, (A \oplus B)^{n-1}] \subseteq [A \oplus B, A^{n-1} \oplus B^{n-1}] \\ &\subseteq [A, A^{n-1}] \oplus [A, B^{n-1}] \oplus [B, A^{n-1}] \oplus [B, B^{n-1}] = A^n \oplus B^n.\end{aligned}$$

Since there are two positive integers p and q such that $A^p = B^q = (0, 1)$, we have $(A \oplus B)^{p+q} \subseteq A^{p+q} \oplus B^{p+q} = (0, 1)$.

In a similar way, we can prove the following theorem.

Theorem 6.5 *The direct sum of two solvable bipolar fuzzy Lie ideals is a solvable bipolar fuzzy Lie ideal.*

Theorem 6.6 *Let $A = (\mu_A^P, \mu_A^N)$ be a bipolar fuzzy Lie ideal in a Lie algebra L. Then $A^n \subseteq [A_n]$ for any $n > 0$, where a bipolar fuzzy subset $[A_n] = ([\mu_{A_n}^P], [\mu_{A_n}^N])$ is defined by*

$$[\mu_{A_n}^P](x) = \sup\{\mu_A^P(a) \mid [x_1, [x_2, [\ldots, [x_n, a] \ldots]]] = x, \quad x_1, \ldots, x_n \in L\},$$
$$[\mu_{A_n}^N](x) = \inf\{\mu_A^N(a) \mid [x_1, [x_2, [\ldots, [x_n, a] \ldots]]] = x, \quad x_1, \ldots, x_n \in L\}.$$

Proof It is enough to prove that $\ll A, A^{n-1} \gg \subseteq [A_n]$. We prove it by induction on n. For n=1 and $x \in L$, we have

$$\begin{aligned}\ll \mu_A^P, \mu_A^P \gg (x) &= \sup\{\min(\mu_A^P(a), \mu_A^P(b)) \mid [a,b]=x\}\\ &\geqslant \sup\{\mu_A^P(b) \mid [a,b]=x, a \in L\} = [\mu_{A_1}](x),\end{aligned}$$

$$\begin{aligned}\ll \mu_A^N, \mu_A^N \gg (x) &= \inf\{\max(\mu_A^N(a), \mu_A^N(b)) \mid [a,b]=x\}\\ &\leqslant \inf\{\mu_A^N(b) : [a,b]=x, a \in L\} = [\mu_{A_1}^N](x).\end{aligned}$$

For $n > 1$,

$$\begin{aligned}&\ll \mu_A^P, \mu_A^{P(n-1)} \gg (x) = \sup\{\min(\mu^P(a), \mu_A^{p(n-1)}(b)) \mid [a,b]=x\}\\ &= \sup\{\min(\mu_A^P(a), [\mu_A^P(b), \mu_A^{P(n-2)}(b)]) \mid [a,b]=x\}\\ &\geqslant \sup\{\min(\mu_A^P(a), \sup\{\ll \mu_A^P, \mu_A^{P(n-2)} \gg (b_i) \mid b = \sum \alpha_i b_i\}) \mid [a,b]=x\}\\ &\geqslant \sup\{\min(\mu^P(a), \sup\{[\mu_{A_{n-1}}^P](b_i) \mid b = \sum \alpha_i b_i\}) \mid [a,b]=x\}\\ &\geqslant \sup\{\min(\mu_A^P(a), [\mu_{A_{n-1}}^P](b_i)) \mid \sum \alpha_i [a,b_i]=x\}\\ &\geqslant \sup\{\min(\mu_A^P(a), \sup\{\mu_{A_{n-1}}^P(c_i) \mid b_i = \sum \beta_i c_i\}) \mid \sum \alpha_i [a,b_i]=x\}\\ &\geqslant \sup\{\min(\mu_A^P(a), \mu_{A_{n-1}}^P(c_i)) \mid \sum \gamma_i [a,c_i]=x\}\\ &\geqslant \sup\{\min(\mu_A^P(a), \sup\{\mu_A^P(d_i)) \mid [x_1,[x_2,[\ldots,[x_{n-1},d_i]\ldots]]]=c_i\} \mid \sum \gamma_i [a,c_i]=x\}\\ &\geqslant \sup\{\min(\mu_A^P(a), \mu_A^P(d_i)) \mid \sum \gamma_i [a,[x_1,[x_2,[\ldots,[x_{n-1},d_i]\ldots]]]]=x\}\\ &\geqslant \sup\{\mu_{A_n}^P(d_i) \mid \sum \gamma_i [a,[x_1,[x_2,[\ldots,[x_{n-1},d_i]\ldots]]]]=x\} \geqslant [\mu_{A_n}^P](x),\end{aligned}$$

$$\begin{aligned}&\ll \mu_A^N, \mu_A^{N(n-1)} \gg (x) = \inf\{\max(\mu^N(a), \mu_A^{N(n-1)}(b)) \mid [a,b]=x\}\\ &= \inf\{\max(\mu_A^N(a), [\mu_A^N(b), \mu_A^{N(n-2)}(b)]) \mid [a,b]=x\}\\ &\leqslant \inf\{\max(\mu_A^N(a), \inf\{\ll \mu^N, \mu_A^{N(n-2)} \gg (b_i) \mid b = \sum \alpha_i b_i\}) \mid [a,b]=x\}\\ &\leqslant \inf\{\max(\mu_A^N(a), \inf\{[\mu_{A_{n-1}}^N](b_i) \mid b = \sum \alpha_i b_i\}) \mid [a,b]=x\}\\ &\leqslant \inf\{\max(\mu_A^N(a), [\mu_{A_{n-1}}^N](b_i)) \mid \sum \alpha_i [a,b_i]=x\}\\ &\leqslant \inf\{\max(\mu_A^N(a), \inf\{\mu_{A_{n-1}}^N(c_i) \mid b_i = \sum \beta_i c_i\}) \mid \sum \alpha_i [a,b_i]=x\}\\ &\leqslant \inf\{\max(\mu_A^N(a), \mu_{A_{n-1}}^N(c_i)) \mid \sum \gamma_i [a,c_i]=x\}\\ &\leqslant \inf\{\max(\mu_A^N(a), \inf\{\mu_A^N(d_i)) \mid [x_1,[x_2,[\ldots,[x_{n-1},d_i]\ldots]]]=c_i\} \mid \sum \gamma_i [a,c_i]=x\}\\ &\leqslant \inf\{\max(\mu_A^N(a), \mu_A^N(d_i)) \mid \sum \gamma_i [a,[x_1,[x_2,[\ldots,[x_{n-1},d_i]\ldots]]]]=x\}\\ &\leqslant \inf\{\mu_{A_n}^N(d_i) \mid \sum \gamma_i [a,[x_1,[x_2,[\ldots,[x_{n-1},d_i]\ldots]]]]=x\} \leqslant [\mu_{A_n}^N](x).\end{aligned}$$

This complete the proof.

Theorem 6.7 *If for a bipolar fuzzy Lie ideal $A = (\mu_A^P, \mu_A^N)$, there exists a positive integer n such that*

$$(a\overline{d}x_1 \circ a\overline{d}x_2 \circ \cdots \circ a\overline{d}x_n)(\mu_A^P) = 0,$$

$$(\overline{ad}x_1 \circ \overline{ad}x_2 \circ \cdots \circ \overline{ad}x_n)(\mu_A^N) = 1,$$

for all $x_1, \ldots, x_n \in L$, then A is nilpotent.

Proof For $x_1, \ldots, x_n \in L$ and $x(\neq 0) \in L$, we have

$$(\overline{ad}x_1 \circ \cdots \circ \overline{ad}x_n)(\mu_A^P)(x) = \sup\{\mu_A^P(a) \mid [x_1, [x_2, [\ldots, [x_n, a]\ldots]]] = x\} = 0,$$

$$(\overline{ad}x_1 \circ \cdots \circ \overline{ad}x_n)(\mu_A^N)(x) = \inf\{\mu_A^N(a) \mid [x_1, [x_2, [\ldots, [x_n, a]\ldots]]] = x\} = 1.$$

Thus, $[A_n] = (0, 1)$. From Theorem 6.6, it follows that $A^n = (0, 1)$. Hence, $A = (\mu_A^P, \mu_A^N)$ is a nilpotent bipolar fuzzy Lie ideal.

The mapping $K : L \times L \to F$ defined by $K(x, y) = Tr(adx \circ ady)$, where Tr is the *trace* of a linear homomorphism, is a symmetric bilinear form which is called the *Killing form*. It is not difficult to see that this form satisfies the identity $K([x, y], z) = K(x, [y, z])$. The form K can be naturally extended to $\overline{K} : J^{L\times L} \to J^F$ defined by putting

$$\overline{K}(\mu_A^P)(\beta) = \sup\{\mu_A^P(x, y) \mid Tr(adx \circ ady) = \beta\},$$
$$\overline{K}(\mu_A^N)(\beta) = \inf\{\mu_A^N(x, y) \mid Tr(adx \circ ady) = \beta\}$$

The Cartesian product of two bipolar fuzzy sets $A = \left(\mu_A^P, \mu_A^N\right)$ and $B = \left(\mu_B^P, \mu_B^N\right)$ is defined as

$$(\mu_A^P \times \mu_B^P)(x, y) = \min(\mu_A^P(x), \mu_B^P(y)),$$
$$(\mu_A^N \times \mu_B^N)(x, y) = \max(\mu_A^N(x), \mu_B^N(y)).$$

Similarly, we define

$$\overline{K}(\mu_A^P \times \mu_B^P)(\beta) = \sup\{\min(\mu_A^P(x), \mu_B^P(y)) \mid Tr(adx \circ ady) = \beta\},$$
$$\overline{K}(\mu_A^N \times \mu_B^N)(\beta) = \inf\{\max(\mu_A^N(x), \mu_B^N(y)) \mid Tr(adx \circ ady) = \beta\}.$$

Theorem 6.8 *Let $A = (\mu_A^P, \mu_A^N)$ be a bipolar fuzzy Lie ideal of Lie algebra L. Then $\overline{K}(\mu_A^P \times 1_{(\alpha x)}) = \alpha \odot \overline{K}(\mu_A^P \times 1_x)$ and $\overline{K}(\mu_A^N \times 0_{(\alpha x)}) = \alpha \odot \overline{K}(\mu_A^N \times 0_x)$ for all $x \in L$, $\alpha \in \mathbb{F}$.*

Proof If $\alpha = 0$, then for $\beta = 0$ we have

$$\begin{aligned}\overline{K}(\mu_A^P \times 1_0)(0) &= \sup\{\min(\mu_A^P(x), 1_0(y)) \mid Tr(adx \circ ady) = 0\}\\ &\geqslant \min(\mu_A^P(0), 1_0(0)) = 0,\\ \overline{K}(\mu_A^N \times 0_0)(0) &= \inf\{\max(\mu_A^N(x), 0_0(y)) : Tr(adx \circ ady) = 0\}\\ &\leqslant \max(\mu_A^N(0), 0_0(0)) = 1.\end{aligned}$$

For $\beta \neq 0$ $Tr(adx \circ ady) = \beta$ means that $x \neq 0$ and $y \neq 0$. So,

$$\overline{K}(\mu_A^P \times 1_0)(\beta) = \sup\{\min(\mu_A^P(x), 1_0(y)) \mid Tr(adx \circ ady) = \beta\} = 0,$$
$$\overline{K}(\mu_A^N \times 0_0)(\beta) = \inf\{\max(\mu_A^N(x), 0_0(y)) \mid Tr(adx \circ ady) = \beta\} = 1.$$

If $\alpha \neq 0$, then for arbitrary β, we obtain

$$\begin{aligned}
\overline{K}(\mu_A^P \times 1_{\alpha x})(\beta) &= \sup\{\min(\mu_A^P(y), 1_{\alpha x}(z)) \mid Tr(ady \circ adz) = \beta\} \\
&= \sup\{\min(\mu_A^P(y), \alpha \odot 1_x(z)) \mid Tr(ady \circ adz) = \beta\} \\
&= \sup\{\min(\mu_A^P(y), 1_x(\alpha^{-1}z)) \mid \alpha Tr(ady \circ ad(\alpha^{-1}z)) = \beta\} \\
&= \sup\{\min(\mu_A^P(y), 1_x(\alpha^{-1}z)) \mid Tr(ady \circ ad(\alpha^{-1}z)) = \alpha^{-1}\beta\} \\
&= \overline{K}(\mu_A^P \times 1_x)(\alpha^{-1}\beta) = \alpha \odot \overline{K}(\mu_A^P \times 1_x)(\beta),
\end{aligned}$$

$$\begin{aligned}
\overline{K}(\mu_A^N \times 0_{\alpha x})(\beta) &= \inf\{\max(\mu_A^N(y), 0_{\alpha x}(z)) \mid Tr(ady \circ adz) = \beta\} \\
&= \inf\{\max(\mu_A^N(y), \alpha \odot 0_x(z)) \mid Tr(ady \circ adz) = \beta\} \\
&= \inf\{\max(\mu_A^N(y), 0_x(\alpha^{-1}z)) \mid \alpha Tr(ady \circ ad(\alpha^{-1}z)) = \beta\} \\
&= \inf\{\max(\mu_A^N(y), 0_x(\alpha^{-1}z)) \mid Tr(ady \circ ad(\alpha^{-1}z)) = \alpha^{-1}\beta\} \\
&= \overline{K}(\mu_A^N \times 1_x)(\alpha^{-1}\beta) = \alpha \odot \overline{K}(\mu_A^N \times 0_x)(\beta).
\end{aligned}$$

This completes the proof.

Theorem 6.9 *Let $A = (\mu_A^P, \mu_A^N)$ be a bipolar fuzzy Lie ideal of a Lie algebra L. Then, $\overline{K}(\mu_A^P \times 1_{(x+y)}) = \overline{K}(\mu_A^P \times 1_x) \oplus \overline{K}(\mu_A^P \times 1_y)$ and $\overline{K}(\mu_A^P \times 0_{(x+y)}) = \overline{K}(\mu_A^P \times 0_x) \oplus \overline{K}(\mu_A^P \times 0_y)$ for all $x, y \in L$.*

Proof Indeed,

$$\begin{aligned}
\overline{K}(\mu_A^P \times 1_{(x+y)})(\beta) &= \sup\{\min(\mu_A^P(z), 1_{x+y}(u)) \mid Tr(adz \circ adu) = \beta\} \\
&= \sup\{\mu_A^P(z) \mid Tr(adz \circ ad(x+y)) = \beta\} \\
&= \sup\{\mu_A^P(z) \mid Tr(adz \circ adx) + Tr(adz \circ ady) = \beta\}
\end{aligned}$$

$$\begin{aligned}
&= \sup\{\min(\mu_A^P(z), \min(1_x(v), 1_y(w))) \mid Tr(adz \circ adv) + Tr(adz \circ adw) = \beta\} \\
&= \sup\{\min(\sup\{\min(\mu_A^P(z), 1_x(v)) \mid Tr(adz \circ adv) = \beta_1\}, \\
&\qquad \sup\{\min(\mu_A^P(z), 1_y(w)) \mid Tr(adz \circ adw) = \beta_2\} \mid \beta_1 + \beta_2 = \beta)\} \\
&= \sup\{\min(\overline{K}(\mu_A^P \times 1_x)(\beta_1), \overline{K}(\mu_A^P \times 1_y)(\beta_2)) \mid \beta_1 + \beta_2 = \beta\} \\
&= \overline{K}(\mu_A^P \times 1_x) \oplus \overline{K}(\mu_A^P \times 1_y)(\beta),
\end{aligned}$$

$$\begin{aligned}
&\overline{K}(\mu_A^N \times 0_{(x+y)})(\beta) = \inf\{\max(\mu_A^N(z), 0_{x+y}(u)) \mid Tr(adz \circ adu) = \beta\} \\
&= \inf\{\mu_A^N(z) \mid Tr(adz \circ ad(x+y)) = \beta\} \\
&= \inf\{\mu_A^N(z) \mid Tr(adz \circ adx) + Tr(adz \circ ady) = \beta\} \\
&= \inf\{\max(\mu_A^N(z), \max(0_x(v), 0_y(w))) \mid Tr(adz \circ adv) + Tr(adz \circ adw) = \beta\} \\
&= \inf\{\max(\inf\{\max(\mu_A^N(z), 0_x(v)) \mid Tr(adz \circ adv) = \beta_1\}, \\
&\qquad \inf\{\max(\mu_A^N(z), 0_y(w) \mid Tr(adz \circ adw) = \beta_2\} \mid \beta_1 + \beta_2 = \beta)\} \\
&= \inf\{\max(\overline{K}(\mu_A^N \times 0_x)(\beta_1), \overline{K}(\mu_A^N \times 0_y)(\beta_2)) \mid \beta_1 + \beta_2 = \beta\} \\
&= \overline{K}(\mu_A^N \times 0_x) \oplus \overline{K}(\mu_A^N \times 0_y)(\beta).
\end{aligned}$$

This completes the proof.

As a consequence of the above two theorems, we obtain

Corollary 6.1 *For each bipolar fuzzy Lie ideal* $A = (\mu_A^P, \mu_A^N)$ *and all* $x, y \in L$, $\alpha, \beta \in \mathbb{F}$, *we have*

$$\overline{K}(\mu_A^P \times 1_{(\alpha x+\beta y)}) = \alpha \odot \overline{K}(\mu_A^P \times 1_x) \oplus \beta \odot \overline{K}(\mu_A^P \times 1_y),$$
$$\overline{K}(\mu_A^N \times 0_{(\alpha x+\beta y)}) = \alpha \odot \overline{K}(\mu_A^N \times 0_x) \oplus \beta \odot \overline{K}(\mu_A^N \times 0_y).$$

6.3 Bipolar Fuzzy Lie Sub-superalgebras

Definition 6.16 Let $V = V_{\bar{0}} \oplus V_{\bar{1}}$ be a $\mathbb{Z}_2$-graded vector space. Suppose that $A_{\bar{0}} = (\mu_{A_{\bar{0}}}^P, \mu_{A_{\bar{0}}}^N)$ and $A_{\bar{1}} = (\mu_{A_{\bar{1}}}^P, \mu_{A_{\bar{1}}}^N)$ are bipolar fuzzy vector subspaces of $V_{\bar{0}}, V_{\bar{1}}$, respectively. We define $A'_{\bar{0}} = (\mu_{A'_{\bar{0}}}^P, \mu_{A'_{\bar{0}}}^N)$ where

$$\mu_{A'_{\bar{0}}}^P(x) = \begin{cases} \mu_{A_{\bar{0}}}^P(x) & x \in V_{\bar{0}} \\ 0 & x \notin V_{\bar{0}} \end{cases}, \quad \mu_{A'_{\bar{0}}}^N(x) = \begin{cases} \mu_{A_{\bar{0}}}^N(x) & x \in V_{\bar{0}} \\ 0 & x \notin V_{\bar{0}} \end{cases}$$

and define $A'_{\bar{1}} = (\mu_{A'_{\bar{1}}}^P, \mu_{A'_{\bar{1}}}^N)$ where

$$\mu_{A'_{\bar{1}}}^P(x) = \begin{cases} \mu_{A_{\bar{1}}}^P(x) & x \in V_{\bar{1}} \\ 0 & x \notin V_{\bar{1}} \end{cases}, \quad \mu_{A'_{\bar{1}}}^N(x) = \begin{cases} \mu_{A_{\bar{1}}}^N(x) & x \in V_{\bar{1}} \\ 0 & x \notin V_{\bar{1}} \end{cases}$$

Then, $A'_{\bar{0}} = (\mu_{A'_{\bar{0}}}^P, \mu_{A'_{\bar{0}}}^N)$ and $A'_{\bar{1}} = (\mu_{A'_{\bar{1}}}^P, \mu_{A'_{\bar{1}}}^N)$ are the bipolar fuzzy vector subspaces of V. Moreover, we have $A'_{\bar{0}} \cap A'_{\bar{1}} = (\mu_{A'_{\bar{0}} \cap A'_{\bar{1}}}^P, \mu_{A'_{\bar{0}} \cap A'_{\bar{1}}}^N)$, where

$$\mu_{A'_{\bar{0}} \cap A'_{\bar{1}}}^P(x) = \mu_{A'_{\bar{0}}}^P(x) \wedge \mu_{A'_{\bar{1}}}^P(x) = \begin{cases} 1 & x = 0 \\ 0 & x \neq 0 \end{cases}, \quad \mu_{A'_{\bar{0}} \cap A'_{\bar{1}}}^N(x) = \mu_{A'_{\bar{0}}}^N(x) \vee \mu_{A'_{\bar{1}}}^N(x) = \begin{cases} -1 & x = 0 \\ 0 & x \neq 0 \end{cases}.$$

So $A'_{\bar{0}} + A'_{\bar{1}}$ is the direct sum and denoted by $A_{\bar{0}} \oplus A_{\bar{1}}$. If $A = (\mu_A^P, \mu_A^N)$ is a bipolar fuzzy vector subspace of V and $A = A_{\bar{0}} \oplus A_{\bar{1}}$, then $A = (\mu_A^P, \mu_A^N)$ is called a $\mathbb{Z}_2$-*graded bipolar fuzzy vector subspace* of V.

Definition 6.17 Let $A = (\mu_A^P, \mu_A^N)$ be a bipolar fuzzy set of $\mathscr{L}$. Then, $A = (\mu_A^P, \mu_A^N)$ is called a *bipolar fuzzy Lie sub-superalgebra* of $\mathscr{L}$, if it satisfies the following conditions:

(1) $A = (\mu_A^P, \mu_A^N)$ is a $\mathbb{Z}_2$-graded bipolar fuzzy vector subspace,

(2) $\mu_A^P([x, y]) \geq \mu_A^P(x) \wedge \mu_A^P(y)$ and $\mu_A^N([x, y]) \leq \mu_A^N(x) \vee \mu_A^N(y)$.

If the condition (2) is replaced by (3) $\mu_A^P([x, y]) \geq \mu_A^P(x) \vee \mu_A^P(y)$ and $\mu_A^N([x, y]) \leq \mu_A^N(x) \wedge \mu_A^N(y)$, then $A = (\mu_A^P, \mu_A^N)$ is called a *bipolar fuzzy Lie ideal* of $\mathscr{L}$.

Example 6.5 Let $\mathscr{L} = \mathscr{L}_{\bar{0}} \oplus \mathscr{L}_{\bar{1}}$, where $\mathscr{L}_{\bar{0}} = \langle e \rangle$, $\mathscr{L}_{\bar{1}} = \langle a_1, \ldots, a_n, b_1, \ldots, b_n \rangle$ and $[a_i, b_i] = e, i = 1, 2, \ldots n$, the remaining brackets being zero. Then, $\mathscr{L}$ is Lie superalgebra.

Define $A_{\bar{0}} = (\mu_{A_{\bar{0}}}^P, \mu_{A_{\bar{0}}}^N)$ where $\mu_{A_{\bar{0}}}^P : \mathscr{L}_{\bar{0}} \to [0, 1]$ by $\mu_{A_{\bar{0}}}^P(x) = \begin{cases} 0.7 & x \in \mathscr{L}_{\bar{0}} \setminus \{0\} \\ 1 & x = 0 \end{cases}$,

$\mu_{A_{\bar{0}}}^N : \mathscr{L}_{\bar{0}} \to [-1, 0]$ by $\mu_{A_{\bar{0}}}^N(x) = \begin{cases} -0.2 & x \in \mathscr{L}_{\bar{0}} \setminus \{0\} \\ -1 & x = 0 \end{cases}$.

Define $A_{\bar{1}} = (\mu_{A_{\bar{1}}}^P, \mu_{A_{\bar{1}}}^N)$ where

$\mu_{A_{\bar{1}}}^P : \mathscr{L}_{\bar{1}} \to [0, 1]$ by $\mu_{A_{\bar{1}}}^P(x) = \begin{cases} 0.5 & x \in \mathscr{L}_{\bar{1}} \setminus \{0\} \\ 1 & x = 0 \end{cases}$,

$\mu_{A_{\bar{1}}}^N : \mathscr{L}_{\bar{1}} \to [-1, 0]$ by $\mu_{A_{\bar{1}}}^N(x) = \begin{cases} -0.4 & x \in \mathscr{L}_{\bar{1}} \setminus \{0\} \\ -1 & x = 0 \end{cases}$.

Define $A = (\mu_A^P, \mu_A^N)$ by $A = A_{\bar{0}} \oplus A_{\bar{1}}$. Then $A = (\mu_A^P, \mu_A^N)$ is a bipolar fuzzy ideal of $\mathscr{L}$.

Definition 6.18 For any $t \in [0, 1]$ and fuzzy subset μ^P of $\mathscr{L}$, the set $U(\mu^P, t) = \{x \in \mathscr{L} | \mu^P(x) \geq t\}$ (resp. $L(\mu^N, t) = \{x \in \mathscr{L} | \mu^N(x) \leq t\}$) is called an upper (resp. lower) t-level cut of μ^P.

The proofs of the following theorems are omitted.

Theorem 6.10 *If $A = (\mu_A^P, \mu_A^N)$ is a bipolar fuzzy Lie sub-superalgebra (resp. bipolar fuzzy ideal) of $\mathscr{L}$, then the sets $U(\mu_A^P, t)$ and $L(\mu_A^N, t)$ are Lie sub-superalgebras (resp. ideals) of $\mathscr{L}$ for every $t \in Im\mu_A^P \cap Im\mu_A^N$.*

Theorem 6.11 *If $A = (\mu_A^P, \mu_A^N)$ is a bipolar fuzzy set of $\mathscr{L}$ such that all nonempty level sets $U(\mu_A^P, t)$ and $L(\mu_A^N, t)$ are Lie sub-superalgebras (resp. ideals) of $\mathscr{L}$, then $A = (\mu_A^P, \mu_A^N)$ is an bipolar fuzzy Lie sub-superalgebra (resp. bipolar fuzzy ideal) of $\mathscr{L}$.*

Theorem 6.12 *If $A = (\mu_A^P, \mu_A^N)$ and $B = (\mu_B^P, \mu_B^N)$ are bipolar fuzzy Lie sub-superalgebras (resp. bipolar fuzzy ideals) of $\mathscr{L}$, then so is $A + B = (\mu_{A+B}^P, \mu_{A+B}^N)$.*

Theorem 6.13 *If $A = (\mu_A^P, \mu_A^N)$ and $B = (\mu_B^P, \mu_B^N)$ are bipolar fuzzy Lie sub-superalgebras (resp. bipolar fuzzy ideals) of $\mathscr{L}$, then so is $A \cap B = (\mu_{A\cap B}^P, \mu_{A\cap B}^N)$.*

Proposition 6.3 *Let $\varphi : \mathscr{L} \to \mathscr{L}'$ be a Lie homomorphism. If $A = (\mu_A^P, \mu_A^N)$ is a bipolar fuzzy Lie sub-superalgebra (resp. bipolar fuzzy ideal) of $\mathscr{L}'$, then the bipolar fuzzy set $\varphi^{-1}(A)$ of $\mathscr{L}$ is also a bipolar fuzzy Lie sub-superalgebra (resp. bipolar fuzzy ideal).*

Proof Since φ preserves the grading, we have $\varphi(x) = \varphi(x_{\bar{0}} + x_{\bar{1}}) = \varphi(x_{\bar{0}}) + \varphi(x_{\bar{1}}) \in \mathscr{L}'_{\bar{0}} \oplus \mathscr{L}'_{\bar{1}}$, for $x = x_{\bar{0}} + x_{\bar{1}} \in \mathscr{L}$. We define $\varphi^{-1}(A)_{\bar{0}} = (\mu^P_{\varphi^{-1}(A)_{\bar{0}}}, \mu^N_{\varphi^{-1}(A)_{\bar{0}}})$ where $\mu^P_{\varphi^{-1}(A)_{\bar{0}}} = \varphi^{-1}(\mu^P_{A_{\bar{0}}}), \mu^N_{\varphi^{-1}(A)_{\bar{0}}} = \varphi^{-1}(\mu^N_{A_{\bar{0}}})$ and define $\varphi^{-1}(A)_{\bar{1}} = (\mu^P_{\varphi^{-1}(A)_{\bar{1}}}, \mu^N_{\varphi^{-1}(A)_{\bar{1}}})$ where $\mu^P_{\varphi^{-1}(A)_{\bar{1}}} = \varphi^{-1}(\mu^P_{A_{\bar{1}}}), \mu^N_{\varphi^{-1}(A)_{\bar{1}}} = \varphi^{-1}(\mu^N_{A_{\bar{1}}})$. By Lemma 6.1, we have that they are bipolar fuzzy subspaces of $\mathscr{L}_{\bar{0}}, \mathscr{L}_{\bar{1}}$, respectively.

Then, we define $\varphi^{-1}(A)'_{\bar{0}} = (\mu^P_{\varphi^{-1}(A)'_{\bar{0}}}, \mu^N_{\varphi^{-1}(A)'_{\bar{0}}})$, where $\mu^P_{\varphi^{-1}(A)'_{\bar{0}}} = \varphi^{-1}(\mu^P_{A'_{\bar{0}}})$, $\mu^N_{\varphi^{-1}(A)'_{\bar{0}}} = \varphi^{-1}(\mu^N_{A'_{\bar{0}}})$, and $\varphi^{-1}(A)'_{\bar{1}} = (\mu^P_{\varphi^{-1}(A)'_{\bar{1}}}, \mu^N_{\varphi^{-1}(A)'_{\bar{1}}})$, where $\mu^P_{\varphi^{-1}(A)'_{\bar{1}}} = \varphi^{-1}(\mu^P_{A'_{\bar{1}}})$, $\mu^N_{\varphi^{-1}(A)'_{\bar{1}}} = \varphi^{-1}(\mu^N_{A'_{\bar{1}}})$.

Clearly,

$$\mu^P_{\varphi^{-1}(A)'_{\bar{0}}}(x) = \begin{cases} \mu^P_{\varphi^{-1}(A)_{\bar{0}}}(x) & x \in \mathscr{L}_{\bar{0}} \\ 0 & x \notin \mathscr{L}_{\bar{0}} \end{cases}, \mu^N_{\varphi^{-1}(A)'_{\bar{0}}}(x) = \begin{cases} \mu^N_{\varphi^{-1}(A)'_{\bar{0}}}(x) & x \in \mathscr{L}_{\bar{0}} \\ 0 & x \notin \mathscr{L}_{\bar{0}} \end{cases},$$

and

$$\mu^P_{\varphi^{-1}(A)'_{\bar{1}}}(x) = \begin{cases} \mu^P_{\varphi^{-1}(A)'_{\bar{1}}}(x) & x \in \mathscr{L}_{\bar{1}} \\ 0 & x \notin \mathscr{L}_{\bar{1}} \end{cases}, \mu^N_{\varphi^{-1}(A)'_{\bar{1}}}(x) = \begin{cases} \mu^N_{\varphi^{-1}(A)'_{\bar{1}}}(x) & x \in \mathscr{L}_{\bar{1}} \\ 0 & x \notin \mathscr{L}_{\bar{1}} \end{cases}.$$

These show that $\varphi^{-1}(A)'_{\bar{0}}$ and $\varphi^{-1}(A)'_{\bar{1}}$ are the extensions of $\varphi^{-1}(A)_{\bar{0}}$ and $\varphi^{-1}(A)_{\bar{1}}$.

For $0 \neq x \in \mathscr{L}$, we have

$$\begin{aligned} \mu^P_{\varphi^{-1}(A)'_{\bar{0}}}(x) \wedge \mu^P_{\varphi^{-1}(A)'_{\bar{1}}}(x) &= \varphi^{-1}(\mu^P_{A'_{\bar{0}}})(x) \wedge \varphi^{-1}(\mu^P_{A'_{\bar{1}}})(x) \\ &= \mu^P_{A'_{\bar{0}}}(\varphi(x)) \wedge \mu^P_{A'_{\bar{1}}}(\varphi(x)) = 0 \end{aligned}$$

and

$$\begin{aligned} \mu^N_{\varphi^{-1}(A)'_{\bar{0}}}(x) \vee \mu^N_{\varphi^{-1}(A)'_{\bar{1}}}(x) &= \varphi^{-1}(\mu^N_{A'_{\bar{0}}})(x) \vee \varphi^{-1}(\mu^N_{A'_{\bar{1}}})(x) \\ &= \mu^N_{A'_{\bar{0}}}(\varphi(x)) \vee \mu^N_{A'_{\bar{1}}}(\varphi(x)) = 0. \end{aligned}$$

Let $x \in \mathscr{L}$. We have

$$\begin{aligned} \mu^P_{\varphi^{-1}(A)'_{\bar{0}}+\varphi^{-1}(A)'_{\bar{1}}}(x) &= \sup_{x=a+b} \{\mu^P_{\varphi^{-1}(A)'_{\bar{0}}}(a) \wedge \mu^P_{\varphi^{-1}(A)'_{\bar{1}}}(b)\} \\ &= \sup_{x=a+b} \{\varphi^{-1}(\mu^P_{A'_{\bar{0}}})(a) \wedge \varphi^{-1}(\mu^P_{A'_{\bar{1}}})(b)\} \\ &= \sup_{x=a+b} \{\mu^P_{A'_{\bar{0}}}(\varphi(a)) \wedge \mu^P_{A'_{\bar{1}}}(\varphi(b))\} \\ &= \sup_{\varphi(x)=\varphi(a)+\varphi(b)} \{\mu^P_{A'_{\bar{0}}}(\varphi(a)) \wedge \mu^P_{A'_{\bar{1}}}(\varphi(b))\} \\ &= \mu^P_{A'_{\bar{0}}+A'_{\bar{1}}}(\varphi(x)) = \mu^P_A(\varphi(x)) = \mu^P_{\varphi^{-1}(A)}(x) \end{aligned}$$

and

$$\begin{aligned}
\mu^N_{\varphi^{-1}(A)'_{\bar{0}}+\varphi^{-1}(A)'_{\bar{1}}}(x) &= \inf_{x=a+b}\{\mu^N_{\varphi^{-1}(A)'_{\bar{0}}}(a) \vee \mu^N_{\varphi^{-1}(A)'_{\bar{1}}}(b)\} \\
&= \inf_{x=a+b}\{\varphi^{-1}(\mu^N_{A'_{\bar{0}}})(a) \vee \varphi^{-1}(\mu^N_{A'_{\bar{1}}})(b)\} \\
&= \inf_{x=a+b}\{\mu^N_{A'_{\bar{0}}}(\varphi(a)) \vee \mu^N_{A'_{\bar{1}}}(\varphi(b))\} \\
&= \inf_{\varphi(x)=\varphi(a)+\varphi(b)}\{\mu^N_{A'_{\bar{0}}}(\varphi(a)) \vee \mu^N_{A'_{\bar{1}}}(\varphi(b))\} \\
&= \mu^N_{A'_{\bar{0}}+A'_{\bar{1}}}(\varphi(x)) = \mu^N_A(\varphi(x)) = \mu^N_{\varphi^{-1}(A)}(x).
\end{aligned}$$

So $\varphi^{-1}(A) = \varphi^{-1}(A)_{\bar{0}} \oplus \varphi^{-1}(A)_{\bar{1}}$ is a $\mathbb{Z}_2$-graded bipolar fuzzy vector subspace of $\mathscr{L}$.

Let $x, y \in \mathscr{L}$. Then,

(1) $\mu^P_{\varphi^{-1}(A)}([x, y]) = \mu^P_A(\varphi([x, y])) = \mu^P_A([\varphi(x), \varphi(y)]) \geq \mu^P_A(\varphi(x)) \wedge \mu^P_A(\varphi(y)) = \mu^P_{\varphi^{-1}(A)}(x) \wedge \mu^P_{\varphi^{-1}(A)}(y)$, and $\mu^N_{\varphi^{-1}(A)}([x, y]) = \mu^N_A(\varphi([x, y])) = \mu^N_A([\varphi(x),\varphi(y)]) \leq \mu^N_A(\varphi(x)) \vee \mu^N_A(\varphi(y)) = \mu^N_{\varphi^{-1}(A)}(x) \vee \mu^N_{\varphi^{-1}(A)}(y)$, thus $\varphi^{-1}(A)$ is a bipolar fuzzy Lie sub-superalgebra.

(2) $\mu^P_{\varphi^{-1}(A)}([x, y]) = \mu^P_A(\varphi([x, y])) = \mu^P_A([\varphi(x), \varphi(y)]) \geq \mu^P_A(\varphi(x)) \vee \mu^P_A(\varphi(y)) = \mu^P_{\varphi^{-1}(A)}(x) \vee \mu^P_{\varphi^{-1}(A)}(y)$, and $\mu^N_{\varphi^{-1}(A)}([x, y]) = \mu^N_A(\varphi([x, y])) = \mu^N_A([\varphi(x),\varphi(y)]) \leq \mu^N_A(\varphi(x)) \wedge \mu^N_A(\varphi(y)) = \mu^N_{\varphi^{-1}(A)}(x) \wedge \mu^N_{\varphi^{-1}(A)}(y)$, thus $\varphi^{-1}(A)$ is a bipolar fuzzy ideal. □

Proposition 6.4 *Let $\varphi : \mathscr{L} \to \mathscr{L}'$ be a Lie homomorphism. If $A = (\mu^P_A, \mu^N_A)$ is a bipolar fuzzy Lie sub-superalgebra of $\mathscr{L}$, then the bipolar fuzzy set $\varphi(A)$ is also a bipolar fuzzy Lie sub-superalgebra of $\mathscr{L}'$.*

Proof. Since $A = (\mu^P_A, \mu^N_A)$ is a bipolar fuzzy Lie sub-superalgebra of $\mathscr{L}$, we have $A = A_{\bar{0}} \oplus A_{\bar{1}}$ where $A_{\bar{0}} = (\mu^P_{A_{\bar{0}}}, \mu^N_{A_{\bar{0}}})$, $A_{\bar{1}} = (\mu^P_{A_{\bar{1}}}, \mu^N_{A_{\bar{1}}})$ are bipolar fuzzy vector subspaces of $\mathscr{L}_{\bar{0}}, \mathscr{L}_{\bar{1}}$, respectively. We define $\varphi(A)_{\bar{0}} = (\mu^P_{\varphi(A)_{\bar{0}}}, \mu^N_{\varphi(A)_{\bar{0}}})$ where $\mu^P_{\varphi(A)_{\bar{0}}} = \varphi(\mu^P_{A_{\bar{0}}}), \mu^N_{\varphi(A)_{\bar{0}}} = \varphi(\mu^N_{A_{\bar{0}}}), \varphi(A)_{\bar{1}} = (\mu^P_{\varphi(A)_{\bar{1}}}, \mu^N_{\varphi(A)_{\bar{1}}})$ where $\mu^P_{\varphi(A)_{\bar{1}}} = \varphi(\mu^P_{A_{\bar{1}}})$, $\mu^N_{\varphi(A)_{\bar{1}}} = \varphi(\mu^N_{A_{\bar{1}}})$. By Lemma 6.5, $\varphi(A)_{\bar{0}}$ and $\varphi(A)_{\bar{1}}$ are bipolar fuzzy subspaces of $\mathscr{L}_{\bar{0}}, \mathscr{L}_{\bar{1}}$, respectively. And extend them to $\varphi(A)'_{\bar{0}}, \varphi(A)'_{\bar{1}}$, we define $\varphi(A)'_{\bar{0}} = (\mu^P_{\varphi(A)'_{\bar{0}}}, \mu^N_{\varphi(A)'_{\bar{0}}})$ where $\mu^P_{\varphi(A)'_{\bar{0}}} = \varphi(\mu^P_{A'_{\bar{0}}}), \mu^N_{\varphi(A)'_{\bar{0}}} = \varphi(\mu^N_{A'_{\bar{0}}})$ and $\varphi(A)'_{\bar{1}} = (\mu^P_{\varphi(A)'_{\bar{1}}}, \mu^N_{\varphi(A)'_{\bar{1}}})$ where $\mu^P_{\varphi(A)'_{\bar{1}}} = \varphi\left(\mu^P_{A'_{\bar{1}}}\right), \mu^N_{\varphi(A)'_{\bar{1}}} = \varphi\left(\mu^N_{A'_{\bar{1}}}\right)$. Clearly,

$$\mu^P_{\varphi(A)'_{\bar{0}}}(x) = \begin{cases} \mu^P_{\varphi(A)_{\bar{0}}}(x) & x \in \mathscr{L}_{\bar{0}} \\ 0 & x \notin \mathscr{L}_{\bar{0}} \end{cases}, \mu^N_{\varphi(A)'_{\bar{0}}}(x) = \begin{cases} \mu^N_{\varphi(A)_{\bar{0}}}(x) & x \in \mathscr{L}_{\bar{0}} \\ 0 & x \notin \mathscr{L}_{\bar{0}} \end{cases},$$
$$\mu^P_{\varphi(A)'_{\bar{1}}}(x) = \begin{cases} \mu^P_{\varphi(A)_{\bar{1}}}(x) & x \in \mathscr{L}_{\bar{1}} \\ 0 & x \notin \mathscr{L}_{\bar{1}} \end{cases}, \mu^N_{\varphi(A)'_{\bar{1}}}(x) = \begin{cases} \mu^N_{\varphi(A)_{\bar{1}}}(x) & x \in \mathscr{L}_{\bar{1}} \\ 0 & x \notin \mathscr{L}_{\bar{1}} \end{cases}.$$

For $0 \neq x \in \mathscr{L}'$, then

$$\begin{aligned}
\mu^P_{\varphi(A)'_{\bar{0}}}(x) \wedge \mu^P_{\varphi(A)'_{\bar{1}}}(x) &= \varphi(\mu^P_{A'_{\bar{0}}})(x) \wedge \varphi(\mu^P_{A'_{\bar{1}}})(x) \\
&= \sup_{x=\varphi(a)} \{\mu^P_{A'_{\bar{0}}}(a)\} \wedge \sup_{x=\varphi(a)} \{\mu^P_{A'_{\bar{1}}}(a)\} \\
&= \sup_{x=\varphi(a)} \{\mu^P_{A'_{\bar{0}}}(a) \wedge \mu^P_{A'_{\bar{1}}}(a)\} = 0, \\
\mu^N_{\varphi(A)'_{\bar{0}}}(x) \vee \mu^N_{\varphi(A)'_{\bar{1}}}(x) &= \varphi(\mu^N_{A'_{\bar{0}}})(x) \vee \varphi(\mu^N_{A'_{\bar{1}}})(x) \\
&= \inf_{x=\varphi(a)} \{\mu^N_{A'_{\bar{0}}}(a)\} \vee \inf_{x=\varphi(a)} \{\mu^N_{A'_{\bar{1}}}(a)\} \\
&= \inf_{x=\varphi(a)} \{\mu^N_{A'_{\bar{0}}}(a) \vee \mu^N_{A'_{\bar{1}}}(a)\} = 0.
\end{aligned}$$

Let $y \in \mathscr{L}'$. We have

$$\begin{aligned}
\mu^P_{\varphi(A)'_{\bar{0}}+\varphi(A)'_{\bar{1}}}(y) &= \sup_{y=a+b} \{\mu^P_{\varphi(A)'_{\bar{0}}}(a) \wedge \mu^P_{\varphi(A)'_{\bar{1}}}(b)\} \\
&= \sup_{y=a+b} \{\varphi(\mu^P_{A'_{\bar{0}}})(a) \wedge \varphi(\mu^P_{A'_{\bar{1}}})(b)\} \\
&= \sup_{y=a+b} \{ \sup_{a=\varphi(m)} \{\mu^P_{A'_{\bar{0}}}(m)\} \wedge \sup_{b=\varphi(n)} \{\mu^P_{A'_{\bar{1}}}(n)\}\} \\
&= \sup_{y=\varphi(x)} \{ \sup_{x=m+n} \{\mu^P_{A'_{\bar{0}}}(m) \wedge \mu^P_{A'_{\bar{1}}}(n)\}\} \\
&= \sup_{y=\varphi(x)} \{(\mu^P_{A'_{\bar{0}}+A'_{\bar{1}}})(x)\} = \sup_{y=\varphi(x)} \{\mu^P_A(x)\} = \mu^P_{\varphi(A)}(y), \\
\mu^N_{\varphi(A)'_{\bar{0}}+\varphi(A)'_{\bar{1}}}(y) &= \inf_{y=a+b} \{\mu^N_{\varphi(A)'_{\bar{0}}}(a) \vee \mu^N_{\varphi(A)'_{\bar{1}}}(b)\} \\
&= \inf_{y=a+b} \{\varphi(\mu^N_{A'_{\bar{0}}})(a) \vee \varphi(\mu^N_{A'_{\bar{1}}})(b)\} \\
&= \inf_{y=a+b} \{ \inf_{a=\varphi(m)} \{\mu^N_{A'_{\bar{0}}}(m)\} \vee \inf_{b=\varphi(n)} \{\mu^N_{A'_{\bar{1}}}(n)\}\} \\
&= \inf_{y=\varphi(x)} \{ \inf_{x=m+n} \{\mu^N_{A'_{\bar{0}}}(m) \vee \mu^N_{A'_{\bar{1}}}(n)\}\} \\
&= \inf_{y=\varphi(x)} \{(\mu^N_{A'_{\bar{0}}+A'_{\bar{1}}})(x)\} = \inf_{y=\varphi(x)} \{\mu^N_A(x)\} = \mu^N_{\varphi(A)}(y).
\end{aligned}$$

So $\varphi(A) = \varphi(A)_{\bar{0}} \oplus \varphi(A)_{\bar{1}}$ is a $\mathbb{Z}_2$-graded bipolar fuzzy vector subspace.

Let $x, y \in \mathscr{L}'$. It is enough to show $\mu^P_{\varphi(A)}([x, y]) \geq \mu^P_{\varphi(A)}(x) \wedge \mu^P_{\varphi(A)}(y)$ and $\mu^N_{\varphi(A)}([x, y]) \leq \mu^N_{\varphi(A)}(x) \vee \mu^N_{\varphi(A)}(y)$. Suppose that $\mu^P_{\varphi(A)}([x, y]) < \mu^P_{\varphi(A)}(x) \wedge \mu^P_{\varphi(A)}(y)$, we have $\mu^P_{\varphi(A)}([x, y]) < \mu^P_{\varphi(A)}(x)$ and $\mu^P_{\varphi(A)}([x, y]) < \mu^P_{\varphi(A)}(y)$. We choose a number $t \in [0, 1]$ such that $\mu^P_{\varphi(A)}([x, y]) < t < \mu^P_{\varphi(A)}(x)$ and $\mu^P_{\varphi(A)}([x, y]) < t < \mu^P_{\varphi(A)}(y)$. Then, there exist $a \in \varphi^{-1}(x)$, $b \in \varphi^{-1}(y)$ such that $\mu^P_A(a) > t$, $\mu^P_A(b) > t$. Since $\varphi([a, b]) = [\varphi(a), \varphi(b)] = [x, y]$, we have $\mu^P_{\varphi(A)}([x, y]) = \sup_{[x,y]=\varphi([a,b])} \{\mu^P_A([a, b])\} \geq \mu^P_A([a, b]) \geq \mu^P_A(a) \wedge \mu^P_A(b) > t > \mu^P_{\varphi(A)}([x, y])$. This is a contradiction.

Suppose that $\mu^N_{\varphi(A)}([x,y]) > \mu^N_{\varphi(A)}(x) \vee \mu^N_{\varphi(A)}(y)$, we have $\mu^N_{\varphi(A)}([x,y]) > \mu^N_{\varphi(A)}(x)$ and $\mu^N_{\varphi(A)}([x,y]) > \mu^N_{\varphi(A)}(y)$. We choose a number $t \in [-1,0]$ such that $\varphi(\mu^N)([x,y]) > t > \mu^N_{\varphi(A)}(x)$ and $\mu^N_{\varphi(A)}([x,y]) > t > \mu^N_{\varphi(A)}(y)$. Then, there exist $a \in \varphi^{-1}(x), b \in \varphi^{-1}(y)$ such that $\mu^N_A(a) < t, \mu^N_A(b) < t$. Since $\varphi([a,b]) = [\varphi(a), \varphi(b)] = [x,y]$, we have $\mu^N_{\varphi(A)}([x,y]) = \inf\limits_{[x,y]=\varphi([a,b])} \{\mu^N_A([a,b])\} \le \mu^N_A([a,b]) \le \mu^N_A(a) \vee \mu^N_A(b) < t < \mu^N_{\varphi(A)}([x,y])$. This is a contradiction.

Therefore, $\varphi(A)$ is a bipolar fuzzy Lie sub-superalgebra of $\mathscr{L}'$. □

We state the following results without proofs.

Proposition 6.5 *Let $\varphi : \mathscr{L} \to \mathscr{L}'$ be a surjective Lie homomorphism. If $A = (\mu^P_A, \mu^N_A)$ is a bipolar fuzzy ideal of $\mathscr{L}$, then the bipolar fuzzy set $\varphi(A)$ is also a bipolar fuzzy ideal of $\mathscr{L}'$.*

Theorem 6.14 *Let $\varphi : \mathscr{L} \to \mathscr{L}'$ be a surjective Lie homomorphism. If $A = (\mu^P_A, \mu^N_A)$ and $B = (\mu^P_B, \mu^N_B)$ are bipolar fuzzy ideals of $\mathscr{L}$, then $\varphi(A+B) = \varphi(A) + \varphi(B)$.*

6.4 Bipolar Fuzzy Bracket Product

Definition 6.19 For any bipolar fuzzy sets $A = (\mu^P_A, \mu^N_A)$ and $B = (\mu^P_B, \mu^N_B)$ of $\mathscr{L}$, we define the *bipolar fuzzy bracket product* $[A,B] = (\mu^P_{[A,B]}, \mu^N_{[A,B]})$ where

$$\mu^P_{[A,B]}(x) = \begin{cases} \sup\limits_{x=\sum\limits_{i\in\mathscr{L}} \alpha_i[x_i,y_i]} \{\min\limits_{i\in\mathscr{L}}\{\mu^P_A(x_i) \wedge \mu^P_B(y_i)\}\} & \text{where } \alpha_i \in F, x_i, y_i \in \mathscr{L} \\ 0 & \text{if } x \text{ is not expressed as } x = \sum\limits_{i\in\mathscr{L}} \alpha_i[x_i,y_i] \end{cases}$$

and

$$\mu^N_{[A,B]}(x) = \begin{cases} \inf\limits_{x=\sum\limits_{i\in\mathscr{L}} \alpha_i[x_i,y_i]} \{\max\limits_{i\in\mathscr{L}}\{\mu^N_A(x_i) \vee \mu^N_B(y_i)\}\} & \text{where } \alpha_i \in \mathbb{F}, x_i, y_i \in \mathscr{L} \\ 0 & \text{if } x \text{ is not expressed as } x = \sum\limits_{i\in\mathscr{L}} \alpha_i[x_i,y_i] \end{cases}.$$

Lemma 6.6 *Let $A_1 = (\mu^P_{A_1}, \mu^N_{A_1})$, $A_2 = (\mu^P_{A_2}, \mu^N_{A_2})$, $B_1 = (\mu^P_{B_1}, \mu^N_{B_1})$ and $B_2 = (\mu^P_{B_2}, \mu^N_{B_2})$ be bipolar fuzzy sets of $\mathscr{L}$ such that $A_1 \subseteq A_2$, $B_1 \subseteq B_2$. Then, $[A_1, B_1] \subseteq [A_2, B_2]$. In particular, if $A = (\mu^P_A, \mu^N_A)$ and $B = (\mu^P_B, \mu^N_B)$ are bipolar fuzzy sets of $\mathscr{L}$, then $[A_1, B] \subseteq [A_2, B]$ and $[A, B_1] \subseteq [A, B_2]$.*

Lemma 6.7 *Let $A_1 = (\mu^P_{A_1}, \mu^N_{A_1})$, $A_2 = (\mu^P_{A_2}, \mu^N_{A_2})$, $B_1 = (\mu^P_{B_1}, \mu^N_{B_1})$, $B_2 = (\mu^P_{B_2}, \mu^N_{B_2})$ and $A = (\mu^P_A, \mu^N_A)$, $B = (\mu^P_B, \mu^N_B)$ be any bipolar fuzzy vector subspaces of $\mathscr{L}$. Then, $[A_1 + A_2, B] = [A_1, B] + [A_2, B]$ and $[A, B_1 + B_2] = [A, B_1] + [A, B_2]$.*

Lemma 6.8 *Let* $A = (\mu_A^P, \mu_A^N)$ *and* $B = (\mu_B^P, \mu_B^N)$ *be bipolar fuzzy vector subspaces of* $\mathscr{L}$*. Then, for any* $\alpha, \beta \in \mathbb{F}$*, we have* $[\alpha A, B] = \alpha[A, B]$ *and* $[A, \beta B] = \beta[A, B]$.

Theorem 6.15 *Let* $A_1 = (\mu_{A_1}^P, \mu_{A_1}^N)$, $A_2 = (\mu_{A_2}^P, \mu_{A_2}^N)$, $B_1 = (\mu_{B_1}^P, \mu_{B_1}^N)$, $B_2 = (\mu_{B_2}^P, \mu_{B_2}^N)$ *and* $A = (\mu_A^P, \mu_A^N)$, $B = (\mu_B^P, \mu_B^N)$ *be bipolar fuzzy vector subspaces of* $\mathscr{L}$*. Then, for any* $\alpha, \beta \in \mathbb{F}$*, we have*

$$[\alpha A_1 + \beta A_2, B] = \alpha[A_1, B] + \beta[A_2, B]$$

$$[A, \alpha B_1 + \beta B_2] = \alpha[A, B_1] + \beta[A, B_2]$$

Proof The results follow from Lemmas 6.7 and 6.8. □

Lemma 6.9 *Let* $A = (\mu_A^P, \mu_A^N)$ *and* $B = (\mu_B^P, \mu_B^N)$ *be any two bipolar fuzzy vector subspaces of* $\mathscr{L}$*. Then,* $[A, B]$ *is a bipolar fuzzy vector subspace of* $\mathscr{L}$*.*

Let $A = (\mu_A^P, \mu_A^N)$ and $B = (\mu_B^P, \mu_B^N)$ be $\mathbb{Z}_2$-graded bipolar fuzzy vector subspaces of $\mathscr{L}$. Then, $A = A_{\bar{0}} \oplus A_{\bar{1}}$, $B = B_{\bar{0}} \oplus B_{\bar{1}}$, where $A_{\bar{0}}$, $B_{\bar{0}}$ are bipolar fuzzy vector subspaces of $\mathscr{L}_{\bar{0}}$ and $A_{\bar{1}}$, $B_{\bar{1}}$ are bipolar fuzzy vector subspaces of $\mathscr{L}_{\bar{1}}$. We define:

$[A_{\bar{0}}, B_{\bar{0}}] = (\mu_{[A_{\bar{0}}, B_{\bar{0}}]}^P, \mu_{[A_{\bar{0}}, B_{\bar{0}}]}^N)$, where

$$\mu_{[A_{\bar{0}}, B_{\bar{0}}]}^P(x) = \sup_{x = \sum_{i \in \mathscr{L}} \alpha_i [x_i, y_i]} \{\min_{i \in \mathscr{L}} \{\mu_{A_{\bar{0}}}^P(x_i) \wedge \mu_{B_{\bar{0}}}^P(y_i)\}\}$$

and

$$\mu_{[A_{\bar{0}}, B_{\bar{0}}]}^N(x) = \inf_{x = \sum_{i \in \mathscr{L}} \alpha_i [x_i, y_i]} \{\max_{i \in \mathscr{L}} \{\mu_{A_{\bar{0}}}^N(x_i) \vee \mu_{B_{\bar{0}}}^N(y_i)\}\},$$

for $x_i \in \mathscr{L}_{\bar{0}}$ and $y_i \in \mathscr{L}_{\bar{0}}$,

$[A_{\bar{0}}, B_{\bar{1}}] = (\mu_{[A_{\bar{0}}, B_{\bar{1}}]}^P, \mu_{[A_{\bar{0}}, B_{\bar{1}}]}^N)$, where

$$\mu_{[A_{\bar{0}}, B_{\bar{1}}]}^P(x) = \sup_{x = \sum_{i \in \mathscr{L}} \alpha_i [x_i, y_i]} \{\min_{i \in \mathscr{L}} \{\mu_{A_{\bar{0}}}^P(x_i) \wedge \mu_{B_{\bar{1}}}^P(y_i)\}\}$$

and

$$\mu_{[A_{\bar{0}}, B_{\bar{1}}]}^N(x) = \inf_{x = \sum_{i \in \mathscr{L}} \alpha_i [x_i, y_i]} \{\max_{i \in \mathscr{L}} \{\mu_{A_{\bar{0}}}^N(x_i) \vee \mu_{B_{\bar{1}}}^N(y_i)\}\},$$

for $x_i \in \mathscr{L}_{\bar{0}}$ and $y_i \in \mathscr{L}_{\bar{1}}$,

$[A_{\bar{1}}, B_{\bar{0}}] = (\mu_{[A_{\bar{1}}, B_{\bar{0}}]}^P, \mu_{[A_{\bar{1}}, B_{\bar{0}}]}^N)$, where

$$\mu^P_{[A_{\bar{1}},B_{\bar{0}}]}(x) = \sup_{x=\sum\limits_{i\in\mathscr{L}}\alpha_i[x_i,y_i]} \{\min_{i\in\mathscr{L}}\{\mu^P_{A_{\bar{1}}}(x_i) \wedge \mu^P_{B_{\bar{0}}}(y_i)\}\}$$

and

$$\mu^N_{[A_{\bar{1}},B_{\bar{0}}]}(x) = \inf_{x=\sum\limits_{i\in\mathscr{L}}\alpha_i[x_i,y_i]} \{\max_{i\in\mathscr{L}}\{\mu^N_{A_{\bar{1}}}(x_i) \vee \mu^N_{B_{\bar{0}}}(y_i)\}\},$$

for $x_i \in \mathscr{L}_{\bar{1}}$ and $y_i \in \mathscr{L}_{\bar{0}}$,

$[A_{\bar{1}}, B_{\bar{1}}] = (\mu^P_{[A_{\bar{1}},B_{\bar{1}}]}, \mu^N_{[A_{\bar{1}},B_{\bar{1}}]})$, where

$$\mu^P_{[A_{\bar{1}},B_{\bar{1}}]}(x) = \sup_{x=\sum\limits_{i\in\mathscr{L}}\alpha_i[x_i,y_i]} \{\min_{i\in\mathscr{L}}\{\mu^P_{A_{\bar{1}}}(x_i) \wedge \mu^P_{B_{\bar{1}}}(y_i)\}\}$$

and

$$\mu^N_{[A_{\bar{1}},B_{\bar{1}}]}(x) = \inf_{x=\sum\limits_{i\in\mathscr{L}}\alpha_i[x_i,y_i]} \{\max_{i\in\mathscr{L}}\{\mu^N_{A_{\bar{1}}}(x_i) \vee \mu^N_{B_{\bar{1}}}(y_i)\}\},$$

for $x_i \in \mathscr{L}_{\bar{1}}$ and $y_i \in \mathscr{L}_{\bar{1}}$.

Note that $[A_{\bar{0}}, B_{\bar{0}}], [A_{\bar{1}}, B_{\bar{1}}]$ are bipolar fuzzy sets of $\mathscr{L}_{\bar{0}}$ and $[A_{\bar{0}}, B_{\bar{1}}], [A_{\bar{1}}, B_{\bar{0}}]$ are bipolar fuzzy sets of $\mathscr{L}_{\bar{1}}$.

Lemma 6.10 *Let $A = (\mu^P_A, \mu^N_A)$ and $B = (\mu^P_B, \mu^N_B)$ be any two $\mathbb{Z}_2$-graded bipolar fuzzy vector subspaces of $\mathscr{L}$. Then*

$[A, B]_{\bar{0}} := [A_{\bar{0}}, B_{\bar{0}}] + [A_{\bar{1}}, B_{\bar{1}}]$ is a bipolar fuzzy vector subspace of $\mathscr{L}_{\bar{0}}$,

$[A, B]_{\bar{1}} := [A_{\bar{0}}, B_{\bar{1}}] + [A_{\bar{1}}, B_{\bar{0}}]$ is a bipolar fuzzy vector subspace of $\mathscr{L}_{\bar{1}}$ and

$[A, B]$ is a $\mathbb{Z}_2$-graded bipolar fuzzy vector subspace of $\mathscr{L}$.

Proof Since $[A_{\bar{0}}, B_{\bar{0}}]$ and $[A_{\bar{1}}, B_{\bar{1}}]$ are bipolar fuzzy vector subspaces of $\mathscr{L}_{\bar{0}}$ by Lemma 6.9, we can get that $[A, B]_{\bar{0}} := [A_{\bar{0}}, B_{\bar{0}}] + [A_{\bar{1}}, B_{\bar{1}}]$ is a bipolar fuzzy vector subspace of $\mathscr{L}_{\bar{0}}$ by Lemma 6.10. Similarly, $[A, B]_{\bar{1}} := [A_{\bar{0}}, B_{\bar{1}}] + [A_{\bar{1}}, B_{\bar{0}}]$ is a bipolar fuzzy vector subspace of $\mathscr{L}_{\bar{1}}$. We define $[A, B]'_{\bar{0}} := [A'_{\bar{0}}, B'_{\bar{0}}] + [A'_{\bar{1}}, B'_{\bar{1}}]$ and $[A, B]'_{\bar{1}} := [A'_{\bar{0}}, B'_{\bar{1}}] + [A'_{\bar{1}}, B'_{\bar{0}}]$.

Let $x \in \mathscr{L}_{\bar{0}}$. We have

$$\begin{aligned}
\mu^P_{[A,B]'_{\bar{0}}}(x) &= (\mu^P_{[A'_{\bar{0}},B'_{\bar{0}}]+[A'_{\bar{1}},B'_{\bar{1}}]})(x) \\
&= \sup_{x=a+b} \{\mu^P_{[A'_{\bar{0}},B'_{\bar{0}}]}(a) \wedge \mu^P_{[A'_{\bar{1}},B'_{\bar{1}}]}(b)\} \\
&= \sup_{x=a+b} \{ \sup_{a=\sum\limits_{i\in\mathscr{L}}\alpha_i[k_i,l_i]} \{\min_{i\in\mathscr{L}}\{\mu^P_{A'_{\bar{0}}}(k_i) \wedge \mu^P_{B'_{\bar{0}}}(l_i)\}\} \wedge \\
&\qquad \sup_{b=\sum\limits_{i\in\mathscr{L}}\beta_i[m_i,n_i]} \{\min_{i\in\mathscr{L}}\{\mu^P_{A'_{\bar{1}}}(m_i) \wedge \mu^P_{B'_{\bar{1}}}(n_i)\}\}
\end{aligned}$$

$$\begin{aligned}
&= \sup_{x=a+b}\{\sup_{a=\sum\limits_{i\in\mathscr{L}}\alpha_i[k_i,l_i]}\{\min_{i\in\mathscr{L}}\{\mu^P_{A_{\bar 0}}(k_i)\wedge\mu^P_{B_{\bar 0}}(l_i)\}\}\wedge\\
&\quad \sup_{b=\sum\limits_{i\in\mathscr{L}}\beta_i[m_i,n_i]}\{\min_{i\in\mathscr{L}}\{\mu^P_{A_{\bar 1}}(m_i)\wedge\mu^P_{B_{\bar 1}}(n_i)\}\}\\
&= \sup_{x=a+b}\{\mu^P_{[A_{\bar 0},B_{\bar 0}]}(a)\wedge\mu^P_{[A_{\bar 1},B_{\bar 1}]}(b)\}\\
&= (\mu^P_{[A_{\bar 0},B_{\bar 0}]+[A_{\bar 1},B_{\bar 1}]})(x) = \mu^P_{[A,B]_{\bar 0}}(x)
\end{aligned}$$

and

$$\begin{aligned}
\mu^N_{[A,B]'_{\bar 0}}(x) &= (\mu^N_{[A'_{\bar 0},B'_{\bar 0}]+[A'_{\bar 1},B'_{\bar 1}]})(x)\\
&= \inf_{x=a+b}\{\mu^N_{[A'_{\bar 0},B'_{\bar 0}]}(a)\vee\mu^N_{[A'_{\bar 1},B'_{\bar 1}]}(b)\}\\
&= \inf_{x=a+b}\{\inf_{a=\sum\limits_{i\in\mathscr{L}}\alpha_i[k_i,l_i]}\{\max_{i\in\mathscr{L}}\{\mu^N_{A'_{\bar 0}}(k_i)\vee\mu^N_{B'_{\bar 0}}(l_i)\}\}\vee\\
&\quad \inf_{b=\sum\limits_{i\in\mathscr{L}}\beta_i[m_i,n_i]}\{\max_{i\in\mathscr{L}}\{\mu^N_{A'_{\bar 1}}(m_i)\vee\mu^N_{B'_{\bar 1}}(n_i)\}\}\\
&= \inf_{x=a+b}\{\inf_{a=\sum\limits_{i\in\mathscr{L}}\alpha_i[k_i,l_i]}\{\max_{i\in\mathscr{L}}\{\mu^N_{A_{\bar 0}}(k_i)\vee\mu^N_{B_{\bar 0}}(l_i)\}\}\vee\\
&\quad \inf_{b=\sum\limits_{i\in\mathscr{L}}\beta_i[m_i,n_i]}\{\max_{i\in\mathscr{L}}\{\mu^N_{A_{\bar 1}}(m_i)\vee\mu^N_{B_{\bar 1}}(n_i)\}\}\\
&= \inf_{x=a+b}\{\mu^N_{[A_{\bar 0},B_{\bar 0}]}(a)\vee\mu^N_{[A_{\bar 1},B_{\bar 1}]}(b)\}\\
&= (\mu^N_{[A_{\bar 0},B_{\bar 0}]+[A_{\bar 1},B_{\bar 1}]})(x) = \mu^N_{[A,B]_{\bar 0}}(x).
\end{aligned}$$

Let $x\notin\mathscr{L}_{\bar 0}$. Then, $\mu^P_{[A,B]'_{\bar 0}}(x)=0$ and $\mu^N_{[A,B]'_{\bar 0}}(x)=-1$.

Similarly, for $x\in\mathscr{L}_{\bar 1}$, we have $\mu^P_{[A,B]'_{\bar 1}}(x)=\mu^P_{[A,B]_{\bar 1}}(x)$ and $\mu^N_{[A,B]'_{\bar 1}}(x)=\mu^N_{[A,B]_{\bar 1}}(x)$, for $x\notin\mathscr{L}_{\bar 1}$, we have $\mu^P_{[A,B]'_{\bar 1}}(x)=0$ and $\mu^N_{[A,B]'_{\bar 1}}(x)=-1$. Then, $[A,B]'_{\bar 0}$ and $[A,B]'_{\bar 1}$ are the extensions of $[A,B]_{\bar 0}$ and $[A,B]_{\bar 1}$.

Clearly, $[A,B]'_{\bar 0}\cap[A,B]'_{\bar 1}=(\mu^P_{[A,B]'_{\bar 0}\cap[A,B]'_{\bar 1}},\mu^N_{[A,B]'_{\bar 0}\cap[A,B]'_{\bar 1}})$, where

$$\mu^P_{[A,B]'_{\bar 0}\cap[A,B]'_{\bar 1}}(x)=\mu^P_{[A,B]'_{\bar 0}}(x)\wedge\mu^P_{[A,B]'_{\bar 1}}(x)=\begin{cases}1 & x=0\\ 0 & x\neq 0\end{cases},$$

$$\mu^N_{[A,B]'_{\bar 0}\cap[A,B]'_{\bar 1}}(x)=\mu^N_{[A,B]'_{\bar 0}}(x)\vee\mu^N_{[A,B]'_{\bar 1}}(x)=\begin{cases}-1 & x=0\\ 0 & x\neq 0\end{cases}.$$

Let $x \in \mathscr{L}$. We have

$$\begin{aligned}[A, B](x) &= [A'_{\bar{0}} + A'_{\bar{1}}, B'_{\bar{0}} + B'_{\bar{1}}](x) \\ &= ([A'_{\bar{0}}, B'_{\bar{0}}] + [A'_{\bar{1}}, B'_{\bar{1}}] + [A'_{\bar{0}}, B'_{\bar{1}}] + [A'_{\bar{1}}, B'_{\bar{0}}])(x) \\ &= ([A, B]'_{\bar{0}} + [A, B]'_{\bar{1}})(x).\end{aligned}$$

Hence, $[A, B] = [A, B]_{\bar{0}} \oplus [A, B]_{\bar{1}}$ is a $\mathbb{Z}_2$-graded bipolar fuzzy vector subspace.□

Lemma 6.11 *Let $A = (\mu_A^P, \mu_A^N)$ and $B = (\mu_B^P, \mu_B^N)$ be any two $\mathbb{Z}_2$-graded bipolar fuzzy vector subspaces of $\mathscr{L}$. Then, $[A, B] = [B, A]$.*

The following theorem is our main theorem in this section. The proof is base on Lemma 6.11. The left is similar to intuitionistic fuzzy ideal of Lie superalgebras.

Theorem 6.16 *Let $A = (\mu_A^P, \mu_A^N)$ and $B = (\mu_B^P, \mu_B^N)$ be any two bipolar fuzzy ideals of $\mathscr{L}$. Then $[A, B]$ is also a bipolar fuzzy ideal of $\mathscr{L}$.*

6.5 Solvable Bipolar Fuzzy Ideals and Nilpotent Bipolar Fuzzy Ideals

Definition 6.20 Let $A = (\mu_A^P, \mu_A^N)$ be a bipolar fuzzy ideal of $\mathscr{L}$. Define inductively a sequence of bipolar fuzzy ideals of $\mathscr{L}$ by $A^{(0)} = A$, $A^{(1)} = [A^{(0)}, A^{(0)}]$, $A^{(2)} = [A^{(1)}, A^{(1)}], \ldots, A^{(n)} = [A^{(n-1)}, A^{(n-1)}]$, then $A^{(n)}$ is called the nth derived bipolar fuzzy ideal of $\mathscr{L}$. In which, $A^{(i+1)} = (\mu^P_{A^{(i+1)}}, \mu^N_{A^{(i+1)}})$ where

$$\mu^P_{A^{(i+1)}}(x) = \begin{cases} \sup\limits_{x=\sum_{j\in\mathscr{L}} \alpha_j[x_j, y_j]} \{\min_{j\in\mathscr{L}}\{\mu^P_{A^{(i)}}(x_j) \wedge \mu^P_{A^{(i)}}(y_j)\}\} & \text{where } \alpha_j \in \mathbb{F}, x_j, y_j \in \mathscr{L} \\ 0 & \text{if } x \text{ is not expressed as } x = \sum_{j\in\mathscr{L}} \alpha_j[x_j, y_j] \end{cases},$$

and

$$\mu^N_{A^{(i+1)}}(x) = \begin{cases} \inf\limits_{x=\sum_{j\in\mathscr{L}} \alpha_j[x_j, y_j]} \{\max_{j\in\mathscr{L}}\{\mu^N_{A^{(i)}}(x_j) \vee \mu^N_{A^{(i)}}(y_j)\}\} & \text{where } \alpha_j \in \mathbb{F}, x_j, y_j \in \mathscr{L} \\ 0 & \text{if } x \text{ is not expressed as } x = \sum_{j\in\mathscr{L}} \alpha_j[x_j, y_j] \end{cases}.$$

From the definition, we can get $\mu^P_{A^{(0)}} \supseteq \mu^P_{A^{(1)}} \supseteq \mu^P_{A^{(2)}} \supseteq \cdots \supseteq \mu^P_{A^{(n)}} \supseteq \cdots$ and $\mu^N_{A^{(0)}} \subseteq \mu^N_{A^{(1)}} \subseteq \mu^N_{A^{(2)}} \subseteq \cdots \subseteq \mu^N_{A^{(n)}} \subseteq \cdots$.

Definition 6.21 Let $A^{(n)}$ be as above. Define: $\eta^{(n)} = \sup\{\mu^P_{A^{(n)}}(x) : 0 \neq x \in \mathscr{L}\}$ and $\kappa^{(n)} = \inf\{\mu^N_{A^{(n)}}(x) : 0 \neq x \in \mathscr{L}\}$. Then it is clear that $\eta^{(0)} \geq \eta^{(1)} \geq \eta^{(2)} \geq \cdots \geq \eta^{(n)} \geq \cdots$ and $\kappa^{(0)} \leq \kappa^{(1)} \leq \kappa^{(1)} \leq \cdots \leq \kappa^{(n)} \leq \cdots$.

Definition 6.22 A bipolar fuzzy ideal $A = (\mu^P_A, \mu^N_A)$ of $\mathscr{L}$ is called a *solvable bipolar fuzzy ideal*, if there is a positive integer n such that $\eta^{(n)} = 0$ and $\kappa^{(n)} = 0$. So it is a solvable bipolar fuzzy ideal; then, there is positive integer n such that $\mu^P_{A^{(n)}} = 1_0$ and $\mu^N_{A^{(n)}} = (-1)_0$.

Example 6.6 The Lie superalgebra N is in Example 6.5.

We define $A_{\bar{0}} = (\mu^P_{A_{\bar{0}}}, \mu^N_{A_{\bar{0}}})$ where $\mu^P_{A_{\bar{0}}}(x) = 1, \mu^N_{A_{\bar{0}}}(x) = -1$ for all $x \in N_{\bar{0}}$. Then, it is a bipolar fuzzy subspace of $N_{\bar{0}}$. Let $x \in N_{\bar{1}}$. Then, $x = k_1a_1 + k_2a_2 + k_3b_1 + k_4b_2$, for $k_i \neq 0$ and $i = 1, 2, 3, 4$. We define $A_{\bar{1}} = (\mu^P_{A_{\bar{1}}}, \mu^N_{A_{\bar{1}}})$ where $\mu^P_{A_{\bar{1}}}(x) = \mu^P_{A_{\bar{1}}}(a_1) \wedge \mu^P_{A_{\bar{1}}}(a_2) \wedge \mu^P_{A_{\bar{1}}}(b_1) \wedge \mu^P_{A_{\bar{1}}}(b_2)$, in which $\mu^P_{A_{\bar{1}}}(a_1) = 0.2$, $\mu^P_{A_{\bar{1}}}(a_2) = 1$, $\mu^P_{A_{\bar{1}}}(b_1) = 0.1$, $\mu^P_{A_{\bar{1}}}(b_2) = 1$, $\mu^P_{A_{\bar{1}}}(0) = 1$, and $\mu^N_{A_{\bar{1}}}(x) = \mu^N_{A_{\bar{1}}}(a_1) \vee \mu^N_{A_{\bar{1}}}(a_2) \vee \mu^N_{A_{\bar{1}}}(b_1) \vee \mu^N_{A_{\bar{1}}}(b_2)$, in which $\mu^N_{A_{\bar{1}}}(a_1) = -0.7, \mu^N_{A_{\bar{1}}}(a_2) = -1, \mu^N_{A_{\bar{1}}}(b_1) = -0.9$, $\mu^N_{A_{\bar{1}}}(b_2) = -1$, $\mu^N_{A_{\bar{1}}}(0) = -1$. Then, A is a bipolar fuzzy subspace of $N_{\bar{1}}$.

Let $x \in N$. Then, $x = ke + k_1a_1 + k_2a_2 + k_3b_1 + k_4b_2$ for $k, k_i \neq 0$ and $i = 1, 2, 3, 4$. We define $A = (\mu^P_A, \mu^N_A)$ where $\mu^P_A(x) = \mu^P_A(e) \wedge \mu^P_A(a_1) \wedge \mu^P_A(a_2) \wedge \mu^P_A(b_1) \wedge \mu^P_A(b_2)$, in which $\mu^P_A(e) = 1$, $\mu^P_A(a_1) = 0.2$, $\mu^P_A(a_2) = 1$, $\mu^P_A(b_1) = 0.1$, $\mu^P_A(b_2) = 1$, $\mu^P_A(0) = 1$ and $\mu^N_A(x) = \mu^N_B(e) \vee \mu^N_A(a_1) \vee \mu^N_A(a_2) \vee \mu^N_A(b_1) \vee \mu^N_A(b_2)$, in which $\mu^N_A(e) = -1$, $\mu^N_A(a_1) = -0.7$, $\mu^N_A(a_2) = -1$, $\mu^N_A(b_1) = -0.9$, $\mu^N_A(b_2) = -1$, $\mu^N_A(0) = -1$. Then, $A = A_{\bar{0}} \oplus A_{\bar{1}}$ is a bipolar fuzzy ideal of N.

Let $A^{(0)} = A$. Note that $[a_i, b_i] = e$ and the other brackets are zero. Then, $\mu^P_{A^{(0)}}(x) = 0.1$, $\mu^N_{A^{(0)}}(x) = -0.7$. We define $A^{(1)} = [A^{(0)}, A^{(0)}]$. If $x \in N_1$, then x can not be expressed as $x = \sum \alpha_i[x_i, y_i], x_i, y_i \in N$, so $\mu^P_{A^{(1)}}(x) = 0$, $\mu^N_{A^{(1)}}(x) = 0$. If $x \in N_0$, then x can be expressed as $x = \alpha_1[a_1, b_1] + \alpha_2[a_2, b_2]$, $\alpha_1, \alpha_2 \in k$. We calculate

$$\mu^P_{A^{(1)}}(x) = \sup_{x=\sum_{i=1,2} \alpha_i[a_i,b_i]} \{\min_{i=1,2}\{\mu^P_{A^{(0)}}(a_i) \wedge \mu^P_{A^{(0)}}(b_i)\}\} = 0.1,$$

$$\mu^N_{A^{(1)}}(x) = \inf_{x=\sum_{i=1,2} \alpha_i[a_i,b_i]} \{\max_{i=1,2}\{\mu^N_{A^{(0)}}(a_i) \vee \mu^N_{A^{(0)}}(b_i)\}\} = -1.$$

Define $A^{(2)} = [A^{(1)}, A^{(1)}]$, we calculate

$$\mu^P_{A^{(2)}}(x) = \sup_{x=\sum_{i=1,2} \alpha_i[a_i,b_i]} \{\min_{i=1,2}\{\mu^P_{A^{(1)}}(a_i) \wedge \mu^P_{A^{(1)}}(b_i)\}\} = 0,$$

$$\mu^N_{A^{(2)}}(x) = \inf_{x=\sum_{i=1,2} \alpha_i[a_i,b_i]} \{\max_{i=1,2}\{\mu^N_{A^{(1)}}(a_i) \vee \mu^N_{A^{(1)}}(b_i)\}\} = 0.$$

So $\eta^{(0)} \geq \eta^{(1)} \geq \eta^{(2)} = 0$ and $\kappa^{(0)} \leq \kappa^{(1)} \leq \kappa^{(2)} = 0$. These show that A is a solvable bipolar fuzzy ideal of N.

Following from the definition of solvable bipolar fuzzy ideals, we can easily get.

Lemma 6.12 *Let $A = (\mu_A^P, \mu_A^N)$ be a bipolar fuzzy Lie ideal of $\mathscr{L}$. Then, $A = (\mu_A^P, \mu_A^N)$ is a solvable bipolar fuzzy ideal if and only if there is a positive integer n such that $\mu^P_{A^{(m)}} = 1_0, \mu^N_{A^{(m)}} = (-1)_0$ for all $m \geq n$.*

Theorem 6.17 *Homomorphic images of solvable bipolar fuzzy ideals are also solvable bipolar fuzzy Lie ideals.*

Proof Let $\varphi : \mathscr{L} \to \mathscr{L}'$ be a homomorphism of Lie superalgebra, and assume that $A = (\mu_A^P, \mu_A^N)$ is a bipolar fuzzy ideal of $\mathscr{L}$. Let $\varphi(A) = B$, i.e, $\mu_B^P = \mu^P_{\varphi(A)}, \mu_B^N = \mu^N_{\varphi(A)}$. We prove $\mu^P_{\varphi(A^{(n)})} = \mu^P_{B^{(n)}}$ and $\mu^N_{\varphi(A^{(n)})} = \mu^N_{B^{(n)}}$ by induction on n, where n is any positive integer. Indeed, let $y \in \mathscr{L}'$. Consider $n = 1$,

$$\begin{aligned}
\mu^P_{\varphi(A^{(1)})}(y) = \mu^P_{\varphi([A,A])}(y) &= \sup_{y=\varphi(x)} \{\mu^P_{[A,A]}(x)\} \\
&= \sup_{y=\varphi(x)} \{ \sup_{x=\sum_{i\in\mathscr{L}} \alpha_i[x_i,y_i]} \{\min_{i\in\mathscr{L}}(\mu_A^P(x_i) \wedge \mu_A^P(y_i))\}\} \\
&= \sup_{y=\sum_{i\in\mathscr{L}} \alpha_i\varphi[x_i,y_i]} \{\min_{i\in\mathscr{L}}(\mu_A^P(x_i) \wedge \mu_A^P(y_i))\} \\
&= \sup_{y=\sum_{i\in\mathscr{L}} \alpha_i[a_i,b_i]} \{\min_{i\in\mathscr{L}}(\mu_A^P(x_i) \wedge \mu_A^P(y_i)) : \varphi(x_i) = a_i, \varphi(y_i) = b_i\} \\
&= \sup_{\sum_{i\in\mathscr{L}} \alpha_i[a_i,b_i]=y} \{\min_{i\in\mathscr{L}}(\mu_B^P(a_i) \wedge \mu_B^P(b_i))\} \\
&= \mu^P_{[B,B]}(y) = \mu^P_{B^{(1)}}(y),
\end{aligned}$$

and

$$\begin{aligned}
\mu^N_{\varphi(A^{(1)})}(y) = \mu^N_{\varphi([A,A])}(y) &= \inf_{y=\varphi(x)} \{\mu^N_{[A,A]}(x)\} \\
&= \inf_{y=\varphi(x)} \{ \inf_{x=\sum_{i\in\mathscr{L}} \alpha_i[x_i,y_i]} \{\max_{i\in\mathscr{L}}(\mu_A^N(x_i) \vee \mu_A^N(y_i))\}\} \\
&= \inf_{y=\sum_{i\in\mathscr{L}} \alpha_i\varphi[x_i,y_i]} \{\max_{i\in\mathscr{L}}(\mu_A^N(x_i) \vee \mu_A^N(y_i))\} \\
&= \inf_{y=\sum_{i\in\mathscr{L}} \alpha_i[a_i,b_i]} \{\max_{i\in\mathscr{L}}(\mu_A^N(x_i) \vee \mu_A^N(y_i)) : \varphi(x_i) = a_i, \varphi(y_i) = b_i\} \\
&= \inf_{\sum_{i\in\mathscr{L}} \alpha_i[a_i,b_i]=y} \{\max_{i\in\mathscr{L}}(\mu_B^N(a_i) \vee \mu_B^N(b_i))\}
\end{aligned}$$

$$= \mu^N_{[B,B]}(y) = \mu^N_{B^{(1)}}(y).$$

These prove the case of $n = 1$. Suppose that the case of $n-1$ is true, then $\mu^P_{\varphi(A^{(n)})} = \mu^P_{\varphi([A^{(n-1)},A^{(n-1)}])} = \mu^P_{[\varphi(A^{(n-1)}),\varphi(A^{(n-1)})]} = \mu^P_{[B^{(n-1)},B^{(n-1)}]} = \mu^P_{B^{(n)}}$ and $\mu^N_{\varphi(A^{(n)})} = \mu^N_{\varphi([A^{(n-1)},A^{(n-1)}])} = \mu^N_{[\varphi(A^{(n-1)}),\varphi(A^{(n-1)})]} = \mu^N_{[B^{(n-1)},B^{(n-1)}]} = \mu^N_{B^{(n)}}$. Let m be a positive integer such that $\mu^P_{A^{(m)}} = 1_0$ and $\mu^N_{A^{(m)}} = (-1)_0$. Then, for any $0 \neq y \in \mathscr{L}'$, we get $\mu^P_{B^{(m)}}(y) = \mu^P_{\varphi(A^{(m)})}(y) = \sup\limits_{y=\varphi(x)} \{1_0(x)\} = 0$, $\mu^N_{B^{(m)}}(y) = \varphi(\mu^N_{A^{(m)}})(y) = \inf\limits_{y=\varphi(x)} \{(-1)_0(x)\} = 0$. So $\mu^P_{B^{(m)}} = 1_0$ and $\mu^N_{B^{(m)}} = (-1)_0$. □

Let $A = (\mu^P_A, \mu^N_A)$ be a bipolar fuzzy ideal of $\mathscr{L}$ and J be an ideal of $\mathscr{L}$. We can prove that A/J is a bipolar fuzzy ideal of $\mathscr{L}/J$.

Theorem 6.18 *Let $A = (\mu^P_A, \mu^N_A)$ be a bipolar fuzzy ideal of $\mathscr{L}$ and A/J be a solvable bipolar fuzzy ideal of $\mathscr{L}/J$. If $B = (\mu^P_B, \mu^N_B)$ is a solvable bipolar fuzzy ideal of $\mathscr{L}$ and is also a bipolar fuzzy ideal of $A = (\mu^P_A, \mu^N_A)$ such that $B(J) = A(J)$, then $A = (\mu^P_A, \mu^N_A)$ is solvable.*

Proof Let φ be the canonical projection from $\mathscr{L}$ to $\mathscr{L}/J$. From the proof of Theorem 6.17, we get $\mu^P_{\varphi(A^{(n)})} = \mu^P_{(A/J)^{(n)}}$ and $\mu^N_{\varphi(A^{(n)})} = \mu^N_{(A/J)^{(n)}}$. Since A/J is solvable, there exists n such that $\mu^P_{(A/J)^{(n)}} = 1_0$ and $\mu^N_{(A/J)^{(n)}} = (-1)_0$.

For $0 \neq \bar{y} \in \mathscr{L}/J$, we have $\sup\limits_{m\in\varphi^{-1}(\bar{y})} \{\mu^P_{A^{(n)}}(m)\} = \mu^P_{\varphi(A^{(n)})}(\bar{y}) = \mu^P_{(A/J)^{(n)}}(\bar{y}) = 0$ and $\inf\limits_{m\in\varphi^{-1}(\bar{y})} \{\mu^N_{A^{(n)}}(m)\} = \mu^N_{\varphi(A^{(n)})}(\bar{y}) = \mu^N_{(A/J)^{(n)}}(\bar{y}) = 0$. Notice that $m \in \mathscr{L}$ and $m \neq 0$; we get $\mu^P_{A^{(n)}}(m) = 0$ and $\mu^N_{A^{(n)}}(m) = 0$.

For $\bar{y} = 0$, we have $\sup\limits_{m\in\varphi^{-1}(0)} \{\mu^P_{A^{(n)}}(m)\} = \mu^P_{\varphi(A^{(n)})}(0) = 1$ and $\inf\limits_{m\in\varphi^{-1}(0)} \{\mu^N_{A^{(n)}}(m)\} = \mu^N_{\varphi(A^{(n)})}(0) = -1$. Since $\varphi^{-1}(0) = J$ and $B(J) = A(J)$, we have $\mu^P_{B^{(n)}}(J) = \mu^P_{A^{(n)}}(J)$ and $\mu^N_{B^{(n)}}(J) = \mu^N_{A^{(n)}}(J)$. For any $x \in J$, B is solvable, then there exists n such that $\mu^P_{B^{(n)}} = 1_0$ and $\mu^N_{B^{(n)}} = (-1)_0$, we have $\mu^P_{A^{(n)}} = 1_0$ and $\mu^N_{A^{(n)}} = (-1)_0$.

Hence, for any $x \in \mathscr{L}$, we always have that $\mu^P_{A^{(n)}} = 1_0$ and $\mu^N_{A^{(n)}} = (-1)_0$, which imply that $A = (\mu^P_A, \mu^N_A)$ is solvable. □

Lemma 6.13 *Let $A = (\mu^P_A, \mu^N_A)$ and $B = (\mu^P_B, \mu^N_B)$ be bipolar fuzzy ideals of $\mathscr{L}$. Then, $(A \oplus B)^{(n)} = A^{(n)} \oplus B^{(n)}$.*

Proof Let $0 \neq x \in \mathscr{L}$. Then, we have $[A, B] = (\mu^P_{[A,B]}, \mu^P_{[A,B]})$, where $\mu^P_{[A,B]}(x) = \sup\limits_{x=\sum\limits_{i\in\mathscr{L}} \alpha_i[x_i,y_i]\ x\in\mathscr{L}} \{\min(\mu^P_A(x_i) \wedge \mu^P_B(y_i))\} \leq \mu^P_A(x) \wedge \mu^P_B(x) = 0$,

$$\mu^N_{[A,B]}(x) = \inf_{x=\sum\limits_{i\in\mathscr{L}} \alpha_i[x_i,y_i]\ x\in\mathscr{L}} \{\max(\mu^N_A(x_i) \vee \mu^N_B(y_i))\} \geq \mu^N_A(x) \vee \mu^N_B(x) = 0.$$

So $\mu^P_{[A,B]} = 1_0$ and $\mu^N_{[A,B]} = (-1)_0$. Consequently, for any positive integer a, b, we have $\mu^P_{[A^{(a)},B^{(b)}]} = 1_0$ and $\mu^N_{[A^{(a)},B^{(b)}]} = (-1)_0$. We prove the lemma by induction on n. Let $n = 1$. Then,

$(A \oplus B)^{(1)} = [A \oplus B, A \oplus B] = [A, A] \oplus [A, B] \oplus [B, A] \oplus [B, B] = A^{(1)} \oplus B^{(1)}$.

Suppose that the case of $n - 1$ is true, then

$$\begin{aligned}(A \oplus B)^{(n)} &= [(A \oplus B)^{(n-1)}, (A \oplus B)^{(n-1)}]\\ &= [A^{(n-1)} \oplus B^{(n-1)}, A^{(n-1)} \oplus B^{(n-1)}]\\ &= A^{(n)} \oplus B^{(n)}.\end{aligned}$$

So we get $(A \oplus B)^{(n)} = A^{(n)} \oplus B^{(n)}$. □

Theorem 6.19 *Direct sum of any solvable bipolar fuzzy Lie ideals is also a solvable bipolar Lie ideal.*

Proof Let $A = (\mu^P_A, \mu^N_A)$ and $B = (\mu^P_B, \mu^N_B)$ be solvable bipolar fuzzy ideals. Then, there exist positive integers m, n such that $\mu^P_{A^{(m)}} = 1_0$, $\mu^N_{A^{(m)}} = (-1)_0$ and $\mu^P_{B^{(n)}} = 1_0$, $\mu^N_{B^{(n)}} = (-1)_0$. Since $(A \oplus B)^{(m+n)} = A^{(m+n)} \oplus B^{(m+n)}$, we have $\mu^P_{(A\oplus B)^{(m+n)}} = \mu^P_{A^{(m+n)}\oplus B^{(m+n)}} = 1_0$ and $\mu^N_{(A\oplus B)^{(m+n)}} = \mu^N_{A^{(m+n)}\oplus B^{(m+n)}} = (-1)_0$. So $A \oplus B$ is a solvable bipolar fuzzy Lie ideal. □

Definition 6.23 Let $A = (\mu^P_A, \mu^N_A)$ be a bipolar fuzzy ideal of $\mathscr{L}$. Define inductively a sequence of bipolar fuzzy ideals of $\mathscr{L}$ by $A^0 = A,\ \ A^1 = [A, A^0],\ \ A^2 = [A, A^1], \ldots, A^n = [A, A^{n-1}] \ldots$, which is called the descending central series of a bipolar fuzzy ideal $A = (\mu^P_A, \mu^N_A)$ of $\mathscr{L}$. We get $\mu^P_{A^0} \supseteq \mu^P_{A^1} \supseteq \mu^P_{A^2} \supseteq \cdots \supseteq \mu^P_{A^n} \supseteq \cdots$ and $\mu^N_{A^0} \subseteq \mu^N_{A^1} \subseteq \mu^N_{A^2} \subseteq \cdots \subseteq \mu^N_{A^n} \subseteq \ldots$

Definition 6.24 For any bipolar fuzzy Lie ideal $A = (\mu^P_A, \mu^N_A)$, define $\eta^n = \sup\{\mu^P_{A^n}(x) : 0 \neq x \in \mathscr{L}\}$ and $\kappa^n = \inf\{\mu^N_{A^n}(x) : 0 \neq x \in \mathscr{L}\}$, for any positive integer n. The bipolar fuzzy ideal is called a nilpotent bipolar fuzzy ideal, if there is a positive integer m such that $\eta^m = 0$ and $\kappa^m = 1$, or equivalently, $\mu^P_{A^m} = 1_0$ and $\mu^N_{A^m} = (-1)_0$.

Example 6.7 Let us take the basis h, e, f of $\mathfrak{sl}(1|1)$ as follows:

$$h = \begin{pmatrix} 1 & 0 \\ 0 & 1 \end{pmatrix}, e = \begin{pmatrix} 0 & 1 \\ 0 & 0 \end{pmatrix}, f = \begin{pmatrix} 0 & 0 \\ 1 & 0 \end{pmatrix}. \tag{6.1}$$

Then, h is an even element, and e and f are odd element. Their bracket products are as follows: $[e, f] = [f, e] = h$, the other brackets $= 0$. Then, $\mathfrak{sl}(1|1)$ is a three-dimensional Lie superalgebra.

Define $A_{\bar{0}} = (\mu^P_{A_{\bar{0}}}, \mu^N_{A_{\bar{0}}}) : \mathfrak{sl}(1|1)_{\bar{0}} \to [-1, 1]$ where

$$\mu^P_{A_{\bar{0}}}(x) = \begin{cases} 0.6 & x = h \\ 1 & \text{otherwise} \end{cases}, \mu^N_{A_{\bar{0}}}(x) = \begin{cases} -0.4 & x = h \\ -1 & \text{otherwise} \end{cases}$$

Define $A_{\bar{1}} = (\mu^P_{A_{\bar{1}}}, \mu^N_{A_{\bar{1}}}) : \mathfrak{sl}(1|1)_{\bar{1}} \to [-1, 1]$ where

$$\mu^P_{A_{\bar{1}}}(x) = \begin{cases} 0.3 & x = e \\ 0.5 & x = f \\ 1 & \text{otherwise} \end{cases}, \mu^N_{A_{\bar{1}}}(x) = \begin{cases} -0.7 & x = e \\ -0.5 & x = f \\ -1 & \text{otherwise} \end{cases}$$

Define $A = (\mu^P_A, \mu^N_A) : \mathfrak{sl}(1|1) \to [-1, 1]$ where $\mu^P_A(x) = \mu^P_{A_{\bar{0}}}(x_{\bar{0}}) \wedge \mu^P_{A_{\bar{1}}}(x_{\bar{1}})$ and $\mu^N_A(x) = \mu^N_{A_{\bar{0}}}(x_{\bar{0}}) \vee \mu^N_{A_{\bar{1}}}(x_{\bar{1}})$. Then, A is a bipolar fuzzy ideal of $\mathfrak{sl}(1|1)$.

Let $A^0 = A$. We define $A^1 = [A, A^0]$, then if $x \in \mathfrak{sl}(1|1)_{\bar{1}}$, x can not be expressed as $x = \sum \alpha_i [x_i, y_i]$, $x_i, y_i \in \mathfrak{sl}(1|1)$, and then, $\mu^P_{A^1}(x) = 0, \mu^N_{A^1}(x) = 0$. If $x \in \mathfrak{sl}(1|1)_{\bar{0}}$, $x = \alpha[e, f]$, $\alpha \in \mathbb{F}$, then $\mu^P_{A^1}(x) = \sup\{\mu^P_A(e) \wedge \mu^P_{A^0}(f)\} = 0.3$ and $\mu^N_{A^1}(x) = \inf\{\mu^N_A(e) \vee \mu^N_{A^0}(f)\} = -0.5$.

Define $A^2 = [A, A^1]$, we calculate if $x \in \mathfrak{sl}(1|1)_{\bar{1}}$, $\mu^P_{A^2}(x) = 0, \mu^N_{A^2}(x) = 0$. If $x \in \mathfrak{sl}(1|1)_{\bar{0}}$, $\mu^P_{A^2}(x) = \sup\{\mu^P_A(e) \wedge \mu^P_{A^1}(f)\} = 0$ and $\mu^N_{A^2}(x) = \inf\{\mu^N_A(e) \vee \mu^N_{A^1}(f)\} = 0$, then we get $\eta^0 \geq \eta^1 \geq \eta^2 = 0$ and $\kappa^0 \leq \kappa^1 \leq \kappa^2 = 0$. So A is a nilpotent bipolar fuzzy Lie ideal of $\mathfrak{sl}(1|1)$.

Theorem 6.20 *Homomorphic images of nilpotent bipolar fuzzy ideals are also nilpotent bipolar fuzzy Lie ideals. Direct sum of nilpotent bipolar fuzzy ideals is also a nilpotent bipolar fuzzy ideal.*

Theorem 6.21 *If $A = (\mu^P_A, \mu^N_A)$ is a nilpotent bipolar fuzzy ideal of $\mathscr{L}$, then it is solvable.*

Chapter 7
m-Polar Fuzzy Lie Ideals

In this chapter, we present concepts of m-polar fuzzy Lie subalgebras and m-polar fuzzy Lie ideals. We discuss the homomorphisms between the Lie subalgebras of a Lie algebra and their relationship between the domains and the codomains of the m-polar fuzzy subalgebras under these homomorphisms. We also describe nilpotent m-polar fuzzy Lie ideals and solvable m-polar fuzzy Lie ideals.

7.1 Introduction

Fuzzy set theory deals with real-life data incorporating vagueness. Zhang [147] extended the theory of fuzzy sets to bipolar fuzzy sets, which register the bipolar behavior of objects. Nowadays, analysts believe that the world is moving toward multipolarity. Therefore, it comes as no surprise that multipolarity in data and information plays a vital role in various fields of science and technology. In neurobiology, multipolar neurons in brain gather a great deal of information from other neurons. In information technology, multipolar technology can be exploited to operate large-scale systems. Based on this motivation, Chen et al. [40] further generalized bipolar fuzzy set theory by introducing the theory of m-polar fuzzy sets. In an m-polar fuzzy set, the membership value of an element belongs to $[0, 1]^m$ which represents all the m different properties of an element. This is more suited for a number of real-world problems, where data come from n agents ($n \geq 2$); hence, multipolar information arises and cannot be properly represented by any existing type of structures. Considering graphic structures, m-polar fuzzy sets can be used to describe the relationship among several individuals. In particular, m-polar fuzzy sets are shown to be useful in adapting accurate problems if it is necessary to make judgements with a group of agreements.

Definition 7.1 An *m-polar fuzzy set* C on a nonempty set X is a mapping $C : X \to [0, 1]^m$. The membership value of every element $x \in X$ is denoted by $C(x) =$

M. Akram, *Fuzzy Lie Algebras*, Infosys Science Foundation Series,
https://doi.org/10.1007/978-981-13-3221-0_7

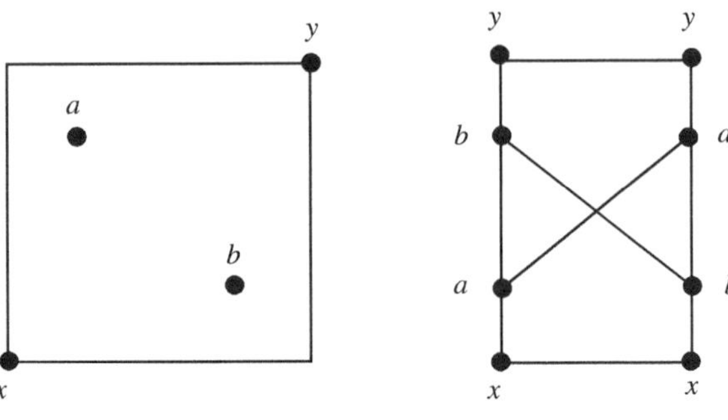

Fig. 7.1 Order relation when $m = 2$

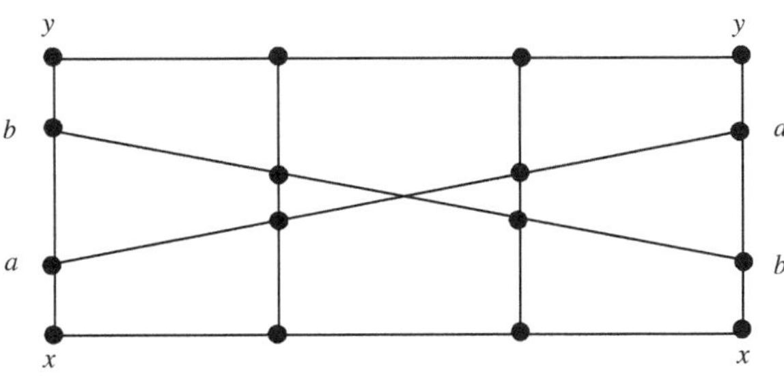

Fig. 7.2 Order relation when $m = 4$

$(P_1 \circ C(x), P_2 \circ C(x), \ldots, P_m \circ C(x))$, where $P_i \circ C : [0, 1]^m \to [0, 1]$ is defined as the ith projection mapping.

Note that $[0, 1]^m$ (mth power of $[0, 1]$) is considered as a partially ordered set with the point-wise order $\leq$, where m is an arbitrary ordinal number (we make an appointment that $m = \{n|n < m\}$ when $m > 0$), $\leq$ is defined by $x \leq y \Leftrightarrow P_i(x) \leq P_i(y)$ for each $i \in m$ ($x, y \in [0, 1]^m$), and $P_i : [0, 1]^m \to [0, 1]$ is the ith projection mapping ($i \in m$). $\mathbf{1} = (1, 1, \ldots, 1)$ is the greatest value, and $\mathbf{0} = (0, 0, \ldots, 0)$ is the smallest value in $[0, 1]^m$. $m\mathscr{F}(X)$ is the power set of all m-polar fuzzy subsets on X.

(i) When $m = 2$, $[0, 1]^2$ is the ordinary closed unit square in $\mathbb{R}^2$, the Euclidean plane. The righter (respectively, the upper), the point in this square, the larger it is. Let $x = (0, 0) = \mathbf{0}$ (the smallest element of $[0, 1]^2$), $a = (0.35, 0.85)$, $b = (0.85, 0.35)$, and $y = (1, 1) = \mathbf{1}$ (the largest element of $[0, 1]^2$). Then $x \leq c \leq y$, $\forall c \in [0, 1]^2$ (especially, $x \leq a \leq y$ and $x \leq b \leq y$ hold). It is easy to note that $a \not\leq b \not\leq a$ because $P_0(a) = 0.35 < 0.85 = P_0(b)$ and $P_1(a) = 0.85 > 0.35 = P_1(b)$ hold. The "order relation $\leq$" on $[0, 1]^2$ can be described in at least two ways. It can be seen in Fig. 7.1,

(ii) When $m = 4$, the order relation can be seen in Fig. 7.2.

Example 7.1 Let $X = \{T_1, T_2, T_3, T_4, T_5\}$ be a set of companies which may have different repute in the market due to its annual profit, market power, and price control of its product. These are multipolar information which are fuzzy in nature. Let C be a 3-polar fuzzy set on X. The degree of membership of each company is shown in Table 7.1.

Table 7.1 3-polar fuzzy set C

Company x	Profit $P_1 \circ C$	Market power $P_2 \circ C$	Price control $P_3 \circ C$
T_1	0.7	0.4	0.9
T_2	0.5	0.3	0.6
T_3	0.9	0.7	0.5
T_4	0.8	0.7	0.6
T_5	0.6	0.3	0.6

The membership value of T_1 is $(0.7, 0.4, 0.9)$ which shows that T_1 has 70% annual profit, 40% power in business market, and 90% price control of its product. The fuzzy strategies in Table 7.1 can be represented by a 3-polar fuzzy set as: $C = \{(T_1, 0.7, 0.4, 0.9),\ (T_2, 0.5, 0.3, 0.6),\ (T_3, 0.9, 0.7, 0.5),\ (T_4, 0.8, 0.7, 0.6),\ (T_5, 0.6, 0.3, 0.6)\}$.

Definition 7.2 Let C and D be two m-polar fuzzy sets on X. Then, the operations $C \cup D$, $C \cap D$, $C \subseteq D$, and $C = D$ are defined as follows:

1. $P_i \circ (C \cup D)(x) = \sup\{P_i \circ C(x), P_i \circ D(x)\} = P_i \circ C(x) \vee P_i \circ D(x)$,
2. $P_i \circ (C \cap D)(x) = \inf\{P_i \circ C(x), P_i \circ D(x)\} = P_i \circ C(x) \wedge P_i \circ D(x)$,
3. $C \subseteq D$ if and only if $P_i \circ C(x) \leq P_i \circ D(x)$,
4. $C = D$ if and only if $P_i \circ C(x) = P_i \circ D(x)$

for all $x \in X$, for each $1 \leq i \leq m$.

Definition 7.3 An m-polar fuzzy set C on a group X is called an *m-polar fuzzy subgroup* if the following conditions are satisfied:

(1) $C(xy) \geq \inf(C(x), C(y))$,
(2) $C(x^{-1}) \geq C(x)$.

That is,

(1) $p_i \circ C(xy) \geq \min(p_i \circ C(x), p_i \circ C(y))$,
(2) $p_i \circ C(x^{-1}) \geq p_i \circ C(x)$

for all $x, y \in X$, $i = 1, 2, 3, \ldots, m$.

7.2 m-Polar Fuzzy Lie Subalgebras

Definition 7.4 Let V be vector space over field $\mathbb{F}$. An m-polar fuzzy set C on V is called an *m-polar fuzzy subspace* if the following conditions are satisfied:

(1) $C(x + y) \geq C(x) \wedge C(y)$,
(2) $C(\alpha x) \geq C(x)$, for all $x, y \in V$ and $\alpha \in \mathbb{F}$.

That is,

(1) $p_i \circ C(x + y) \geq \inf(p_i \circ C(x), p_i \circ C(y))$,
(2) $p_i \circ C(\alpha x) \geq p_i \circ C(x)$

for all $x, y \in V$ and $\alpha \in \mathbb{F}$, $i = 1, 2, 3, \ldots, m$.

Definition 7.5 Let L be Lie algebra. An m-polar fuzzy set C on L is called an *m-polar fuzzy Lie subalgebra* if the following conditions are satisfied:

(1) $C(x + y) \geq C(x) \wedge C(y)$,
(2) $C(\alpha x) \geq C(x)$,
(3) $C([x, y]) \geq C(x) \wedge C(y)$ for all $x, y \in L$ and $\alpha \in \mathbb{F}$.

That is,

(1) $p_i \circ C(x + y) \geq \inf(p_i \circ C(x), p_i \circ C(y))$,
(2) $p_i \circ C(\alpha x) \geq p_i \circ C(x)$,
(3) $p_i \circ C([x, y]) \geq \inf(p_i \circ C(x), p_i \circ C(y))$

for all $x, y \in L$ and $\alpha \in \mathbb{F}$, $i = 1, 2, 3, \ldots, m$.

Example 7.2 Let $\Re^3 = \{(x, y, z) | x, y, z \in \mathbb{R}\}$ be the set of all three-dimensional real vectors. Then, $\Re^3$ with the bracket $[\cdot, \cdot]$ defined as the usual cross product, i.e., $[x, y] = x \times y$, forms a real Lie algebra. We also define an m-polar fuzzy set $C : \Re^3 \to [0, 1]^m$ by

$$C(x, y, z) = \begin{cases} (0.6, 0.6, \ldots, 0.6) \text{ if } x = y = z = 0, \\ (0.2, 0.2, \ldots, 0.2) \text{ otherwise.} \end{cases}$$

By routine computations, we can verify that the above m-polar fuzzy set C is an m-polar fuzzy Lie subalgebra and Lie ideal of the Lie algebra $\Re^3$.

Proposition 7.1 *Every m-polar fuzzy Lie ideal is an m-polar fuzzy Lie subalgebra.*

We note here that the converse of Proposition 7.1 does not hold in general as can be seen in the following example.

Example 7.3 Consider $\mathbb{F} = \mathbb{R}$. Let $L = \Re^3 = \{(x, y, z) : x, y, z \in \mathbb{R}\}$ be the set of all three-dimensional real vectors which forms a Lie algebra and define

$$\Re^3 \times \Re^3 \to \Re^3$$

$$[x, y] \to x \times y,$$

where $\times$ is the usual cross product. We define an m-polar fuzzy set $C : \Re^3 \to [0, 1]^m$ by

$$C(x, y, z) = \begin{cases} (1, 1, \ldots, 1) & \text{if } x = y = z = 0, \\ (0.5, 0.5, \ldots, 0.5) & \text{if } x \neq 0, y = z = 0, \\ (0, 0, \ldots, 0) & \text{otherwise.} \end{cases}$$

Then, C is an m-polar fuzzy Lie subalgebra of L but C is not an m-polar fuzzy Lie ideal of L since

$$C([(1,0,0)\ (1,1,1)]\ ) = C(0,-1,1) = (0,0,\ldots,0),$$

$$C(1,0,0) = (0.5,0.5,\ldots,0.5)$$

That is,

$$C([(1,0,0)\ (1,1,1)]\ ) \not\geq C(1,0,0).$$

Theorem 7.1 *Let C be an m-polar fuzzy Lie subalgebra in a Lie algebra L. Then, C is an m-polar fuzzy Lie subalgebra of L if and only if the nonempty upper s-level cut $C_{[s]} = U(C,s) = \{x \in L \mid C(x) \geq s\}$ is a Lie subalgebra of L, for all $s \in [0,1]^m$.*

Example 7.4 Consider the group algebra $\mathbb{C}[S_3]$, where S_3 is the symmetric group. Then, $\mathbb{C}[S_3]$ assumes the structure of a Lie algebra via the bracket (commutator) operation.

Clearly, the linear span of the elements $\hat{g} = g - g^{-1}$ for $g \in S_3$ is the subalgebra of $\mathbb{C}[S_3]$, which is also known as Plesken Lie algebra and denoted by $L(S_3)_{\mathbb{C}}$. It is easy to see that $L(S_3)_{\mathbb{C}} = Span_{\mathbb{C}}\{\widehat{(1,2,3)}\}$ and $\widehat{(1,2,3)} = (1,2,3) - (1,3,2)$.

We define an m-polar fuzzy set $C : L(S_3)_{\mathbb{C}} \to [0,1]^m$ by

$$C(g) = \begin{cases} (t_1, t_2, \ldots, t_m), & g = \gamma(1,2,3) - \gamma(1,3,2), \text{ where } \gamma \in \mathbb{C}, g \in \mathbb{C}[S_3] \\ (s_1, s_2, \ldots, s_m), & \text{otherwise , where } s_i < t_i \end{cases}$$

By routine calculations, we have $\{g \in \mathbb{C}[S_3] : C(g) > (s_1, s_2, \ldots, s_m)\} = L(S_3)_{\mathbb{C}}$. Then, we see that $L(S_3)_{\mathbb{C}}$ can be realized $C_{[s]}$ as an upper s_i-level cut and C is an m-polar fuzzy Lie ideal of $L(S_3)_{\mathbb{C}}$.

Definition 7.6 Let C and D be two m-polar fuzzy sets of L. We define the *sup-inf-product* $[CD]$ of C and D as follows: for all $x, y, z \in L$

$$[CD](x) = \begin{cases} \sup_{x=[yz]}\{\inf(C(y), D(z))\} \\ \mathbf{0}, \text{ if x} \neq \text{[yz].} \end{cases}$$

Let C and D be m-polar fuzzy Lie subalgebras of the Lie algebra L. Then, $[CD]$ may not be an m-polar fuzzy Lie subalgebra of L as this can be seen in the following counterexample:

Example 7.5 Let $\{e_1, e_2, \ldots, e_8\}$ be a basis of a vector space over a field $\mathbb{F}$. Then, it is not difficult to see that, by putting:

$$[e_1, e_2] = e_5, \quad [e_1, e_3] = e_6, \quad [e_1, e_4] = e_7, \quad [e_1, e_5] = -e_8,$$

$$[e_2, e_3] = e_8, \quad [e_2, e_4] = e_6, \quad [e_2, e_6] = -e_7, \quad [e_3, e_4] = -e_5,$$

$$[e_3, e_5] = -e_7, \quad [e_4, e_6] = -e_8, \quad [e_i, e_j] = -[e_j, e_i]$$

and $[e_i, e_j] = 0$ for all $i \leq j$, we can obtain a Lie algebra over a field $\mathbb{F}$. The following fuzzy sets

$$C(x) := \begin{cases} (1, 1, \ldots, 1) \text{ if } x \in \{0, e_1, e_5, e_6, e_7, e_8\}, \\ (0, 0, \ldots, 0) \text{ otherwise,} \end{cases}$$

$$D(x) := \begin{cases} (1, 1, \ldots, 1) & \text{if } x = 0, \\ (0.5, 0.5, \ldots, 0.5) & \text{if } x \in \{e_2, e_5, e_6, e_7, e_8\}, \\ (0, 0, \ldots, 0) & \text{otherwise,} \end{cases}$$

are clearly fuzzy Lie subalgebras of a Lie algebra L. Thus, C and D are m-polar fuzzy Lie subalgebras of L because the level Lie subalgebras

$$U(C; (1, 1, \ldots, 1)) =< e_1, e_5, e_6, e_7, e_8 >,$$

$$U(D; (0.5, 0.5, \ldots, 0.5)) =< e_2, e_5, e_6, e_7, e_8 >$$

are Lie subalgebras of L. But $[CD]$ is not an m-polar fuzzy Lie subalgebra because the following condition does not hold:

$$[CD](e_7 + e_8) \geq \inf\{[CD](e_7), [CD](e_8)\}.$$

$$(1) \quad [CD](e_7) = \sup \begin{cases} \inf\{C(e_1), D(e_4)\} = (0, 0, \ldots, 0), & e_7 = [e_1, e_4], \\ \inf\{C(e_2), D(e_6)\} = (0, 0, \ldots, 0), & e_7 = -[e_2, e_6], \\ \inf\{C(e_3), D(e_5)\} = (0, 0, \ldots, 0), & e_7 = -[e_3, e_5], \\ \inf\{C(e_4), D(e_1)\} = (0, 0, \ldots, 0), & e_7 = -[e_4, e_1], \\ \inf\{C(e_6), D(e_2)\} = (0.5, 0.5, \ldots, 0.5), & e_7 = [e_6, e_2], \\ \inf\{C(e_5), D(e_3)\} = (0, 0, \ldots, 0), & e_7 = [e_5, e_3]. \end{cases}$$

Thus, $[CD](e_7) = (0.5, 0.5, \ldots, 0.5)$.
(2) By using similar arguments, we can show that $[CD](e_8) = (0.5, 0.5, \ldots, 0.5)$.
(3) $[CD](e_7 + e_8) = \sup\{(i) - (vi)\}$
(i) if $e_7 + e_8 = [e_1(e_4 - e_5)]$, then $\inf\{C(e_1), D(e_4 - e_5)\} = \inf\{C(e_1), D(e_4), D(e_5)\} = (0, 0, \ldots, 0)$, since $D(e_4) = (0, 0, \ldots, 0)$, and if $e_7 + e_8 = [(e_5 - e_4)e_1]$, then, $\inf\{C(e_5 - e_4), D(e_1)\} = \inf\{C(e_5), D(e_4), D(e_1)\} = (0, 0, \ldots, 0)$, since $C(e_4) = (0, 0, \ldots, 0)$. By using similar method, we can also obtain the following numerical results:
(ii) If $e_7 + e_8 = [e_2(e_3 - e_6)]$, then $\inf(C(e_2), D(e_3 - e_6)) = (0, 0, \ldots, 0)$.
(iii) If $e_7 + e_8 = [e_3(-e_2 - e_5)]$, then $\inf(C(e_3), D(e_2 - e_5)) = (0, 0, \ldots, 0)$.
(iv) If $e_7 + e_8 = [e_4(-e_1 - e_6)]$, then $\inf(C(e_4), D(-e_3 - e_1)) = (0, 0, \ldots, 0)$.
(v) If $e_7 + e_8 = [e_5(-e_3 - e_1)]$, then $\inf(C(e_5), D(-e_3 - e_1)) = (0, 0, \ldots, 0)$.
(vi) If $e_7 + e_8 = [e_6(-e_2 - e_4)]$, then $\inf(C(e_6), D(-e_2 - e_4)) = (0, 0, \ldots, 0)$.

Thus, $[CD](e_7 + e_8) =$

$\sup\{(0, 0, \ldots, 0), (0, 0, \ldots, 0), (0, 0, \ldots, 0), (0, 0, \ldots, 0), (0, 0, \ldots, 0), (0, 0, \ldots, 0)\} = (0, 0, \ldots, 0).$

Hence, we have proved that

$$[CD](e_7 + e_8) \ngeq \inf\{[CD](e_7), [CD](e_8)\}.$$

We now refine the product of two m-polar fuzzy Lie subalgebras C and D of L to an extended form.

Definition 7.7 Let C and D be two m-polar fuzzy sets of L. Then, we define the *sup-inf-product* $\ll CD \gg$ of C and D as follows, for all $x, y, z \in L$

$$\ll CD \gg (x) = \begin{cases} \sup_{x=\sum_{i=1}^{n}[x_i y_i]}\{\inf_{i\in\mathbb{N}}\{\inf(C(x_i), D(y_i))\}\} \\ \mathbf{0}, \ \text{if } x \neq \sum_{i=1}^{n}[x_i y_i]. \end{cases}$$

From the definitions of $[CD]$ and $\ll CD \gg$, we can easily see that $[CD] \subseteq \ll CD \gg$ and $[CD] \neq \ll CD \gg$ hold generally even if C and D are both m-polar fuzzy Lie subalgebras of L, and in this case, $\ll CD \gg$ is also an m-polar fuzzy Lie subalgebra of L.

Theorem 7.2 *Let C be an m-polar fuzzy Lie subalgebra of Lie algebra L. Define a binary relation $\sim$ on L by $x \sim y$ if and only if $C(x - y) = C(0)$ for all $x, y \in L$. Then $\sim$ is a congruence relation on L.*

Proof We first prove that "$\sim$" is an equivalent relation. We only need to show the transitivity of "$\sim$" because the reflectivity and symmetricity of "$\sim$" hold trivially. Let $x, y, z \in L$. If $x \sim y$ and $y \sim z$, then $C(x - y) = C(0)$, $C(y - z) = C(0)$. Hence, it follows that

$$C(x - z) = C(x - y + y - z) \geq \inf(C(x - y), C(y - z)) = C(0).$$

Consequently, $x \sim z$. We now verify that "$\sim$" is a congruence relation on L. For this purpose, we let $x \sim y$ and $y \sim z$. Then, $C(x - y) = C(0)$, $C(y - z) = C(0)$. Now, for $x_1, x_2, y_1, y_2 \in L$, we have

$$\begin{aligned} C((x_1 + x_2) - (y_1 + y_2)) &= C((x_1 - y_1) + (x_2 - y_2)) \\ &\geq \inf(C(x_1 - y_1), C(x_2 - y_2)) = C(0), \\ C(\alpha x_1 - \alpha y_1) &= C(\alpha(x_1 - y_1)) \geq C(x_1 - y_1) = C(0), \\ C([x_1, x_2] - [y_1, y_2]) &= C([x_1 - y_1], [x_2 - y_2]) \\ &\geq \inf\{C(x_1 - y_1), C(x_2 - y_2)\} = C(0). \end{aligned}$$

That is, $x_1 + x_2 \sim y_1 + y_2$, $\alpha x_1 \sim \alpha y_1$ and $[x_1, x_2] \sim [y_1, y_2]$. Thus, "$\sim$" is indeed a congruence relation on L.

Definition 7.8 Let C be an m-polar fuzzy set on a set L. An *m-polar fuzzy relation* on C is an m-polar fuzzy set D of $L \times L$ such that $D(xy) \leq \inf(C(x), C(y))\, \forall x, y \in L$.

Definition 7.9 Let C and D be m-polar fuzzy sets on a set L. If C is an m-polar fuzzy relation on a set L, then C is said to be an *m-polar fuzzy relation* on D if $C(x, y) \leq \inf(D(x), D(y))$ for all $x, y \in L$.

Theorem 7.3 *Let C and D be two m-polar fuzzy Lie subalgebras of a Lie algebra L. Then, $C \times D$ is an m-polar fuzzy Lie subalgebra of $L \times L$.*

Proof Let $x = (x_1, x_2)$ and $y = (y_1, y_2) \in L \times L$. Then

$$\begin{aligned}
(C \times D)(x + y) &= (C \times D)((x_1, x_2) + (y_1, y_2))\\
&= (C \times D)(x_1 + y_1, x_2 + y_2)\\
&= \inf(C(x_1 + y_1), D(x_2 + y_2))\\
&\geq \inf(\inf(C(x_1), C(y_1)), T(D(x_2), D(y_2)))\\
&= \inf(\inf(C(x_1), D(x_2)), \inf(C(y_1), D(y_2)))\\
&= \inf((C \times D)(x_1, x_2), (C \times D)(y_1, y_2))\\
&= \inf((C \times D)(x), (C \times D)(y)),
\end{aligned}$$

$$\begin{aligned}
(C \times D)(\alpha x) &= (C \times D)(\alpha(x_1, x_2)) = (C \times D)(\alpha x_1, \alpha x_2)\\
&= \inf(C(\alpha x_1), D(\alpha x_2)) \geq \inf(C(x_1), D(x_2))\\
&= (C \times D)(x_1, x_2) = (C \times D)(x),
\end{aligned}$$

$$\begin{aligned}
(C \times D)([x, y]) &= (C \times D)([(x_1, x_2), (y_1, y_2)])\\
&\geq \inf(\inf(C(x_1), D(x_2)), \inf(C(y_1), D(y_2)))\\
&= \inf((C \times D)(x_1, x_2), (C \times D)(y_1, y_2))\\
&= \inf((C \times D)(x), (C \times D)(y)).
\end{aligned}$$

This shows that $C \times D$ is an m-polar fuzzy Lie subalgebra of $L \times L$.

Definition 7.10 Let L_1 and L_2 be two Lie algebras over a field $\mathbb{F}$. Then, a linear transformation $f : L_1 \to L_2$ is called a *Lie homomorphism* if $f([x, y]) = [f(x), f(y)]$ holds for all $x, y \in L_1$.

For the Lie algebras L_1 and L_2, it can be easily observed that if $f : L_1 \to L_2$ is a Lie homomorphism and C is an m-polar fuzzy Lie subalgebra of L_2, then the m-polar fuzzy set $f^{-1}(C)$ of L_1 is also an m-polar fuzzy Lie subalgebra.

Definition 7.11 Let L_1 and L_2 be two Lie algebras. Then, a Lie homomorphism $f : L_1 \to L_2$ is said to have a natural extension $f : J^{L_1} \to J^{L_2}$ defined by for all $C \in J^{L_1}, y \in L_2$:

$$f(C)(y) = \sup\{C(x) : x \in f^{-1}(y)\}.$$

We now call these sets the homomorphic images of the m-polar fuzzy set C.

Theorem 7.4 *The homomorphic image of an m-polar fuzzy Lie subalgebra is still an m-polar fuzzy Lie subalgebra of its codomain.*

Proof Let $y_1, y_2 \in L_2$. Then

$$\{x \mid x \in f^{-1}(y_1 + y_2)\} \supseteq \{x_1 + x_2 \mid x_1 \in f^{-1}(y_1)\ and, x_2 \in f^{-1}(y_2)\}.$$

Now, we have

$$\begin{aligned} f(C)(y_1 + y_2) &= \sup\{C(x) \mid x \in f^{-1}(y_1 + y_2)\} \\ &\geq \{C(x_1 + x_2), \mid x_1 \in f^{-1}(y_1), and, x_2 \in f^{-1}(y_2)\} \\ &\geq \sup\{\inf\{C(x_1), C(x_2)\}, \mid x_1 \in f^{-1}(y_1), and, x_2 \in f^{-1}(y_2)\} \\ &= \inf\{\sup\{C(x_1), \mid x_1 \in f^{-1}(y_1)\}, \sup\{C(x_2), \mid x_2 \in f^{-1}(y_2)\}\} \\ &= \inf\{f(C)(y_1), f(C)(y_2)\}. \end{aligned}$$

For $y \in L_2$ and $\alpha \in \mathbb{F}$, we have

$$\{x \mid x \in f^{-1}(\alpha y)\} \supseteq \{\alpha x, \mid x \in f^{-1}(y)\}.$$

$$\begin{aligned} f(C)(\alpha y) &= \sup\{C(\alpha x) \mid x \in f^{-1}(y)\} \\ &\geq \{C(\alpha x), \mid x \in f^{-1}(\alpha y)\} \\ &\geq \sup\{C(x), \mid x \in f^{-1}(y)\} \\ &= f(C)(y), \end{aligned}$$

If $y_1, y_2 \in L_2$, then

$$\{x \mid x \in f^{-1}([y_1, y_2])\} \supseteq \{[x_1, x_2] | x_1 \in f^{-1}(y_1),,\ x_2 \in f^{-1}(y_2)\}.$$

Now

$$\begin{aligned} f(C)([y_1, y_2]) &= \sup\{C(x) \mid x \in f^{-1}([y_1, y_2])\} \\ &\geq \{C([x_1, x_2]), \mid x_1 \in f^{-1}(y_1), and, x_2 \in f^{-1}(y_2)\} \\ &\geq \sup\{\inf\{C(x_1), C(x_2)\}, \mid x_1 \in f^{-1}(y_1), and, x_2 \in f^{-1}(y_2)\} \\ &= \inf\{\sup\{C(x_1), \mid x_1 \in f^{-1}(y_1)\}, \sup\{C(x_2), \mid x_2 \in f^{-1}(y_2)\}\} \\ &= \inf\{f(C)(y_1), f(C)(y_2)\}. \end{aligned}$$

Thus, $f(C)$ is a fuzzy Lie algebra of L_2.

Theorem 7.5 *Let $f : L_1 \to L_2$ be a surjective Lie homomorphism. If C and D are m-polar fuzzy Lie subalgebras of L_1, then $f(\ll CD \gg) = \ll f(C)f(D) \gg$.*

Proof Assume that $f(\ll CD \gg) < \ \ll f(C)f(D) \gg$. Now, we choose a number $t \in [0, 1]$ such that $f(\ll CD \gg)(x) \ < t < \ \ll f(C)f(D) \gg (x)$. Then, there exist $y_i, z_i \in L_2$ such that $x = \sum_{i=1}^{n}[y_i z_i]$ with $f(C)(y_i) > t$ and $f(D)(z_i) > t$. Since f is surjective, there exists $y \in L_1$ such that $f(y) = x$ and $y = \sum_{i=1}^{n}[a_i b_i]$ for some $a_i \in f^{-1}(y_i)$, $b_i \in f^{-1}(z_i)$ with $f(a_i) = y_i$, $f(b_i) = z_i$, $C(a_i) > t$, and $D(b_i) > t$. Since

$$f(\sum_{i=1}^{n}[a_i b_i]) = \sum_{i=1}^{n} f([a_i b_i]) = \sum_{i=1}^{n}[f(a_i) f(b_i)] = \sum_{i=1}^{n}[y_i z_i] = x,$$

$f(\ll CD \gg)(x) > t$. This is a contradiction. Similarly, for the case $f(\ll CD \gg) > \ll f(C)f(D) \gg$, we can also obtain a contradiction. Hence, $f(\ll CD \gg) = \ll f(C)f(D) \gg$.

Definition 7.12 Let C and D be m-polar fuzzy subalgebras of L. Then, C is said to be of the same type of D if there exists $f \in Aut(L)$ such that $C = D \circ f$, i.e., $C(x) = D(f(x))$ for all $x \in L$.

Theorem 7.6 *Let C and D be two m-polar fuzzy subalgebras of L. Then, C is an m-polar fuzzy subalgebra having the same type of D if and only if C is isomorphic to D.*

Proof We only need to prove the necessity part because the sufficiency part is trivial. Let C be an m-polar fuzzy subalgebra having the same type of D. Then, there exists $\phi \in Aut(L)$ such that

$$C(x) = D(\phi(x)) \ \forall x \in L.$$

Let $f : C(L) \to D(L)$ be a mapping defined by $f(C(x)) = D(\phi(x))$ for all $x \in L$, that is,

$$f(C(x)) = D(\phi(x)) \ \forall x \in L.$$

Then, it is clear that f is surjective. Also, f is injective because if $f(C(x)) = f(C(y))$ for all $x, y \in L$, then $D(\phi(x)) = D(\phi(y))$ and hence $C(x) = D(y)$. Finally, f is a homomorphism because for $x, y \in L$,

$$f(C(x + y)) = D(\phi(x + y)) = D(\phi(x) + \phi(y)),$$

$$f(C(\alpha x)) = D(\phi(\alpha x)) = \alpha D(\phi(x)),$$

$$f(C([x, y])) = D(\phi([x, y])) = D([\phi(x), \phi(y)]\ ).$$

Hence, C is isomorphic to D. This completes the proof.

7.3 m-Polar Fuzzy Lie Ideals

Definition 7.13 Let L be a Lie algebra. An m-polar fuzzy set C on L is called an *m-polar fuzzy Lie ideal* if the following conditions are satisfied:

(1) $C(x+y) \geq \inf(C(x), C(y))$,
(2) $C(\alpha x) \geq C(x)$,
(3) $C([x, y]) \geq C(x)$ for all $x, y \in L$ and $\alpha \in \mathbb{F}$.

That is,

(1) $p_i \circ C(x+y) \geq \inf(p_i \circ C(x), p_i \circ C(y))$,
(2) $p_i \circ C(\alpha x) \geq p_i \circ C(x)$,
(3) $p_i \circ C([x, y]) \geq p_i \circ C(x)$

for all $x, y \in L$ and $\alpha \in \mathbb{F}$, $i = 1, 2, 3, \ldots, m$.

Example 7.6 Let $\Re^3 = \{(x, y, z) : x, y, z \in \mathbb{R}\}$ be the set of all three-dimensional real vectors. Then, $\Re^3$ with the bracket $[\cdot, \cdot]$ defined as the usual cross product, i.e., $[x, y] = x \times y$, forms a real Lie algebra. We also define an m-polar fuzzy set $C : \Re^3 \to [0, 1]^m$ by

$$C(x, y, z) = \begin{cases} (0.8, 0.8, \ldots, 0.8) \text{ if } x = y = z = 0, \\ (0.1, 0.1, \ldots, 0.1) \text{ otherwise.} \end{cases}$$

By routine computations, we can verify that the above m-polar fuzzy set C is an m-polar fuzzy Lie ideal of the Lie algebra $\Re^3$.

Example 7.7 A subalgebra $sl_2(\mathbb{C})$ of all 2×2 matrices with trace 0 is an ideal of $gl_2(\mathbb{C})$. The bases of $sl_2(\mathbb{C})$ are: $h = \begin{pmatrix} 0 & 1 \\ 0 & 0 \end{pmatrix}$, $f = \begin{pmatrix} 0 & 0 \\ 1 & 0 \end{pmatrix}$, and $e = \begin{pmatrix} 1 & 0 \\ 0 & -1 \end{pmatrix}$. The commutators are $[e, f] = h$, $[h, f] = -2f$, and $[h, e] = 2e$.

We define an m-polar fuzzy set $C : gl_2(\mathbb{C}) \to [0, 1]^m$ by

$$C(g) = \begin{cases} (1, 1, \ldots, 1), & g \in sl_n(\mathbb{C}) \\ (0, 0, \ldots, 0), & \text{otherwise.} \end{cases}$$

By routine computations, we see that C is an m-polar fuzzy ideal.

We state the following theorem without its proof.

Theorem 7.7 *Let C be an m-polar fuzzy Lie ideal in a Lie algebra L. Then, C is an m-polar fuzzy Lie ideal of L if and only if the nonempty upper s-level cut $C_{[s]} = \{x \in L \mid C(x) \geq s\}$ is Lie ideal of L, for all $s = (s_1, s-2, \ldots, s_m) \in [0, 1]^m$.*

Example 7.8 Consider the group algebra $\mathbb{C}[S_3]$, where S_3 is the symmetric group. Then, $\mathbb{C}[S_3]$ assumes the structure of a Lie algebra via the bracket (commutator) operation.

Clearly, the linear span of the elements $\hat{g} = g - g^{-1}$ for $g \in S_3$ is the subalgebra of $\mathbb{C}[S_3]$, which is also known as Plesken Lie algebra and denoted by $L(S_3)_{\mathbb{C}}$. It is easy to see that $L(S_3)_{\mathbb{C}} = Span_{\mathbb{C}}\{\widehat{(1,2,3)}\}$ and $\widehat{(1,2,3)} = (1,2,3) - (1,3,2)$.

We define an m-polar fuzzy set $C : L(S_3)_{\mathbb{C}} \to [0,1]^m$ by

$$C(g) = \begin{cases} (t_1, t_2, \ldots, t_m), & g = \gamma(1,2,3) - \gamma(1,3,2), \text{ where } \gamma \in \mathbb{C}, g \in \mathbb{C}[S_3] \\ (s_1, s_2, \ldots, s_m), & \text{otherwise , where } s_i < t_i \end{cases}$$

By routine calculations, we have $\{g \in \mathbb{C}[S_3] : C(g) > (s_1, s_2, \ldots, s_m)\} = L(S_3)_{\mathbb{C}}$. Then, we see that $L(S_3)_{\mathbb{C}}$ can be realized $C_{[s]}$ as an upper s_i-level cut and C is an m-polar fuzzy Lie ideal of $L(S_3)_{\mathbb{C}}$.

Definition 7.14 Let $C \in J^L$, and an m-polar fuzzy subspace of L generated by C will be denoted by $[C]$. It is the intersection of all m-polar fuzzy subspaces of L containing C. For all $x \in L$, we define:

$$[C](x) = \sup\left\{\inf C(x_i) \mid x = \sum \alpha_i x_i, \alpha_i \mathbb{F}, x_i \in L\right\}.$$

Definition 7.15 Let $f : L_1 \to L_2$ be a homomorphism of Lie algebras which has an extension $f : J^{L_1} \to J^{L_2}$ defined by:

$$f(C)(y) = \sup\{C(x), x \in f^{-1}(y)\}.$$

for all $C \in J^{L_1}$, $y \in L_2$. Then, $f(C)$ is called the *homomorphic image* of C.

Proposition 7.2 *Let $f : L_1 \to L_2$ be a homomorphism of Lie algebras, and let C be an m-polar fuzzy Lie ideal of L_1.*

Then, (i) *$f(C)$ is an m-polar fuzzy Lie ideal of L_2,*
(ii) $f([C]) \supseteq [f(C)]$.

Proposition 7.3 *If C and D are m-polar fuzzy Lie ideals in L, then $[C, D]$ is an m-polar fuzzy Lie ideal of L.*

Theorem 7.8 *Let C_1, C_2, D_1, D_2 be m-polar fuzzy Lie ideals in L such that $C_1 \subseteq C_2$ and $D_1 \subseteq D_2$, then $[C_1, D_1] \subseteq [C_2, D_2]$.*

Proof Indeed,

$$\begin{aligned} \ll C_1, D_1 \gg (x) &= \sup\{\inf(C_1(a), D_1(b)) \mid a, b \in L_1, [a,b] = x\} \\ &\geqslant \sup\{\inf(C_2(a), D_2(b)) \mid a, b \in L_1, [a,b] = x\} \\ &= \ll C_2, D_2 \gg (x). \end{aligned}$$

Hence $[C_1, D_1] \subseteq [C_2, D_2]$.

Let C be an m-polar fuzzy Lie ideal in L. Putting

$$C^0 = C,\ C^1 = [C, C_0],\ C^2 = [C, C_1],\ \ldots,\ C^n = [C, C^{n-1}]$$

we obtain a descending series of an m-polar fuzzy Lie ideals

$$C^0 \supseteq C^1 \supseteq C^2 \supseteq \cdots \supseteq C^n \supseteq \cdots$$

and a series of m-polar fuzzy sets D^n such that

$$D^n = \sup\{C^n(x) \mid 0 \neq x \in L\}.$$

Definition 7.16 An m-polar fuzzy Lie ideal C is called *nilpotent* if there exists a positive integer n such that $C^n = \mathbf{0}$.

Theorem 7.9 *A homomorphic image of a nilpotent m-polar fuzzy Lie ideal is a nilpotent m-polar fuzzy Lie ideal.*

Proof Let $f : L_1 \to L_2$ be a homomorphism of Lie algebras, and let C be a nilpotent m-polar fuzzy Lie ideal in L_1. Assume that $f(C) = D$. We prove by induction that $f(C^n) \supseteq D^n$ for every natural n. First, we claim that $f([C, C]) \supseteq [f(C), f(C)] = [D, D]$. Let $y \in L_2$, then

$$\begin{aligned}
f(\ll C, C \gg)(y) &= \sup\{\ll C, C \gg)(x) \mid f(x) = y\} \\
&= \sup\{\sup\{\inf(C(a), C(b)) \mid a, b \in L_1, [a, b] = x, f(x) = y\}\} \\
&= \sup\{\inf(C(a), C(b)) \mid a, b \in L_1, [a, b] = x, f(x) = y\} \\
&= \sup\{\inf(C(a), C(b)) \mid a, b \in L_1, [f(a), f(b)] = y\} \\
&= \sup\{\inf(C(a), C(b)) \mid a, b \in L_1, f(a) = u, f(b)] = v, [u, v] = y\} \\
&\geqslant \sup\left\{\inf\left(\sup_{a\in f^{-1}(u)} C(a), \sup_{b\in f^{-1}(v)} C(b)\right) \mid [u, v] = y\right\} \\
&= \sup\{\inf(f(C)(u), f(C)(v)) \mid [u, v] = y\} = \ll f(C), f(C) \gg (y),
\end{aligned}$$

Thus

$$f([C, C]) \supseteq f(\ll C, C \gg) \supseteq \ll f(C), f(C) \gg = [f(C), f(C)]\,.$$

For $n > 1$, we get

$$f(C^n) = f([C, C^{n-1}]) \supseteq [f(C), f(C^{n-1})] \supseteq [D, D^{n-1}] = D^n.$$

Let m be a positive integer such that $C^m = \mathbf{0}$. Then, for $0 \neq y \in L_2$ we have

$$D^m(y) \leqslant f(\mu^P_{C^n})(y) = f(0)(y) = \sup\{0(a) \mid f(a) = y\} = \mathbf{0}.$$

This completes the proof.

Let C be an m-polar fuzzy Lie ideal in L. Putting

$$C^{(0)} = C,\ C^{(1)} = [C^{(0)}, C^{(0)}],\ C^{(2)} = [C^{(1)}, C^{(1)}], \ldots, C^{(n)} = [C^{(n-1)}, C^{(n-1)}]$$

we obtain series

$$C^{(0)} \subseteq C^{(1)} \subseteq C^{(2)} \subseteq \cdots \subseteq C^{(n)} \subseteq \cdots$$

of m-polar fuzzy Lie ideals and a series of m-polar fuzzy sets $D^{(n)}$ such that

$$D^n = \sup\{C^n(x) \mid 0 \neq x \in L\}.$$

Definition 7.17 An m-polar fuzzy Lie ideal C is called *solvable* if there exists a positive integer n such that $D^{(n)} = \mathbf{0}$.

Proposition 7.4 *A nilpotent m-polar fuzzy Lie ideal is solvable.*

Proof It is enough to prove that $C^{(n)} \subseteq C^n$ for all positive integers n. We prove it by induction on n and by the use of Theorem 7.8:

$$C^{(1)} = [C, C] = C^1, \quad C^{(2)} = [C^{(1)}, C^{(1)}] \subseteq [C, C^{(1)}] = C^2.$$

$$C^{(n)} = [C^{(n-1)}, C^{(n-1)}] \subseteq [C, C^{(n-1)}] \subseteq [C, C^{(n-1)}] = C^n.$$

This completes the proof.

Definition 7.18 Let C and D be two m-polar fuzzy Lie ideals of a Lie algebra L. The sum $C \oplus D$ is called a *direct sum* if $C \cap D = \mathbf{0}$.

Theorem 7.10 *The direct sum of two nilpotent m-polar fuzzy Lie ideals is also a nilpotent m-polar fuzzy Lie ideal.*

Proof Suppose that C and D are two m-polar fuzzy Lie ideals such that $C \cap D = \mathbf{0}$. We claim that $[C, D] = \mathbf{0}$. Let $x(\neq 0) \in L$, then

$$\ll C, D \gg (x) = \sup\{\inf(C(a), D(b)) \mid [a, b] = x\} \leqslant \inf(C(x), D(x)) = \mathbf{0}.$$

This proves our claim. Thus, we obtain $[C^m, D^n] = \mathbf{0}$ for all positive integers m, n. Now, we again claim that $(C \oplus D)^n \subseteq C^n \oplus D^n$ for positive integer n. We prove this claim by induction on n. For $n = 1$,

$$(C \oplus D)^1 = [C \oplus D, C \oplus D] \subseteq [C, C] \oplus [C, D] \oplus [D, C] \oplus [D, D] = C^1 \oplus D^1.$$

Now for $n > 1$,

$$\begin{aligned}(C \oplus D)^n &= [C \oplus D, (C \oplus D)^{n-1}] \subseteq [C \oplus D, C^{n-1} \oplus D^{n-1}] \\ &\subseteq [C, C^{n-1}] \oplus [C, D^{n-1}] \oplus [D, C^{n-1}] \oplus [D, D^{n-1}] = C^n \oplus D^n.\end{aligned}$$

Since there are two positive integers p and q such that $C^p = D^q = \mathbf{0}$, we have $(C \oplus D)^{p+q} \subseteq C^{p+q} \oplus D^{p+q} = \mathbf{0}$.

In a similar way, we can prove the following theorem.

Theorem 7.11 *The direct sum of two solvable m-polar fuzzy Lie ideals is a solvable m-polar fuzzy Lie ideal.*

Theorem 7.12 *Let C be an m-polar fuzzy Lie ideal in a Lie algebra L. Then, $C^n \subseteq [C_n]$ for any $n > 0$, where an m-polar fuzzy subset $[C_n]$ is defined by*

$$[C_n](x) = \sup\{C(a) \mid [x_1, [x_2, [\ldots, [x_n, a]\ldots]]] = x, \quad x_1, \ldots, x_n \in L\}.$$

Proof It is enough to prove that $\ll C, C^{n-1} \gg \subseteq [C_n]$. We prove it by induction on n. For $n = 1$ and $x \in L$, we have

$$\begin{aligned} \ll C, C \gg (x) &= \sup\{\inf(C(a), C(b)) \mid [a, b] = x\} \\ &\geqslant \sup\{C(b) \mid [a, b] = x, a \in L\} = [C_1](x). \end{aligned}$$

For $n > 1$,

$$\begin{aligned} &\ll C, C^{(n-1)} \gg (x) = \sup\{\inf(C(a), C^{(n-1)}(b)) \mid [a, b] = x\} \\ &= \sup\{\inf(C(a), [C(b), C^{(n-2)}(b)]\,) \mid [a, b] = x\} \\ &\geqslant \sup\{\inf(C(a), \sup\{\ll C, C^{(n-2)} \gg (b_i) \mid b = \textstyle\sum \alpha_i b_i\}) \mid [a, b] = x\} \\ &\geqslant \sup\{\inf(C(a), \sup\{[C_{n-1}](b_i) \mid b = \textstyle\sum \alpha_i b_i\}) \mid [a, b] = x\} \\ &\geqslant \sup\{\inf(C(a), [C_{n-1}](b_i)) \mid \textstyle\sum \alpha_i [a, b_i] = x\} \\ &\geqslant \sup\{\inf(C(a), \sup\{C_{n-1}(c_i) \mid b_i = \textstyle\sum \beta_i c_i\}) \mid \textstyle\sum \alpha_i [a, b_i] = x\} \\ &\geqslant \sup\{\inf(C(a), C_{n-1}(c_i)) \mid \textstyle\sum \gamma_i [a, c_i] = x\} \\ &\geqslant \sup\{\inf(C(a), \sup\{C(d_i)) \mid [x_1, [x_2, [\ldots, [x_{n-1}, d_i]\ldots]]] = c_i\} \mid \textstyle\sum \gamma_i [a, c_i] = x\} \\ &\geqslant \sup\{\inf(C(a), C(d_i)) \mid \textstyle\sum \gamma_i [a, [x_1, [x_2, [\ldots, [x_{n-1}, d_i]\ldots]]]] = x\} \\ &\geqslant \sup\{C_n(d_i) \mid \textstyle\sum \gamma_i [a, [x_1, [x_2, [\ldots, [x_{n-1}, d_i]\ldots]]]] = x\} \geqslant [C_n](x). \end{aligned}$$

This completes the proof.

Theorem 7.13 *If for an m-polar fuzzy Lie ideal C there exists a positive integer n such that*

$$(\overline{ad}x_1 \circ \overline{ad}x_2 \circ \cdots \circ \overline{ad}x_n)(C) = \mathbf{0}.$$

for all $x_1, \ldots, x_n \in L$, then C is nilpotent.

Proof For $x_1, \ldots, x_n \in L$ and $x(\neq 0) \in L$, we have

$$(\overline{ad}x_1 \circ \cdots \circ \overline{ad}x_n)(C)(x) = \sup\{C(a) \mid [x_1, [x_2, [\ldots, [x_n, a]\ldots]]] = x\} = \mathbf{0}.$$

Thus $[C_n] = \mathbf{0}$. From Theorem 7.12, it follows that $C^n = \mathbf{0}$. Hence, C is a nilpotent m-polar fuzzy Lie ideal.

The mapping $K : L \times L \to \mathbb{F}$ defined by $K(x, y) = Tr(adx \circ ady)$, where Tr is the *trace* of a linear homomorphism, is a symmetric bilinear form which is called the *Killing form*. It is not difficult to see that this form satisfies the identity $K([x, y], z) = K(x, [y, z])$. The form K can be naturally extended to $\overline{K} : J^{L\times L} \to J^{\mathbb{F}}$ defined by putting

$$\overline{K}(C)(\beta) = \sup\{C(x, y) \mid Tr(adx \circ ady) = \beta\}.$$

The Cartesian product of two m-polar fuzzy sets C and D is defined as

$$(C \times D)(x, y) = \inf(C(x), D(y)).$$

Similarly, we define

$$\overline{K}(C \times D)(\beta) = \sup\{\inf(C(x), D(y)) \mid Tr(adx \circ ady) = \beta\}.$$

Theorem 7.14 *Let C be an m-polar fuzzy Lie ideal of Lie algebra L. Then, $\overline{K}(C \times 1_{(\alpha x)}) = \alpha \odot \overline{K}(C \times 1_x)$ for all $x \in L$, $\alpha \in \mathbb{F}$.*

Proof If $\alpha = 0$, then for $\beta = 0$ we have

$$\begin{aligned}\overline{K}(C \times 1_0)(0) &= \sup\{\inf(C(x), 1_0(y)) \mid Tr(adx \circ ady) = 0\}\\ &\geqslant \inf(C(0), 1_0(0)) = \mathbf{0}.\end{aligned}$$

For $\beta \neq 0$ $Tr((adx \circ ady) = \beta$ means that $x \neq 0$ and $y \neq 0$. So,

$$\overline{K}(C \times 1_0)(\beta) = \sup\{\inf(C(x), 1_0(y)) \mid Tr((adx \circ ady) = \beta\} = \mathbf{0}.$$

If $\alpha \neq 0$, then for arbitrary β we obtain

$$\begin{aligned}\overline{K}(C \times 1_{\alpha x})(\beta) &= \sup\{\inf(C(y), 1_{\alpha x}(z)) \mid Tr(ady \circ adz) = \beta\}\\ &= \sup\{\inf(C(y), \alpha \odot 1_x(z)) \mid Tr(ady \circ adz) = \beta\}\\ &= \sup\{\inf(C(y), 1_x(\alpha^{-1}z)) \mid \alpha Tr(ady \circ ad(\alpha^{-1}z)) = \beta\}\\ &= \sup\{\inf(C(y), 1_x(\alpha^{-1}z)) \mid Tr(ady \circ ad(\alpha^{-1}z)) = \alpha^{-1}\beta\}\\ &= \overline{K}(C \times 1_x)(\alpha^{-1}\beta) = \alpha \odot \overline{K}(C \times 1_x)(\beta).\end{aligned}$$

This completes the proof.

Theorem 7.15 *Let C be an m-polar fuzzy Lie ideal of a Lie algebra L. Then, $\overline{K}(C \times 1_{(x+y)}) = \overline{K}(C \times 1_x) \oplus \overline{K}(C \times 1_y)$ and $\overline{K}(C \times 0_{(x+y)}) = \overline{K}(C \times 0_x) \oplus \overline{K}(C \times 0_y)$ for all $x, y \in L$.*

Proof Indeed,

$$\begin{aligned}\overline{K}(C \times 1_{(x+y)})(\beta) &= \sup\{\inf(C(z), 1_{x+y}(u)) \mid Tr(adz \circ adu) = \beta\}\\ &= \sup\{C(z) \mid Tr(adz \circ ad(x + y)) = \beta\}\\ &= \sup\{C(z) \mid Tr(adz \circ adx) + Tr(adz \circ ady) = \beta\}\end{aligned}$$

$$
\begin{aligned}
&= \sup\{\inf(C(z), \inf(1_x(v), 1_y(w))) \mid Tr(adz \circ adv) + Tr(adz \circ adw) = \beta\} \\
&= \sup\{\inf(\sup\{\inf(C(z), 1_x(v)) \mid Tr(adz \circ adv) = \beta_1\}, \\
&\qquad \sup\{\inf(C(z), 1_y(w)) \mid Tr(adz \circ adw) = \beta_2\} \mid \beta_1 + \beta_2 = \beta)\} \\
&= \sup\{\inf(\overline{K}(C \times 1_x)(\beta_1), \ \overline{K}(C \times 1_y)(\beta_2)) \mid \beta_1 + \beta_2 = \beta\} \\
&= \overline{K}(C \times 1_x) \oplus \overline{K}(C \times 1_y)(\beta).
\end{aligned}
$$

This completes the proof.

We conclude that:

Corollary 7.1 *For each m-polar fuzzy Lie ideal C and all $x, y \in L$, $\alpha, \beta \in \mathbb{F}$, we have*

$$\overline{K}(C \times 1_{(\alpha x + \beta y)}) = \alpha \odot \overline{K}(C \times 1_x) \oplus \beta \odot \overline{K}(C \times 1_y).$$

Chapter 8
Fuzzy Soft Lie Algebras

In this chapter, we present certain concepts, including soft intersection Lie algebras, fuzzy soft Lie algebras, $(\in_\alpha, \in_\alpha \vee q_\beta)$-fuzzy soft Lie subalgebras, bipolar fuzzy soft Lie algebras and $(\in, \in \vee q)$-bipolar fuzzy soft Lie algebras.

8.1 Introduction

8.1.1 Soft Sets

There are many real-life problems in various fields, including social sciences, physical sciences, and life sciences, that contain uncertain and vague data. A lot of mathematical theories, including probability theory, Zadeh's fuzzy set theory, and Pawlak's rough set theory, are very useful for the purpose of handling different types of uncertain data. Molodtsov [103] pointed out some drawbacks of these theories. To overcome these difficulties, he introduced the idea of soft sets. The theory of soft sets is playing a very important role in many fields, including data analysis, and decision-making. Maji et al. [98] introduced some fundamental algebraic operations for soft sets. Ali et al. [3] presented some new operations for soft sets. Based on the idea of parametrization, a soft set gives a series of approximate descriptions of a complicated object from various different aspects. Each approximate description has two parts, namely predicate a subset of the universe. More specifically, we can define the notion of soft set in the following way: Let X be the universe of discourse and E be the universe of all possible parameters related to the objects in X. Each parameter is a word or a sentence. In most cases, parameters are considered to be attributes, characteristics, or properties of objects in X. The pair (X, E) is also known as a *soft universe*. The power set of X is denoted by $P(X)$.

M. Akram, *Fuzzy Lie Algebras*, Infosys Science Foundation Series,
https://doi.org/10.1007/978-981-13-3221-0_8

Definition 8.1 A pair $F_A = (F, A)$ is called *soft set* over X, where $A \subseteq E$, F is a set-valued function $F : A \to P(X)$. In other words, a soft set over X is a parameterized family of subsets of X. For any $x \in A$, $F(x)$ may be considered as set of x-approximate elements of soft set (F, A). A soft set F_A over the universe X can be represented by the set of ordered pairs

$$F_A = \{(x, F(x)) \mid x \in E, F(x) \in P(X)\}.$$

By means of parametrization, a soft set produces a series of approximate descriptions of a complicated object being perceived from various points of view. It is apparent that a soft set $F_A = (F, A)$ over a universe X can be viewed as a parameterized family of subsets of X. For any parameter $\varepsilon \in A$, the subset $F(\varepsilon) \subseteq X$ may be interpreted as the set of *ε-approximate elements*.

Example 8.1 Let $X = \{1, 2, 3, \ldots, 10\}$ be a set of first ten positive integers and $E = \{e_1, e_2, e_3, e_4, e_5\}$ be the set of parameters, where
e_1 stands for the parameter "divisibility by 2"
e_2 stands for the parameter "divisibility by 3"
e_3 stands for the parameter "divisibility by 4"
e_4 stands for the parameter "divisibility by 5"
e_5 stands for the parameter "divisibility by prime numbers."

If $A = \{e_1, e_2, e_3, e_4\}$, then the soft set (F, A) is given by

$$S = \{F(e_1), F(e_2), F(e_3), F(e_4)\},$$

where

$$F(e_1) = \{2, 4, 6, 8, 10\}, F(e_2) = \{3, 6, 9\}, F(e_3) = \{4, 8\}, F(e_4) = \{5, 10\}.$$

Thus, the soft set (F, A) is a parameterized family of subsets of X. The tabular arrangement of the soft set (F, A) is (Table 8.1)

Table 8.1 Tabular arrangement of the soft set

Parameters	1	2	3	4	5	6	7	8	9	10
e_1	0	1	0	1	0	1	0	1	0	1
e_2	0	0	1	0	0	1	0	0	1	0
e_3	0	0	0	1	0	0	0	1	0	0
e_4	0	0	0	0	1	0	0	0	0	1

Example 8.2 Let $X = \{h_1, h_2, h_3, h_4, h_5, h_6\}$ be the set of houses under consideration, and let $E = \{e_1, e_2, e_3, e_4, e_5\}$ be the set of all parameters, where
e_1 stands for the parameter "expansive"

e_2 stands for the parameter "wooden"
e_3 stands for the parameter "low cost"
e_4 stands for the parameter "cheaper"
e_5 stands for the parameter "green surroundings."

Let $A = \{e_1, e_2, e_3\}$ be the set parameters for selection of a house, then the soft set (F, A) is given by

$$S = \{F(e_1), F(e_2), F(e_3)\},$$

where $F(e_1) = \{h_1, h_3\}$ = expansive houses
$F(e_2) = \{h_1, h_3, h_6\}$ = wooden houses
$F(e_3) = \{h_2, h_4\}$ = low-cost houses.
The tabular arrangement of the soft set (F, A) is (Table 8.2)

Table 8.2 Tabular arrangement of the soft set

Parameters	h_1	h_2	h_3	h_4	h_5	h_6
e_1	1	0	1	0	0	0
e_2	1	0	1	0	0	1
e_3	0	1	0	1	0	0

Example 8.3 Suppose a soft set (F, A) describes attractiveness of the shirts which the authors are going to wear. Here X = the set of all shirts under consideration $=\{x_1, x_2, x_3, x_4, x_5\}$ and A ={colorful, bright, cheap, warm} $= \{e_1, e_2, e_3, e_4\}$. $F(e_1) = \{x_1, x_5\}$, $F(e_2) = \{x_2, x_4\}$, $F(e_3) = \{x_2, x_5\}$ and $F(e_4) = \{x_1, x_2, x_5\}$. So, the soft set (F, A) is a subfamily $\{F(e_1), F(e_2), F(e_3), F(e_4)\}$ of $P(X)$, which represents the attractiveness of shirts w. r. t the given parameters.

Definition 8.2 Let F_A and G_B be two soft sets over a common universe X. F_A is a said to be *soft subset* of G_B, denoted by $F_A \tilde{\subseteq} G_B$, if $F(x) \subseteq G(x)$ for all $x \in E$.

8.1.2 Fuzzy Soft Sets

A soft set is a mapping from parameter to the crisp subset of universe. However, the situation may be more complicated in real world because of the fuzzy characters of the parameters. In fuzzy soft sets, the soft set theory is extended to a fuzzy one, and the fuzzy membership is used to describe parameter approximate elements of fuzzy soft set. Maji et al. [97] extended the idea of soft sets and introduced the hybrid model called fuzzy soft sets. By using this definition of fuzzy soft sets, many interesting applications of soft set theory have been expanded by some researchers. Roy and Maji [116] gave some applications of fuzzy soft sets. Som [122] defined soft relation and fuzzy soft relation on the theory of soft sets.

Definition 8.3 A pair (f, A) is called a *fuzzy soft set* over X, where f is a mapping given by $f : A \to \mathscr{P}(X)$, $\mathscr{P}(X) = \mathscr{I}^X$, $\mathscr{I} = [0, 1]$. In general, for every $\varepsilon \in A$, $f(\varepsilon) = f_\varepsilon$ is a fuzzy set of X and it is called *fuzzy value set* of parameter ε. The set of all fuzzy soft sets over X with parameters from E is called a *fuzzy soft class*, and it is denoted by $\mathscr{F}\mathscr{S}(X, E)$.

Definition 8.4 Let X be a universe and E a set of attributes. Then the pair (X, E) denotes the collection of all fuzzy soft sets on X with attributes from E and is called *fuzzy soft class*.

Example 8.4 Let $X = \{a, b, c, d\}$ be a set of houses under considerations. Let

$$E = \{\text{very costly, costly, beautiful, in green surrounding, cheap}\}$$

be a set of parameters. Take $A, B \subset E$ as $A = \{\text{very costly, costly, cheap}\}$ and $B = \{\text{beautiful, in green surrounding}\}$. Then
$f(\text{very costly}) = \{(\text{a}, 0.3), (\text{b}, 0.8), (\text{c}, 0.1), (\text{d}, 0.3)\}$
$f(\text{costly}) = \{(\text{a}, 0.5), (\text{b}, 1), (\text{c}, 0.2)(\text{d}, 0.1)\}$
$f(\text{cheap}) = \{(\text{a}, 0.5), (\text{b}, 0.4), (\text{c}, 0.3), (\text{d}, 0.1)\}$
Thus, fuzzy soft set (f,A) over X "describe cost of the houses."
$g(\text{beautiful}) = \{(\text{a}, 0.2), (\text{b}, 0.3), (\text{c}, 0.4), (, 0.5)\}$
$g(\text{in green surrounding}) = \{(\text{a}, 0.4), (\text{b}, 0.2), (\text{c}, 0.4), (\text{d}, 0.1)\}$.
Hence, fuzzy soft set (g, B) over X describes "attractiveness of the houses." Clearly, (f, A) and (g, B) are fuzzy soft sets in a fuzzy soft class (X, E).

Definition 8.5 A fuzzy soft set (f, A) over X is called a *null fuzzy soft set*, denoted by Φ, if for all $\varepsilon \in A$, $f(\varepsilon)$ is the null fuzzy set $\underline{0}$ of X, where $\underline{0}(x) = 0$ for all $x \in X$. A fuzzy soft set (g, A) over X is called a *whole (absolute) fuzzy soft set*, denoted by $\mathbb{X}$, if for all $\varepsilon \in A$, $g(\varepsilon)$ is the whole fuzzy set $\overline{1}$ of X, where $\overline{1}(x) = 1$ for all $x \in X$.

Example 8.5 Let $X = \{c_1, c_2, c_3, c_4\}$ be a set of four cars under consideration and $E = \{e_1, e_2, e_3\}$ be a set of parameters, where
e_1 denotes costly,
e_2 denotes beautiful,
e_3 denotes fuel efficient.
(i) Let $A = \{e_2, e_3\} \subset E$. Then
$f(e_2) = \{(c_1, 0.4), (c_2, 0.3), (c_3, 0.5), (c_4, 0.7)\}$
$f(e_3) = \{(c_1, 0.5), (c_2, 0.2), (c_3, 0.4), (c_4, 0.1)\}$
Hence, $(f, A) = \{(e_2, f(e_2)), (e_3, f(e_3))\}$ is a fuzzy soft set over universe X.
(ii) Let $B = \{e_1, e_3\} \subset E$. Then
$(g, B) = \{g(e_1) = \{(c_1, 0), (c_2, 0), (c_3, 0), (c_4, 0)\}, g(e_3) = \{(c_1, 0), (c_2, 0), (c_3, 0), (c_4, 0)\}\}$ is a null fuzzy soft set over X.
$(h, B) = \{h(e_1) = \{c_1, 1), (c_2, 1), (c_3, 1), (c_4, 1)\}, h(e_3) = \{c_1, 1), (c_2, 1), (c_3, 1), (c_4, 1)\}\}$ is an absolute fuzzy soft set.

Definition 8.6 Let (f, A) be a fuzzy soft set over X. For each $t \in [0, 1]$, the set $(f, A)^t = (f^t, A)$ is called a *t-level soft set* of (f, A), where $f^t_\varepsilon = \{x \in X \mid f_\varepsilon(x) \geq t\}$ for all $\varepsilon \in A$. Clearly, $(f, A)^t$ is a soft set over X.

Definition 8.7 Let f and g be any two fuzzy subsets of X. Then the *product* $f \circ g$ is a fuzzy subset of X defined by

$$(f \circ g)(z) = \begin{cases} \bigvee_{z=[x,y]}(f(x) \wedge g(y)) & \textit{if there exist } x,\ y \in X \textit{ such that } z = [x, y], \\ 0 & \textit{otherwise.} \end{cases}$$

Definition 8.8 Let f and g be any two fuzzy subsets of X. Then the *sum* $f + g$ is a fuzzy subset of X defined by

$$(f + g)(z) = \begin{cases} \bigvee_{z=[x,y]}(f(x) \wedge f(y) \wedge g(x) \wedge g(y)) \text{ for } z = [x, y], \\ 0 \qquad\qquad\qquad\qquad\qquad\qquad\qquad\qquad \textit{otherwise.} \end{cases}$$

Definition 8.9 The *extended product* of two fuzzy soft sets (f, A) and (g, B) over X is a fuzzy soft set, denoted by $(f \circ g, C)$, where $C = A \cup B$ and

$$(f \circ g)(\varepsilon) = \begin{cases} f(\varepsilon) & \text{if } \varepsilon \in A - B, \\ g(\varepsilon) & \text{if } \varepsilon \in B - A, \\ f(\varepsilon) \circ g(\varepsilon) & \text{if } \varepsilon \in A \cap B, \end{cases}$$

for all $\varepsilon \in C$. *This is denoted by* $(f \circ g, C) = (f, A)\tilde{\circ}(g, B)$.

Definition 8.10 If $A \cap B \neq \emptyset$, then the *restricted product* (h,C) of two fuzzy soft sets (f,A) and (g,B) over X is defined as the fuzzy soft set, $(h,A \cap B)$ denoted by $(f,A)\, o_R\, (g,B)$ where $h(\varepsilon) = f(\varepsilon) \circ g(\varepsilon)$, for all $\varepsilon \in A \cap B$. Here $f(\varepsilon) \circ g(\varepsilon)$ is the product of two fuzzy subsets of X.

8.1.3 Bipolar Fuzzy Soft Sets

Bipolar fuzzy sets and soft sets are two different methods for representing uncertainty and vagueness. The concept of hybrid model called bipolar fuzzy soft set was originally proposed by Yang and Li [132]. Further, Abdullah et al. [1] discussed bipolar fuzzy soft sets and its applications in decision-making problem. Let $BF(X)$ denotes the family of all bipolar fuzzy sets in X.

Definition 8.11 Let X be an initial universe and $A \subseteq E$ be a set of parameters. A pair (f, A) is called a *bipolar fuzzy soft set* over X, where f is a mapping given by $f : A \to BF(X)$. A bipolar fuzzy soft set is a parameterized family of bipolar fuzzy subsets of X. For any $\varepsilon \in A$, f_ε is referred to as the set of ε-*approximate* elements of the bipolar fuzzy soft set (f, A), which is actually a bipolar fuzzy set on X and can be written as

$$f_\varepsilon = \{\langle x, (\mu^P_{f_\varepsilon}(x), \mu^N_{f_\varepsilon}(x))\rangle \mid x \in X\},$$

where $\mu^P_{f_\varepsilon}(x)$ denotes the degree of x keeping the parameter ε, $\mu^N_{f_\varepsilon}(x)$ denotes the degree of x keeping the nonparameter ε.

Example 8.6 Let $X=\{c_1, c_2, c_3, c_4\}$ be the set of four cars under consideration and $E=\{e_1$=Costly, e_2=Beautiful, e_3=Fuel Efficient, e_4=Modern Technology $\}$ be the set of parameters and $A=\{e_1, e_2, e_3\}\subseteq E$. Then,

$$(f, A) = \left\{ \begin{array}{l} f(e_1) = \left\{ \begin{array}{c} (c_1, 0.1, -0.5), (c_2, 0.3, -0.6), \\ (c_3, 0.4, -0.2), (c_4, 0.7, -0.2) \end{array} \right\}, \\ f(e_2) = \left\{ \begin{array}{c} (c_1, 0.3, -0.5), (c_2, 0.4, -0.2), \\ (c_3, 0.5, -0.2), (c_4, 0.4, -0.2) \end{array} \right\}, \\ f(e_3) = \left\{ \begin{array}{c} (c_1, 0.8, -0.11), (c_2, 0.3, -0.6), \\ (c_3, 0.4, -0.3), (c_4, 0.6, -0.2) \end{array} \right\} \end{array} \right\}$$

Definition 8.12 Let (f, A) and (g, B) be two bipolar fuzzy soft sets over X. We say that (f, A) is a *bipolar fuzzy soft subset* of (g, B) and write $(f, A) \Subset (g, B)$ if $A \subseteq B$ and $f(\varepsilon) \subseteq g(\varepsilon)$ for $\varepsilon \in A$. (f, A) and (g, B) are said to be *bipolar fuzzy soft equal sets* and write $(f, A) = (g, B)$ if $(f, A) \Subset (g, B)$ and $(g, B) \Subset (f, A)$.

According to [132] for any two bipolar fuzzy soft sets (f, A) and (g, B) over X we define

- the *extended intersection* $(h, C) = (f, A)\tilde{\cap}(g, B)$, where $C = A \cup B$ and

$$h(\varepsilon) = \left\{ \begin{array}{ll} f_\varepsilon & \text{if } \varepsilon \in A - B, \\ g_\varepsilon & \text{if } \varepsilon \in B - A, \\ f_\varepsilon \cap g_\varepsilon & \text{if } \varepsilon \in A \cap B, \end{array} \right.$$

- the *extended union* $(h, C) = (f, A)\tilde{\cup}(g, B)$, where $C = A \cup B$ and

$$h(\varepsilon) = \left\{ \begin{array}{ll} f_\varepsilon & \text{if } \varepsilon \in A - B, \\ g_\varepsilon & \text{if } \varepsilon \in B - A, \\ f_\varepsilon \cup g_\varepsilon & \text{if } \varepsilon \in A \cap B, \end{array} \right.$$

- the operation $(f,A) \wedge (g,B) = (h, A \times B)$, where $h(a, b) = h(a) \cap g(b)$ for all $(a,b) \in A \times B$.

8.2 Soft Intersection Lie Algebras

Definition 8.13 Let L be a Lie algebra. Let $F_A = (F, A)$ be a soft set over L. Then F_A is called a *soft Lie subalgebra (resp. soft Lie ideal)* over L if $F(x)$ is a Lie subalgebra (resp. Lie ideal) of a Lie algebra L for all $x \in A$.

Example 8.7 Let $\mathfrak{R}^3 = \{(x, y, z)|x, y, z \in \mathbb{R}\}$ be the set of all three-dimensional real vectors. Then $\mathfrak{R}^3$ with the bracket $[.,.]$ defined as the usual cross product, i.e., $[x, y] = x \times y = (x_2y_3 - x_3y_2, x_3y_1 - x_1y_3, x_1y_2 - x_2y_1)$ forms a real Lie algebra over the field R. Now we define a soft set $\langle F, \mathfrak{R}^3\rangle$ as $F : \mathfrak{R}^3 \longrightarrow P(\mathfrak{R}^3)$ by $F((0, 0, 0)) =$

$\{(0,0,0)\}$, $F(x,0,0) = \{(0,0,0), (x,0,0) : x \neq 0\}$ and $F(x,y,z) = \Re^3$. By routine computations, it is easy to see that $(F, \Re^3)$ is soft Lie subalgebra but not soft Lie ideal of $\Re^3$.

Example 8.8 Let V be a vector space over a field $\mathbb{F}$ such that $dim(V) = 5$. Let $V = \{e_1, e_2, \ldots, e_5\}$ be a basis of a vector space over a field $\mathbb{F}$ with Lie brackets as follows:

$$[e_1, e_2] = e_3, \quad [e_1, e_3] = e_5, \quad [e_1, e_4] = e_5, \quad [e_1, e_5] = 0,$$

$$[e_2, e_3] = e_5, \quad [e_2, e_4] = 0, \quad [e_2, e_5] = 0, \quad [e_3, e_4] = 0,$$

$$[e_3, e_5] = 0, \quad [e_4, e_5] = 0, \quad [e_i, e_j] = -[e_j, e_i]$$

and $[e_i, e_j] = 0$ for all $i = j$. Then V is a Lie algebra over $\mathbb{F}$. Let (F, V) be soft over V and defined by

$$F(x) = \begin{cases} \langle e_7 \rangle \text{ if } x = e_1 \\ \langle e_8 \rangle \text{ if } x = e_2, e_3 \\ \langle e_7, e_8 \rangle \text{ if } x = e_4, e_5 \\ V \text{ otherwise.} \end{cases}$$

Routine computations show that (F, V) is a soft Lie ideal over V.

Proposition 8.1 *If F_A and F_B are soft Lie subalgebras (resp. soft Lie ideals) over L, then $F_A \tilde{\wedge} F_B$ and $F_A \tilde{\cap} F_B$ are soft Lie subalgebras (resp. soft Lie ideals) over L.*

Definition 8.14 Let F_A be a soft Lie subalgebra (resp. soft Lie ideal) over L.

(i) F_A is called the *trivial soft Lie subalgebra (resp. soft Lie ideal)* over L if $F(x) = \{\underline{0}\}$ for all $x \in A$,
(ii) F_A is called the *whole soft Lie subalgebra (resp. soft Lie ideal)* over L if $F(x) = L$ for all $x \in A$.

Definition 8.15 Let L_1, L_2 be two Lie algebras and $\varphi : L_1 \to L_2$ a mapping of Lie algebras. If F_A and G_B are soft sets over L_1 and L_2, respectively, then $\varphi(F_A)$ is a soft set over L_2 where $\varphi(F) : E \to P(L_2)$ is defined by $\varphi(F)(x) = \varphi(F(x))$ for all $x \in E$ and $\varphi^{-1}(G_B)$ is a soft set over L_1 where $\varphi^{-1}(G) : E \to P(L_1)$ is defined by $\varphi^{-1}(G)(y) = \varphi^{-1}(G(y))$ for all $y \in E$.

Proposition 8.2 *Let $\varphi : L_1 \to L_2$ be an onto homomorphism of Lie algebras.*

(i) *If F_A is a soft Lie algebra over L_1, then $\varphi(F_A)$ is a soft Lie algebra over L_2,*
(ii) *If F_B is a soft Lie algebra over L_2, then $\varphi^{-1}(F_B)$ is a soft Lie algebra over L_1 if it is non-null.*

Theorem 8.1 *Let $f : L_1 \to L_2$ be a homomorphism of Lie algebras. Let F_A and G_B be two soft Lie algebras over L_1 and L_2, respectively.*

(a) *If $F(x) = \ker(\varphi)$ for all $x \in A$, then $\varphi(F_A)$ is the trivial soft Lie algebra over L_2,*

(b) *If φ is onto and F_A is whole, then $\varphi(F_A)$ is the whole soft Lie algebra over L_2,*
(c) *If $G(y) = \varphi(L_1)$ for all $y \in B$, then $\varphi^{-1}(G_B)$ is the whole soft Lie algebra over L_1,*
(d) *If φ is injective and G_B is trivial, then $\varphi^{-1}(G_B)$ is the trivial soft Lie algebra over L_1.*

We now introduce the concept of soft intersection Lie subalgebras (resp. soft intersection Lie ideals.

Definition 8.16 Let $L = E$ be a Lie algebra and let A be a subset of L. Let F_A be a soft set over X. Then, F_A is called a *soft intersection Lie subalgebra* over X if it satisfies the following conditions:

(a) $F(x + y) \supseteq F(x) \cap F(y)$,
(b) $F(mx) \supseteq F(x)$,
(c) $F([x, y]) \supseteq F(x) \cap F(y)$

for all $x, y \in A$, $m \in \mathbb{F}$. A soft set F_A is called a *soft intersection Lie ideal* over X if it satisfies (a), (b) and the following condition:

(d) $F([x, y]) \supseteq F(x)$

for all $x, y \in A$.

Example 8.9 Assume that $X=\mathbb{Z}$ is the universal set. Let $E = \Re^2 = \{(x, y)|x, y \in \mathbb{R}\}$ be the set of all two-dimensional real vectors. Then $\Re^2$ with the bracket [., .] defined as the usual cross product, i.e., $[x, y] = x \times y = xy - yx$ forms a real Lie algebra over the field $\mathbb{R}$. Let $A = \{(0, 0), (0, x), x \neq 0\}$ be a subset of E. Let F_A be a soft set over X. Then $F(0, 0) = \mathbb{Z}$ and $F(0, x) = \{-2, -1, 0, 1, 2\}$. It is easy to see that F_A is a soft intersection Lie algebra (resp. soft intersection Lie ideal) over X.

From now on, we will always assume $L = E$ unless otherwise specified.

The following propositions are obvious.

Proposition 8.3 *Let L be a Lie algebra and let A be Lie subalgebra (resp. Lie ideal) of L. If F_A is a soft intersection Lie subalgebra (resp. soft intersection Lie ideal) over X, then $F(\underline{0}) \supseteq F(x)$ for all $x \in A$.*

Proposition 8.4 *Let L be a Lie algebra and let A be Lie subalgebra (resp. Lie ideal) of L. If F_A is a soft intersection Lie algebra (resp. soft intersection Lie ideal) over X, then $F(-x) = F(x)$ for all $x \in A$.*

Proposition 8.5 *Let L be a Lie algebra and let A and B be Lie subalgebras (resp. Lie ideals) of L. If F_A and G_B are soft intersection Lie subalgebras (resp. soft intersection Lie ideals) over X, then $F_A \tilde{\wedge} G_B$ is a soft intersection Lie subalgebras (resp. soft intersection Lie ideal) over X, where $F_A \tilde{\wedge} G_B$ is defined by $F_A \tilde{\wedge} G_B(x, y) = F(x) \cap G(y)$ for all $(x, y) \in A \times B$.*

Proof Let $(x_1, y_1), (x_2, y_2) \in A \times B$ and $m \in \mathbb{F}$. Then

$$\begin{aligned}
(F_A \tilde{\wedge} G_B)((x_1, y_1) + (x_2, y_2)) &= (F_A \tilde{\wedge} G_B)((x_1 + x_2, y_1 + y_2)) \\
&= F(x_1 + x_2) \cap G(y_1 + y_2) \\
&\supseteq (F(x_1) \cap F(x_2)) \cap (G(y_1) \cap G(y_2)) \\
&= (F(x_1) \cap G(y_1)) \cap (F(x_2) \cap G(y_2)) \\
&= (F_A \tilde{\wedge} G_B)(x_1, y_1) \cap (F_A \tilde{\wedge} G_B)(x_2, y_2), \\
(F_A \tilde{\wedge} G_B)(m(x_1, y_1)) &= (F_A \tilde{\wedge} G_B)(mx_1, my_1) \\
&= F(mx_1) \cap G(my_1) \\
&\supseteq F(x_1) \cap G(y_1) \\
&= (F_A \tilde{\wedge} G_B)(x_1, y_1), \\
(F_A \tilde{\wedge} G_B)([(x_1, y_1), (x_2, y_2)]) &= (F_A \tilde{\wedge} G_B)([x_1, x_2], [y_1, y_2]) \\
&= F([x_1, x_2]) \cap G([y_1, y_2]) \\
&\supseteq ([F(x_1), F(x_2)] \cap [G(y_1), G(y_2)]) \\
&= [F(x_1), G(y_1)] \cap [F(x_2), G(y_2)] \\
&= (F_A \tilde{\wedge} G_B)[x_1, y_1] \cap (F_A \tilde{\wedge} G_B)[x_2, y_2], \\
(F_A \tilde{\wedge} G_B)([(x_1, y_1), (x_2, y_2)]) &= (F_A \tilde{\wedge} G_B)([x_1, x_2], [y_1, y_2]) \\
&= F([x_1, x_2]) \cap G([y_1, y_2]) \\
&\supseteq F(x_1) \cap G(y_1) \\
&= (F_A \tilde{\wedge} G_B)[x_1, y_1].
\end{aligned}$$

Hence, $F_A \tilde{\wedge} G_B$ is a soft intersection Lie subalgebra (resp. soft intersection Lie ideal) over X.

Theorem 8.2 *Let $\{(F_i)_{A_i} \mid i \in \Lambda\}$ be a family of soft intersection Lie subalgebras (resp. soft intersection Lie ideals) over X. Then $\tilde{\bigwedge}_{i \in \Lambda}(F_i)_{A_i}$ is a soft intersection Lie subalgebra (resp. soft intersection Lie ideal) over X.*

Proposition 8.6 *Let L be a Lie algebra and let A be a Lie subalgebra (resp. Lie ideal) of L. If F_A and G_A are soft intersection Lie subalgebras (resp. soft intersection Lie ideals) over X, then $F_A \tilde{\cap} G_A$ is a soft intersection Lie algebra (resp. soft intersection Lie ideal) over X, where $F_A \tilde{\cap} G_A$ is defined by $F_A \tilde{\cap} G_A(x) = F(x) \cap G(x)$ for all $x \in A$.*

Proof Let $x, y \in A$ and $m \in \mathbb{F}$. Then

$$\begin{aligned}
(F_A \tilde{\cap} G_A)(x + y) &= F(x + y) \cap G(x + y) \\
&\supseteq (F(x) \cap F(y)) \cap (G(x) \cap G(y)) \\
&= (F(x) \cap G(x)) \cap (F(y) \cap G(y)) \\
&= (F_A \tilde{\cap} G_A)(x) \cap (F_A \tilde{\cap} G_A)(y), \\
(F_A \tilde{\cap} G_A)(mx) &= F(mx) \cap G(mx) \supseteq F(x) \cap G(x)
\end{aligned}$$

$$
\begin{aligned}
&= (F_A \tilde{\cap} G_A)(x),\\
(F_A \tilde{\cap} G_A)([x, y]) &= F([x, y]) \cap G([x, y])\\
&\supseteq (F(x) \cap F(y)) \cap (G(x) \cap G(y))\\
&= (F(x) \cap G(x)) \cap (F(y) \cap G(y))\\
&= (F_A \tilde{\cap} G_A)(x) \cap (F_A \tilde{\cap} G_A)(y),\\
(F_A \tilde{\cap} G_A)([x, y]) &= F([x, y]) \cap G([x, y])\\
&\supseteq F(x) \cap G(x)\\
&= (F_A \tilde{\cap} G_A)(x).
\end{aligned}
$$

Hence, $F_A \tilde{\cap} G_A$ is a soft intersection Lie subalgebras (resp. soft intersection Lie ideal) over X.

Theorem 8.3 *Let $\{(F_i)_{A_i} \mid i \in \Lambda\}$ be a family of soft intersection Lie subalgebras (resp. soft intersection Lie ideals) over X. Then $\tilde{\bigcap}_{i \in \Lambda}(F_i)_{A_i}$ is a soft intersection Lie subalgebra (resp. soft intersection Lie ideal) over X.*

Proposition 8.7 *Let L be a Lie algebra and let A and B be Lie subalgebras (resp. Lie ideals) of L. If F_A and G_B are soft intersection Lie subalgebras (resp. soft intersection Lie ideals) over X, then $F_A \tilde{\times} G_B$ is a soft intersection Lie algebra (resp. soft intersection Lie ideal) over X, where $F_A \tilde{\times} G_B$ is defined by $F_A \tilde{\times} G_B(x, y) = F(x) \times G(y)$ for all $(x, y) \in A \times B$.*

Proof Let $(x_1, y_1), (x_2, y_2) \in A \times B$ and $m \in \mathbb{F}$. Then

$$
\begin{aligned}
(F_A \tilde{\times} G_B)((x_1, y_1) + (x_2, y_2)) &= (F_A \tilde{\times} G_B)(x_1 + x_2, y_1 + y_2)\\
&= F(x_1 + x_2) \times G(y_1 + y_2)\\
&\supseteq (F(x_1) \cap F(x_2)) \times (G(y_1) \cap G(y_2))\\
&= (F(x_1) \times G(y_1)) \cap (F(x_2) \times G(y_2))\\
&= (F_A \tilde{\times} G_B)(x_1, y_1) \cap (F_A \tilde{\times} G_B)(x_2, y_2),\\
(F_A \tilde{\times} G_B)(m(x_1, y_1)) &= (F_A \tilde{\times} G_B)(mx_1, my_1)\\
&= F(mx_1) \times G(my_1)\\
&\supseteq F(x_1) \times G(y_1)\\
&= (F_A \tilde{\times} G_B)(x_1, y_1),\\
(F_A \tilde{\times} G_B)([(x_1, y_1), (x_2, y_2)]) &= (F_A \tilde{\times} G_B)([x_1, x_2], [y_1, y_2])\\
&= F([x_1, x_2]) \times G([y_1, y_2])\\
&\supseteq (F(x_1) \cap F(x_2)) \times (G(y_1) \cap G(y_2))\\
&= (F(x_1) \times G(y_1)) \cap (F(x_2) \times G(y_2))\\
&= (F_A \tilde{\times} G_B)(x_1, y_1) \cap (F_A \tilde{\times} G_B)(x_2, y_2),
\end{aligned}
$$

$$\begin{aligned}(F_A\tilde{\times}G_B)([(x_1, y_1), (x_2, y_2)]) &= (F_A\tilde{\times}G_B)([x_1, x_2], [y_1, y_2]) \\ &= F([x_1, x_2]) \times G([y_1, y_2]) \\ &\supseteq F(x_1) \times G(y_1) \\ &= (F(x_1) \times G(y_1)) \\ &= (F_A\tilde{\times}G_B)(x_1, y_1).\end{aligned}$$

Hence, $F_A\tilde{\times}G_B$ is a soft intersection Lie subalgebras (resp. soft intersection Lie ideal) over X.

Theorem 8.4 *Let $\{(F_i)_{A_i} \mid i \in \Lambda\}$ be a family of soft intersection Lie subalgebras (resp. soft intersection Lie ideals) over X. Then $\tilde{\prod}_{i\in\Lambda}(F_i)_{A_i}$ is a soft intersection Lie subalgebra (resp. soft intersection Lie ideal) over X.*

Proposition 8.8 *Let L be a Lie algebra and let A, B and C be Lie subalgebras (resp. Lie ideals) of L. If F_A, G_B and F_C are soft intersection Lie subalgebras (resp. soft intersection Lie ideals) over X, $F_A\tilde{\subseteq}G_B$ and $F_C\tilde{\subseteq}G_B$, then $F_A\tilde{\cap}F_C\tilde{\subseteq}G_B$ over X.*

Proof Straightforward.

Definition 8.17 Let F_A and G_B be two soft sets over the common universe X and let φ be a function from A to B. Then, soft image of F_A under φ denoted by $\varphi(F_A)$ is a soft set over X by

$$\varphi(F)(y) = \begin{cases} \bigcup\{F(x) \mid x \in A \text{ and } \varphi(x) = y\} & \text{if } \varphi^{-1}(y) \neq \emptyset, \\ \emptyset & \text{otherwise} \end{cases}$$

for all $y \in B$, and *soft pre-image* (or soft inverse image) of G_B under φ denoted by $\varphi^{-1}(G_B)$ is a soft set over X by $\varphi^{-1}(G)(x) = G(\varphi(x))$ for all $x \in A$.

Proposition 8.9 *Let L be a Lie algebra and let A be Lie ideal of L. If F_A is a soft intersection Lie subalgebra (resp. soft intersection Lie ideal) over X, then $A_F = \{x \in A \mid F(x) = F(\underline{0})\}$ is a soft intersection Lie subalgebra (resp. soft intersection Lie ideal) over X.*

Theorem 8.5 *Let L be a Lie algebra and A and B Lie ideals of L. Let φ be a Lie homomorphism from A to B. If G_B is soft intersection Lie subalgebra (resp. soft intersection Lie ideal) over X. Then $\varphi^{-1}(G_B)$ is a soft intersection Lie subalgebra (resp. soft intersection Lie ideal) over X.*

Proof Let $x, y, z \in A$ and $m \in \mathbb{F}$. Then

$$\begin{aligned}\varphi^{-1}(G_B)(x + y) = G_B(\varphi(x + y)) &= G_B(\varphi(x) + \varphi(y)) \\ &\supseteq G(\varphi(x)) \cap G(\varphi(y)) \\ &= \varphi^{-1}(G)(x) \cap \varphi^{-1}(G)(y),\end{aligned}$$

$$\begin{aligned}
\varphi^{-1}(G_B)(mx) &= G_B(\varphi(mx)) \supseteq G(\varphi(x)) = \varphi^{-1}(G_B)(x),\\
\varphi^{-1}(G_B)([x,y]) &= G(\varphi([x,y])) = G([\varphi(x),\varphi(y)]\,)\\
&\supseteq G(\varphi(x)) \cap G(\varphi(y))\\
&= \varphi^{-1}(G)(x) \cap \varphi^{-1}(G)(y),\\
\varphi^{-1}(G_B)([x,y]) &= G_B(\varphi([x,y]) = G_B([\varphi(x),\varphi(y)]\,)\\
&\supseteq G(\varphi(x)) = \varphi^{-1}(G_B)(x).
\end{aligned}$$

Hence, $\varphi^{-1}(G_B)$ is a soft intersection Lie subalgebra (resp. soft intersection Lie ideal) over X.

Theorem 8.6 *Let L be a Lie algebra and A and B Lie ideals of L, and let φ be a Lie isomorphism from A to B. If F_A is soft intersection Lie subalgebra (resp. soft intersection Lie ideal) over X, then $\varphi(F_A)$ is a soft intersection Lie subalgebra (resp. soft intersection Lie ideal) over X.*

Proof Since φ is surjective, there exist $x, y \in A$ such that $a = \varphi(x)$ and $b = \varphi(y)$ for all $a, b \in B$. Then

$$\begin{aligned}
\varphi(F_A)(x+y) &= \cup\{F(z) \mid z \in A, \varphi(z) = a+b\}\\
&= \cup\{F(x+y) \mid x, y \in A, \varphi(x) = a,\ \varphi(y) = b\}\\
&\supseteq \cup\{F(x) \cap F(y) \mid x, y \in A, \varphi(x) = a,\ \varphi(y) = b\}\\
&= (\cup\{F(x) \mid x \in A, \varphi(x) = a\}) \cap (\cup\{F(y) \mid y \in A, \varphi(y) = b\})\\
&= \varphi(F_A)(a) \cap \varphi(F_A)(b),\\
\varphi(F_A)(mx) &= \cup\{F(z) \mid z \in A, \varphi(z) = ma\}\\
&= \cup\{F(mx) \mid x \in A, \varphi(x) = a\}\\
&\supseteq \cup\{F(x) \mid x \in A, \varphi(x) = a\}\\
&= \varphi(F_A)(a),\\
\varphi(F)([x,y]) &= \cup\{F(z) \mid z \in A, \varphi(z) = [a,b]\}\\
&= \cup\{F_A([x,y]) \mid x, y \in A, \varphi(x) = a,\ \varphi(y) = b\}\\
&\supseteq \cup\{F(x) \cap F(y) \mid x, y \in A, \varphi(x) = a,\ \varphi(y) = b\}\\
&= (\cup\{F(x) \mid x \in A, \varphi(x) = a\}) \cap (\cup\{F(y) \mid y \in A, \varphi(y) = b\})\\
&= \varphi(F_A)(a) \cap \varphi(F_A)(b),\\
\varphi(F_A)([x,y]) &= \cup\{F(z) \mid z \in A, \varphi(z) = [a,b]\}\\
&= \cup\{F([x,y]) \mid x, y \in A, \varphi(x) = a,\ \varphi(y) = b\}\\
&\supseteq \cup\{F(x) \mid x \in A, \varphi(x) = a\}\\
&= \cup\{F(x) \mid x \in A, \varphi(x) = a\}\\
&= \varphi(F_A)(a).
\end{aligned}$$

Hence, $\varphi(F_A)$ is a soft intersection Lie subalgebra (resp. soft intersection Lie ideal) over X.

8.3 Fuzzy Soft Lie Algebras

Definition 8.18 Let (f, A) be a fuzzy soft set over L. Then (f, A) is said to be a *fuzzy soft Lie subalgebra* over L if $f(x)$ is a fuzzy Lie subalgebra of L for all $x \in A$, that is, a fuzzy soft set (f, A) on L is called a *fuzzy soft Lie subalgebra* of L if

(a) $f_\varepsilon(x + y) \geq \min\{f_\varepsilon(x), f_\varepsilon(y)\}$,
(b) $f_\varepsilon(mx) \geq f_\varepsilon(x)$,
(c) $f_\varepsilon([x, y]) \geq \min\{f_\varepsilon(x), f_\varepsilon(y)\}$

hold for all $x, y \in L$ and $m \in \mathbb{F}$.

Example 8.10 Let $\Re^2 = \{(x, y)|x, y \in \mathbb{R}\}$ be the set of all two-dimensional real vectors. Then $\Re^2$ with $[x, y] = x \times y$ is a real Lie algebra. Let $\mathbb{N}$ and $\mathbb{Z}$ denote the set of all natural numbers and the set of all integers, respectively. Define $f : \mathbb{Z} \to [0, 1]^{\Re^2}$ by $f(n) = f_n : \Re^2 \to [0, 1]$ for all $n \in \mathbb{Z}$,

$$f_n(x) = \begin{cases} 0.6 & \text{if } x = (0, 0) = \underline{0}, \\ 0.2 & \text{if } x = (0, a), a \neq 0, \\ 0 & \text{otherwise.} \end{cases}$$

By routine computations, we can easily check that $(f, \mathbb{Z})$ is a fuzzy soft Lie subalgebra of $\Re^2$.

The following propositions are obvious.

Proposition 8.10 *Let (f, A) be a fuzzy soft Lie subalgebra of L, then*

(i) $f_\varepsilon(\underline{0}) \geq f_\varepsilon(x) \quad \forall\, x \in L$,
(ii) $f_\varepsilon([x, y]) = f_\varepsilon(-[y, x]) = f_\varepsilon([y, x]) \ \forall x, y \in L$.

Proposition 8.11 *Let (f, A) and (g, B) be fuzzy soft Lie subalgebras of L, then $(f, A)\tilde{\cap}(g, B)$ is a fuzzy soft Lie subalgebra of L.*

Proposition 8.12 *Let (f, A) and (g, B) be fuzzy soft Lie subalgebras over L, then $(f, A) \wedge (g, B)$ is a fuzzy soft Lie subalgebra over L.*

Proposition 8.13 *Let (f, A) and (g, B) be fuzzy soft Lie subalgebras over L. If $A \cap B = \emptyset$, then $(f, A)\tilde{\cup}(g, B)$ is a fuzzy soft Lie subalgebra of L.*

Proposition 8.14 *Let (f, A) be a fuzzy soft Lie subalgebra over L and let $\{(h_i, B_i) \mid i \in I\}$ be a nonempty family of fuzzy soft Lie subalgebras of (f, A). Then*

(i) $\tilde{\cap}_{i\in I}(h_i, B_i)$ *is a fuzzy soft Lie subalgebra of* (f, A),
(ii) $\bigwedge_{i\in I}(h_i, B_i)$ *is a fuzzy soft Lie subalgebra of* $\bigwedge_{i\in I}(f, A)$,
(iii) *If* $B_i \cap B_j = \emptyset$ *for all* $i, j \in I$, *then* $\tilde{\bigvee}_{i\in I}(H_i, B_i)$ *is a fuzzy soft Lie subalgebra of* $\tilde{\bigvee}_{i\in I}(f, A)$.

Theorem 8.7 *Let (f, A) be a fuzzy soft set over L. (f, A) is a fuzzy soft Lie subalgebras if and only if $(f, A)^t$ is a soft Lie subalgebra over L for each $t \in [0, 1]$.*

Proof Suppose that (f, A) is a fuzzy soft Lie subalgebra. For each $t \in [0, 1]$, $\varepsilon \in A$ and $x_1, x_2 \in f_\varepsilon^t$, then $f_\varepsilon(x_1) \geq t$ and $f_\varepsilon(x_2) \geq t$. From Definition 8.18, it follows that f_ε is a fuzzy Lie subalgebra over L. Thus

$$f_\varepsilon(x_1 + x_2) \geq \min(f_\varepsilon(x_1), f_\varepsilon(x_2)), f_\varepsilon(x_1 + x_2) \geq t,$$

$$f_\varepsilon(mx_1) \geq f_\varepsilon(x_1), f_\varepsilon(mx_1) \geq t,$$

$$f_\varepsilon([x_1, x_2]) \geq \min(f_\varepsilon(x_1), f_\varepsilon(x_2)), f_\varepsilon([x_1, x_2]) \geq t.$$

This implies that $x_1 + x_2, mx_1, [x_1, x_2] \in f_\varepsilon^t$, that is, f_ε^t is a Lie subalgebra over L. According to Definition 8.6, $(f, A)^t$ is a soft Lie subalgebra over L for each $t \in [0, 1]$. Conversely, assume that $(f, A)^t$ is a soft Lie subalgebra over L for each $t \in [0, 1]$. For each $\varepsilon \in A$ and $x_1, x_2 \in G$, let $t = \min\{f_\varepsilon(x_1), f_\varepsilon(x_2)\}$, then $x_1, x_2 \in f_\varepsilon^t$. Since f_ε^t is a Lie subalgebra over L, then $x_1 + x_2, mx_1, [x_1, x_2] \in f_\varepsilon^t$. This means that

$$f_\varepsilon(x_1 + x_2) \geq \min(f_\varepsilon(x_1), f_\varepsilon(x_2)).$$

$$f_\varepsilon(mx_1) \geq f_\varepsilon(x_1),$$

$$f_\varepsilon([x_1, x_2]) \geq \min(f_\varepsilon(x_1), f_\varepsilon(x_2)),$$

that is, f_ε^t is a fuzzy Lie subalgebra over L. According to Definition 8.18, (f, A) is a fuzzy soft Lie subalgebra over L. This completes the proof.

Definition 8.19 Let $\phi : X \to Y$ and $\psi : A \to B$ be two functions, A and B are parametric sets from the crisp sets X and Y, respectively. Then the pair (ϕ, ψ) is called a *fuzzy soft function* from X to Y.

Definition 8.20 Let (f, A) and (g, B) be two fuzzy soft sets on X and Y, respectively, and let (ϕ, ψ) be a fuzzy soft function from X to Y.
(1) The image of (f, A) under the fuzzy soft function (ϕ, ψ), denoted by $(\phi, \psi)(f, A)$, is the fuzzy soft set on Y defined by $(\phi, \psi)(f, A) = (\phi(f), \psi(A))$, where for all $k \in \psi(A)$, $y \in Y$

$$\phi(f)_k(y) = \begin{cases} \bigvee_{\phi(x)=y} \bigvee_{\psi(a)=k} f_a(x) & \text{if } x \in \psi^{-1}(y), \\ 0 & \text{otherwise.} \end{cases}$$

(2) The pre-image of (g, B) under the fuzzy soft function (ϕ, ψ), denoted by $(\phi, \psi)^{-1}(g, B)$, is the fuzzy soft set on X defined by $(\phi, \psi)^{-1}(g, B) = (\phi^{-1}(g), \psi^{-1}(B))$ where $\phi^{-1}(g)_a(x) = g_{\psi(a)}(\phi(x))$, for all $a \in \psi^{-1}(A)$, for all $x \in X$.

Definition 8.21 Let (ϕ, ψ) be a fuzzy soft function from X to Y. If ϕ is a homomorphism from X to Y, then (ϕ, ψ) is said to be *fuzzy soft homomorphism*. If ϕ is an isomorphism from X to Y and ψ is one-to-one mapping from A onto B, then (ϕ, ψ) is said to be fuzzy soft isomorphism.

Theorem 8.8 *Let (g, B) be a fuzzy soft Lie algebra on L_2 and let (ϕ, ψ) be a fuzzy soft homomorphism from L_1 to L_2. Then $(\phi, \psi)^{-1}(g, B)$ is a fuzzy soft Lie algebra on L_1.*

Proof Let $x_1, x_2 \in L_1$, $m \in \mathbb{F}$ then

$$\begin{aligned}
\phi^{-1}(g_\varepsilon)(x_1+x_2) &= g_{\psi(\varepsilon)}(\phi(x_1+x_2)) = g_{\psi(\varepsilon)}(\phi(x_1)+\phi(x_2))\\
&\geq \min\{g_{\psi(\varepsilon)}(\phi(x_1)), g_{\psi(\varepsilon)}(\phi(x_2))\}\\
&= \min\{\phi^{-1}(g_\varepsilon)(x_1), \phi^{-1}(g_\varepsilon)(x_2)\},\\
\phi^{-1}(g_\varepsilon)(mx_1) &= g_{\psi(\varepsilon)}(\phi(mx_1)) = g_{\psi(\varepsilon)}(m\phi(x_1))\\
&\geq g_{\psi(\varepsilon)}(\phi(x_1)) = \phi^{-1}(g_\varepsilon)(x_1),\\
\phi^{-1}(g_\varepsilon)([x_1, x_2]) &= g_{\psi(\varepsilon)}(\phi([x_1, x_2])) = g_{\psi(\varepsilon)}([\phi(x_1), \phi(x_2)]\ )\\
&\geq \min\{g_{\psi(\varepsilon)}(\phi(x_1)), g_{\psi(\varepsilon)}(\phi(x_2))\}\\
&= \min\{\phi^{-1}(g_\varepsilon)(x_1), \phi^{-1}(g_\varepsilon)(x_2)\}.
\end{aligned}$$

Hence, $(\phi, \psi)^{-1}(g, B)$ is a fuzzy soft Lie algebra on L_1.

Definition 8.22 Let (f, A) be a fuzzy soft Lie subalgebra on L. Define a sequence of fuzzy soft Lie subalgebras on L by putting $f^0 = f$ and $f^n = [f^{n-1}, f^{n-1}]$ for $n > 0$. If there exists a positive integer n such that $f^n = 0$, then a fuzzy soft Lie subalgebra (f, A) is called *solvable*.

Theorem 8.9 *Homomorphic image of a solvable fuzzy Lie subalgebra is a solvable fuzzy soft Lie subalgebra.*

Proof Let $f : L_1 \to L_2$ be a homomorphism of Lie algebras. Suppose that (f, A) is a solvable fuzzy Lie subalgebra in L_1. We prove by induction on n that $\phi(f^n) \supseteq [\phi(f)]^{\,n}$, where n is any positive integer. First we claim that $\phi([f_\varepsilon, f_\varepsilon]) \supseteq [\phi(f_\varepsilon), \phi(f_\varepsilon)]$. Let $y \in L_2$, then

$$\begin{aligned}
&\phi(\ll f_\varepsilon, f_\varepsilon \gg)(y) = \sup\{\ll f_\varepsilon, f_\varepsilon \gg (x) \mid \phi(x) = y\}\\
&= \sup\{\sup\{\min\{f_\varepsilon(a), f_\varepsilon(b)\} \mid a, b \in L_1, [a, b] = x, \phi(x) = y\}\}\\
&= \sup\{\min\{f_\varepsilon(a), f_\varepsilon(b)\} \mid a, b \in L_1, [a, b] = x, \phi(x) = y\}\\
&= \sup\{\min\{f_\varepsilon(a), f_\varepsilon(b)\} \mid a, b \in L_1, [\phi(a), \phi(b)]\ = x\}\\
&= \sup\{\min\{f_\varepsilon(a), f_\varepsilon(b)\} \mid a, b \in L_1, \phi(a) = u, \phi(b)]\ = v, [u, v] = y\}\\
&\geq \sup\{\min\{\sup_{a\in\phi^{-1}(u)} f_\varepsilon(a), \sup_{b\in\phi^{-1}(v)} f_\varepsilon(b)\} \mid [u, v] = y\}\\
&= \sup\{\min(\phi(f_\varepsilon)(u), \phi(f_\varepsilon)(v)) \mid [u, v] = y\} = \ll \phi(f_\varepsilon), \phi(f_\varepsilon) \gg (y).
\end{aligned}$$

Thus,

$$\phi([f_e, f_e]) \supseteq [\phi(f_e), \phi(f_e)].$$

Now for $n > 1$, we get
$\phi(f_e^n) = \phi([f_e^{n-1}, f_e^{n-1}]) \supseteq [\phi(f_e^{n-1}), \phi(f_e^{n-1})] \quad \supseteq [(\phi(f_\varepsilon))^{n-1}, (\phi(f_\varepsilon))^{n-1}] = (\phi(f_\varepsilon))^n$. This completes the proof.

Definition 8.23 Let (f, A) be a fuzzy soft Lie subalgebra on L and let $f_n = [f, f_{n-1}]$ for $n > 0$, where $f_0 = f$. If there exists a positive integer n such that $f_n = 0$ then f is called *nilpotent*.

Theorem 8.10 *Homomorphic image of a nilpotent fuzzy soft Lie subalgebra is a nilpotent fuzzy soft Lie subalgebra.*

Theorem 8.11 *If (f, A) is a nilpotent fuzzy soft Lie subalgebra, then it is solvable.*

Definition 8.24 Let (f, A) be a fuzzy soft set over L. Then (f, A) is said to be an $(\in, \in \vee q)$-*fuzzy soft Lie subalgebra* over L if $f(x)$ is an $(\in, \in \vee q)$-fuzzy Lie subalgebra of L for all $x \in A$.

Definition 8.25 Given a fuzzy set μ in L and $A \subseteq [0, 1]$, we define two set-valued functions $f : A \to P(L)$ and $f_q : A \to P(L)$ by

$$f(t) = \{x \in L \mid x_t \in \mu\}, \quad f_q(t) = \{x \in L \mid x_t q\mu \in \mu\}$$

for all $t \in A$, respectively. Then (f, A) and (f_q, A) are soft sets over L, which are called an $\in$-soft set and a q-soft set over L, respectively.

Example 8.11 Let V be a vector space over a field $\mathbb{F}$ such that $dim(V) = 5$. Let $\{e_1, e_2, \ldots, e_5\}$ be a basis of a vector space over a field $\mathbb{F}$ with Lie brackets as follows:

$$[e_1, e_2] = e_3, \quad [e_1, e_3] = e_5, \quad [e_1, e_4] = e_5, \quad [e_1, e_5] = 0,$$

$$[e_2, e_3] = e_5, \quad [e_2, e_4] = 0, \quad [e_2, e_5] = 0, \quad [e_3, e_4] = 0,$$

$$[e_3, e_5] = 0, \quad [e_4, e_5] = 0, \quad [e_i, e_j] = -[e_j, e_i]$$

and $[e_i, e_j] = 0$ for all $i = j$. Then V is a Lie algebra over $\mathbb{F}$. We define a fuzzy set $\mu : V \to [0, 1]$ by

$$\mu(x) := \begin{cases} 0.6 \text{ if } x = \underline{0}, \\ 0.3 \text{ otherwise} . \end{cases}$$

Then μ is an $(\in, \in \vee q)$-fuzzy Lie subalgebra of L. Take $A = (0, 0.5]$ and let (f, A) be an $\in$-soft set over L. Then $f(t) = V$ if $t \in (0, 0.3]$, which is Lie subalgebras of L. Hence, (f, A) is a soft Lie algebra over L.

Proposition 8.15 *Let μ be a fuzzy set in a Lie algebra L and let (f, A) be an $\in$-soft set on L with A =(0, 1]. Then (f, A) is a soft L-algebra on L if and only if μ is a fuzzy L-subalgebra of L.*

Proof Assume that (f, A) is a soft Lie algebra on L. If μ is not a fuzzy Lie subalgebra of L, then there exist $a, b \in L$ such that $\mu(a+b) < \min(\mu(a), \mu(b))$. Take $t \in A$ such that $\mu(a+b) < t \leq \min(\mu(a), \mu(b))$. Then $a_t \in \mu$ and $b_t \in \mu$ but $(a+b)_{\min(t,t)}=(a+b)_t \notin \mu$. Hence, $a, b \in f(t)$, but $a+b \notin f(t)$, a contradiction. Thus, $\mu(x+y) \geq \min(\mu(x), \mu(y))$ for all $x, y \in L$. The verification of other conditions is similar.

Conversely, suppose that μ is a fuzzy Lie subalgebra of L. Let $t \in \mu$ and $x, y \in f(t)$. Then x_t and $y_t \in \mu$. It follows from Definition 8.25 that $(x+y)_t = (x+y)_{\min(t,t)} \in \mu$ so that $x+y \in f(t)$. The verification of other conditions is similar. Hence, $f(t)$ is a Lie subalgebra of L, i.e., (f, A) is a soft Lie algebra on L.

Proposition 8.16 *Let μ be a fuzzy set in a Lie algebra L and let (f_q, A) be a q-soft set over L with $A = (0, 1]$. Then (f_q, A) is a soft L-algebra over L if and only if μ is a fuzzy L-subalgebra of L.*

Proof Suppose that μ is a fuzzy Lie subalgebra of L. Let $t \in A$ and $x, y \in f_q(t)$. Then $xt_q\mu$ and $y_tq\mu$, i.e., $\mu(x)+t > 1$ and $\mu(y)+t > 1$. It follows from Definition 8.18 that

$$\mu(x+y)+t \geq \min(\mu(x), \mu(y))+t = \min(\mu(x)+t, \quad \mu(y)+t) > 1$$

so that $(x+y)_tq\mu$, i.e., $x+y \in f_q(t)$. Likewise, $mx \in f_q(t), [x, y] \in f_q(t)$. Hence, $f_q(t)$ is a Lie subalgebra of L for all $t \in A$, and so (f_q, A) is a soft Lie algebra over L.

The proof of converse part is obvious. This completes the proof.

Proposition 8.17 *Let μ be a fuzzy set in a Lie algebra L and let (f, A) be an $\in$-soft set on L with $A =$ (0.5, 1]. Then the following assertions are equivalent:*

(A) *(f, A) is a soft Lie algebra over L,*
(B) (1) $\min(\mu(x+y), 0.5) \geq \min(\mu(x), \mu(y))$,
(2) $\min(\mu(mx), 0.5) \geq \mu(x)$,
(3) $\min(\mu([x, y]), 0.5) \geq \min(\mu(x), \mu(y))$

$x, y \in L, m \in \mathbb{F}$.

Proposition 8.18 *Let μ be a fuzzy set in a Lie algebra L and let (f, A) be an $\in$-soft set on L with $A =$ (0, 0.5]. Then the following assertions are equivalent:*

(i) *μ is an $(\in, \in \vee q)$-fuzzy Lie subalgebra of L,*
(ii) *(f, A) is a soft Lie algebra over L.*

Proof Assume that μ is an $(\in, \in \vee q)$-fuzzy Lie subalgebra of L. Let $t \in A$ and $x, y \in f(t)$. Then $x_t \in \mu$ and $y_t \in \mu$ or equivalently $\mu(x) \geq t$ and $\mu(y) \geq t$. It follows from Proposition 8.17 that

$$\mu(x+y) \geq \min(\mu(x), \mu(y), 0.5) \geq \min(t, 0.5) = t,$$

$$\mu(mx) \geq \min(\mu(x), 0.5) \geq \min(t, 0.5) = t,$$

$$\mu([x, y]) \geq \min(\mu(x), \mu(y), 0.5) \geq \min(t, 0.5) = t,$$

so that $(x + y)_t \in \mu$, $(mx)_t \in \mu$ and $[x, y]_t \in \mu$ or equivalently $x + y, mx, [x, y] \in f(t)$. Hence, (f, A) is a soft Lie algebra on L.

Conversely, suppose that (ii) is valid. If there exist $a, b \in L$ such that

$$\mu(a + b) < \min(\mu(a), \mu(b), 0.5),$$

then we take $t \in (0, 1)$ such that

$$\mu(a + b) < t \leq \min(\mu(a), \mu(b), 0.5).$$

Thus, $t \leq 0.5$, $a_t \in \mu$ and $b_t \in \mu$, that is, $a \in f(t)$ and $b \in f(t)$. Since $f(t)$ is a Lie subalgebra of L, it follows that $a + b \in f(t)$ for all $t \leq 0.5$ so that $(a + b)_t \in \mu$ or equivalently $\mu(a + b) \geq t$ for all $t \leq 0.5$, a contradiction. Verification of other conditions is similar. Hence,

$$\mu(x + y) \geq \min(\mu(x), \mu(y), 0.5) \text{ for all } x, y \in L,$$

$$\mu(mx) \geq \min(\mu(x), 0.5) \text{ for all } x \in L,\ m \in \mathbb{F},$$

$$\mu([x, y]) \geq \min(\mu(x), \mu(y), 0.5) \text{ for all } x, y \in L.$$

It follows from Definition 8.18 that μ is an $(\in, \in \vee q)$-fuzzy Lie subalgebra of L.

Proposition 8.19 *Let H be a Lie subalgebra of a Lie algebra L and let (f, A) be a soft set over L. If $A = (0, 0.5]$, then there exists an $(\in, \in \vee q)$-fuzzy Lie subalgebra μ of L such that*

$$f(t) = \{x \in H \mid x_t \in \mu\} = H\ \forall\ t \in A.$$

Proof Obvious.

Proposition 8.20 *Let μ be a fuzzy set in a Lie algebra L and let (f_q, A) be a q-soft set over L with $A = (0, 1]$. Then the following assertions are equivalent:*

(i) *μ is a fuzzy Lie subalgebra of L,*
(ii) *$(f_q(t) \neq \emptyset \rightarrow f_q(t))$ is a Lie subalgebra of L for all $t \in A$.*

Proof Assume that μ is a fuzzy Lie subalgebra of L. Let $t \in A$ be such that $f_q(t) \neq \emptyset$. Let $x, y \in L$ and $m \in \mathbb{F}$ be such that $x + y \in f_q(t)$, $mx \in f_q(t)$ and $[x, y] \in f_q(t)$. Then $(x + y)_t q \mu$, $(mx)_{\in} \mu_t q$ and $[x, y]_t q \mu$ or equivalently, $\mu(x + y) + t > 1$, $\mu(mx) + t > 1$ and $\mu([x, y]) + t > 1$. Using Definition 8.18, we have

$$\mu(x + y) + t \geq \min(\mu(x), \mu(y)) + t = \min(\mu(x) + t, \mu(y) + t) > 1,$$

$$\mu(mx) + t \geq \mu(x) + t > 1,$$

$$\mu([x, y]) + t \geq \min(\mu(x), \mu(y)) + t = \min(\mu(x) + t, \mu(y) + t) > 1,$$

and so $(x + y)_t q\mu$, $(mx)_t q\mu$, $[x, y]_t q\mu$, i.e., $x + y, mx, [x, y] \in f_q(t)$. Thus, $f_q(t)$ is a Lie subalgebra of L.

Conversely, assume that (ii) is valid. Suppose there exist $a, b \in L$ such that $\mu(a + b) < \min(\mu(a), \mu(b))$. Then $\mu(a + b) + s \leq 1 < \min(\mu(a), \mu(b)) + s$ for some $s \in A$. It follows that $(a)_s q\mu$ and $b_s q\mu$, i.e., $a \in f_q(s)$ and $b \in f_q(s)$. Since $f_q(s)$ is a Lie subalgebra of L, we get $a + b \in f_q(s)$, and so $(a + b)_s q\mu$ or equivalently $\mu(a + b) + s > 1$, a contradiction. Thus, $\mu(x + y) \geq \min(\mu(x), \mu(y))$ for all $x, y \in L$. The verification for other conditions is similar. Hence, μ is a fuzzy Lie subalgebra of L.

8.4 $(\in_\alpha, \in_\alpha \vee q_\beta)$-Fuzzy Soft Lie Subalgebras

We now present generalized fuzzy soft Lie subalgebras and describe some of their properties.

Definition 8.26 A fuzzy set μ in a Lie algebra L is called an *$(\in_\alpha, \in_\alpha \vee q_\beta)$-fuzzy soft Lie subalgebra* of L if it satisfies the following conditions:

(d) $x_r, y_s \in_\alpha f_\varepsilon \rightarrow (x + y)_{\min(r,s)} \in_\alpha \vee q_\beta f_\varepsilon$,
(e) $x_r \in_\alpha f_\varepsilon \rightarrow (mx)_r \in_\alpha \vee q_\beta f_\varepsilon$,
(f) $x_r, y_s \in_\alpha f_\varepsilon \rightarrow [x, y]_{\min(r,s)} \in_\alpha \vee q_\beta f_\varepsilon$

for all $x, y \in L, m \in \mathbb{F}, r, s \in (\alpha, 1]$.

Example 8.12 Let $\Re^3 = \{(x, y, z)|x, y, z \in \mathbb{R}\}$ be the set of all three-dimensional real vectors. Then $\Re^3$ with $[x, y] = x \times y$ is a real Lie algebra. Let $E = (0.2, 0.6]$. Define a fuzzy soft set (f, A) over L as follows.

$$f_\varepsilon(x) = \begin{cases} e & \text{if } x = (0, 0, 0), \\ 0.2 & \text{otherwise.} \end{cases}$$

Then it is now routine to verify that (f, A) is an $(\in_{0.2}, \in_{0.2} \vee q_{0.6})$-fuzzy soft Lie subalgebra of $\Re^3$.

For any fuzzy soft set (f, A) over a Lie algebra L, $e \in A$ and $r \in (\alpha, 1]$, denote $f_{e_r} = \{x \in G|x_r \in_\alpha f_\varepsilon\}, (f_\varepsilon)_r = \{x \in L|x_r\, q_\beta f_\varepsilon\}$, and $[f_\varepsilon]_r = \{x \in L|x_r \in_\alpha \vee q_\beta f_\varepsilon\}$.

Theorem 8.12 *Let L be a Lie algebra and (f, A) a fuzzy soft set on L. Then*

(1) *(f, A) is an $(\in_\alpha, \in_\alpha \vee q_\beta)$-fuzzy soft Lie subalgebra on L if and only if nonempty subset f_{ε_r} is a Lie subalgebra of L for all $\varepsilon \in A$ and $r \in (\alpha, \beta]$,*
(2) *If $2\beta = 1 + \alpha$, then (f, A) is an $(\in_\alpha, \in_\alpha \vee q_\beta)$-fuzzy soft Lie subalgebra on L if and only if nonempty subset $(f_\varepsilon)_r$ is a Lie subalgebra of L for all $\varepsilon \in A$ and $r \in (\beta, 1]$,*

(3) *(f, A) is an $(\in_\alpha, \in_\alpha \vee q_\beta)$-fuzzy soft Lie subalgebra on L if and only if nonempty subset $[f_\varepsilon]_r$ is a Lie subalgebra of L for all $\varepsilon \in A$ and $r \in (\alpha, \min\{2\beta - \alpha, 1\}]$.*

Proof We prove only (2) and (3).

(2) Assume that $2\beta = 1 + \alpha$. Let (f, A) be an $(\in_\alpha, \in_\alpha \vee q_\beta)$-fuzzy soft Lie subalgebra on L and assume that $(f_\varepsilon)_r \neq \emptyset$ for some $e \in A$ and $r \in (\beta, 1]$. Let $x, y \in (f_\varepsilon)_r$. Then $x_r q_\beta f_\varepsilon$ and $y_r q_\beta f_\varepsilon$, that is, $f_\varepsilon(x) + r > 2\beta$ and $f_\varepsilon(y) + r > 2\beta$. Since (f, A) is an (α, β)-fuzzy soft Lie subalgebra on L, we have $\min\{f_\varepsilon(x + y), \alpha\} \geq \min\{f_\varepsilon(x), f_\varepsilon(y), \beta\}$. Thus, by $r > \beta$,
$\min\{f_\varepsilon(x + y) + r, \alpha + r\} = \min\{f_\varepsilon(x + y), \alpha\} + r \geq \min\{f_\varepsilon(x), f_\varepsilon(y), \beta\} + r$ $= \min\{f_\varepsilon(x) + r, f_\varepsilon(y) + r, \beta + r\} > 2\beta$. From $r \leq 1 = 2\beta - \alpha$, that is, $r + \alpha \leq 2\beta$, we have $f_\varepsilon(x + y) + r > 2\beta$ and so $x + y \in (f_\varepsilon)_r$. Similarly, we can show that $mx \in (f_\varepsilon)_r$, $[x, y] \in (f_\varepsilon)_r$. Therefore, $(f_\varepsilon)_r$ is a Lie subalgebra of L.

Conversely, assume that the given conditions hold. If there exist $e \in A$ and $x, y \in L$ such that $\min\{f_\varepsilon(x + y), \alpha\} < \min\{f_\varepsilon(y), \beta\}$. Take $r = 2\beta - \min\{f_\varepsilon(x + y), \alpha\}$. Then $r \in (\beta, 1]$, $f_\varepsilon(x + y) \leq 2\beta - r$, $f_\varepsilon(x) > \min\{G(e)(x + y), \alpha\} = 2\beta - r$, $f_\varepsilon(y) > \min\{G(e)(x + y), \alpha\} = 2\beta - r$, that is, $x \in (f_\varepsilon)_r$, $y \in (f_\varepsilon)_r$ but $x + y \notin (f_\varepsilon)_r$, a contradiction. The verification for other conditions is similar. Hence, (f, A) satisfies above condition. Therefore, (F, A) is an $(\in_\alpha, \in_\alpha \vee q_\beta)$-fuzzy soft Lie subalgebra on L.

(3) Let (f, A) be an $(\in_\alpha, \in_\alpha \vee q_\beta)$-fuzzy soft Lie subalgebra on L and assume that $[f_\varepsilon]_r \neq \emptyset$ for some $\varepsilon \in A$ and $r \in (\alpha, \min\{2\beta - \alpha, 1\}]$. Let $x, y \in [f_\varepsilon]_r$. Then $x_r \in_\alpha \vee q_\beta f_\varepsilon$ and $y_r \in_\alpha \vee q_\beta f_\varepsilon$, that is, $f_\varepsilon(x) \geq r > \alpha$ or $f_\varepsilon(x) > 2\beta - r \geq 2\beta - (2\beta - \alpha) = \alpha$ and $f_\varepsilon(y) \geq r > \alpha$ or $f_\varepsilon(y) > 2\beta - r \geq 2\beta - (2\beta - \alpha) = \alpha$. Since (f, A) is an $(\in_\alpha, \in_\alpha \vee q_\beta)$-fuzzy soft Lie subalgebra on L, we have $\min\{f_\varepsilon(x + y), \alpha\} \geq \min\{f_\varepsilon(x), f_\varepsilon(y), \beta\}$ and so $f_\varepsilon(x + y) \geq \min\{f_\varepsilon(x), f_\varepsilon(y), \beta\}$ since $\alpha < \min\{f_\varepsilon(x), f_\varepsilon(y), \beta\}$ in any case.

We now consider the following cases.

Case 1: $r \in (\alpha, \beta]$. Then $2\beta - r \geq \beta \geq r$. If $f_\varepsilon(x) \geq r$ and $f_\varepsilon(y) \geq r$ or $f_\varepsilon(x) > 2\beta - r$ and $f_\varepsilon(y) > 2\beta - r$, then $f_\varepsilon(x + y) \geq \min\{f_\varepsilon(x), f_\varepsilon(y), \beta\} \geq r$. Hence, $(x + y)_r \in_\alpha f_\varepsilon$.

Case 2: $r \in (\beta, \min\{2\beta - \alpha, 1\}]$. Then $r > \beta > 2\beta - r$. If $f_\varepsilon(x) \geq r$ and $f_\varepsilon(y) \geq r$ or $f_\varepsilon(x) > 2\beta - r$ $f_\varepsilon(y) > 2\beta - r$, then $f_\varepsilon(x + y) \geq \min\{f_\varepsilon(x), f_\varepsilon(y), \beta\} > 2\beta - r$. Hence, $(x + y)_r q_\beta f_\varepsilon$. Thus, in any case, $(x + y)_r \in_\alpha \vee q_\beta f_\varepsilon$, that is, $x + y \in [f_\varepsilon]_r$. Similarly, we can show $mx \in [f_\varepsilon]_r$, $[x, y] \in [f_\varepsilon]_r$. Therefore, $[f_\varepsilon]_r$ is a Lie subalgebra.

Conversely, assume that the given conditions hold. If there exist $\varepsilon \in A$ and $x, y \in L$ such that $\min\{f_\varepsilon(x + y), \alpha\} < r = \min\{f_\varepsilon(x), f_\varepsilon(y), \beta\}$. Then $x_r \in_\alpha f_\varepsilon$, $y_r \in_\alpha f_\varepsilon$ but $(x + y)_r \overline{\in_\alpha \vee q_\beta} f_\varepsilon$, that is, $x \in [f_\varepsilon]_r$, $y \in [f_\varepsilon]_r$ but $x + y \notin [f_\varepsilon]_r$, a contradiction. The verification for other conditions is similar. Hence, (f, A) satisfies above condition. Therefore, (F, A) is an $(\in_\alpha, \in_\alpha \vee q_\beta)$-fuzzy soft Lie subalgebra on L.

Corollary 8.1 *Let L be a Lie algebra and $\alpha, \alpha', \beta, \beta' \in [0, 1]$ such that $\alpha < \beta$, $\alpha' < \beta'$, $\alpha < \alpha'$ and $\beta' < \beta$. Then any $(\in_\alpha, \in_\alpha \vee q_\beta)$-fuzzy Lie subalgebra of L is an $(\in_{\alpha'}, \in_{\alpha'} \vee q_{\beta'})$-fuzzy soft Lie subalgebra on L.*

We state the following Lemmas without their proofs.

Lemma 8.1 *Let L be a Lie algebra and (f, A) and (g, B) fuzzy soft sets on L. If (f, A) and (g, B) are $(\in_\alpha, \in_\alpha \vee q_\beta)$-fuzzy soft Lie subalgebras on L, then so are $(f, A) \Cap (g, B)$ and $(f, A)\tilde{\cap}(g, B)$.*

Lemma 8.2 *Let L be a Lie algebra and (f, A) and (g, B) fuzzy soft sets on L. If (f, A) and (g, B) are $(\in_\alpha, \in_\alpha \vee q_\beta)$-fuzzy soft Lie subalgebras on L, then so are $(f, A) \Cup (g, B)$ and $(f, A)\tilde{\cup}(g, B)$.*

Denote by $\mathbb{FSI}(L, E)$ the set of all $(\in_\alpha, \in_\alpha \vee q_\beta)$-fuzzy soft Lie subalgebras on L.

Theorem 8.13 *Let L be a Lie algebra. $(\mathbb{FSI}(L, E), \tilde{\cup}, \Cap)$ is a complete distributive lattice under the ordering relation $\Subset_{(\alpha,\beta)}$.*

Proof For any $(f, A), (g, B) \in \mathbb{FSI}(L, E)$, by Lemmas 8.1, 8.2, $(f, A)\tilde{\cup}(g, B) \in \mathbb{FSI}(L, E)$ and $(f, A) \Cap (g, B) \in \mathbb{FSI}(L, E)$. It is obvious that $(f, A)\tilde{\cup}(g, B)$ and $(f, A) \Cap (g, B)$ are the least upper bound and the greatest lower bound of (f, A) and (g, B), respectively. There is no difficulty in replacing $\{(f, A), (g, B)\}$ with an arbitrary family of $\mathbb{FSI}(L, E)$ and so $(\mathbb{FSI}(L, E), \tilde{\cup}, \Cap)$ is a complete lattice. Now we prove that the following distributive law

$$(f, A) \Cap ((g, B)\tilde{\cup}(h, C)) = ((f, A) \Cap (g, B))\tilde{\cup}((f, A) \Cap (h, C))$$

holds for all $(f, A), (g, B), (h, C) \in \mathbb{FSI}(L, E)$. Suppose that $(f, A) \Cap ((g, B)\tilde{\cup} (h, C)) = (J, A \cap (B \cup C))$, $((f, A) \Cap (G, B))\tilde{\cup}((f, A) \Cap (h, C)) = (K, (A \cap B) \cup (A \cap C)) = (K, A \cap (B \cup C))$.

Now for any $\varepsilon \in A \cap (B \cup C)$, it follows that $\varepsilon \in A$ and $\varepsilon \in B \cup C$. We consider the following cases. Case 1: $\varepsilon \in A$, $\varepsilon \notin B$ and $\varepsilon \in C$. Then $J_\varepsilon = f_\varepsilon \cap h_\varepsilon = K_\varepsilon$.

Case 2: $\varepsilon \in A$, $\varepsilon \in B$ and $\varepsilon \notin C$. Then $J_\varepsilon = f_\varepsilon \cap g_\varepsilon = K_\varepsilon$.

Case 3: $\varepsilon \in A$, $\varepsilon \in B$ and $\varepsilon \in C$. Then $J_\varepsilon = f_\varepsilon \cap (g_\varepsilon \cup h_\varepsilon) = (f_\varepsilon \cap g_\varepsilon) \cup (f_\varepsilon \cap h_\varepsilon) = K_\varepsilon$.

Therefore, J and $\mathbb{F}$ are the same operators, and so $(h, A) \Cap ((g, B)\tilde{\cup}(h, C)) = ((f, A) \Cap (g, B))\tilde{\cup}((f, A) \Cap (h, C))$. It follows that $(f, A) \Cap ((g, B)\tilde{\cup}(h, C)) =_{(\alpha,\beta)} ((f, A) \Cap (g, B))\tilde{\cup}((f, A) \Cap (h, C))$. This completes the proof.

We state the following theorem without its proof.

Theorem 8.14 $(\mathbb{FSI}(L, E), \Cup, \tilde{\cap})$ *is a complete distributive lattice under the ordering relation $\Subset'_{(\alpha,\beta)}$, where for any $(f, A), (g, B) \in \mathbb{FSI}(L, E)$, $(f, A) \Subset'_{(\alpha,\beta)} (g, B)$ if and only if $B \subseteq A$ and $f_\varepsilon \subseteq \vee q_{(\alpha,\beta)} g_\varepsilon$ for any $\varepsilon \in B$.*

8.5 Bipolar Fuzzy Soft Lie Algebras

Definition 8.27 Let (f, A) be a bipolar fuzzy soft set over L. Then (f, A) is said to be a *bipolar fuzzy soft Lie subalgebra* over L if $f(x)$ is a bipolar fuzzy Lie subalgebra of L for all $x \in A$, that is, a bipolar fuzzy soft set (f, A) over L is called a *bipolar fuzzy soft Lie subalgebra* of L if the following conditions are satisfied:

(1) $\mu^P_{f_\varepsilon}(x+y) \geq \min\{\mu^P_{f_\varepsilon}(x), \mu^P_{f_\varepsilon}(y)\}$,
(2) $\mu^N_{f_\varepsilon}(x+y) \leqslant \max\{\mu^N_{f_\varepsilon}(x), \mu^N_{f_\varepsilon}(y)\}$,
(3) $\mu^P_{f_\varepsilon}(mx) \geq \mu^P_{f_\varepsilon}(x), \quad \mu^N_{f_\varepsilon}(mx) \leqslant \mu^N_{f_\varepsilon}(x)$,
(4) $\mu^P_{f_\varepsilon}([x, y]) \geq \min\{\mu^P_{f_\varepsilon}(x), \mu^P_{f_\varepsilon}(y)\}$,
(5) $\mu^N_{f_\varepsilon}([x, y]) \leqslant \max\{\mu^N_{f_\varepsilon}(x), \mu^N_{f_\varepsilon}(y)\}$

for all $x, y \in L$ and $m \in \mathbb{F}$.

Example 8.13 The real vector space $\Re^2$ with $[x, y] = x \times y$ is a real Lie algebra. Let $\mathbb{N}$ and $\mathbb{Z}$ denote the set of all natural numbers and the set of all integers, respectively. By routine computations, we can easily check that $(f, \mathbb{Z})$, where $f : \mathbb{Z} \to ([0, 1] \times [-1, 0])^{\Re^2}$ with $f(n) = (\mu^P_{f_n}, \mu^N_{f_n}) : \Re^2 \to [0, 1] \times [-1, 0]$ for all $n \in \mathbb{Z}$,

$$\mu^P_{f_n}(x) = \begin{cases} 0.6 & \text{if } x = (0,0) = \underline{0}, \\ 0.2 & \text{if } x = (0, a), a \neq 0, \\ 0 & \text{otherwise}, \end{cases}$$

$$\mu^N_{f_n}(x) = \begin{cases} -0.3 & \text{if } x = (0,0) = \underline{0}, \\ -0.2 & \text{if } x = (0, a), a \neq 0, \\ -1 & \text{otherwise}, \end{cases}$$

is a bipolar fuzzy soft Lie subalgebra of $\Re^2$.

We state the following propositions without their proofs.

Proposition 8.21 *Let (f, A) be a bipolar fuzzy soft Lie subalgebra on L, then*

(i) $\mu^P_{f_\varepsilon}(\underline{0}) \geq \mu^P_{f_\varepsilon}(x), \quad \mu^N_{f_\varepsilon}(\underline{0}) \leqslant \mu^N_{f_\varepsilon}(x)$,
(ii) $\mu^P_{f_\varepsilon}([x, y]) = \mu^P_{f_\varepsilon}(-[y, x]) = \mu^P_{f_\varepsilon}([y, x])$,
(iii) $\mu^N_{f_\varepsilon}([x, y]) = \mu^N_{f_\varepsilon}(-[y, x]) = \mu^N_{f_\varepsilon}([y, x])$

for all $x, y \in L$.

Proposition 8.22 *Let (f, A) and (g, B) be bipolar fuzzy soft Lie subalgebras over L, then $(f, A)\widetilde{\cap}(g, B)$ and $(f, A) \wedge (g, B)$ are bipolar fuzzy soft Lie subalgebras over L. If $A \cap B = \emptyset$, then also $(f, A)\widetilde{\cup}(g, B)$ is a bipolar fuzzy soft Lie subalgebra.* □

Proposition 8.23 *Let (f, A) be a bipolar fuzzy soft Lie subalgebra over L and let $\{(h_i, B_i) \mid i \in I\}$ be a nonempty family of bipolar fuzzy soft Lie subalgebras of (f, A). Then*

(a) $\widetilde{\bigcap}_{i\in I}(h_i, B_i)$ *is a bipolar fuzzy soft Lie subalgebra of* (f, A),
(b) $\bigwedge_{i\in I}(h_i, B_i)$ *is a bipolar fuzzy soft Lie subalgebra of* $\bigwedge_{i\in I}(f, A)$,
(c) *If* $B_i \cap B_j = \emptyset$ *for all* $i, j \in I, i \neq j$, *then* $\widetilde{\bigvee}_{i\in I}(H_i, B_i)$ *is a bipolar fuzzy soft Lie subalgebra of* $\widetilde{\bigvee}_{i\in I}(f, A)$.

Definition 8.28 Let (f, A) be a bipolar fuzzy soft set over X. For each $s \in [0, 1]$, $t \in [-1, 0]$, the set $(f, A)^{(s,t)} = (f^{(s,t)}, A)$, where

$$(f, A)_{\varepsilon}^{(s,t)} = \{x \in X \mid \mu_{f_\varepsilon}^P(x) \geq s, \quad \mu_{f_\varepsilon}^N(x) \leqslant t\} \text{ for all } \varepsilon \in A,$$

is called an (s, t)*-level soft set* of (f, A). Clearly, $(f, A)^{(s,t)}$ is a soft set over X.

Theorem 8.15 *Let* (f, A) *be a bipolar fuzzy soft set over* L. (f, A) *is a bipolar fuzzy soft Lie subalgebra if and only if* $(f, A)^{(s,t)}$ *is a soft Lie subalgebra over* L *for each* $s \in [0, 1]$, $t \in [-1, 0]$.

Proof Suppose that (f, A) is a bipolar fuzzy soft Lie subalgebra. Then for each $s \in [0, 1], t \in [-1, 0], \varepsilon \in A$ and $x_1, x_2 \in (f, A)_{\varepsilon}^{(s,t)}$ we have $\mu_{f_\varepsilon}^P(x_1) \geq s, \mu_{f_\varepsilon}^P(x_2) \geq s$, and $\mu_{f_\varepsilon}^N(x_1) \leqslant t, \mu_{f_\varepsilon}^N(x_2) \leqslant t$. From Definition 8.28, it follows that $(f, A)_{\varepsilon}^{(s,t)}$ is a bipolar fuzzy Lie subalgebra over L. Thus, $\mu_{f_\varepsilon}^P(x_1 + x_2) \geq \min(\mu_{f_\varepsilon}^P(x_1), \mu_{f_\varepsilon}^P(x_2))$, $\mu_{f_\varepsilon}^P(x_1 + x_2) \geq s$, $\mu_{f_\varepsilon}^N(x_1 + x_2) \leqslant \max(\mu_{f_\varepsilon}^N(x_1), \mu_{f_\varepsilon}^N(x_2))$, $\mu_{f_\varepsilon}^N(x_1 + x_2) \leqslant t$. This implies that $x_1 + x_2 \in f_{\varepsilon}^{s}$. The verification for other conditions is similar. Thus, $(f, A)^{(s,t)}$ is a soft Lie subalgebra over L for each $s \in [0, 1], t \in [-1, 0]$.

Conversely, assume that $(f, A)^{(s,t)}$ is a soft Lie subalgebra over L for each $s \in [0, 1], t \in [-1, 0]$. For each $\varepsilon \in A$ and $x_1, x_2 \in (f, A)_{\varepsilon}^{(s,t)}$, let $s = \min\{\mu_{f_\varepsilon}^P(x_1), \mu_{f_\varepsilon}^P(x_2)\}$ and let $t = \max\{\mu_{f_\varepsilon}^N(x_1), \mu_{f_\varepsilon}^N(x_2)\}$, then $x_1, x_2 \in (f, A)_{\varepsilon}^{(s,t)}$. Since $(f, A)_{\varepsilon}^{(s,t)}$ is a Lie subalgebra over L, then $x_1 + x_2 \in (f, A)_{\varepsilon}^{(s,t)}$. This means that $\mu_{f_\varepsilon}^P(x_1 + x_2) \geq \min(\mu_{f_\varepsilon}^P(x_1), \mu_{f_\varepsilon}^P(x_2))$ and $\mu_{f_\varepsilon}^N(x_1 + x_2) \leqslant \max(\mu_{f_\varepsilon}^N(x_1), \mu_{f_\varepsilon}^N(x_2))$. The verification for other conditions is similar. Thus, according to Definition 8.27, (f, A) is a bipolar fuzzy soft Lie subalgebra over L. This completes the proof.

Definition 8.29 Let $\phi : L_1 \to L_2$ and $\psi : A \to B$ be two functions, A and B are parametric sets from the crisp sets L_1 and L_2, respectively. Then the pair (ϕ, ψ) is called a *bipolar fuzzy soft function* from L_1 to L_2.

Definition 8.30 Let (f, A) and (g, B) be two bipolar fuzzy soft sets over L_1 and L_2, respectively, and let (ϕ, ψ) be a bipolar fuzzy soft function from L_1 to L_2.

The *image of* (f, A) under the bipolar fuzzy soft function (ϕ, ψ), denoted by $(\phi, \psi)(f, A)$, is the bipolar fuzzy soft set on L_2 defined by $(\phi, \psi)(f, A) = (\phi(f), \psi(A))$, where for all $k \in \psi(A)$, $y \in L_2$

$$\mu_{\phi(f)_k}^P(y) = \begin{cases} \bigvee_{\phi(x)=y} \bigvee_{\psi(a)=k} f_a(x) & \text{if } x \in \psi^{-1}(y), \\ 1 & \text{otherwise,} \end{cases}$$

$$\mu^N_{\phi(f)_k}(y) = \begin{cases} \bigwedge_{\phi(x)=y} \bigwedge_{\psi(a)=k} f_a(x) & \text{if } x \in \psi^{-1}(y), \\ -1 & \text{otherwise.} \end{cases}$$

The *pre-image of* (g, B) under the bipolar fuzzy soft function (ϕ, ψ), denoted by $(\phi, \psi)^{-1}(g, B)$, is the bipolar fuzzy soft set over L_1 defined by $(\phi, \psi)^{-1}(g, B) = (\phi^{-1}(g), \psi^{-1}(B))$, where for all $a \in \psi^{-1}(A)$ for all $x \in L_1$,

$$\mu^P_{\phi^{-1}(g)_a}(x) = \mu^P_{g_{\psi(a)}}(\phi(x)), \quad \mu^N_{\phi^{-1}(g)_a}(x) = \mu^N_{g_{\psi(a)}}(\phi(x)).$$

Definition 8.31 Let (ϕ, ψ) be a bipolar fuzzy soft function from L_1 to L_2. If ϕ is a homomorphism from L_1 to L_2 then (ϕ, ψ) is said to be a *bipolar fuzzy soft homomorphism.* If ϕ is a isomorphism from L_1 to L_2 and ψ is one-to-one mapping from A onto B then (ϕ, ψ) is said to be a *bipolar fuzzy soft isomorphism.*

Theorem 8.16 *Let (g, B) be a bipolar fuzzy soft Lie subalgebra over L_2 and let (ϕ, ψ) be a bipolar fuzzy soft homomorphism from L_1 to L_2. Then $(\phi, \psi)^{-1}(g, B)$ is a bipolar fuzzy soft Lie subalgebra over L_1.*

Proof Let $x_1, x_2 \in L_1$, then

$$\begin{aligned} \phi^{-1}(\mu^P_{g_\varepsilon})(x_1 + x_2) &= \mu^P_{g_{\psi(\varepsilon)}}(\phi(x_1 + x_2)) = \mu^P_{g_{\psi(\varepsilon)}}(\phi(x_1) + \phi(x_2)) \\ &\geq \min\{\mu^P_{g_{\psi(\varepsilon)}}(\phi(x_1)), \mu^P_{g_{\psi(\varepsilon)}}(\phi(x_2))\} \\ &= \min\{\phi^{-1}(\mu^P_{g_\varepsilon})(x_1), \phi^{-1}(\mu^P_{g_\varepsilon})(x_2)\}, \end{aligned}$$

$$\begin{aligned} \phi^{-1}(\mu^N_{g_\varepsilon})(x_1 + x_2) &= \mu^N_{g_{\psi(\varepsilon)}}(\phi(x_1 + x_2)) = \mu^N_{g_{\psi(\varepsilon)}}(\phi(x_1) + \phi(x_2)) \\ &\leqslant \max\{\mu^N_{g_{\psi(\varepsilon)}}(\phi(x_1)), \mu^N_{g_{\psi(\varepsilon)}}(\phi(x_2))\} \\ &= \max\{\phi^{-1}(\mu^N_{g_\varepsilon})(x_1), \phi^{-1}(\mu^N_{g_\varepsilon})(x_2)\}. \end{aligned}$$

The verification for other conditions is similar, and hence, we omit the detail. Hence, $(\phi, \psi)^{-1}(g, B)$ is a bipolar fuzzy soft Lie subalgebra over L_1.

Note that $(\phi, \psi)(f, A)$ may not be a bipolar fuzzy soft Lie subalgebra over L_2.

8.6 $(\in, \in \vee q)$-Bipolar Fuzzy Soft Lie Algebras

Let $c \in G$ be fixed. If $\gamma \in (0, 1]$ and $\delta \in [-1, 0)$ are two real numbers, then $c(\gamma, \delta) = \langle x, c_\gamma, c_\delta \rangle$ is called a *bipolar fuzzy point* in G, where γ(resp, δ) is the positive degree of membership (resp, negative degree of membership) of $c(\gamma, \delta)$ and $c \in G$ is the support of $c(\gamma, \delta)$. Let $c(\gamma, \delta)$ be a bipolar fuzzy in G and let $A = \langle x, \mu^P_A, \mu^N_A \rangle$ be a bipolar fuzzy in G. Then $c(\gamma, \delta)$ is said to belong to A, written $c(\gamma, \delta) \in A$ if $\mu^P_A(c) \geq \gamma$ and $\mu^N_A(c) \leqslant \delta$. We say that $c(\gamma, \delta)$ is quasicoincident with A, written $c(\gamma, \delta) q A$, if $\mu^P_A(c) + \gamma > 1$ and $\mu^N_A(c) + \delta < -1$. To say that $c(\gamma, \delta) \in \vee q A$ (resp, $c(\gamma, \delta) \in \wedge q A$) means that $c(\gamma, \delta) \in A$ or $c(\gamma, \delta) q A$ (resp, $c(\gamma, \delta) \in A$ and $c(\gamma, \delta) q A$) and $c(\gamma, \delta) \overline{\in \vee q} A$ means that $c(\gamma, \delta) \in \vee q A$ does not hold.

Definition 8.32 A bipolar fuzzy set $A = (\mu_A^P, \mu_A^N)$ in L is called an $(\in, \in \vee q)$-*bipolar fuzzy Lie subalgebra* of L if it satisfies the following conditions:

(a) $x(s_1, t_1), y(s_2, t_2) \in A \Rightarrow (x + y)(\min(s_1, s_2), \max(t_1, t_2)) \in \vee q A$,
(b) $x(s, t) \in A \Rightarrow (mx)(s, t) \in \vee q A$,
(c) $x(s_1, t_1), y(s_2, t_2) \in A \Rightarrow ([x, y])(\min(s_1, s_2), \max(t_1, t_2)) \in \vee q A$

for all $x, y \in L, m \in \mathbb{F}, s, s_1, s_2 \in (0, 1],\ t, t_1, t_2 \in [-1, 0)$.

Example 8.14 Let $\mathfrak{R}^3$ be as in Example 8.12. We define a bipolar fuzzy set $A : \mathfrak{R}^3 \to [0, 1] \times [-1, 0]$ by

$$\mu_A^P(x) := \begin{cases} 1 & \text{if } x = (0, 0, 0), \\ 0.4 & \text{otherwise}, \end{cases}$$

$$\mu_A^N(x) := \begin{cases} 0 & \text{if } x = (0, 0, 0), \\ -0.2 & \text{otherwise}. \end{cases}$$

By routine computations, it is easy to see that A is not an $(\in, \in \vee q)$-bipolar fuzzy Lie subalgebra of L.

Theorem 8.17 *A bipolar fuzzy set A in a Lie algebra L is an* $(\in, \in \vee q)$*-bipolar fuzzy Lie subalgebra of L if and only if*

- $\mu_A^P(x + y) \geq \min(\mu_A^P(x), \mu_A^P(y), 0.5), \mu_A^N(x + y) \leqslant \max(\mu_A^N(x), \mu_A^N(y), -0.5)$,
- $\mu_A^P(mx) \geq \min(\mu_A^P(x), 0.5),\ \ \mu_A^N(mx) \leqslant \max(\mu_A^N(x), -0.5)$,
- $\mu_A^P([x, y]) \geq \min(\mu_A^P(x), \mu_A^P(y), 0.5),\ \mu_A^N([x, y]) \leqslant \max(\mu_A^N(x), \mu_A^N(y), -0.5)$

hold for all $x, y \in L,\ m \in \mathbb{F}$.

Theorem 8.18 *A bipolar fuzzy set A of a Lie algebra of L is an* $(\in, \in \vee q)$*-bipolar fuzzy Lie subalgebra of L if and only if for all* $s \in (0.5, 1],\ t \in [-1, -0.5)$ *each nonempty* $A_{(s,t)}$ *is a Lie subalgebra of L.*

Proof Assume that A is an $(\in, \in \vee q)$-bipolar fuzzy Lie subalgebra of L and let $s \in (0.5, 1], t \in [-1, -0.5)$. If $x, y \in A_{(s,t)}$, then $\mu_A^P(x) \geq s$ and $\mu_A^P(y) \geq s$, $\mu_A^N(x) \leq t$ and $\mu_A^N(y) \leq t$. Thus, $\mu_A^P(x + y) \geq \min(\mu_A^P(x), \mu_A^P(y), 0.5) \geq \min(s, 0.5) = s$ and $\mu_A^N(x + y) \leqslant \max(\mu_A^N(x), \mu_A^N(y), -0.5) \leqslant \max(t, -0.5) = t$, so $x + y \in A_{(s,t)}$. The verification for other conditions is similar. The proof of converse part is obvious.

Theorem 8.19 *If A is a bipolar fuzzy set in a Lie algebra L, then* $A_{(s,t)}$ *is a Lie subalgebra of L if and only if*

- $\max(\mu_A^P(x + y), 0.5) \geq \min(\mu_A^P(x), \mu_A^P(y))$,
  $\min(\mu_A^N(x + y), -0.5) \leqslant \max(\mu_A^N(x), \mu_A^N(y))$,
- $\max(\mu_A^P(mx), 0.5) \geq \min(\mu_A^P(x))$,
  $\min(\mu_A^N(mx), -0.5) \leqslant \max(\mu_A^N(x))$,
- $\max(\mu_A^P([x, y]), 0.5) \geq \min(\mu_A^P(x), \mu_A^P(y))$,
  $\min(\mu_A^N([x, y]), -0.5) \leqslant \max(\mu_A^N(x), \mu_A^N(y))$

for all $x, y \in L,\ m \in \mathbb{F}$.

Definition 8.33 Let (f, A) be a bipolar fuzzy soft set over a Lie algebra L. Then (f, A) is called an $(\in, \in_\alpha \vee q_\beta)$-*bipolar fuzzy soft Lie subalgebra* if $f(\alpha)$ is an $(\in, \in_\alpha \vee q_\beta)$-bipolar fuzzy Lie subalgebra of L for all $\alpha \in A$.

Theorem 8.20 *Let (f, A) and (g, B) be two $(\in, \in_\alpha \vee q_\beta)$-bipolar fuzzy soft Lie subalgebras over a Lie algebra L. Then $(f, A) \wedge (g, B)$ is an $(\in, \in_\alpha \vee q_\beta)$-bipolar fuzzy soft Lie subalgebra over L.*

Proof By the definition, we can write $(f, A) \wedge (g, B) = (h, C)$, where $C = A \times B$ and $h(\alpha, \beta) = f(\alpha) \cap g(\beta)$ for all $(\alpha, \beta) \in C$. Now for any $(\alpha, \beta) \in C$, since (f, A) and (g, B) are $(\in, \in_\alpha \vee q_\beta)$-bipolar fuzzy soft Lie subalgebras over L, we have both $f(\alpha)$ and $g(\beta)$ are $(\in, \in_\alpha \vee q_\beta)$-bipolar fuzzy Lie subalgebras of L. Thus, $h(\alpha, \beta) = f(\alpha) \cap g(\beta)$ is an $(\in, \in_\alpha \vee q_\beta)$-bipolar fuzzy Lie subalgebra of L. Hence, $(f, A) \wedge (g, B)$ is an $(\in, \in_\alpha \vee q_\beta)$-bipolar fuzzy soft Lie subalgebra over L.

Theorem 8.21 *Let (f, A) and (g, B) be two $(\in, \in_\alpha \vee q_\beta)$-bipolar fuzzy soft Lie subalgebras over a Lie algebra L. Then $(f, A)\tilde{\cap}(g, B)$ is an $(\in, \in_\alpha \vee q_\beta)$-bipolar fuzzy soft Lie subalgebra over L.*

Proof We have $(f, A)\tilde{\cap}(g, B) = (h, C)$, where $C = A \cup B$ and

$$h(\varepsilon) = \begin{cases} f(\varepsilon) & \text{if } \varepsilon \in A - B, \\ g(\varepsilon) & \text{if } \varepsilon \in B - C, \\ f(\varepsilon) \cap g(\varepsilon) & \text{if } \varepsilon \in A \cap B. \end{cases}$$

for all $\alpha \in C$.

Now for any $\alpha \in C$, we consider the following cases.

Case 1: $\alpha \in A - B$. Then $h(\alpha) = f(\alpha)$ is an $(\in, \in_\alpha \vee q_\beta)$-bipolar fuzzy Lie subalgebra of L since (f, A) is an $(\in, \in_\alpha \vee q_\beta)$-bipolar fuzzy soft Lie subalgebra over L.

Case 2: $\alpha \in B - A$. Then $h(\alpha) = g(\alpha)$ is an $(\in, \in_\alpha \vee q_\beta)$-bipolar fuzzy Lie subalgebra of L since (g, B) is an $(\in, \in_\alpha \vee q_\beta)$-bipolar fuzzy soft Lie subalgebra over L.

Case 3: $\alpha \in A \cap B$. Then $h(\alpha) = f(\alpha) \cap g(\alpha)$ is an $(\in, \in_\alpha \vee q_\beta)$-bipolar fuzzy Lie subalgebra of L by the assumption. Thus, in any case, $h(\alpha)$ is an $(\in, \in_\alpha \vee q_\beta)$-bipolar fuzzy Lie subalgebra of L. Therefore, $(f, A)\tilde{\cap}(g, B)$ is an $(\in, \in_\alpha \vee q_\beta)$-bipolar fuzzy soft Lie subalgebra over L.

Theorem 8.22 *Let (f, A) and (g, B) be two $(\in, \in_\alpha \vee q_\beta)$-bipolar fuzzy soft Lie subalgebras over a Lie algebra L. If $A \cap B \neq \emptyset$, then $(f, A)\tilde{\cap}(g, B)$ is an $(\in, \in_\alpha \vee q_\beta)$-bipolar fuzzy soft Lie subalgebra over L.*

Proof $(f, A)\tilde{\cap}(g, B) = (h, C)$, where $C = A \cap B$ and $h(\alpha) = f(\alpha) \cap g(\alpha)$ for all $\alpha \in C$. Now for any $\alpha \in C$, since (f, A) and (g, B) are $(\in, \in_\alpha \vee q_\beta)$-bipolar fuzzy soft Lie subalgebras over L, we have both $f(\alpha)$ and $g(\alpha)$ are $(\in, \in_\alpha \vee q_\beta)$-bipolar fuzzy Lie subalgebras of L. Thus, $h(\alpha) = f(\alpha) \cap g(\alpha)$ is an $(\in, \in_\alpha \vee q_\beta)$-bipolar fuzzy Lie subalgebra of L. Therefore, $(f, A)\tilde{\cap}(g, B)$ is an $(\in, \in_\alpha \vee q_\beta)$-bipolar fuzzy soft Lie subalgebra over L.

Theorem 8.23 *Let (f, A) be an $(\in, \in_\alpha \vee q_\beta)$-bipolar fuzzy soft Lie subalgebra over L and let $\{(h_i, B_i) \mid i \in I\}$ be a nonempty family of $(\in, \in_\alpha \vee q_\beta)$-bipolar fuzzy soft Lie subalgebras of (f, A). Then*

(a) $\widetilde{\bigcap}_{i\in I}(h_i, B_i)$ *is an $(\in, \in_\alpha \vee q_\beta)$-bipolar fuzzy soft Lie subalgebra of (f, A),*
(b) $\bigwedge_{i\in I}(h_i, B_i)$ *is an $(\in, \in_\alpha \vee q_\beta)$-bipolar fuzzy soft Lie subalgebra of $\bigwedge_{i\in I}(f, A)$,*
(c) *If $B_i \cap B_j = \emptyset$ for all $i, j \in I$, then $\widetilde{\bigvee}_{i\in I}(H_i, B_i)$ is an $(\in, \in_\alpha \vee q_\beta)$-bipolar fuzzy soft Lie subalgebra of $\widetilde{\bigvee}_{i\in I} f, A)$.*

Theorem 8.24 *Let (f, A) and (g, B) be two $(\in, \in_\alpha \vee q_\beta)$-bipolar fuzzy soft Lie subalgebras over a Lie algebra L. If A and B are disjoint, then $(f, A)\widetilde{\cup}(g, B)$ is an $(\in, \in_\alpha \vee q_\beta)$-bipolar fuzzy soft Lie subalgebra over L.*

Chapter 9
Rough Fuzzy Lie Algebras

In this chapter, we apply the concept of hybrid model rough fuzzy set to Lie algebras. We present the notions of rough fuzzy Lie subalgebras, rough fuzzy Lie ideals, fuzzy rough Lie subalgebras, and rough intuitionistic fuzzy Lie subalgebras. We illustrate these concepts with examples. We also describe some of their properties.

9.1 Introduction

Most of the real-world problems ranging from engineering to medical and medical to social fields involve uncertainty in data. Zadeh [139] was the first one to introduce fuzzy set theory. The idea of fuzzy sets is welcomed because it deals with uncertain and vague information. Pawlak [111] introduced the concept of rough set. He was a Polish mathematician (citizen of Poland) and computer scientist. Rough means approximate or inexact. Rough set theory expresses vagueness in terms of a boundary region of a set not in terms of membership function as in fuzzy set. The idea of rough set theory is useful to study the intelligence systems containing inexact, uncertain, or incomplete information. It is a mathematical approach to imprecise knowledge. Rough set theory expresses vagueness by means of a boundary region of a set. The emptiness of boundary region of a set shows that this is a crisp set, and nonemptiness shows that this is a rough set. A subset of a universe in rough set theory is expressed by two approximations which are known as lower and upper approximations. Equivalence classes are the basic building blocks in rough set theory, for upper and lower approximations. Dubois and Prade [59] considered rough fuzzy sets and fuzzy rough sets, and they concluded that these two hybrid models are different approaches to handle vagueness. They reported that these are not opposite theories, but obtain beneficial results.

Definition 9.1 Let X be a nonempty finite universe and R an equivalence relation on X. A pair (X, R) is called a *Pawlak approximation space*. Let Y be a subset of X; then, the lower and upper approximations of Y are defined as follows:

M. Akram, *Fuzzy Lie Algebras*, Infosys Science Foundation Series,
https://doi.org/10.1007/978-981-13-3221-0_9

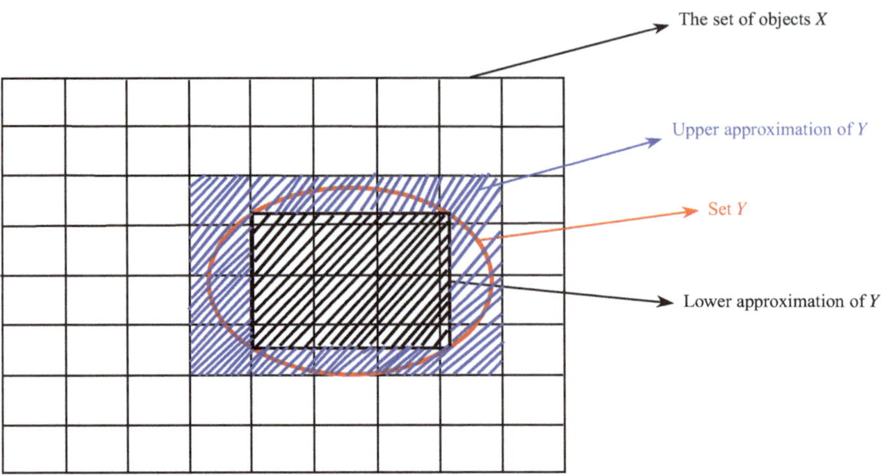

Fig. 9.1 Diagram of a rough set

$$\underline{R}(Y) = \{x \in X : [x]_R \subseteq Y\},$$
$$\overline{R}(Y) = \{x \in X : [x]_R \cap Y \neq \phi\},$$

where

$$[x]_R = \{y \in X : (x, y) \in R\}$$

denotes equivalence class of R containing x. $\underline{R}$ and $\overline{R}$ are called the *lower and upper approximations operators*, respectively. The pair $(\underline{R}(Y), \overline{R}(Y))$ is called a *Pawlak rough set*.

The graphical representation of rough set is shown in Fig. 9.1

Example 9.1 Let $X = \{1, 2, 3, 4, 5, 6\}$ be a universe and $R = \{\{1, 5\}, \{2, 3\}, \{4, 6\}\}$ an equivalence relation on X. Let $Y = \{2, 3, 5\}$. Then
$[1]_R = \{1, 5\} = [5]_R \nsubseteq Y$ but $[1]_R \cap Y \neq \emptyset \neq [5]_R \cap Y$
$[2]_R = \{2, 3\} = [3]_R \subseteq Y$ but $[3]_R \cap Y \neq \emptyset \neq [2]_R \cap Y$
$[4]_R = \{4, 6\} = [6]_R \nsubseteq Y$ but $[4]_R \cap Y = \emptyset$
Hence, $\underline{R}(Y) = \{2, 3\}$ and
$\overline{R}(Y) = \{1, 2, 3, 5\}$
$bn(X) = \overline{R}(Y) - \underline{R}(Y) = \{1, 5\} \neq \emptyset$
Thus, $(\underline{R}(Y), \overline{R}(Y))$ is a rough set w.r.t R.

Definition 9.2 Let X be a universe and R an equivalence relation on X. The lower and upper approximations of a fuzzy set $S \in F(X)$, denoted by $\underline{R}S$ and $\overline{R}S$, respectively, are defined as fuzzy sets in X such that

$$(\underline{R}S)(x) = \bigwedge_{y \in X}((1 - R(x, y)) \vee S(y)) = \bigwedge_{y \in [x]_R} S(y),$$

$$(\overline{R}S)(x) = \bigvee_{y \in X}(R(x, y) \wedge S(y)) = \bigvee_{y \in [x]_R} S(y),$$

for all $x \in X$. The pair $RS = (\underline{R}S, \overline{R}S)$ is called *rough fuzzy set*.

Example 9.2 Let $X = \{a, b, c\}$ be a set of universe and R an equivalence relation on X defined by

$$R = \begin{pmatrix} 1 & 1 & 0 \\ 1 & 1 & 0 \\ 0 & 0 & 1 \end{pmatrix}.$$

R can also be written as

$$R = \{\{a, b\}, \{c\}\}$$
$$R = \{(a, a), (a, b), (b, a), (b, b), (c, c)\}.$$

Let $A \in F(X)$ be defined as $A = \{(a, 0.2), (b, 0.4), (c, 0.8)\}$. Then by definition of rough fuzzy sets, we have

$$\begin{aligned}
(\underline{R}A)(a) &= (0 \vee 0.2) \wedge (0 \vee 0.4) \wedge (1 \vee 0.8) \\
&= 0.2 \wedge 0.4 \wedge 1 = 0.2. \\
(\underline{R}A)(b) &= (0 \vee 0.2) \wedge (0 \vee 0.4) \wedge (1 \vee 0.8) \\
&= 0.2 \wedge 0.4 \wedge 1 = 0.2. \\
(\underline{R}A)(c) &= (1 \vee 0.2) \wedge (1 \vee 0.4) \wedge (0 \vee 0.8) \\
&= 1 \wedge 1 \wedge 0.8 = 0.8. \\
(\overline{R}A)(a) &= (1 \wedge 0.2) \vee (1 \wedge 0.4) \vee (0 \wedge 0.8) \\
&= 0.2 \vee 0.4 \vee 0 = 0.4.
\end{aligned}$$

$$\begin{aligned}
(\overline{R}A)(b) &= (1 \wedge 0.2) \vee (1 \wedge 0.4) \vee (0 \wedge 0.8) \\
&= 0.2 \vee 0.4 \vee 0 = 0.4. \\
(\overline{R}A)(c) &= (0 \wedge 0.2) \vee (0 \wedge 0.4) \vee (1 \wedge 0.8) \\
&= 0 \vee 0 \vee 0.8 = 0.8.
\end{aligned}$$

Thus,

$$\underline{R}A = \{(a, 0.2), (b, 0.2), (c, 0.8)\},$$

$$\overline{R}A = \{(a, 0.4), (b, 0.4), (c, 0.8)\}.$$

Hence, $(\underline{R}A, \overline{R}A)$ is rough fuzzy set.

Definition 9.3 Consider the approximation of a fuzzy set $F = (U(S,t))_t$, $t \in [0, 1]$, in an approximation space (X, R), where R is an *equivalence relation*. For each t-level set F_t, we have a rough set:

Reference set: $U(S, t)$,

Lower approximation: $\underline{R}U(S,t) = \{x \in X \mid [x]_R \subseteq U(S,t)\}$,
Upper approximation: $\overline{R}U(S,t) = \{x \in X \mid [x]_R \cap U(S,t) \neq \emptyset\}$.

That is, $(\underline{R}U(S,t), \overline{R}U(S,t)) = (\underline{R}S, \overline{R}S)_t$ is a rough set with reference set $U(S, t)$. For the family of t-*level sets*, we have a family of lower and upper approximations, $\underline{R}U(S,t)$ and $\overline{R}U(S,t)$, $t \in [0, 1]$.

Properties. Let X be a universe, and let R be an equivalence relation on X. The lower and upper approximations of fuzzy sets $S, T \in F(X)$ have the following properties:

1. $\underline{R}(X) = X = \overline{R}(X)$,
2. $\underline{R}(\emptyset) = \emptyset = \overline{R}(\emptyset)$,
3. $\overline{R}(S \cup T) = \overline{R}(S) \cup \overline{R}(T)$,
4. $\underline{R}(S \cup T) \supseteq \underline{R}(S) \cup \underline{R}(T)$,
5. $S \subseteq T \longrightarrow \underline{R}(S) \subseteq \underline{R}(T)$ and $\overline{R}(S) \subseteq \overline{R}(T)$,
6. $\underline{R}(S \cap T) = \underline{R}(S) \cap \underline{R}(T)$,
7. $\overline{R}(S \cap T) \subseteq \overline{R}(S) \cap \overline{R}(T)$,
8. $\underline{R}(\sim S) = \sim \overline{R}(S)$, $\overline{R}(\sim S) = \sim \underline{R}(S)$,
9. $\underline{R}(\underline{R}(S)) = \overline{R}(\underline{R}(S)) = \underline{R}(S)$, $\overline{R}(\overline{R}(S)) = \underline{R}(\overline{R}(S)) = \overline{R}(S)$,

where $\sim S$ denotes the complement of S.

Definition 9.4 Let X be a nonempty set; a fuzzy set R on $X \times X$ is called a *fuzzy equivalence relation*, if it satisfies:

1. $\mu_R(x, x) = 1$ (reflexivity),
2. $\mu_R(x, y) = \mu_R(y, x)$ (symmetry),
3. $\mu_R(x, z) \geq \min\{\mu_R(x, y), \mu_R(y, z)\}$ (max-min transitivity).

Then, the pair (X, R) is called a fuzzy approximation space, and the *approximation operators* are defined by (for any $A \in F(X)$, where $F(X)$ denotes the fuzzy power set): For all $x \in X$,

$$\underline{apr}_R(A)(x) = \inf\{\max\{\mu_A(y), 1 - \mu_R(x, y)\} \mid y \in X\},$$

$$\overline{apr}_R(A)(x) = \sup\{\min\{\mu_A(y), \mu_R(x, y)\} \mid y \in X\}.$$

Fuzzy set $\underline{apr}_R(A)$ is called a lower approximation of A, and $\overline{apr}_R(A)$ is called an upper approximation of A. The pair $(\underline{apr}_R(A), \overline{apr}_R(A))$ is called *fuzzy rough approximation of* A, and it is also called a *fuzzy rough set*.

Example 9.3 Let $X = \{a, b, c\}$ be a set of universe and R a fuzzy relation on X defined by

$$R = \begin{pmatrix} 1 & 0.2 & 0.7 \\ 0.2 & 1 & 0.2 \\ 0.7 & 0.2 & 1 \end{pmatrix}.$$

That is,
$\mu_R(a, a) = \mu_R(b, b) = \mu_R(c, c) = 1.$ (reflexivity)
$\mu_R(a, b) = \mu_R(b, a),\ \ \mu_R(b, c) = \mu_R(c, b),\ \ \mu_R(a, c) = \mu_R(c, a).$ (symmetry)
Now, we check max-min transitivity of R:

$$R^2 = \begin{pmatrix} 1 & 0.2 & 0.7 \\ 0.2 & 1 & 0.2 \\ 0.7 & 0.2 & 1 \end{pmatrix} \begin{pmatrix} 1 & 0.2 & 0.7 \\ 0.2 & 1 & 0.2 \\ 0.7 & 0.2 & 1 \end{pmatrix} = \begin{pmatrix} 1 & 0.2 & 0.7 \\ 0.2 & 1 & 0.2 \\ 0.7 & 0.2 & 1 \end{pmatrix} = R.$$

Thus, R is max-min transitive. Hence, R is a fuzzy equivalence relation on X. Let $A = \{(a, 0.2), (b, 0.4), (c, 0.8)\}$ be a fuzzy set on X, and then by definition of fuzzy rough sets, we have

$$\underline{apr}_R(A)(a) = 0.2 \wedge 0.8 \wedge 0.8 = 0.2,$$

$$\overline{apr}_R(A)(a) = 0.2 \vee 0.2 \vee 0.7 = 0.7.$$

Similarly,

$$\underline{apr}_R(A)(b) = 0.4, \qquad \underline{apr}_R(A)(c) = 0.3,$$
$$\overline{apr}_R(A)(b) = 0.4, \qquad \overline{apr}_R(A)(c) = 0.8.$$

Thus,

$$\underline{apr}_R(A) = \{(a, 0.2), (b, 0.4), (c, 0.3)\},$$

$$\overline{apr}_R(A) = \{(a, 0.7), (b, 0.4), (c, 0.8)\}.$$

Hence, $(\underline{apr}_R(A), \overline{apr}_R(A))$ is fuzzy rough set.

Properties. Let (X, R) be a fuzzy approximation space. Then, the lower and upper approximation operators $\underline{apr}_R(A)$ and $\overline{apr}_R(A)$ satisfy the following properties for any $A, B \in F(X)$,

1. $\underline{apr}_R(\phi) = \overline{apr}_R(\phi) = \phi$,
2. $\underline{apr}_R(X) = \overline{apr}_R(X) = X$,
3. $\underline{apr}_R(A) = \sim \overline{apr}_R(\sim A)$,
4. $A \subseteq B \Rightarrow \underline{apr}_R(A) \subseteq \underline{apr}_R(B)$,
5. $\underline{apr}_R(A \cup B) \supseteq \underline{apr}_R(A) \cup \underline{apr}_R(B)$,

6. $\underline{apr}_R(A \cap B) = \underline{apr}_R(A) \cap \underline{apr}_R(B)$,
7. $\overline{apr}_R(A) = \sim \underline{apr}_R(\sim A)$,
8. $A \subseteq B \Rightarrow \overline{apr}_R(A) \subseteq \overline{apr}_R(B)$,
9. $\overline{apr}_R(A \cup B) = \overline{apr}_R(A) \cup \overline{apr}_R(B)$,
10. $\overline{apr}_R(A \cap B) \subseteq \overline{apr}_R(A) \cap \overline{apr}_R(B)$.

Remark 9.1 A *rough set* is the approximation of a crisp set in a crisp approximation space. It is a pair of crisp set. A *rough fuzzy set* is derived from the approximation of a fuzzy set in a crisp approximation space. It is a pair of fuzzy sets in which all elements in the same equivalence class have the same membership. The membership of an element is determined by the original memberships of all those elements equivalent to that element. A *fuzzy rough set* is derived from the approximation of a fuzzy set in a fuzzy approximation space. It is a pair of fuzzy sets in which the membership of an element is determined by the degrees of similarity of all those elements in the set.

9.2 Rough Fuzzy Lie Ideals

Definition 9.5 Let $(\underline{R}S, \overline{R}S)$ be a rough fuzzy set over L. A pair $(\underline{R}S, \overline{R}S)$ is called a *rough fuzzy Lie subalgebra* over L if $(\underline{R}S)(x)$ and $(\overline{R}S)(x)$ are fuzzy Lie subalgebras of L for all $x \in L$. Equivalently, a rough fuzzy set $(\underline{R}S, \overline{R}S)$ over L is called a *rough fuzzy Lie subalgebra* of L if the following conditions are satisfied:

1.
$$\begin{cases} (\underline{R}S)(x+y) \geq \min\{(\underline{R}S)(x), (\underline{R}S)(y)\}, \\ (\overline{R}S)(x+y) \geq \min\{\overline{R}S(x), \overline{R}S(y)\}, \end{cases}$$

2.
$$\begin{cases} (\underline{R}S)(\alpha x) \geq (\underline{R}S)(x), \\ (\overline{R}S)(\alpha x) \geq (\overline{R}S)(x), \end{cases}$$

3.
$$\begin{cases} (\underline{R}S)([x,y]) \geq \min\{(\underline{R}S)(x), (\underline{R}S)(y)\}, \\ (\overline{R}S)([x,y]) \geq \min\{(\overline{R}S)(x)(\overline{R}S)(y)\}, \end{cases}$$

for all $x, y \in L, \alpha \in \mathbb{F}$.

Definition 9.6 A rough fuzzy set $(\underline{R}S, \overline{R}S)$ on L is called a *rough fuzzy Lie ideal* if it satisfies the conditions (1) and (2) of Definition 9.5 and the following

(4)

$$\begin{cases} (\underline{R}S)([x, y]) \geq (\underline{R}S)(x) \\ (\overline{R}S)([x, y]) \geq \overline{R}S(x) \end{cases}$$

for all $x, y \in L$.

Remark 9.2 From condition (2) of Definition 9.5, it follows that:

1. $(\underline{R}S)(0) \geq (\underline{R}S)(x)$, $(\overline{R}S)(0) \geq (\overline{R}S)(x)$.
2. $(\underline{R}S)(-x) \geq (\underline{R}S)(x)$, $(\overline{R}S)(-x) \geq (\overline{R}S)(x)$.
3. $(\underline{R}S)([x, y]) = (\underline{R}S)(-[y, x]) = (\underline{R}S)([y, x])$, $(\overline{R}S)([x, y]) = (\overline{R}S)(-[y, x]) = (\overline{R}S)([y, x])$

for all $x, y \in L$.

Proposition 9.1 *Every rough fuzzy Lie ideal is a rough fuzzy Lie subalgebra.*

The converse of the Proposition 9.1 is not true, in general, as it can be seen in the following example.

Example 9.4 Consider $\mathbb{F} = \mathbb{R}$. Let $\Re^3 = \{(r_1, r_2, r_3) : r_1, r_2, r_3 \in \mathbb{R}\}$ be the set of all three-dimensional real vectors which consist of elements of the form:

1. $o = (r_1, r_2, r_3)$ such that $r_1 = r_2 = 0 = r_3$,
2. $y = (r_1, r_2, r_3)$ such that $r_1 \neq 0, r_2 = r_3 = 0$, and
3. $z = (r_1, r_2, r_3)$, otherwise.

Then, it is clear that $\Re^3$ endowed with the operation defined by $[r, r'] = r \times r'$ forms a real Lie algebra. Let $S = \{(o, 0.9), (y, 0.6), (z, 0.2)\}$ be a fuzzy set on $\{o, y, z\}$ and R an equivalence relation as shown in Table 9.1.

The lower and upper fuzzy approximations $\underline{R}S$ and $\overline{R}S$ are given as:

$$(\underline{R}S)(r) = \begin{cases} 0.6, & \text{if } r = o, \\ 0.6, & \text{if } r = y, \\ 0.2, & \text{if } r = z, \end{cases} \qquad (\overline{R}S)(r) = \begin{cases} 0.9, & \text{if } r = o, \\ 0.9, & \text{if } r = y, \\ 0.2, & \text{if } r = z, \end{cases}$$

Then, $(\underline{R}S, \overline{R}S)$ is a rough fuzzy Lie subalgebra of L but $(\underline{R}S, \overline{R}S)$ is not a rough fuzzy Lie ideal of L since

$$(\underline{R}S)([(1, 0, 0), (1, 1, 1)]) = (\underline{R}S)(0, -1, 1) = 0.2 \text{ and } (\underline{R}S)(1, 0, 0) = 0.6,$$

Table 9.1 Equivalence relation of $\Re^3$

R	o	y	z
o	1	1	1
y	1	1	0
z	1	1	1

that is,

$$(\underline{R}S)([(1,0,0),(1,1,1)]) = 0.2 \not\geq (\underline{R}S)(1,0,0) = 0.6.$$

Proposition 9.2 *Let $(\underline{R}S,\ \overline{R}S)$ be a rough fuzzy Lie subalgebra in a Lie algebra L. Then, $(\underline{R}S,\ \overline{R}S)$ is a rough fuzzy Lie subalgebra of L if and only if the nonempty t-level cut $(\underline{R}S,\ \overline{R}S)_t$ is a rough Lie subalgebra of L when $U(S,t)$ is proper subset, where $t \in [0,\ 1]$.*

Proposition 9.3 *Let $(\underline{R}S,\ \overline{R}S)$ be a rough fuzzy Lie ideal in a Lie algebra L. Then, $(\underline{R}S,\ \overline{R}S)$ is a rough fuzzy Lie ideal of L if and only if the set $(\underline{R}S,\ \overline{R}S)_t$ is a rough Lie ideal of L when $U(S,t)$ is proper subset, where $t \in [0,\ 1]$.*

Definition 9.7 Let $A = (\underline{R}S,\ \overline{R}S)$ and $B = (\underline{R}T,\ \overline{R}T)$ be rough fuzzy Lie algebras L. We define the *sup-min-product* $\ll AB \gg = (\ll \underline{A}\ \underline{B} \gg, \ll \overline{A}\ \overline{B} \gg) = ((\underline{R}S)(\underline{R}T),\ (\overline{R}S)(\overline{R}T))$ of A and B as follows: for all $x, y, z \in L$

$$\ll \underline{A}\ \underline{B} \gg (x) = \begin{cases} \sup_{x=\sum_{i=1}^{n}[x_i y_i]} \min((\underline{R}S)(x_i), \underline{R}T(y_i)), & \text{if } x = \sum_{i=1}^{n}[x_i y_i], \\ 0, & x \neq \sum_{i=1}^{n}[x_i y_i], \end{cases}$$

$$\ll \overline{A}\ \overline{B} \gg (x) = \begin{cases} \sup_{x=\sum_{i=1}^{n}[x_i y_i]} \min((\overline{R}S)(x_i), \overline{R}T(y_i)), & \text{if } x = \sum_{i=1}^{n}[x_i y_i], \\ 0, & x \neq \sum_{i=1}^{n}[x_i y_i]. \end{cases}$$

Proposition 9.4 *If $(\underline{R}S,\ \overline{R}S)$ and $(\underline{R}T,\ \overline{R}T)$ are rough fuzzy Lie algebras of a Lie algebra L, then the functions $(\underline{R}S,\ \overline{R}S)\tilde{\cap}(\underline{R}T,\ \overline{R}T)$ and $(\underline{R}S,\ \overline{R}S) \wedge (\underline{R}T,\ \overline{R}T)$: $L \to [0,\ 1]$ are rough fuzzy Lie algebra of L, where*

$$(\underline{R}S \wedge \underline{R}T)(x) = \min\{(\underline{R}S)(x),\ (\underline{R}T)(x)\},$$
$$(\overline{R}S \wedge \overline{R}T)(x) = \min\{(\overline{R}S)(x),\ (\overline{R}T)(x)\}.$$

Also, $(\underline{R}S,\ \overline{R}S)\tilde{\cup}(\underline{R}T,\ \overline{R}T) : L \to [0,\ 1]$ is also a rough fuzzy Lie algebra of L.

Proof Let $x, y \in L$ and $\alpha \in \mathbb{F}$. Then

$$\begin{aligned}
(\underline{R}S \wedge \underline{R}T)(x+y) &= \min\{(\underline{R}S)(x+y),\ (\underline{R}T)(x+y)\} \\
&\geq \min\{\min\{(\underline{R}S)(x),\ (\underline{R}S)(y)\},\ \min\{(\underline{R}T)(x),\ (\underline{R}T)(y)\}\} \\
&= \min\{\min\{(\underline{R}S)(x),\ (\underline{R}T)(x)\},\ \min\{(\underline{R}S)(y),\ \underline{R}T(y)\}\} \\
&= \min\{(\underline{R}S \wedge \underline{R}T)(x),\ (\underline{R}S \wedge \underline{R}T)(y)\}, \\
(\underline{R}S \wedge \underline{R}T)(\alpha x) &= \min\{\underline{R}S(\alpha x),\ (\underline{R}T)(\alpha x)\} \geq \min\{(\underline{R}S)(x),\ (\underline{R}T)(x)\} \\
&= (\underline{R}S \wedge \underline{R}T)(x), \\
(\underline{R}S \wedge \underline{R}T)([x,y]) &= \min\{(\underline{R}S)([x,y]),\ (\underline{R}T)([x,y])\} \geq \min\{(\underline{R}S)(x),\ (\underline{R}T)(x)\}
\end{aligned}$$

$$= (\underline{R}S \wedge \underline{R}T)(x).$$

Hence, $(\underline{R}S \vee \underline{R}T)$ is a fuzzy Lie algebra of L. Similarly, $(\overline{R}S \vee \overline{R}T)$ is a fuzzy Lie algebra of L. Hence, $(\underline{R}S,\ \overline{R}S) \wedge (\underline{R}T,\ \overline{R}T)$ is a rough fuzzy Lie algebra of L. Similarly, we can prove other cases.

Proposition 9.5 *If* $\{(\underline{R}S_i,\ \overline{R}S_i)|i \in I\}$ *is a family of rough fuzzy Lie algebras of Lie algebras L, then*

1. $\tilde{\bigwedge\limits_{i\in I}}(\underline{R}S_i,\ \overline{R}S_i)$ *is a rough fuzzy Lie subalgebra of L.*
2. $\tilde{\bigcap\limits_{i\in I}}(\underline{R}S_i,\ \overline{R}S_i)$ *is a rough fuzzy Lie subalgebra of L.*
3. $\tilde{\bigcup\limits_{i\in I}}(\underline{R}S_i,\ \overline{R}S_i)$ *is a rough fuzzy Lie subalgebra of L.*

Proof Since each $\underline{R}S_i$ is a fuzzy Lie algebra, they all must satisfy these conditions:

(i) $\underline{R}S_i(x + y) \geq \min\{\underline{R}S_i(x), \underline{R}S_i(y)\}$,
(ii) $\underline{R}S_i(\alpha x) \geq \underline{R}S_i(x)$,
(iii) $\underline{R}S_i([x, y]) \geq \min\{\underline{R}S_i(x), \underline{R}S_i(y)\}$ for all $x, y \in L$ and $\alpha \in \mathbb{F}$.

by using Definition 9.3 and infimum $\left(\bigwedge \underline{R}S_i\right)(x) = \min\{\underline{R}S_i(x)|i \in I\} = \underline{R}S_k(x)$. Hence, $\left(\bigwedge \underline{R}S_i\right)$ is a fuzzy Lie algebra. Similarly, $\left(\bigwedge \overline{R}S_i\right)$ is a fuzzy Lie algebra. Hence, $\bigwedge(\underline{R}S_i,\ \overline{R}T_i)$ is a rough fuzzy Lie algebra. Similarly, we can prove for other cases.

Proposition 9.6 *If* $\{(\underline{R}S_i,\ \overline{R}S_i)|i \in I\}$ *is a family of rough fuzzy Lie ideal of Lie algebras L, then*

1. $\tilde{\bigwedge\limits_{i\in I}}(\underline{R}S_i,\ \overline{R}S_i)$ *is a rough fuzzy Lie ideal of L,*
2. $\tilde{\bigcap\limits_{i\in I}}(\underline{R}S_i,\ \overline{R}S_i)$ *is a rough fuzzy Lie ideal of L,*
3. $\tilde{\bigcup\limits_{i\in I}}(\underline{R}S_i,\ \overline{R}S_i)$ *is a rough fuzzy Lie ideal of L.*

Proposition 9.7 *Let J be a Lie ideal of a Lie algebra L. If* $(\underline{R}S,\ \overline{R}S)$ *is a rough fuzzy Lie ideal of L, then the rough fuzzy set* $\overline{(\underline{R}S,\ \overline{R}S)} = (\overline{\underline{R}S},\ \overline{\overline{R}S})$ *of* L/J *defined by*

$$(\overline{\underline{R}S})(a + J) = \sup_{x\in J}(\underline{R}S)(a + x) \text{ and } (\overline{\overline{R}S})(a + J) = \sup_{x\in J}(\overline{R}S)(a + x)$$

is a rough fuzzy Lie ideal of the quotient Lie algebra L/J.

Proof Clearly, $\underline{R}S$ is well defined. Let $x + J, y + J \in L/J$; then

$$\begin{aligned}
\overline{R}S((x+J)+(y+J)) &= \overline{R}S((x+y)+J) = \sup_{z\in J}(\underline{R}S)((x+y)+z) \\
&= \sup_{z=s+t\in J}(\underline{R}S)((x+y)+(s+t)) \\
&\geq \sup_{s,t\in J}\min\{(\underline{R}S)(x+s), (\underline{R}S)(y+t)\} \\
&= \min\{\sup_{s\in J}(\underline{R}S)(x+s), \sup_{t\in J}\underline{R}S(\underline{R}S)(y+t)\} \\
&= \min\{\overline{R}S(x+J), \overline{R}S(y+J)\}, \\
\overline{R}S(\alpha(x+J)) &= \overline{R}S(\alpha x+J) = \sup_{z\in J}(\underline{R}S)(\alpha x+z) \\
&\geq \sup_{z\in J}(\underline{R}S)(x+z) = \overline{R}S(x+J), \\
\overline{R}S([x+J, y+J]) &= \overline{R}S([x,y]+J) = \sup_{z\in J}(\underline{R}S)([x,y]+z) \\
&\geq \sup_{z\in J}(\underline{R}S)(x+z) = \overline{R}S(x+J).
\end{aligned}$$

Hence, $\underline{R}S$ is a fuzzy Lie ideal of L/J. Hence, $\overline{R}S$ is a fuzzy Lie ideal of L/J. Therefore, $(\underline{R}S, \overline{R}S)$ is a rough fuzzy Lie ideal of L/J.

Definition 9.8 Let $(\underline{R}S, \overline{R}T)$ and $(\underline{Q}T, \overline{Q}T)$ be rough fuzzy Lie algebras on L_1 and L_2 over the same field $\mathbb{F}$. A *homomorphism* of rough fuzzy Lie algebras $g : (\underline{R}S\ \overline{R}S) \to (\underline{Q}T, \overline{Q}T)$ is a homomorphism $g : L_1 \to L_2$ which satisfies

$$g\Big((\underline{R}S)(l_1)\Big) = (\underline{Q}T)(g(l_1)) \text{ and } g\Big((\overline{R}S)(l_1)\Big) = (\overline{Q}T)(g(l_1)),\ \forall l_1 \in L_1, R(l_1) \in L_2.$$

Proposition 9.8 *Let $g : L_1 \to L_2$ be a monomorphism of Lie algebras. If $(\underline{R}S, \overline{R}S)$ is a rough fuzzy Lie algebra of L_1 and $(\underline{Q}T, \overline{Q}T)$ is the image of $(\underline{R}S, \overline{R}S)$ under g. Then, $(\underline{Q}T, \overline{Q}T)$ is a rough fuzzy Lie algebra of L_2.*

Proof For any $x, y \in L_1$ and $\alpha \in \mathbb{F}$

$$\begin{aligned}
(\underline{Q}T)(g(x+y)) &= g((\underline{R}S)(x+y)) \\
&\geq g((\underline{R}T)(x)) \wedge g((\underline{R}T)(y)) \\
&= (\underline{Q}T)(g(x)) \wedge (\underline{Q}T)(g(y)) \\
(\underline{Q}T)(g(\alpha x)) &= g(\underline{R}S)(\alpha x) \geq g(\underline{R}S)(x) = (\underline{Q}T)(g(x))
\end{aligned}$$

$$\begin{aligned}
(\underline{Q}T)([x,y]) &= g((\underline{R}S)(([x,y]))) \\
&\geq \min\{(\underline{R}S)(g(x)), (\underline{R}S)(g(y))\} \\
&= \min\{g((\underline{Q}T)(x)), g((\underline{R}S)(y))\}.
\end{aligned}$$

Hence, $\underline{Q}T$ is a fuzzy Lie algebra of L_2. Similarly, $\overline{Q}T$ is a fuzzy Lie algebra of L_2. Hence, $(\underline{Q}T, \overline{Q}T)$ is a rough fuzzy Lie algebra of L_2.

Definition 9.9 The *kernal of rough fuzzy Lie algebra homomorphism* is a homomorphism $g : L_1 \to L_2$, that is, defined by $Ker\ R = \{l \in L_1 : R(l) = 0_2\}$. That is,

$$g\Big((\underline{R}S)(l)\Big) = (\underline{Q}T)(0_2), \quad g\Big((\overline{R}T)(l)\Big) = (\overline{Q}T)(0_2),$$

where 0_2 is the identity element in L_2.

Definition 9.10 Let $(\underline{R}S, \overline{R}S)$ and $(\underline{Q}T, \overline{Q}T)$ be rough fuzzy Lie algebras on L_1 and L_2. An *isomorphism* of rough fuzzy Lie algebras $g : (\underline{R}S, \overline{R}S) \to (\underline{Q}T, \overline{Q}T)$ is a bijective homomorphism $g : L_1 \to L_2$ which satisfies $g\Big((\underline{R}S)(l_1)\Big) = (\underline{Q}T)(l_2)$ and $g\Big((\overline{R}S)(l_1)\Big) = (\overline{Q}T)(l_2)), \forall\, l_1 \in L_1, l_2 \in L_2$.

Proposition 9.9 *Let $g : L_1/J \to L_2$ be an epimorphism of Lie algebras with ideal J.*
If $(\underline{Q}T, \overline{Q}T)$ and $(\underline{R}S, \overline{R}S)$ are rough fuzzy isomorphic Lie algebra of L_2 and L_1/J, then there exist an isomorphism $(\underline{Q}T, \overline{Q}T)_t$ and $(\underline{R}S, \overline{R}S)_t$ where $t \in [0, 1]$.

Proof This proof follows from Proposition 9.7 and Definition 9.10.

Definition 9.11 Let $(\underline{R}S, \overline{R}S)$ and $(\underline{R}T, \overline{R}T)$ be rough fuzzy Lie algebras on L_1. An *automorphism* of rough fuzzy Lie algebras $g : (\underline{R}S, \overline{R}S) \to (\underline{R}T, \overline{R}T)$ is a bijective homomorphism $g : L_1 \to L_1$ which satisfies $g\Big((\underline{R}S)(l_1)\Big) = (\underline{R}T)(l_2)$ and $g\Big((\overline{R}S)(l_1)\Big) = (\overline{R}T)(l_2), \forall\, l_1, l_2 \in L_1$.

Proposition 9.10 *Let $g : (\underline{R}S, \overline{R}S) \to (\underline{Q}T, \overline{Q}T)$ be rough fuzzy Lie algebras isomorphism on L_1 to L_2. Then, $(\underline{R}S, \overline{R}S)$ is a rough fuzzy Lie subalgebra of L_1 if and only if the nonempty t-level cut $(\underline{Q}T, \overline{Q}T)_t$ is a rough Lie subalgebras of L_2 when $U(T, t)$ is proper subset, where $t \in [0, 1]$.*

Proof This proof follows from Proposition 9.9.

Proposition 9.11 *Let $f : L_1 \to L_2$ be an epimorphism of Lie algebras. If $(\underline{Q}T, \overline{Q}T)$ is a rough fuzzy Lie algebra of L_2 and $(\underline{R}S, \overline{R}S)$ is the pre-image of $(\underline{Q}T, \overline{Q}T)$ under f, then $(\underline{R}S, \overline{R}S)$ is a rough fuzzy Lie algebra of L_1.*

Proof For any $x, y \in L_1$ and $\alpha \in \mathbb{F}$

$$\begin{aligned}
(\underline{R}S)(x + y) &= (\underline{Q}T)(f(x + y)) = (\underline{Q}T)(f(x) + f(y)) \\
&\geq \min\{(\underline{Q}T)(f(x)), (\underline{Q}T)(f(y))\} \\
&= \min\{(\underline{R}S)(x), (\underline{R}S)(y)\}, \\
(\underline{R}S)(\alpha x) &= (\underline{Q}T)(f(\alpha x)) = (\underline{Q}T)(\alpha f(x)) \\
&\geq (\underline{Q}T)(f(x)) = (\underline{R}S)(x),
\end{aligned}$$

$$\begin{aligned}(\underline{R}S)([x, y]) &= (\underline{Q}T)(f([x, y])) = (\underline{Q}T)([f(x), f(y)])\\ &\geq \min\{(\underline{Q}T)(f(x)), (\underline{Q}T)(f(y))\}\\ &= \min\{(\underline{R}S)(f(x)), (\underline{R}S)(f(y))\}.\end{aligned}$$

Thus, $\underline{R}S$ is a fuzzy Lie algebra of L_1. Similarly, $\overline{R}S$ is a fuzzy Lie algebra of L_1. Hence, $(\underline{R}S, \overline{R}S)$ is a rough fuzzy Lie algebra of L_1.

Proposition 9.12 *Let $f : L_1 \to L_2$ be an epimorphism of Lie ideal. If $(\underline{Q}T, \overline{Q}T)$ is a rough fuzzy Lie ideal of L_2 and $(\underline{R}S, \overline{R}S)$ is the pre-image of $(\underline{Q}T, \overline{Q}T)$ under f. Then $(\underline{R}S, \overline{R}S)$ is a rough fuzzy Lie ideal of L_1.*

Proposition 9.13 *Let $f : L_1 \to L_2$ be a surjective Lie homomorphism. If $(\underline{R}S, \overline{R}S)$ and $(\underline{Q}T, \overline{Q}T)$ are rough fuzzy Lie subalgebras of L_1, then $f(\ll AB \gg) = \ll f(A)\overline{f}(B) \gg$.*

Definition 9.12 Let L_1 and L_2 be two Lie algebras and f be a function of L_1 into L_2. If $(\underline{R}S, \overline{R}S)$ is a rough fuzzy set in L_2, then the *pre-image* of $(\underline{Q}T, \overline{Q}T)$ under f is the rough fuzzy set in L_1 defined by

$$(f^{-1}(\underline{Q}T),\ f^{-1}(\overline{Q}T))(x) = ((\underline{R}S)(f(x)),\ (\overline{R}S)(f(x))) \qquad \forall x \in L_1.$$

Proposition 9.14 *Let $f : L_1 \to L_2$ be an onto homomorphism of Lie algebras. If $(\underline{Q}T, \overline{Q}T)$ is a rough fuzzy Lie algebra of L_2, then $f^{-1}(\underline{Q}T, \overline{Q}T)$ is a rough fuzzy Lie algebra of L_1.*

Proof Let $x_1, x_2 \in L_1$ and $\alpha \in \mathbb{F}$; then

$$\begin{aligned}f^{-1}(\underline{Q}T)(x_1 + x_2) &= (\underline{Q}T)(f(x_1) + f(x_2))\\ &\geq \min\{(\underline{Q}T)(f(x_1)), (\underline{Q}T)(f(x_2))\}\\ &= \min\{f^{-1}(\underline{Q}T)(x_1), f^{-1}(\underline{Q}T)(x_2)\},\\ f^{-1}(\underline{Q}T)(\alpha x_1) &= (\underline{Q}T)(f(\alpha x_1))\\ &\geq (\underline{Q}T)(\alpha f(x_1)) = \alpha f^{-1}(\underline{Q}T)(x_1),\\ f^{-1}(\underline{Q}T)([x, y]) &= (\underline{Q}T)(f([x, y])) = (\underline{Q}T)([f(x), f(y)])\\ &\geq \min\{(\underline{Q}T)(f(x)), (\underline{Q}T)(f(y))\}\\ &= \min\{f^{-1}(\underline{Q}T)(x), f^{-1}(\underline{Q}T)(y)\}.\end{aligned}$$

Hence, $f^{-1}(\underline{Q}T)$ is a fuzzy Lie algebra of L_1. Similarly, $f^{-1}(\overline{Q}T)$ is a fuzzy Lie algebra of L_1. Hence, $f^{-1}(\underline{Q}S, \overline{Q}S)$ is a rough fuzzy Lie algebra of L_1.

Proposition 9.15 *Let $f : (\underline{R}S, \overline{R}S) \to (\underline{Q}T, \overline{Q}T)$ be an onto homomorphism of rough fuzzy Lie ideal. If $(\underline{Q}T, \overline{Q}T)$ is a rough fuzzy Lie ideal of L_2, then $f^{-1}(\underline{Q}T, \overline{Q}T)$ is a rough fuzzy Lie ideal of L_1.*

Proposition 9.16 *Let $f : L_1 \to L_2$ be an onto homomorphism of Lie algebras. If $(\underline{Q}T, \overline{Q}T)$ is a rough fuzzy Lie algebra of L_2, then $f^{-1}((\underline{Q}S, \overline{Q}S)') = (f^{-1}(\underline{Q}S, \overline{Q}S))'$.*

Proof Let $\underline{Q}T$ be a fuzzy Lie algebra of L_2. Then for $x \in L_1$,

$$\begin{aligned} f^{-1}((\underline{Q}T)')(x) &= (\underline{Q}T)'(f(x)) \\ &= 1 - (\underline{Q}T)(f(x)) \\ &= 1 - f^{-1}((\underline{Q}T)')(x) \\ &= (f^{-1}(\underline{Q}T))'(x). \end{aligned}$$

That is, $f^{-1}((\underline{Q}T)') = (f^{-1}(\underline{Q}T))'$. Similarly, for $\overline{Q}S$, we have $f^{-1}((\overline{Q}S)') = (f^{-1}(\overline{Q}S))'$. This completes the proof.

Proposition 9.17 *Let $f : L_1 \to L_2$ be an onto homomorphism of Lie ideal. If $(\underline{Q}S, \overline{Q}S)$ is a rough fuzzy Lie ideal of L_2, then $f^{-1}((\underline{Q}S, \overline{Q}S)') = (f^{-1}(\underline{Q}S, \overline{Q}S))'$.*

Definition 9.13 Let $(\underline{R}S, \overline{R}S)$ be a rough fuzzy set in a Lie algebra L and f a mapping defined on L. Then, the fuzzy set $(\underline{R}S, \overline{R}S)^f$ in $f(L)$ is defined by

$$(\underline{R}S, \overline{R}S)^f(y) = (\underline{R}S^f(y), \overline{R}S^f(y)),$$

That is,

$$\underline{R}S^f(y) = \sup_{x \in f^{-1}(y)} (\underline{R}S)(x) \text{ and } \overline{R}S^f(y) = \sup_{x \in f^{-1}(y)} (\overline{R}S)(x)$$

for every $y \in f(L)$, is called the *image* of $(\underline{R}S, \overline{R}S)$ under f. A rough fuzzy set $(\underline{R}S, \overline{R}S)$ in L has the *sup property* if for any subset $A \subseteq L$, there exists $a_0 \in A$ such that

$$(\underline{R}S, \overline{R}S)(a_0) = \Big((\underline{R}S)(a_0), (\overline{R}S)(a_0)\Big) = \Big(\sup_{a_0 \in A}(\underline{R}S)(a_0), \sup_{a_0 \in A}(\overline{R}S)(a_0)\Big).$$

Proposition 9.18 *A Lie algebra homomorphism image of a rough fuzzy Lie algebra(ideal) having the sup property is a rough fuzzy Lie algebra(ideal).*

Proof Let $f : L_1 \to L_2$ be an epimorphism of L_1 onto L_2 and $\underline{R}S$ be a fuzzy Lie algebra of L_1 with the sup property. Consider $f(x), f(y) \in f(L_1)$. Let $x_0, y_0 \in f^{-1}(f(x))$ be such that

$$(\underline{R}S)(x_0) = \sup_{t \in f^{-1}(f(x))} (\underline{R}S)(t) \text{ and } (\underline{R}S)(y_0) = \sup_{t \in f^{-1}(f(y))} (\underline{R}S)(t),$$

respectively. Then

$$\begin{aligned}
(\underline{Q}T)(f(x)+f(y)) = & \sup_{t\in f^{-1}(f(x)+f(y))} (\underline{R}S)(t) \\
& \geq (\underline{R}S)(x_0+y_0) \\
& \geq \min\{(\underline{R}S)(x_0)+(\underline{R}S)(y_0)\} \\
& = \min\{\sup_{t\in f^{-1}(f(x))} (\underline{R}S)(t), \sup_{t\in f^{-1}(f(y))} (\underline{R}S)(t)\} \\
& = \min\{(\underline{Q}T)(f(x))+(\underline{Q}T)(f(y))\}, \\
(\underline{Q}T)(f(\alpha x)) = & \sup_{t\in f^{-1}(f(\alpha x))} (\underline{R}S)(t) \geq (\underline{R}S)(x_0) \\
& \geq \min\{(\underline{R}S)(x_0)\} = (\underline{Q}T)(f(x)), \\
(\underline{Q}T)([f(x), f(y)]) = & (\underline{Q}T)(f([x,y])) = \sup_{t\in f^{-1}(f([x,y]))} (\underline{R}S)(t) \\
& \geq (\underline{R}S)([x_0, y_0]) \\
& \geq \min\{(\underline{R}S)(x_0), (\underline{R}S)(y_0)\} \\
& = \min\{\sup_{t\in f^{-1}(f(x))} (\underline{R}S)(t), \sup_{t\in f^{-1}(f(y))} (\underline{R}S)(t)\} \\
& = \min\{(\underline{Q}T)(f(x)), (\underline{R}S)(f(y))\}.
\end{aligned}$$

Consequently, $\underline{Q}T$ is a fuzzy Lie algebra of L_2. Similarly, $\overline{Q}T$ is a fuzzy Lie algebra of L_2. This completes the proof.

Definition 9.14 Let L_1 and L_2 be Lie algebras and f a function of $(\underline{R}S, \overline{R}S)$ is a rough fuzzy set in L_1; then, the *image* of $(\underline{R}S, \overline{R}S)$ under f is the rough fuzzy set defined by

$$f(\underline{R}S)(y) = \begin{cases} \sup\{(\underline{R}S)(t) \mid t \in L_1, f(t) = y\}, & \text{if } f^{-1}(y) \neq \emptyset, \\ 1, & \text{otherwise.} \end{cases}$$

$$f(\overline{R}S)(y) = \begin{cases} \sup\{(\overline{R}S)(t) \mid t \in L_1, f(t) = y\}, & \text{if } f^{-1}(y) \neq \emptyset. \\ 1, & \text{otherwise.} \end{cases}$$

Definition 9.15 Let L_1 and L_2 be any sets, and let $f : L_1 \to L_2$ be any function. A rough fuzzy set $(\underline{R}S, \overline{R}S)$ is called f*-invariant* if and only if for $x, y \in L_1$, $f(x) = f(y)$ implies $(\underline{R}S)(x) = y$ and $(\overline{R}S)(x) = y$.

Proposition 9.19 *Let $f : L_1 \to L_2$ be an epimorphism of Lie algebras(ideal). Then, $(\underline{R}S, \overline{R}S)$ is an f-invariant rough fuzzy Lie algebra(ideal) of L_1 if and only if $f(\underline{R}S, \overline{R}S) = (f(\underline{R}S), f(\overline{R}S))$ is a rough fuzzy Lie algebra(ideal) of L_2.*

Proof Let $x, y \in L_2$ and $\alpha \in \mathbb{F}$. Then, there exist $a, b \in L_1$ such that $f(a) = x$, $f(b) = y$, $x + y = f(a + b)$ and $\alpha x = \alpha f(a)$. Since $\underline{R}S$ is f-invariant,

$$\begin{aligned}
f(\underline{R}S)(x+y) =& (\underline{R}S)(a+b) \geq \min\{(\underline{R}S)(a), (\underline{R}S)(b)\} \\
=& \min\{f((\underline{R}S)(x)), f((\underline{R}S)(y))\}, \\
f(\underline{R}S)(\alpha x) =& (\underline{R}S)(\alpha a) \geq (\underline{R}S)(a) \\
=& f((\underline{R}S)(x)), \\
f(\underline{R}S)([x, y]) =& (\underline{R}S)([a, b]) = [(\underline{R}S)(a), (\underline{R}S)(b)] \\
\geq& (\underline{R}S)(a) = f(\underline{R}S)(x).
\end{aligned}$$

Hence, $f(\underline{R}S)$ is a rough fuzzy Lie algebra of L_2.Similarly, $f(\overline{R}S)$ is a rough fuzzy Lie algebra of L_2. This completes the proof.

Conversely, if $f(\underline{R}S)$ is a rough fuzzy Lie algebra of L_2, then for any $x \in L_1$

$$\begin{aligned}
f^{-1}(f(\underline{R}S))(x) =& f(\underline{R}S)(f(x)) = \sup\{(\underline{R}S)(t) \mid t \in L_1, f(t) = f(x)\} \\
=& \sup\{(\underline{R}S)(t) \mid t \in L_1, (\underline{R}S)(t) = (\underline{R}S)(x)\} = (\underline{R}S)(x).
\end{aligned}$$

Hence, $f^{-1}(f(\underline{R}S)) = \underline{R}S$ is a fuzzy Lie algebra by Proposition 9.14. Similarly, $f^{-1}(f(\underline{R}S)) = \underline{R}S$ is a fuzzy Lie algebra. This completes the proof.

Definition 9.16 An algebra C of Lie algebra L is said to be *characteristic* if $f(C) = C$, for all $f \in \text{Aut}(L)$, where $\text{Aut}(L)$ is the set of all automorphisms of L. Rough fuzzy Lie algebra $(\underline{R}S, \overline{R}S)$ of Lie algebra L is said to be *rough fuzzy characteristic* if $(\underline{R}S)(f(x)) = (\underline{R}S)(x)$ and $(\overline{R}S)(f(x)) = (\overline{R}S)(x)$, for all $x \in L$ and $f \in \text{Aut}(L)$.

Lemma 9.1 *Let $(\underline{R}S, \overline{R}S)$ be a rough fuzzy Lie ideal of a Lie algebra L, and let $x \in L$. Then, $S(x) = t$ if and only if $x \in U(\underline{R}S; t)$,$x \in U(\overline{R}S; t)$ and $x \notin U(\underline{R}S; s)$,$x \notin U(\overline{R}S; s)$, for all $s > t$.*

Proposition 9.20 *A rough fuzzy Lie ideal is characteristic if and only if each its level set is a characteristic Lie ideal.*

Definition 9.17 Let $(\underline{R}S, \overline{R}S)$ be a rough fuzzy Lie ideal in L. Define a sequence of rough fuzzy Lie ideals in L putting

$$(\underline{R}S)^0 = \underline{R}S, \ (\overline{R}S)^0 = \overline{R}S$$

and

$$(\underline{R}S)^{n_1} = [(\underline{R}S)^{n_1-1}, \ (\underline{R}S)^{n_1-1}], \ (\overline{R}S)^{n_2} = [(\overline{R}S)^{n_2-1}, \ (\overline{R}S)^{n_2-1}],$$

for $n_1, n_2 > 0$. If there exist positive integers n_1, n_2 such that $(\underline{R}S)^{n_1} = 0$ and $(\underline{R}S)^{n_2} = 0$, then a rough fuzzy Lie ideal $(\underline{R}S, \overline{R}S)$ is called *solvable* .

Proposition 9.21 *Homomorphic image of a solvable rough fuzzy Lie ideal is a solvable rough fuzzy Lie ideal.*

Proof Let $f : L_1 \to L_2$ be a homomorphism of Lie algebras. Suppose that $(\underline{R}S, \overline{R}S)$ is a solvable rough fuzzy Lie algebra in L_1. Taking lower approximation $\underline{R}S$, we prove by induction on n that $f([\underline{R}S, \underbrace{\cdots}_{(n-1)times}, \underline{R}S]) \supseteq [f(\underline{R}S), \underbrace{\cdots}_{(n-1)times}, f(\underline{R}S)]$, where n is any positive integer. First we claim that $f([\underline{R}S, \underline{R}S]) \supseteq [f(\underline{R}S), f(\underline{R}S)]$. Let $y \in L_2$, then

$$\begin{aligned} f(\ll \underline{R}S, \underline{R}S \gg) &= \sup\{\ll \underline{R}S, \underline{R}S \gg (x) | f(x) = y\} \\ &= \sup\{\sup\{\min\{(\underline{R}S)(a), (\underline{R}S)(b)\} | a, b \in L_1, [a, b] = x, f(x) = y\}\} \\ &= \sup\{\min\{(\underline{R}S)(a), (\underline{R}S)(b)\} | a, b \in L_1, [a, b] = x, f(\underline{R}S)(x) = y\} \\ &= \sup\{\min\{(\underline{R}S)(a), (\underline{R}S)(b)\} | a, b \in L_1, [f(\underline{R}S)(a), f(\underline{R}S)(b)] = x\} \\ &= \sup\{\min\{(\underline{R}S)(a), (\underline{R}S)(b)\} | a, b \in L_1, f(\underline{R}S)(a) = u, f(\underline{R}S)(b) = v, [u, v] = y\} \\ &\geq \sup\{\min\{\sup_{a \in f^{-1}(\underline{R}S)(u)} (\underline{R}S)(a), \sup_{b \in f^{-1}(\underline{R}S)(v)} (\underline{R}S)(b)\} | [u, v] = y\} \\ &= \sup\{\min(f(\underline{R}S)(u), f(\underline{R}S)(v)) | [u, v] = y\} = \ll f(\underline{R}S), f(\underline{R}S) \gg (y). \end{aligned}$$

Thus, $f([\underline{R}S, \underline{R}S]) \supseteq (f \ll \underline{R}S, \underline{R}S \gg) \supseteq \ll f(\underline{R}S), f(\underline{R}S) \gg = [f(\underline{R}S), f(\underline{R}S)]$.
Now for $n > 1$, we get

$$\begin{aligned} f((\underline{R}S)^n) &= f([(\underline{R}S)^{n-1}, (\underline{R}S)^{n-1}]) \supseteq [f((\underline{R}S)^{n-1}), f((\underline{R}S)^{n-1})] \\ &\supseteq [(f(\underline{R}S)^{n-1}), (f(\underline{R}S)^{n-1})] = (f(\underline{R}S)^n). \end{aligned}$$

Similarly, we can show for upper approximation. This completes the proof.

Definition 9.18 Let $(\underline{R}S, \overline{R}S)$ be a rough fuzzy Lie algebra in L, and let $\underline{R}S^{n_1} = [\underline{R}S^{n_1-1}, \underline{R}S^{n_1-1}]$ and $\overline{R}S^{n_2} = [\overline{R}S^{n_1-2}, \overline{R}S^{n_2-1}]$ for $n_1, n_2 > 0$, where $\underline{R}S^0 = \underline{R}S$ and $\overline{R}S^0 = \overline{R}S$. If there exist positive integers n_1, n_2 such that $\underline{R}S^{n_1} = 0$ and $\overline{R}S^{n_2} = 0$, then $\underline{R}S$ is called *nilpotent*.

Proposition 9.22 *Homomorphic image of a nilpotent rough fuzzy Lie ideal is a nilpotent rough fuzzy Lie ideal.*

Proposition 9.23 *If $(\underline{R}S, \overline{R}S)$ is a nilpotent rough fuzzy Lie ideal, then it is solvable.*

9.3 Fuzzy Rough Lie Algebras

Definition 9.19 Let $(\underline{apr}_R(A), \overline{apr}_R(A))$ be a fuzzy rough set in L. A pair $(\underline{apr}_R(A), \overline{apr}_R(A))$ is called a *fuzzy rough Lie subalgebra* of L if $\underline{apr}_R(A)$ and $\overline{apr}_R(A)$ are fuzzy Lie subalgebras of L for all $x \in L$. Equivalently, a fuzzy rough set $(\underline{apr}_R(A), \overline{apr}_R(A))$ over L is called a *fuzzy rough Lie subalgebra* of L if the following conditions are satisfied:

(i)

$$\begin{cases} \underline{apr}_R(A)(x+y) \geq \min\{\underline{apr}_R(A)(x), \underline{apr}_R(A)(y)\}, \\ \overline{apr}_R(A)(x+y) \geq \min\{\overline{apr}_R(A)(x), \overline{apr}_R(A)(y)\}, \end{cases}$$

(ii)

$$\begin{cases} \underline{apr}_R(A)(\alpha x) \geq \underline{apr}_R(A)(x), \\ \overline{apr}_R(A)(\alpha x) \geq \overline{apr}_R(A)(x), \end{cases}$$

(iii)

$$\begin{cases} \underline{apr}_R(A)([x, y]) \geq \min\{\underline{apr}_R(A)(x), \underline{apr}_R(A)(y)\}, \\ \overline{apr}_R(A)([x, y]) \geq \min\{\overline{apr}_R(A)(x), \overline{apr}_R(A)(y)\} \end{cases}$$

for all $x, y \in L, \alpha \in \mathbb{F}$.

We illustrate fuzzy rough Lie subalgebra with an example.

Example 9.5 Let $gl_n(\mathbb{R})$ be the general linear Lie algebra L over a field $\mathbb{F} = \mathbb{R}$ which consists of O = null matrices, I_n = identity matrices, and M = otherwise. Let $A = \{(O, 0.9), (I_n, 0.5), (M, 0.3)\}$ a fuzzy set on $\{O, I_n, M\}$ and R a fuzzy equivalence relation as shown in Table 9.2.

The lower and upper fuzzy approximations $\underline{apr}_R(A)$ and $\overline{apr}_R(A)$, respectively, are given as:

$$(\underline{apr}_R(A))(p) = \begin{cases} 0.3, \text{ if } p = O, \\ 0.3, \text{ if } p = I_n, \\ 0.3, \text{ if } p = M, \end{cases} \qquad (\overline{apr}_R(A))(p) = \begin{cases} 0.9, \text{ if } p = O, \\ 0.8, \text{ if } p = I_n, \\ 0.7, \text{ if } p = M. \end{cases}$$

By direct calculations, it is easy to see that it is a fuzzy rough Lie subalgebra of $gl_n(\mathbb{R})$.

Definition 9.20 A fuzzy rough set $(\underline{apr}_R(A), \overline{apr}_R(A))$ on L is called a *fuzzy rough Lie ideal* if it satisfies the conditions (i) and (ii) of Definition 9.19 and the following

(iv)

$$\begin{cases} \underline{apr}_R(A)([x, y]) \geq \underline{apr}_R(A)(x) \\ \overline{apr}_R(A)([x, y]) \geq \overline{apr}_R(A)(x) \end{cases}$$

Table 9.2 Fuzzy equivalence relation R on $gl_n(\mathbb{R})$

R	O	I_n	M
O	1	0.8	0.7
I_n	0.8	1	0.7
M	0.7	0.7	1

Table 9.3 Fuzzy equivalence relation R on $\mathfrak{R}^3$

R	o	i	x	y
o	1	0.8	0.4	0.1
i	0.8	1	0	0
x	0.4	0	1	0.5
y	0.1	0	0.5	1

for all $x, y \in L$.

Proposition 9.24 *Let $(\underline{apr}_R(A), \overline{apr}_R(A))$ be a fuzzy rough Lie subalgebra in a Lie algebra L. Then, $(\underline{apr}_R(A), \overline{apr}_R(A))$ is a fuzzy rough Lie subalgebra of L if and only if the nonempty (s,t)-level cut $(\underline{apr}_R(A), \overline{apr}_R(A))_{(s,t)}$ is a rough Lie subalgebras of L when $X(R,s)$ is a tolerance relation and $X(A,t)$ is proper subset, where $s, t \in [0,1]$.*

We explain Proposition 9.24 with an example.

Example 9.6 Consider $\mathbb{F} = \mathbb{R}$. Let $\mathfrak{R}^3 = \{r = (r_1, r_2, r_3) : r_1, r_2, r_3 \in \mathbb{R}\}$ be the set of all three-dimensional real vectors which consists of elements of the form:

1. $o = (r_1, r_2, r_3)$ such that $r_1 = r_2 = 0 = r_3$,
2. $i = (r_1, r_2, r_3)$ such that $r_1 = 1 = r_2 = r_3$,
3. $x = (r_1, r_2, r_3)$ such that $r_1 \neq 0, r_2 = r_3 = 0$,
4. $y = (r_1, r_2, r_3)$, otherwise.

Then, it is clear that $\mathfrak{R}^3$ endowed with the operation defined by $[r, r'] = r \times r'$ forms a real Lie algebra. Let $A = \{(o, 0.9), (i, 0.4), (x, 0.5), (y, 0.8)\}$ be a fuzzy set on $\{o, y, z\}$, R an equivalence relation as shown in Table 9.3.

The lower and upper fuzzy approximations $\underline{apr}_R(A)$ and $\overline{apr}_R(A)$ are given as:

$$\underline{apr}_R(A)(r) = \begin{cases} 0.4, \text{ if } r = o, \\ 0.4, \text{ if } r = i, \\ 0.5, \text{ if } r = x, \\ 0.5, \text{ if } r = y, \end{cases} \quad , \quad \overline{apr}_R(A)(r) = \begin{cases} 0.9, \text{ if } r = o, \\ 0.8, \text{ if } r = i, \\ 0.5, \text{ if } r = x, \\ 0.8, \text{ if } r = y, \end{cases}$$

Direct calculations show that it is a fuzzy rough Lie subalgebra of $\mathfrak{R}^3$. Take $s = 0.5$ and $t = 0.5$, Reference set : $X(A,t) = \{o, x, y\}$, Tolerance relation is shown in Table 9.4.

Lower approximation: $\underline{apr}_R(A)_{(s,t)} = \{x \in X | [x]_{X(R,s)} \subseteq U(S,t)\} = \{x, y\}$,
Upper approximation: $\overline{apr}_R(A)_{(s,t)} = \{x \in X | [x]_{X(R,s)} \cap X(S,t) \neq \emptyset\} = \{o, i, x, y\} = \mathfrak{R}^3$. The t-level set is a rough Lie algebra.

Proposition 9.25 *Let $g : L_1/J \to L_2$ be an epimorphism of Lie algebras with ideal J.*
If $(\underline{apr}_S(B), \overline{apr}_S(B))$ and $(\underline{apr}_R(A), \overline{apr}_R(A))$ are fuzzy rough isomorphic Lie algebras of L_2 and L_1/J, then there exists an isomorphism $(\underline{apr}_S(B), \overline{apr}_S(B))_{(s_2,t_2)}$ and $(\underline{apr}_R(A), \overline{apr}_R(A))_{(s_1,t_1)}$ where $s_1, s_2, t_1, t_2 \in [0,1]$.

Table 9.4 Tolerance relation $X(R, s)$ on $\mathfrak{R}^3$

$X(R, s)$	o	i	x	y
o	1	1	0	0
i	1	1	0	0
x	0	0	1	1
y	0	0	1	1

Proposition 9.26 *Let $g : (\underline{apr}_R(A), \overline{apr}_R(A)) \to (\underline{apr}_S(B), \overline{apr}_S(B))$ are fuzzy rough Lie algebras isomorphism on L_1 to L_2. Then, $(\underline{apr}_R(A), \overline{apr}_R(A))$ is a fuzzy rough Lie subalgebra of L_1 if and only if the nonempty (s,t)-level cut $(\underline{apr}_S(B), \overline{apr}_S(B))_{(s,t)}$ is a rough Lie subalgebras of L_2 when $X(S, s)$ is an equivalence relation and $X(B, t)$ is proper subset, where $s, t \in [0, 1]$.*

9.4 Rough Intuitionistic Fuzzy Lie Algebras

Definition 9.21 Let X be a nonempty set and R an equivalence relation on X. Let F be an intuitionistic fuzzy set in X with the membership function μ and nonmembership function λ. The lower and the upper approximations $\mu_{(\underline{R}S)}$ and $\mu_{(\overline{R}S)}$, respectively, of the intuitionistic fuzzy set F are intuitionistic fuzzy sets of the quotient set X/R with
(i) membership function defined by

$$\mu_{(\underline{R}S)} = \bigwedge_{y \in [x]_R} \mu_F(y), \ \mu_{(\overline{R}S)} = \bigvee_{y \in [x]_R} \mu_F(y),$$

(ii) and nonmembership function defined by

$$\lambda_{(\underline{R}S)} = \bigvee_{y \in [x]_R} \lambda_F(y), \ \lambda_{(\overline{R}S)} = \bigwedge_{y \in [x]_R} \lambda_F(y).$$

Definition 9.22 Let $(\underline{R}S, \overline{R}S)$ be a rough intuitionistic fuzzy set in L. A pair $(\underline{R}S, \overline{R}S)$ is called a *rough intuitionistic fuzzy Lie subalgebra* of L if $(\underline{R}S)(x)$ and $(\overline{R}S)(x)$ are intuitionistic fuzzy Lie subalgebras of L for all $x \in L$. Equivalently, a rough intuitionistic fuzzy set $(\underline{R}S, \overline{R}S)$ over L is called a *rough intuitionistic fuzzy Lie subalgebra* of L if the following conditions are satisfied:

(i)

$$\begin{cases} \mu_{(\underline{R}S)}(x+y) \geq \min\{\mu_{(\underline{R}S)}(x), \mu_{(\underline{R}S)}(y)\}, \\ \lambda_{(\underline{R}S)}(x+y) \leq \max\{\lambda_{(\underline{R}S)}(x), \lambda_{(\underline{R}S)}(y)\}, \\ \mu_{(\overline{R}S)}(x+y) \geq \min\{\mu_{(\overline{R}S)}(x), \mu_{(\overline{R}S)}(y)\}, \\ \lambda_{(\overline{R}S)}(x+y) \leq \max\{\lambda_{(\overline{R}S)}(x), \lambda_{(\overline{R}S)}(y)\}, \end{cases}$$

Table 9.5 Intuitionistic fuzzy equivalence relation R on $gl_n(\mathbb{R})$

R	O	I_n	M
O	1	1	0
I_n	1	1	0
M	0	0	1

(ii)

$$\begin{cases} \mu_{(\underline{R}S)}(\alpha x) \geq \mu_{(\underline{R}S)}(x), \\ \lambda_{(\underline{R}S)}(\alpha x) \leq \lambda_{(\underline{R}S)}(x), \\ \mu_{(\overline{R}S)}(\alpha x) \geq \mu_{(\overline{R}S)}(x), \\ \lambda_{(\overline{R}S)}(\alpha x) \leq \lambda_{(\overline{R}S)}(x), \end{cases}$$

(iii)

$$\begin{cases} \mu_{(\underline{R}S)}([x, y]) \geq \min\{\mu_{(\underline{R}S)}(x), \mu_{(\underline{R}S)}(y)\}, \\ \lambda_{(\underline{R}S)}([x, y]) \leq \max\{\lambda_{(\underline{R}S)}(x), \lambda_{(\underline{R}S)}(y)\}, \\ \mu_{(\overline{R}S)}([x, y]) \geq \min\{\mu_{(\overline{R}S)}(x), \mu_{(\overline{R}S)}(y)\}, \\ \lambda_{(\overline{R}S)}([x, y]) \leq \max\{\lambda_{(\overline{R}S)}(x), \lambda_{(\overline{R}S)}(y)\} \end{cases}$$

for all $x, y \in L, \alpha \in \mathbb{F}$.

We illustrate rough intuitionistic fuzzy Lie subalgebra with an example.

Example 9.7 Let $gl_n(\mathbb{R})$ be the general linear Lie algebra L over a field $\mathbb{F} = \mathbb{R}$ which consist of O = null matrices, I_n = identity matrices, and M = otherwise and $S = \{(O, 0.9, 0.0), (I_n, 0.5, 0.4), (M, 0.3, 0.6)\}$ an intuitionistic set on $\{O, I_n, M\}$ and R an equivalence relation as shown in Table 9.5.

The lower and upper intuitionistic fuzzy approximations $\underline{R}S$ and $\overline{R}S$, respectively, are given as:

$$(\underline{R}S)(p) = \begin{cases} (0.5, 0.4), \text{ if } p = O, \\ (0.5, 0.4), \text{ if } p = I_n, \\ (0.3, 0.6), \text{ if } p = M, \end{cases} \qquad (\overline{R}S)(p) = \begin{cases} (0.9, 0.0), \text{ if } p = O, \\ (0.9, 0.0), \text{ if } p = I_n, \\ (0.3, 0.6), \text{ if } p = M. \end{cases}$$

Direct calculations show that it is an rough intuitionistic fuzzy Lie subalgebra of $gl_n(\mathbb{R})$.

Definition 9.23 A rough intuitionistic fuzzy set $(\underline{R}S, \overline{R}S)$ on L is called a *rough intuitionistic fuzzy Lie ideal* if it satisfies the conditions (i) and (ii) of Definition 9.22 and the following

(iv)

$$\begin{cases} \mu_{(\underline{R}S)}([x, y]) \geq \mu_{(\underline{R}S)}(x), \lambda_{(\underline{R}S)}([x, y]) \leq \lambda_{(\underline{R}S)}(x), \\ \mu_{(\overline{R}S)}([x, y]) \geq \mu_{(\overline{R}S)}(x), \lambda_{(\overline{R}S)}([x, y]) \leq \lambda_{(\overline{R}S)}(x) \end{cases}$$

for all $x, y \in L$.

Remark 9.3 From condition (ii) of Definition 9.22, it follows that:

1. $\mu_{(\underline{R}S)}(0) \geq \mu_{(\underline{R}S)}(x), \lambda_{(\underline{R}S)}(0) \leq \lambda_{(\underline{R}S)}(x), \mu_{(\overline{R}S)}(0) \geq \mu_{(\overline{R}S)}(x), \lambda_{(\overline{R}S)}(0) \leq \lambda_{(\overline{R}S)}(x)$,
2. $\mu_{(\underline{R}S)}(-x) \geq \mu_{(\underline{R}S)}(x), \lambda_{(\underline{R}S)}(-x) \leq \lambda_{(\underline{R}S)}(x), \mu_{(\overline{R}S)}(-x) \geq \mu_{(\overline{R}S)}(x), \lambda_{(\overline{R}S)}(-x) \leq \lambda_{(\overline{R}S)}(x)$,
3. $\mu_{(\underline{R}S)}([x, y]) = \mu_{(\underline{R}S)}(-[y, x]) = \mu_{(\underline{R}S)}([y, x]), \lambda_{(\underline{R}S)}([x, y]) = \lambda_{(\underline{R}S)}(-[y, x]) = \lambda_{(\underline{R}S)}([y, x]), \mu_{(\overline{R}S)}([x, y]) = \mu_{(\overline{R}S)}(-[y, x]) = \mu_{(\overline{R}S)}([y, x]), \lambda_{(\overline{R}S)}([x, y]) = \lambda_{(\overline{R}S)}(-[y, x]) = \lambda_{(\overline{R}S)}([y, x])$,

for all $x, y \in L$.

Proposition 9.27 *Every rough intuitionistic fuzzy Lie ideal is a rough intuitionistic fuzzy Lie subalgebra.*

Proposition 9.28 *Let* $(\underline{R}S, \overline{R}S)$ *be a rough intuitionistic fuzzy Lie subalgebra in a Lie algebra* L*. Then,* $(\underline{R}S, \overline{R}S)$ *is a rough intuitionistic fuzzy Lie subalgebra of* L *if and only if the nonempty* (t, s)*-level cut* $(\underline{R}S, \overline{R}S)_{(t,s)}$ *is a rough Lie subalgebras of* L *when* $X(S, t)$ *and* $L(R, s)$ *are proper subset, where* $s, t \in [0, 1]$.

We explain Proposition 9.28 with an example.

Example 9.8 Consider $F = \mathbb{R}$. Let $\mathfrak{R}^3 = \{r = (r_1, r_2, r_3) : r_1, r_2, r_3 \in \mathbb{R}\}$ be the set of all three-dimensional real vectors which consists of elements of the form:

1. $o = (r_1, r_2, r_3)$ such that $r_1 = r_2 = 0 = r_3$,
2. $i = (r_1, r_2, r_3)$ such that $r_1 = 1 = r_2 = r_3$,
3. $x = (r_1, r_2, r_3)$ such that $r_1 \neq 0, r_2 = r_3 = 0$,
4. $y = (r_1, r_2, r_3)$, otherwise.

Then, it is clear that $\mathfrak{R}^3$ endowed with the operation defined by $[r, r'] = r \times r'$ forms a real Lie algebra Let $S = \{(o, 0.9, 0), (i, 0.4, 0.3), (x, 0.3, 0.4), (y, 0.8, 0.1)\}$ be an intuitionistic fuzzy set on $\{o, y, z\}$, R an equivalence relation as shown in Table 9.6.

The lower and upper fuzzy approximations $\underline{R}S$ and $\overline{R}S$ are given as:

$$(\underline{R}S)(r) = \begin{cases} (0.4, 0.3), & \text{if } r = o, \\ (0.4, 0.3), & \text{if } r = i, \\ (0.3, 0.4), & \text{if } r = x, \\ (0.3, 0.4), & \text{if } r = y, \end{cases}, \quad (\overline{R}S)(r) = \begin{cases} (0.9, 0.0), & \text{if } r = o, \\ (0.9, 0.0), & \text{if } r = i, \\ (0.8, 0.1), & \text{if } r = x, \\ (0.8, 0.1), & \text{if } r = y. \end{cases}$$

By direct calculations, it is easy to see that it is an intuitionistic fuzzy rough Lie subalgebra of $\mathfrak{R}^3$.

Table 9.6 Intuitionistic fuzzy equivalence relation R on $\mathfrak{R}^3$

R	o	i	x	y
o	1	1	0	0
i	1	1	0	0
x	0	0	1	1
y	0	0	1	1

Take $s = 0.4$ and $t = 0.3$, Reference set: $X(S, s) = \{o, i, y\} = L(S, t)$,

Lower approximation: $(\underline{R}S)_{(s,t)} = \{x \in X | [x]_{U(R,s)} \subseteq U(S, t)\} = \{o, i\}$,
Upper approximation: $(\overline{R}S)_{(s,t)} = \{x \in X | [x]_{U(R,s)} \cap U(S, t) \neq \emptyset\} = \{o, i, x, y\} = \mathfrak{R}^3$.

The t-level set is a rough Lie algebra.

Proposition 9.29 *If* $\{(\underline{R}S_i, \overline{R}S_i) \mid i \in I\}$ *is a family of rough intuitionistic fuzzy Lie algebra of Lie algebras* L*, then*

1. $\tilde{\bigwedge}_{i\in I}(\underline{R}S_i, \overline{R}S_i)$ *is a rough intuitionistic fuzzy Lie subalgebra of* L*,*
2. $\tilde{\bigcap}_{i\in I}(\underline{R}S_i, \overline{R}S_i)$ *is a rough intuitionistic fuzzy Lie subalgebra of* L*,*
3. $\tilde{\bigcup}_{i\in I}(\underline{R}S_i, \overline{R}S_i)$ *is a rough intuitionistic fuzzy Lie subalgebra of* L*.*

Definition 9.24 Let $(\underline{R}S, \overline{R}S)$ and $(\underline{Q}T, \overline{Q}T)$ be rough intuitionistic fuzzy Lie algebras on L_1 and L_2 over the same field $\mathbb{F}$. A *homomorphism* of rough intuitionistic fuzzy Lie algebras $g : (\underline{R}S, \overline{R}S) \to (\underline{Q}T, \overline{Q}T)$ is a homomorphism $g : L_1 \to L_2$, which satisfies

$$g\Big(\mu_{(\underline{R}S)}(l_1)\Big) = \mu_{(\underline{Q}T)}(g(l_1)),\ g\Big(\lambda_{(\underline{R}S)}(l_1)\Big) = \lambda_{(\underline{Q}T)}(g(l_1)),$$

and

$$g\Big(\mu_{(\overline{R}S)}(l_1)\Big) = \mu_{(\overline{Q}T)}(g(l_1)),\ g\Big(\lambda_{(\overline{R}S)}(l_1)\Big) = \lambda_{(\overline{Q}T)}(g(l_1)),$$

for all $l_1 \in L_1$, $R(l_1) \in L_2$.

Definition 9.25 Let L_1 and L_2 be two Lie algebras and f be a function of L_1 into L_2. If $(\underline{R}S, \overline{R}S)$ is a rough intuitionistic fuzzy set in L_2, then the *pre-image* of $(\underline{Q}T, \overline{Q}T)$ under f is the rough intuitionistic fuzzy set in L_1 defined by

$$f^{-1}(\underline{Q}T), (x) = (\mu_{(\underline{R}S)}(f(x)),\ \lambda_{(\underline{R}S)}(f(x)))$$

and

$$f^{-1}(\overline{Q}T), (x) = (\mu_{(\overline{R}S)}(f(x)),\ \lambda_{(\overline{R}S)}(f(x)))$$

for all $x \in L_1$.

Definition 9.26 Let $(\underline{R}S, \overline{R}S)$ be a rough intuitionistic fuzzy set in a Lie algebra L and f a mapping defined on L. Then, the intuitionistic fuzzy set $(\underline{R}S, \overline{R}S)^f$ in $f(L)$ defined by

$$(\underline{R}S, \overline{R}S)^f(y) = (\underline{R}S^f(y), \overline{R}S^f(y)),$$

That is,

$$\underline{R}S^f(y) = \left(\sup_{x\in f^{-1}(y)} \mu_{(\underline{R}S)}(x),\ \inf_{x\in f^{-1}(y)} \lambda_{(\underline{R}S)}(x)\right)$$

and

$$\overline{R}S^f(y) = \left(\sup_{x \in f^{-1}(y)} \mu_{(\overline{R}S)}(x), \inf_{x \in f^{-1}(y)} \lambda_{(\overline{R}S)}(x) \right)$$

for every $y \in f(L)$, is called the *image* of $(\underline{R}S, \overline{R}S)$ under f. A rough intuitionistic fuzzy set $(\underline{R}S, \overline{R}S)$ in L has the *sup property* if for any subset $A \subseteq L$, there exists $a_0 \in A$ such that

$$\underline{R}S(a_0) = \left(\mu_{(\underline{R}S)}(a_0), \lambda_{(\underline{R}S)}(a_0) \right) = \left(\sup_{a_0 \in A} \mu_{(\underline{R}S)}(a_0), \inf_{a_0 \in A} \lambda_{(\underline{R}S)}(a_0) \right),$$

and

$$\overline{R}S(a_0) = \left(\mu_{(\overline{R}S)}(a_0), \lambda_{(\overline{R}S)}(a_0) \right) = \left(\sup_{a_0 \in A} \mu_{(\overline{R}S)}(a_0), \inf_{a_0 \in A} \lambda_{(\overline{R}S)}(a_0) \right).$$

Proposition 9.30 *Let $g : L_1 \to L_2$ be a monomorphism of Lie algebras(Lie ideal). If $(\underline{R}S, \overline{R}S)$ is a rough intuitionistic fuzzy Lie algebra(Lie ideal) of L_1 and $(\underline{Q}T, \overline{Q}T)$ is the image of $(\underline{R}S, \overline{R}S)$ under g, then $(\underline{Q}T, \overline{Q}T)$ is a rough intuitionistic fuzzy Lie algebra(Lie ideal) of L_2.*

Definition 9.27 The *kernal of rough intuitionistic fuzzy Lie algebra homomorphism* is a homomorphism $g : L_1 \to L_2$, that is, defined by $Ker\ R = \{l \in L_1 : R(l) = 0_2\}$. That is,

$$g\left(\mu_{(\underline{R}S)}(l_1)\right) = \mu_{(\underline{Q}T)}(0_2), \ g\left(\lambda_{(\underline{R}S)}(l_1)\right) = \lambda_{(\underline{Q}T)}(0_2),$$

and

$$g\left(\mu_{(\overline{R}S)}(l_1)\right) = \mu_{(\overline{Q}T)}(0_2), \ g\left(\lambda_{(\overline{R}S)}(l_1)\right) = \lambda_{(\overline{Q}T)}(0_2)$$

where 0_2 is the identity element in L_2.

Definition 9.28 Let $(\underline{R}S, \overline{R}S)$ and $(\underline{Q}T, \overline{Q}T)$ be rough intuitionistic fuzzy Lie algebras on L_1 and L_2. An *isomorphism* of rough intuitionistic fuzzy Lie algebras $g : (\underline{R}S, \overline{R}S) \to (\underline{Q}T, \overline{Q}T)$ is a bijective homomorphism $g : L_1 \to L_2$ which satisfies

$$g\left(\mu_{(\underline{R}S)}(l_1)\right) = \mu_{(\underline{Q}T)}(l_2), \ g\left(\lambda_{(\underline{R}S)}(l_1)\right) = \lambda_{(\underline{Q}T)}(l_2),$$

and

$$g\left(\mu_{(\overline{R}S)}(l_1)\right) = \mu_{(\overline{Q}T)}(l_2), \ g\left(\lambda_{(\overline{R}S)}(l_1)\right) = \lambda_{(\overline{Q}T)}(l_2)),$$

for all $l_1 \in L_1, l_2 \in L_2$.

Proposition 9.31 *Let $g : L_1/J \to L_2$ be an epimorphism of Lie algebras with ideal J.*
If $(\underline{Q}T, \overline{Q}T)$ and $(\underline{R}S, \overline{R}S)$ are rough intuitionistic fuzzy isomorphic Lie algebra

of L_2 and L_1/J, then there exist an isomorphism $(\underline{Q}T, \overline{Q}T)_{(t,s)}$ and $(\underline{R}S, \overline{R}S)_{(t,s)}$ where $s, t \in [0, 1]$.

Definition 9.29 Let $(\underline{R}S, \overline{R}S)$ and $(\underline{R}T, \overline{R}T)$ be rough intuitionistic fuzzy Lie algebras on L_1. An *automorphism* of rough intuitionistic fuzzy Lie algebras $g : (\underline{R}S, \overline{R}S) \to (\underline{R}T, \overline{R}T)$ is a bijective homomorphism $g : L_1 \to L_1$ which satisfies

$$g\Big(\mu_{(\underline{R}S)}(l_1)\Big) = \mu_{(\underline{R}T)}(l_2),\ g\Big(\lambda_{(\underline{R}S)}(l_1)\Big) = \lambda_{(\underline{R}T)}(l_2),$$

and

$$g\Big(\mu_{(\overline{R}S)}(l_1)\Big) = \mu_{(\overline{R}T)}(l_2),\ g\Big(\lambda_{(\overline{R}S)}(l_1)\Big) = \lambda_{(\overline{R}T)}(l_2)),$$

$\forall\, l_1, l_2 \in L_1$.

Proposition 9.32 *Let $g : (\underline{R}S, \overline{R}S) \to (\underline{Q}T, \overline{Q}T)$ be rough intuitionistic fuzzy Lie algebras isomorphism on L_1 to L_2. Then, $(\underline{R}S, \overline{R}S)$ is a rough intuitionistic fuzzy Lie subalgebra of L_1 if and only if the nonempty t-level cut $(\underline{Q}T, \overline{Q}T)_{(t,s)}$ is a rough Lie subalgebras of L_2 when $X(T, t)$ and $L(T, s)$ are proper subsets, where $s, t \in [0, 1]$.*

Proposition 9.33 *Let $f : L_1 \to L_2$ be an epimorphism of Lie algebras. If $(\underline{Q}T, \overline{Q}T)$ is a rough intuitionistic fuzzy Lie algebra of L_2 and $(\underline{R}S, \overline{R}S)$ is the pre-image of $(\underline{Q}T, \overline{Q}T)$ under f, then $(\underline{R}S, \overline{R}S)$ is a rough intuitionistic fuzzy Lie algebra of L_1.*

Proposition 9.34 *Let J be a Lie ideal of a Lie algebra L. If $(\underline{R}S, \overline{R}S)$ is a rough intuitionistic fuzzy Lie ideal of L, then the rough intuitionistic fuzzy set $\overline{(\underline{R}S, \overline{R}S)} = (\overline{\underline{R}S}, \overline{\overline{R}S})$ of L/J defined by*

$$(\overline{\mu_{\underline{R}S}})(a+J) = \sup_{x\in J} \mu_{(\underline{R}S)}(a+x),\ (\overline{\lambda_{\underline{R}S}})(a+J) = \inf_{x\in J} \lambda_{(\underline{R}S)}(a+x)$$

and

$$(\overline{\mu_{\overline{R}S}})(a+J) = \sup_{x\in J} \mu_{(\overline{R}S)}(a+x),\ (\overline{\lambda_{\overline{R}S}})(a+J) = \inf_{x\in J} \lambda_{(\overline{R}S)}(a+x)$$

is a rough intuitionistic fuzzy Lie ideal of the quotient Lie algebra L/J.

Proposition 9.35 *Let $f : L_1 \to L_2$ be an onto homomorphism of Lie algebras. If $(\underline{Q}T, \overline{Q}T)$ is a rough intuitionistic fuzzy Lie algebra of L_2, then $f^{-1}(\underline{Q}T, \overline{Q}T)$ is a rough intuitionistic fuzzy Lie algebra of L_1.*

Proposition 9.36 *A Lie algebra homomorphism image of a rough intuitionistic fuzzy Lie algebra(ideal) having the sup property is a rough intuitionistic fuzzy Lie algebra(ideal).*

Chapter 10
Fuzzy n-Lie Algebras

In this chapter, properties of fuzzy subalgebras and ideals of n-ary Lie algebras are described. Methods of construction fuzzy ideals are presented. Connections with various fuzzy quotient n-Lie algebras are proved. We present Pythagorean fuzzy ideals of n-Lie algebras. This chapter is due to [53].

10.1 Introduction

In 1985, Filippov [71] proposed a generalization of the concept of a Lie algebra by replacing the binary operation by n-ary. He defined an n-ary Lie algebra structure on a vector space V as an operation which associates with each n-tuple $(x_1, \ldots, x_n)$ of elements in V another element $[x_1, \ldots, x_n]$ which is n-linear, skew-symmetric:

$$[x_{\sigma(1)}, \ldots, x_{\sigma(n)}] = \operatorname{sign}(\sigma)\, [x_1, \ldots, x_n]$$

and satisfies the *generalized Jacobi identity* (called also the *Filippov identity*)

$$[[x_1, \ldots, x_n], y_2, \ldots, y_n] = \sum_{i=1}^{n} [x_1, \ldots, x_{i-1}, [x_i, y_2, \ldots, y_n], x_{i+1}, \ldots, x_n],$$

where $\sigma \in \mathbb{S}_n$. Now, such structures are also called *n-Lie algebras* or *Filippov algebras*. For $n = 2$ we obtain a classical Lie algebra.

Note that such an n-ary operation, realized on the smooth function algebra of a manifold and additionally assumed to be an n-derivation, is an *n-Poisson structure*. This general concept, however, was not introduced neither by Filippov [71] nor by other mathematicians that time. It was done much later in 1994 by Takhtajan [123] in order to formalize mathematically the n-ary generalization of Hamiltonian mechanics proposed by Nambu [106] at tow places. Apparently Nambu [106] was motivated by some problems of quark dynamics, and the *n-bracket operation* he considered was:

M. Akram, *Fuzzy Lie Algebras*, Infosys Science Foundation Series,
https://doi.org/10.1007/978-981-13-3221-0_10

$$[f_1, \ldots, f_n] := det \begin{pmatrix} \frac{\partial f_1}{\partial x_1} & \cdots & \frac{\partial f_1}{\partial x_n} \\ \vdots & \ddots & \vdots \\ \frac{\partial f_n}{\partial x_1} & \cdots & \frac{\partial f_n}{\partial x_n} \end{pmatrix}$$

where $V = \mathbb{R}[x_1, \ldots, x_n]$ is the vector space of polynomials in n-variables.

Nambu does not mentions that the n-bracket operation satisfies the generalized Jacobi identity but Filippov reports this operation in his paper [71] among other examples of n-Lie algebras. The formal proof is given in [72]. Another interesting example of an n-Lie algebra is the $(n + 1)$-dimensional real Euclidean vector space V with the n-ary vector product $[x_1, x_2, \ldots, x_n]$ of the vectors $x_1, x_2, \ldots, x_n \in V$. Recall that the vector product is a skew-symmetric polylinear function of its variables and, if $e_1, e_2, \ldots, e_{n+1}$ form an orthogonal basis of the space V, then

$$[x_1, x_2, \ldots, x_n] := det \begin{pmatrix} x_{11} & x_{12} & \ldots & x_{1n} & e_1 \\ x_{21} & x_{22} & \ldots & x_{2n} & e_2 \\ \ldots & \ldots & \ldots & \ldots & \ldots \\ x_{n+1\,1} & x_{n+1\,2} & \ldots & x_{n+1\,n} & e_{n+1} \end{pmatrix},$$

where $(x_{1i}, x_{2i}, \ldots, x_{n+1\,i})$ are the coordinates of the vectors x_i, $i = 1, 2, \ldots, n$ (for details and others examples, see [71]).

A nonempty subset S of an n-Lie algebra L is its *subalgebra* if it is a subspace of a vector space L and $[x_1, \ldots, x_n] \in S$ for all $x_1, \ldots, x_n \in S$. A subspace S of an n-Lie algebra L is its an *i-ideal* if $[x_1, \ldots, x_{i-1}, y, x_{i+1}, \ldots, x_n] \in S$ for all $x_1, \ldots, x_{i-1}, x_{i+1}, \ldots, x_n \in L$ and $y \in S$. If S is an i-ideal for every $1 \le i \le n$, then we say that it is an *ideal* of L. Two n-Lie algebras L_1, L_2 over the same field $\mathbb{F}$ are isomorphic if there exists a vector space isomorphism $\varphi : L_1 \longrightarrow L_2$ such that $\varphi([x_1, \ldots, x_n]) = [\varphi(x_1), \ldots, \varphi(x_n)]$ for all $x_1, \ldots, x_n \in L$. Let L be an n-Lie algebra. Fixing in $[x_1, x_2, \ldots, x_n]$ elements $x_2, \ldots, x_{n-1}$, we obtain a new binary operation $\langle x, y\rangle = [x, x_2, \ldots, x_{n-1}, y]$ with the property $\langle x_k, y\rangle = \langle y, x_k\rangle = 0$ for all $k = 2, \ldots, n-1$ and all $y \in L$. It is easily to see that L with respect to this new operation is an arbitrary Lie algebra. It is called a *binary retract*. Fixing various $x_2, \ldots, x_{n-1}$ we obtain various (generally nonisomorphic) retracts. Obviously, any subalgebra (ideal) of an n-Lie algebra is a subalgebra (ideal) of each binary retract of L. The converse is not true. Hence, results obtained for n-Lie algebras are essential generalizations of results proved for Lie algebras.

10.2 Fuzzy Subalgebras and Ideals

Basing on the idea of fuzzification of algebras with one n-ary operation proposed in [65], we present a fuzzification of n-Lie algebras.

Definition 10.1 Let L be an n-Lie algebra. A *fuzzy subalgebra* of L is a fuzzy subspace μ such that

$$\mu([x_1, \ldots, x_n]) \geq \min\{\mu(x_1), \ldots, \mu(x_n)\} \quad \text{for all } x_1, \ldots, x_n \in L.$$

Definition 10.2 Let L be an n-Lie algebra. A *fuzzy ideal* of L is a fuzzy subspace μ such that

$$\mu([x_1, \ldots, x_n]) \geq \mu(x_i) \quad \text{for all } x_1, \ldots, x_n \in L \text{ and } 0 \leq i \leq n.$$

The following facts are obvious. Their proofs are very similar to the proofs of analogous results for fuzzy n-ary systems and fuzzy Lie algebras.

Proposition 10.1 *A fuzzy subspace μ of an n-Lie algebra L is its fuzzy ideal if and only if*

$$\mu([x_1, \ldots, x_n]) \geq \max\{\mu(x_1), \ldots, \mu(x_n)\} \tag{10.1}$$

for all $x_1, \ldots, x_n \in L$.

Proposition 10.2 *If μ is a fuzzy ideal of an n-Lie algebra L, then*

$$L_\mu = \{x \in L \mid \mu(x) = \mu(0)\}$$

is an ideal of L contained in every nonempty level subset of μ.

Proposition 10.3 *Let μ and λ be two fuzzy ideals of an n-Lie algebra L such that $\mu(0) = \lambda(0)$. Then $L_{\mu\cap\lambda} = L_\mu \cap L_\lambda$.*

Theorem 10.1 *Let $\varphi : L \longrightarrow L'$ be an n-Lie algebra homomorphism of an n-Lie algebra L onto an n-Lie algebra L'. Then the following conditions hold:*

(1) if μ is a fuzzy ideal of L, then $\varphi(\mu)$ is a fuzzy ideal of L',
(2) if ν is a fuzzy ideal of L', then $\varphi^{-1}(\nu)$ is a fuzzy ideal of L,
(3) $\varphi^{-1}(U(\nu, t)) = \varphi^{-1}(U(\nu, t))$ for every $t \in [0, 1]$ and every fuzzy ideal ν of L'.

Proposition 10.4 *Let L be an n-Lie algebra. Then the intersection of any family of fuzzy subalgebras (ideals) of L is again a fuzzy subalgebra (ideal) of L.*

It is easy to see that the union of fuzzy subalgebras (ideals) of an n-Lie algebra L is not a fuzzy subalgebra (ideal) of L, in general. But we have the following proposition on the union of fuzzy subalgebras (ideals) of L.

Proposition 10.5 *Let $\{\mu_i : i \in I\}$ be a chain of fuzzy subalgebras (ideals) of an n-Lie algebra L. Then $\bigcup_i \mu_i$ is a fuzzy subalgebra (ideal) of L.*

Theorem 10.2 *For a fuzzy subset μ of an n-Lie algebra L, the following statements are equivalent.*

(1) μ is a fuzzy subalgebra (ideal) of L,
(2) Each nonempty $U(\mu, t)$ is a subalgebra (ideal) of L.

Proof Let μ be a fuzzy ideal of L. Since μ is a fuzzy subspace of L, by Theorem 10.1, each nonempty $U(\mu, t)$ is a subspace of L. Therefore, it is enough to prove that $[\underbrace{L, \ldots, L}_{i-1}, U(\mu, t), \underbrace{L, \ldots, L}_{n-i}] \subseteq U(\mu, t)$. For every $y \in U(\mu, t)$ and $x_1, \ldots, x_n \in L$ we show that $[x_1, \ldots, x_{i-1}, y, x_{i+1}, \ldots, x_n] \in U(\mu, t)$. Since μ is a fuzzy ideal, we have

$$t \leq \mu(y) \leq \mu([x_1, \ldots, x_{i-1}, y, x_{i+1}, \ldots, x_n])$$

and so $[x_1, \ldots, x_{i-1}, y, x_{i+1}, \ldots, x_n] \in U(\mu, t)$.

Conversely, assume that every nonempty $U(\mu, t)$ is an ideal of L. Therefore, $U(\mu, t)$ is a subspace of L and so by Theorem 10.1, μ is a fuzzy subspace of L. Now, for every $y \in L$, we put $t_0 = \mu(y)$. Then, $y \in \overline{U}(\mu, t_0)$. Therefore, for every $x_1, \ldots x_n \in L$ we have $[x_1, \ldots, x_{i-1}, y, x_{i+1}, \ldots, x_n] \in \overline{U}(\mu, t_0)$ which implies that $\mu([x_1, \ldots, x_{i-1}, y, x_{i+1}, \ldots, x_n]) \geq t_0 = \mu(y)$. So, μ is a fuzzy ideal. For subalgebras the proof is analogous.

Proposition 10.6 *Let L be an n-Lie algebra and μ be a fuzzy subalgebra of L. Let $\overline{U}(\mu, t_1)$ and $\overline{U}(\mu, t_2)$ (with $t_1 < t_2$) be any two level subalgebras of μ. Then $\overline{U}(\mu, t_1) = \overline{U}(\mu, t_2)$ if and only if there is no x in L such that $t_1 \leq \mu(x) < t_2$.*

Theorem 10.3 *Let $\{S_\lambda \mid \lambda \in \Lambda\}$, where $\emptyset \neq \Lambda \subseteq [0, 1]$, be a collection of ideals of an n-Lie algebra L such that*

$$\text{(i) } L = \bigcup_{\lambda \in \Lambda} S_\lambda,$$

$$\text{(ii) } t_1 > t_2 \iff S_{t_1} \subset S_{t_2} \quad \textit{for all } t_1, t_2 \in \Lambda.$$

Then μ defined by

$$\mu(x) = \sup\{\lambda \in \Lambda \mid x \in S_\lambda\}$$

is a fuzzy ideal of L.

Proof By Theorem 10.2, it is sufficient to show that every nonempty level $U(\mu, t_1)$ is an ideal of L.

Let $U(\mu, t_1) \neq \emptyset$ for some fixed $t_1 \in [0, 1]$. Then

$$t_1 = \sup\{\lambda \in \Lambda \mid \lambda < t_1\} = \sup\{\lambda \in \Lambda \mid S_{t_1} \subset S_\lambda\}$$

or

$$t_1 \neq \sup\{\lambda \in \Lambda \mid \lambda < t_1\} = \sup\{\lambda \in \Lambda \mid S_{t_1} \subset S_\lambda\}.$$

In the first case, we have $U(\mu, t_1) = \bigcap_{\lambda < t_1} S_\lambda$, because

$$x \in U(\mu, t_1) \iff x \in S_\lambda \text{ for all } \lambda < t_1 \iff x \in \bigcap_{\lambda < t_1} S_\lambda.$$

In the second, there exists $\varepsilon > 0$ such that $(t_1 - \varepsilon, \lambda) \cap \Lambda = \emptyset$. In this case $U(\mu, t_1) = \bigcup_{\lambda \geq t_1} S_\lambda$. Indeed, if $x \in \bigcup_{\lambda \geq t_1} S_\lambda$, then $x \in S_\lambda$ for some $\lambda \geq t_1$, which gives $\mu(x) \geq \lambda \geq t_1$. Thus, $x \in U(\mu, t_1)$, i.e., $\bigcup_{\lambda \geq t_1} S_\lambda \subseteq U(\mu, t_1)$.

Conversely, if $x \notin \bigcup_{\lambda \geq t_1} S_\lambda$, then $x \notin S_\lambda$ for all $\lambda \geq t_1$, which implies $x \notin S_\lambda$ for all $\lambda > t_1 - \varepsilon$, i.e., if $x \in S_\lambda$ then $\lambda \leq t_1 - \varepsilon$. Thus, $\mu(x) \leq t_1 - \varepsilon$. Therefore, $x \notin U(\mu, t_1)$. Hence, $U(\mu, t_1) \subseteq \bigcup_{\lambda \geq t_1} S_\lambda$, and consequently $U(\mu, t_1) = \bigcup_{\lambda \geq t_1} S_\lambda$. This completes our proof.

Theorem 10.4 *Let μ be a fuzzy subset defined on an n-Lie algebra L and let $Im(\mu) = \{t_0, t_1, t_2, \ldots\}$, where $1 \geq t_0 > t_1 > t_2 \ldots \geq 0$. If $S_0 \subset S_1 \subset S_2 \subset \ldots$ are subalgebras (ideals) of L such that $\mu(S_k \setminus S_{k-1}) = t_k$ for $k = 0, 1, 2, \ldots$, where $S_{-1} = \emptyset$, then μ is a fuzzy subalgebra (ideal) of L.*

Proof First consider the case when all S_i are subalgebras. If $[x_1, \ldots, x_n] \in L \setminus \bigcup_k S_k$ then also at least one of $x_1, \ldots, x_n$ is in $L \setminus \bigcup_k S_k$ because in the opposite case $x_1, \ldots, x_n$ and $[x_1, \ldots, x_n]$ will be in some S_k. So, in this case

$$\mu([x_1, \ldots, x_n]) = 0 = \min\{\mu(x_1), \ldots, \mu(x_n)\}.$$

It is clear that for arbitrary elements $x_1, \ldots, x_n \in L$, there exists only one k such that $[x_1, \ldots, x_n] \in S_k \setminus S_{k-1}$ and only one k_i such that $x_i \in S_{k_i} \setminus S_{k_i - 1}$. Thus, $\mu([x_1, \ldots, x_n]) = t_k$, $\mu(x_i) = t_{k_i}$.

Suppose $t_{k_i} > t_k$ for all $i = 1, 2, \ldots, n$. Then, by the assumption, $k_i < k$ and $S_{k_i} \subseteq S_s \subseteq S_{k-1} \subset S_k$, where $s = \max\{k_1, \ldots, k_n\}$. Hence, $x_1, \ldots, x_n \in S_{k-1}$ and, in the consequence, $[x_1, \ldots, x_n] \in S_{k-1}$ because S_{k-1} is an ideal. This is a contradiction. Therefore, there is at least one $t_{k_i} \leq t_k$. In this case $\mu([x_1, \ldots, x_n]) = t_k \geq t_{k_i} \geq \min\{\mu(x_1), \ldots, \mu(x_n)\}$. Since μ also is a fuzzy subspace of a vector space L, it is a fuzzy subalgebra of L.

Now, let all S_i be ideals and let $[x_1, \ldots, x_n] \in S_k \setminus S_{k-1}$ for some $x_1, \ldots, x_n \in L$. Then these $x_1, \ldots, x_n$ are in $L \setminus S_{k-1}$. If not, then there exists $x_i \in S_{k-1}$. But in this case $[x_1, \ldots, x_n] \in S_{k-1}$ because S_{k-1} is an ideal. This is a contradiction. So, all $x_i \in L \setminus S_{k-1}$. Hence, $\max\{\mu(x_1), \ldots, \mu(x_n)\} \leq t_k = \mu([x_1, \ldots, x_n])$. Now, if $[x_1, \ldots, x_n] \in L \setminus \bigcup_k S_k$, then also all $x_1, \ldots, x_n$ are in $L \setminus \bigcup_k S_k$. Thus, $\max\{\mu(x_1), \ldots, \mu(x_n)\} = \mu([x_1, \ldots, x_n])$. This completes the proof that μ is a fuzzy ideal.

Corollary 10.1 *For any chain $S_0 \subset S_1 \subset S_2 \subset \ldots$ of subalgebras (ideals) of an n-Lie algebra L and any chain of reals $1 \geq t_0 > t_1 > \ldots \geq 0$ there exists a fuzzy subalgebra (ideal) μ of L such that $\overline{U(\mu, t_k)} = S_k$.*

Theorem 10.5 *Let $Im(\mu) = \{t_i \mid i \in I\}$ be the image of a fuzzy subalgebra (ideal) μ of an n-Lie algebra L. Then*

(a) there exists a unique $t_0 \in Im(\mu)$ such that $t_0 \geq t_i$ for all $t_i \in Im(\mu)$,

(b) *L is the set-theoretic union of all* $U(\mu, t_i)$, $t_i \in Im(\mu)$,
(c) $\Omega = \{U(\mu, t_i) \mid t_i \in Im(\mu)\}$ *is linearly ordered by inclusion,*
(d) Ω *contains all level subalgebras (ideals) of* μ *if and only if* μ *attains its infimum on all subalgebras (ideals) of* L.

Proof (a) Follows from the fact that $t_0 = \mu(0) \geq \mu(x)$ for all $x \in L$.
(b) If $x \in L$, then $\mu(x) = t_x \in Im(\mu)$. Thus, $x \in \bigcup U(\mu, t_i) \subseteq L$, where $t_i \in Im(\mu)$, which proves (b).
(c) Since $U(\mu, t_i) \subseteq U(\mu, t_j) \Longleftrightarrow t_i \geq t_j$ for $i, j \in I$, then Ω linearly ordered by inclusion.
(d) Suppose that Ω contains all levels of μ. Let S be a subalgebra (ideal) of L. If μ is constant on S, then we are done. Assume that μ is not constant on S. We have two cases: (1) $S = L$ and (2) $S \neq L$. For $S = L$, let $v = \inf\ Im(\mu)$. Then $v \leq t \in Im(\mu)$, i.e., $U(\mu, v) \supseteq U(\mu, t)$ for all $t \in Im(\mu)$. But $U(\mu, t_0) = L \in \Omega$ because Ω contains all levels of μ. Hence, there exists $t' \in Im(\mu)$ such that $U(\mu, t') = L$. It follows that $U(\mu, v) \supset U(\mu, t') = L$ so that $U(\mu, v) = U(\mu, t') = L$ because every level of μ is a subalgebra (resp. ideal) of L.

Now it is sufficient to show that $v = t'$. If $v < t'$, then there exists $t'' \in Im(\mu)$ such that $v \leq t'' < t'$. This implies $U(\mu, t'') \supset U(\mu, t') = L$, which is a contradiction. Therefore, $v = t' \in Im(\mu)$.

In the case $S \neq L$, we consider the fuzzy set μ_S defined by

$$\mu_S(x) = \begin{cases} u \text{ for } x \in S, \\ 0 \text{ for } x \in L \setminus S. \end{cases}$$

Clearly, μ_S is a fuzzy subalgebra (ideal) of L if S is a subalgebra (ideal).

Let

$$J^* = \{i \in I \mid \mu(x) = t_i \text{ for some } x \in S\}.$$

Then $\Omega_S = \{U(\mu, t_i) \mid i \in J^*\}$ contains (by the assumption) all levels of μ_S. This means that there exists $x_0 \in S$ such that $\mu(x_0) = \inf\{\mu_S(x) \mid x \in S\}$, i.e., $\mu(x_0) = \mu_S(x)$ for some $x \in S$. Hence, μ attains its infimum on all subalgebras (ideals) of L.

To prove the converse let $U(\mu, u)$ be a level subalgebra of μ. If $u = t$ for some $t \in Im(\mu)$, then $U(\mu, u) \in \Omega$. If $u \neq t$ for all $t \in Im(\mu)$, then there does not exist $x \in L$ such that $\mu(x) = u$.

Let $S = \{x \in L \mid \mu(x) > u\}$. Obviously, $0 \in S$ and $\mu(x_i) > u$ for all $x_i \in S$. From the fact that μ is a fuzzy subalgebra, we obtain

$$\mu([x_1, \ldots, x_n]) \geq \min\{\mu(x_1),\ \mu(x_2), \ldots, \mu(x_n)\} > u\,,$$

which proves $[x_1, \ldots, x_n] \in S$. Hence, S is a subalgebra. By hypothesis, there exists $y \in S$ such that $\mu(y) = \inf\{\mu(x) \mid x \in S\}$. But $\mu(y) \in Im(\mu)$ implies $\mu(y) = t'$ for some $t' \in Im(\mu)$. Hence, $\inf\{\mu(x) \mid x \in S\} = t' > u$.

Note that there does not exist $z \in L$ such that $u \leq \mu(z) < t'$. This gives $U(\mu, u) = U(\mu, t')$. Hence, $U(\mu, u) \in \Omega$. Thus, Ω contains all level subalgebras of μ.

Theorem 10.6 *If every fuzzy subalgebra (ideal) μ defined on an n-Lie algebra L has a finite number of values, then every descending chain of subalgebras (ideals) of L terminates at finite step.*

Proof Suppose there exists a strictly descending chain

$$S_0 \supset S_1 \supset S_2 \supset \dots$$

of ideals of L which does not terminate at finite step. We prove that μ defined by

$$\mu(x) = \begin{cases} \frac{k}{k+1} & \text{for } x \in S_k \setminus S_{k+1}, \\ 1 & \text{for } x \in \bigcap S_k, \end{cases}$$

where $k = 0, 1, 2, \dots$ and $S_0 = L$, is a fuzzy ideal with an infinite number of values.

If $[x_1, \dots, x_n] \in \bigcap S_k$, then obviously

$$\mu([x_1, \dots, x_n]) = 1 \geq \max\{\mu(x_1),\ \mu(x_2), \dots, \mu(x_n)\}.$$

If $[x_1, \dots, x_n] \notin \bigcap S_k$, then $[x_1, \dots, x_n] \in S_p \setminus S_{p+1}$ for some $p \geq 0$ and there exists at least one $i = 1, 2, \dots, n$ such that $x_i \notin \bigcap S_k$, because $x_1, \dots, x_n \in \bigcap S_k$ implies $[x_1, \dots, x_n] \in \bigcap S_k$.

Let S_m be a maximal ideal of L such that at least one of $x_1, \dots, x_n$ belongs to $S_m \setminus S_{m+1}$. Then $m \leq p$. Indeed, for $m > p$ we have $x_1, x_2, \dots, x_n \in S_m \subseteq S_{p+1} \subset S_p$ and, consequently $[x_1, \dots, x_n] \in S_{p+1}$, which is impossible. Thus, $m \leq p$ and

$$\mu([x_1, \dots, x_n]) = \frac{p}{p+1} \geq \max\{\mu(x_1), \dots, \mu(x_n)\} = \frac{m}{m+1}.$$

This proves that μ is a fuzzy ideal and has an infinite number of different values. This is a contradiction. Hence, every descending chain of ideals terminates at finite step.

For subalgebras, the proof is analogous.

Theorem 10.7 *Every ascending chain of subalgebras (ideals) of an n-Lie algebra L terminates at finite step if and only if the set of values of any fuzzy subalgebra (ideal) of L is a well-ordered subset of* $[0, 1]$.

Proof If the set of values of a fuzzy subalgebra (ideal) μ is not well ordered, then there exists a strictly decreasing sequence $\{t_i\}$ such that $t_i = \mu(x_i)$ for some $x_i \in L$. But in this case, $U(\mu, t_i)$ form a strictly ascending chain of subalgebras (ideals) of L, which is a contradiction.

In order to prove the converse, suppose that there exists a strictly ascending chain $S_1 \subset S_2 \subset S_3 \subset \dots$ of subalgebras (ideals) of L. Then $M = \bigcup\limits_{i \in \mathbb{N}} S_i$ is a subalgebra (ideal) of L and μ defined by

$$\mu(x) = \begin{cases} 0 \text{ for } x \notin M\,, \\ \frac{1}{k} \text{ where } k = \min\{i \mid x \in S_i\} \end{cases}$$

is a fuzzy subalgebra (ideal) on L.

Indeed, for every $x_1, \ldots, x_n \in M$ there exist a minimal number k_i such that $x_i \in S_{k_i}$, and a minimal number p such that $[x_1, \ldots, x_n] \in S_p$. If all S_i are subalgebras, then for $k = \max\{k_1, k_2, \ldots, k_n\}$ all $x_1, \ldots, x_n$ and $[x_1, \ldots, x_n]$ are in S_k. Thus, $k \geq p$. Consequently,

$$\mu([x_1, \ldots, x_n]) = \frac{1}{p} \geq \frac{1}{k} = \min\{\mu(x_1), \mu(x_2), \ldots, \mu(x_n)\}\,.$$

The case when at least one of $x_1, x_2, \ldots, x_n$ is not in M is obvious. Hence, μ is a fuzzy subalgebra.

Now, if all S_i are ideals, then $[x_1, \ldots, x_n] \in S_m$ for $m = \min\{k_1, \ldots, k_n\}$. Thus, $p \leq m$. Hence

$$\mu([x_1, \ldots, x_n]) = \frac{1}{p} \geq \frac{1}{m} = \max\{\mu(x_1), \mu(x_2), \ldots, \mu(x_n)\}\,,$$

which means that in this case μ is a fuzzy ideal.

Since the chain $S_1 \subset S_2 \subset S_3 \subset \ldots$ is not terminating, μ has a strictly descending sequence of values. This contradicts that the set of values of any fuzzy subalgebra (ideal) is well ordered. The proof is complete.

Definition 10.3 A fuzzy subset μ of an n-Lie algebra L is said to be *normal* if $\mu(0) = 1$.

The following lemma is obvious.

Lemma 10.1 *If μ is a fuzzy subalgebra (ideal) of an n-Lie algebra L, then μ^+ defined by*

$$\mu^+(x) = \mu(x) + 1 - \mu(0)$$

is a normal fuzzy subalgebra (ideal) of L.

Corollary 10.2 *Any fuzzy subalgebra (ideal) of an n-Lie algebra L is contained in some normal fuzzy subalgebra (ideal) of it.*

Proof Indeed, $\mu(x) \leq \mu(x) + 1 - \mu(0) = \mu^+(x)$ for every $x \in L$.

Proposition 10.7 *A maximal normal fuzzy subalgebra of an n-Lie algebra L takes only two values:* 0 *and* 1.

Proof If $\mu(x) = 1$ for all $x \in L$, then obviously μ is a maximal normal fuzzy subalgebra of L. If μ is a maximal normal fuzzy subalgebra of L and $0 < \mu(a) < 1$ for some $a \in L$, then a fuzzy subset ν defined by $\nu(x) = \frac{1}{2}\,(\mu(x) + \mu(a))$ is a fuzzy subalgebra of L. Moreover, ν^+ is a nonconstant normal fuzzy subalgebra of L such that $\mu(x) \leq \nu^+(x)$ for every $x \in L$. Thus, μ is not maximal. Obtained contradiction shows that $\mu(a) = 0$ for all $\mu(a) < 1$.

Proposition 10.8 *Let μ be a fuzzy subalgebra (ideal) of an n-Lie algebra L. If $h : [0, \mu(0)] \to [0, 1]$ is an increasing function, then a fuzzy subset μ_h defined on L by $\mu_h(x) = h(\mu(x))$ is a fuzzy subalgebra (ideal). Moreover, μ_h is normal if and only if $h(\mu(0)) = 1$.*

Proof Straightforward.

If μ is a fuzzy subset of an n-Lie algebra L, and f is a function defined on L, then the fuzzy subset ν of $f(L)$ defined by $\nu(y) = \sup_{x \in f^{-1}(y)}\{\mu(x)\}$, for all $y \in f(L)$ is called the *image* of μ under f. Similarly, if ν is a fuzzy subset in $f(L)$, then the fuzzy set $\mu = \nu \circ f$ in L is called *preimage* of ν under f.

Theorem 10.8 *An n-Lie algebra homomorphic preimage of a fuzzy ideal is a fuzzy ideal.*

Proof Let $\varphi : L_1 \longrightarrow L_2$ be an n-Lie algebra homomorphism, and ν be a fuzzy ideal of L_2 and μ be the preimage of ν under φ. Then, as it is not difficult to see, μ is a fuzzy subspace of L and

$$\begin{aligned}\mu([x_1, \ldots, x_{i-1}, y, x_{i+1}, \ldots, x_n]) &= \nu(\varphi([x_1, \ldots, x_{i-1}, y, x_{i+1}, \ldots, x_n])) \\ &= \nu([\varphi(x_1), \ldots, \varphi(x_{i-1}), \varphi(y), \varphi(x_{i+1}), \ldots, \varphi(x_n)]) \\ &\geq \nu(\varphi(y)) = \mu(y),\end{aligned}$$

for all $x_1, \ldots, x_n, y \in L$ and $\alpha \in \mathbb{F}$.

A fuzzy set μ of a set X is said to possess *sup property* if for every nonempty subset S of X, there exists $x_0 \in S$ such that $\mu(x_0) = \sup_{x \in S}\{\mu(x)\}$.

Theorem 10.9 *An n-Lie algebra homomorphism image of a fuzzy ideal having the sup property is a fuzzy ideal.*

Proof Suppose that $\varphi : L_1 \longrightarrow L_2$ is an n-Lie algebra homomorphism, μ is a fuzzy ideal of L_1 with the sup property, and ν is the image of μ under φ. Suppose that $\varphi(x), \varphi(y) \in \varphi(L)$. Let $x_0 \in \varphi^{-1}(\varphi(x))$ and $y_0 \in \varphi^{-1}(\varphi(y))$ be such that $\mu(x_0) = \sup_{t \in \varphi^{-1}(\varphi(x))}\{\mu(t)\}$ and $\mu(y_0) = \sup_{t \in \varphi^{-1}(\varphi(y))}\{\mu(t)\}$, respectively. Then,

$$\begin{aligned}\nu(\varphi(x) + \varphi(y)) &= \sup_{t \in \varphi^{-1}(\varphi(x)+\varphi(y))}\{\mu(t)\} \geq \mu(x_0 + y_0) \geq \min\{\mu(x_0), \mu(y_0)\} \\ &= \min\left\{\sup_{t \in \varphi^{-1}(\varphi(x))}\{\mu(t)\},\ \sup_{t \in \varphi^{-1}(\varphi(y))}\{\mu(t)\}\right\} \\ &= \min\{\nu(\varphi(x),\ \nu(\varphi(y))\},\end{aligned}$$

and

$$\begin{aligned}\nu(\varphi(-x)) &= \sup_{t \in \varphi^{-1}(\varphi(-x))}\{\mu(t)\} \geq \mu(-x_0) = \mu(x_0) = \nu(\varphi(x)), \\ \nu(\alpha\varphi(x)) &= \sup_{t \in \alpha\varphi^{-1}(\varphi(x))}\{\mu(t)\} \geq \mu(\alpha x_0) \geq \mu(x_0) = \nu(\varphi(x)).\end{aligned}$$

Finally, let $\varphi(x_1), \ldots, \varphi(x_n), \varphi(y) \in \varphi(L)$ and let $a_1 \in \varphi^{-1}(\varphi(x_1)), \ldots, a_n \in \varphi^{-1}(\varphi(x_n)), b \in \varphi^{-1}(\varphi(y))$ be such that

$$\mu(a_1) = \sup_{t\in\varphi^{-1}(\varphi(x_1))}\{\mu(t)\}, \ldots, \mu(a_n) = \sup_{t\in\varphi^{-1}(\varphi(x_n))}\{\mu(t)\},$$
$$\mu(b) = \sup_{t\in\varphi^{-1}(\varphi(y))}\{\mu(t)\}.$$

Then,

$$\begin{aligned}&\nu([\varphi(x_1), \ldots, \varphi(x_{i-1}), \varphi(y), \varphi(x_{i+1}), \ldots, \varphi(x_n)])\\ &= \nu(\varphi([\varphi(x_1), \ldots, \varphi(x_{i-1}), \varphi(y), \varphi(x_{i+1}), \ldots, \varphi(x_n)]))\\ &= \sup_{t\in\varphi^{-1}(\varphi([\varphi(x_1),\ldots,\varphi(x_{i-1}),\varphi(y),\varphi(x_{i+1}),\ldots,\varphi(x_n)]))}\{\mu(t)\}\\ &\geq \mu([a_1, \ldots, a_{i-1}, b, a_{i+1}, \ldots, a_n]) \geq \mu(b) = \nu(\varphi(y)).\end{aligned}$$

This proves that ν is a fuzzy ideal of $\varphi(L)$.

10.3 Fuzzy Quotient n-Lie Algebras

If J is an ideal of an n-Lie algebra L, then we can define a new n-Lie algebra on the quotient space L/J with the n-linear map

$$[x_1 + J, \ldots, x_n + J] := [x_1, \ldots, x_n] + J,$$

for all $x_1, \ldots, x_n \in L$.

If J is an ideal of an n-Lie algebra L, then the quotient space L/J is also an n-Lie algebra and is called quotient n-Lie algebra.

Theorem 10.10 *Let L be an n-Lie algebra.*

(1) *Let μ be a fuzzy ideal of L and let $t = \mu(0)$. Then the fuzzy subset μ^* of $L/U(\mu, t)$ defined by $\mu^*(x + U(\mu, t)) = \mu(x)$ for all $x \in L$ is a fuzzy ideal of $L/U(\mu, t)$,*
(2) *If I is an ideal of L and ν is a fuzzy ideal of L/I such that $\nu(x + I) = \nu(I)$ only when $x \in I$, then there exists a fuzzy ideal μ of L such that $U(\mu, t) = I$, where $t = \mu(0)$; and $\nu = \mu^*$.*

Proof (1). Since μ is a fuzzy ideal of L, $U(\mu, t)$ is an ideal of L. Now, μ^* is well defined, because if $x + U(\mu, t) = y + U(\mu, t)$ for $x, y \in L$, then $x - y \in U(\mu, t)$ and so $\mu(x - y) = \mu(0)$. Hence, $\mu(x) = \mu(y)$ which implies that $\mu^*(x + U(\mu, t)) = \mu^*(y + U(\mu, t))$.

Now, we show μ^* is a fuzzy ideal of L. Let $x, y \in L$ and $\alpha \in \mathbb{F}$. Then, we have

$$\begin{aligned}\mu^*((x + U(\mu, t)) + (y + U(\mu, t))) &= \mu^*((x + y) + U(\mu, t)) = \mu(x + y)\\ &\geq \min\{\mu(x), \mu(y)\} = \min\{\mu^*(x + U(\mu, t)), \mu^*(y + U(\mu, t))\},\end{aligned}$$

and

$$\mu^*(-x+U(\mu,t)) = \mu(-x) = \mu(x) = \mu^*(x+U(\mu,t)),$$
$$\mu^*(\alpha(x+U(\mu,t))) = \mu^*(\alpha x+\overline{\mu_t}) = \mu(\alpha x) \geq \mu(x) = \mu^*(x+U(\mu,t)).$$

Finally, for $x_1,\ldots,x_n \in L$, we have

$$\mu^*([x_1+U(\mu,t),\ldots,x_{i-1}+U(\mu,t), y+U(\mu,t), x_{i+1}+U(\mu,t),\ldots,x_n+U(\mu,t)])$$
$$= \mu^*([x_1,\ldots x_{i-1}, y, x_{i+1},\ldots,x_n]+U(\mu,t))$$
$$= \mu([x_1,\ldots x_{i-1}, y, x_{i+1},\ldots,x_n]) \geq \mu(y) = \mu^*(y+U(\mu,t)).$$

(2). We define a fuzzy subset μ of L by $\mu(x) = \nu(x+J)$ for all $x \in L$. A routine computation shows that μ is a fuzzy ideal of L. Now, $U(\mu,t) = J$, because

$$x \in U(\mu,t) \Longleftrightarrow \mu(x) = t = \mu(0) \Longleftrightarrow \nu(x+J) = \nu(J) \Longleftrightarrow x \in J.$$

Finally, $\mu^* = \nu$, since $\mu^*(x+J) = \mu^*(x+U(\mu,t)) = \mu(x) = \nu(x+J)$.

Let μ be any fuzzy ideal of an n-Lie algebra L and let $x \in L$. The fuzzy subset μ^*_x of L defined by $\mu^*_x(a) = \mu(a-x)$ for all $a \in L$ is called the *fuzzy coset* determined by x and μ.

Let J be an ideal of L. If χ_I is the characteristic function of J, then it is easy to see that $(\chi_J)^*_x$ is the characteristic function of $x+J$.

Theorem 10.11 *Let μ be any fuzzy ideal of an n-Lie algebra L. Then the set of all fuzzy cosets of μ in L, i.e., the set $L[\mu] = \{\mu^*_x \mid x \in L\}$, is an n-Lie algebra under the following operations:*

$$\begin{array}{ll} \mu^*_x + \mu^*_y = \mu^*_{x+y} & \text{for all } x, y \in L, \\ \alpha\mu^*_x = \mu^*_{\alpha x} & \text{for all } x \in L,\ \alpha \in \mathbb{F}, \\ [\mu^*_{x_1},\ldots,\ \mu^*_{x_n}] = \mu^*_{[x_1,\ldots,x_n]} & \text{for all } x_1,\ldots,x_n \in L. \end{array}$$

Theorem 10.12 *If μ is any fuzzy ideal of an n-Lie algebra L, then the map $\varphi: L \longrightarrow L[\mu]$ defined by $\varphi(x) = \mu^*_x$ for all $x \in L$ is a homomorphism with kernel $U(\mu,t)$, where $t = \mu(0)$.*

Proof It is easy to see that f is a homomorphism. We show $\mu(x) = \mu(0)$ implies $\mu^*_x = \mu^*_0$. For this, let $a \in L$. Then, $\mu(a) \leq \mu(0) = \mu(x)$. If $\mu(a) < \mu(x)$, then $\mu(a-x) = \mu(a)$, by Lemma 1.8. On the other hand, if $\mu(a) = \mu(x)$, then $a, x \in \{y \in L \mid \mu(y) = \mu(0)\}$. Hence, $\mu(a-x) = \mu(0) = \mu(x) = \mu(a)$. Therefore, in either case, we have shown that $\mu(a-x) = \mu(a)$ for all $a \in L$. Consequently, $\mu^*_x = \mu^*_0$. Also, $\mu^*_x = \mu^*_0$ implies that $\mu(x) = \mu(0)$. Hence, $\mu^*_x = \mu^*_0$ if and only if $\mu(x) = \mu(0)$. Now, we have

$$\ker\varphi = \{x \in L \mid \varphi(x) = \mu^*_0\} = \{x \in L \mid \mu^*_x = \mu^*_0\} = \{x \in L \mid \mu(x) = \mu(0)\} = U(\mu,t),$$

where $t = \mu(0)$.

Theorem 10.13 *Given a homomorphism of n-Lie algebras $\varphi : L \longrightarrow L'$ and fuzzy ideal μ of L and μ' of L' such that $\varphi(\mu) \subseteq \mu'$. Then, there is a homomorphism of n-Lie algebras $\varphi^* : L[\mu] \longrightarrow L'[\mu']$, where $\varphi^*(\mu_x^*) = \mu'^*_{\varphi(x)}$, such that the following diagram is commutative.*

$$\begin{array}{ccc} L & \xrightarrow{\varphi} & L' \\ \downarrow & & \downarrow \\ L[\mu] & \xrightarrow{\varphi^*} & L'[\mu'] \end{array}$$

Proof If $\mu_x^* = \mu_y^*$, then $\mu(x-y) = \mu(0)$. So

$$\mu'(\varphi(x) - \varphi(y)) = \mu'(\varphi(x-y)) = \varphi^{-1}(\mu')(x-y) \geq \mu(x-y) = \mu(0),$$

and so $\mu'(\varphi(x) - \varphi(y)) = \mu(0)$. Hence, $\mu'(\varphi(x)) = \mu'(\varphi(y))$ holds. Thus, φ^* is well defined. It is easily seen that φ^* is a homomorphism.

Let μ be a fuzzy ideal of an n-Lie algebra L. For any $x, y \in L$, we define a binary relation $\sim$ on L by $x \sim y$ if and only if $\mu(x-y) = \mu(0)$. Then $\sim$ is a congruence relation on L. We denote $[x]_\mu$ the equivalence class containing x, and $L/\mu = \{[x]_\mu \mid x \in L\}$ the set of all equivalence classes of L. Then, L/μ is an n-Lie algebra under the following operations:

$$\begin{array}{ll} [x]_\mu + [y]_\mu = [x+y]_\mu & \text{for all } x, y \in L, \\ \alpha[x]_\mu = [cx]_\mu & \text{for all } x \in L,\ \alpha \in \mathbb{F}, \\ [[x_1]_\mu, \ldots\ [x_n]_\mu] = [[x_1, \ldots, x_n]]_\mu & \text{for all } x, y \in L. \end{array}$$

Theorem 10.14 (Fuzzy first isomorphism theorem) *Let $\varphi : L \longrightarrow L'$ be an epimorphism of n-Lie algebras and λ be a fuzzy ideal of L'. Then $L/\varphi^{-1}(\lambda) \cong L'/\lambda$.*

Let J be an ideal and μ a fuzzy ideal of an n-Lie algebra L. If μ is restricted to J, then μ is a fuzzy ideal of J and J/μ is an ideal of L/μ.

Theorem 10.15 (Fuzzy second isomorphism theorem) *Let μ and λ be two fuzzy ideals of an n-Lie algebra L with $\mu(0) = \lambda(0)$. Then $\frac{L_\mu + L_\lambda}{\lambda} \cong \frac{L_\mu}{\mu \cap \lambda}$.*

Theorem 10.16 (Fuzzy third isomorphism theorem) *Let μ and λ be two fuzzy ideals of an n-Lie algebra L with $\lambda \subseteq \mu$ and $\mu(0) = \lambda(0)$. Then $\frac{L/\lambda}{L_\mu/\lambda} \cong L/\mu$.*

10.4 Pythagorean Fuzzy n-Lie Algebras

Pythagorean fuzzy set theory was first proposed by Yager [130] in 2013 as a generalization of intuitionistic fuzzy set ($A = (\mu_A, \lambda_A), forshort$). A Pythagorean fuzzy set is a new evaluation format, which is characterized by the membership function and the nonmembership function satisfying the condition that their square sum not exceed 1. Zhang and Xu [148] provided the detailed mathematical expression for Pythagorean fuzzy set and put forward the concept of Pythagorean fuzzy number. The Pythagorean fuzzy set is more general than the intuitionistic fuzzy set because the space of Pythagorean fuzzy set's membership degree is greater than the space of intuitionistic fuzzy set's membership degree. For instance, when a decision maker gives the evaluation information whose membership degree is 0.4 and nonmembership degree is 0.9, it can be known that the intuitionistic fuzzy number fails to address this issue because $0.4 + 0.9 > 1$. However $(0.4)^2 + (0.9)^2 < 1$, that is, the Pythagorean fuzzy number is capable of representing this evaluation information. For this case, the Pythagorean fuzzy set shows its wider applicability than the intuitionistic fuzzy set. Pythagorean fuzzy sets as a novel evaluation have been prosperously applied in various fields, such as the internet stocks investment, the service quality of domestic airline [148].

Definition 10.4 Let X be a nonempty set. A Pythagorean fuzzy set P in X is expressed as the following mathematical symbol:

$$P = \{\langle x, \mu_P(x), \lambda_P(x)\rangle | x \in X\},$$

characterized by a membership function $\mu_P : X \to [0, 1]$ and a nonmembership function $\lambda_P : X \to [0, 1]$ such that $0 \le \mu_P^2(x) + \lambda_P^2(x) \le 1$ for all $x \in X$. Moreover, for all $x \in X$, $\pi_P(x) = \sqrt{1 - \mu_P^2(x) - \lambda_P^2(x)}$ is called a Pythagorean fuzzy index or degree of hesitancy of x in P.

For computational convenience, $P = (\mu_P, \lambda_P)$ is called a Pythagorean fuzzy number, where $\mu_P, \lambda_P \in [0, 1]$, $\mu_P^2 + \lambda_P^2 \le 1$, and $\pi_P = \sqrt{1 - \mu_P^2 - \lambda_P^2}$.

Definition 10.5 Let $P = (\mu_P, \lambda_P)$, $P_1 = (\mu_{P_1}, \lambda_{P_1})$, and $P_2 = (\mu_{P_2}, \lambda_{P_2})$ be three Pythagorean fuzzy numbers, then

1. $P_1 \ge P_2$ if and only if $\mu_{P_1} \ge \mu_{P_2}$ and $\lambda_{P_1} \le \lambda_{P_2}$,
2. $P^c = (\lambda_P, \mu_P)$,
3. $P_1 \cup P_2 = (\max\{\mu_{P_1}, \mu_{P_2}\}, \min\{\lambda_{P_1}, \lambda_{P_2}\})$,
4. $P_1 \cap P_2 = (\min\{\mu_{P_1}, \mu_{P_2}\}, \max\{\lambda_{P_1}, \lambda_{P_2}\})$,
5. $P_1 \oplus P_2 = \left(\sqrt{\mu_{P_1}^2 + \mu_{P_2}^2 - \mu_{P_1}^2\mu_{P_2}^2}, \lambda_{P_1}\lambda_{P_2}\right)$,
6. $P_1 \otimes P_2 = \left(\mu_{P_1}\mu_{P_2}, \sqrt{\lambda_{P_1}^2 + \lambda_{P_2}^2 - \lambda_{P_1}^2\lambda_{P_2}^2}\right)$.

The difference between intuitionistic fuzzy number and Pythagorean fuzzy number is shown in Fig. 10.1.

Definition 10.6 A Pythagorean fuzzy set $P = (\mu_P, \lambda_P)$ on vector space V is called a *Pythagorean fuzzy subspace* if the following conditions are satisfied:

(a) $\mu_P(x + y) \ge \min(\mu_P(x), \mu_P(y))$ and $\lambda_P(x + y) \le \max(\lambda_P(x), \lambda_P(y))$,
(b) $\mu_P(\alpha x) \ge \mu_P(x)$ and $\lambda_P(\alpha x) \le \lambda_P(x)$

for all $x, y \in V$ and $\alpha \in \mathbb{F}$.
Obviously, $\mu_P^2(x) + \lambda_P^2(x) \le 1$ for all $x \in V$.

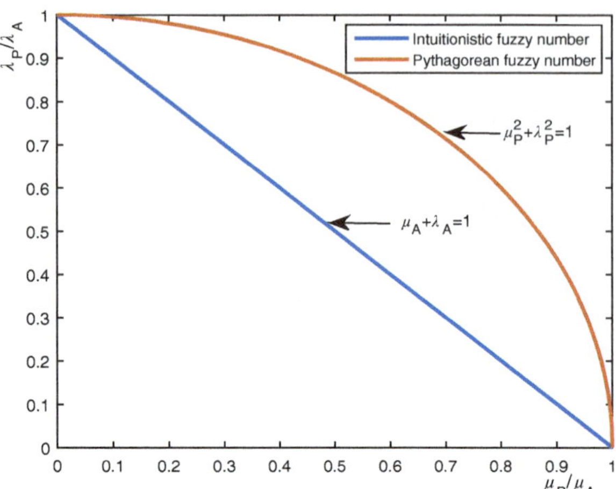

Fig. 10.1 Representation of intuitionistic fuzzy number and Pythagorean fuzzy number

From (b) it follows that:

(c) $\mu_P(0) \geq \mu_P(x), \quad \lambda_P(0) \leq \lambda_P(x)$,
(d) $\mu_P(-x) \geq \mu_P(x), \quad \lambda_P(-x) \leq \lambda_P(x)$.

Example 10.1 Let $V = \Re^3 = \{(x, y, z) : x, y, z \in \mathbb{R}\}$ be the set of all three-dimensional real vectors over field $\mathbb{R}$, which forms a vector space. We define a Pythagorean fuzzy set $P = (\mu_P, \lambda_P) : \Re^3 \to [0, 1] \times [0, 1]$ by

$$\mu_P(x, y, z) = \begin{cases} 1 & \text{if } x = y = z = 0, \\ 0.7 & \text{if } x \neq 0, y = z = 0, \\ 0 & \text{otherwise}, \end{cases} \quad \lambda_P(x, y, z) = \begin{cases} 0 & \text{if } x = y = z = 0, \\ 0.6 & \text{if } x \neq 0, y = z = 0, \\ 1 & \text{otherwise}. \end{cases}$$

By direct calculations, it is easy to see that $P = (\mu_P, \lambda_P)$ is a Pythagorean fuzzy subspace of V.

Definition 10.7 Let L be an n-Lie algebra. A *Pythagorean fuzzy subalgebra* of L is a Pythagorean fuzzy subspace P such that

$$\mu_P([x_1, \ldots, x_n]) \geq \min\{\mu_P(x_1), \ldots, \mu_P(x_n)\},$$

$$\lambda_P([x_1, \ldots, x_n]) \leq \max\{\lambda_P(x_1), \ldots, \lambda_P(x_n)\}$$

for all $x_1, \ldots, x_n \in L$.

Definition 10.8 Let L be an n-Lie algebra. A *Pythagorean fuzzy ideal* of L is a Pythagorean fuzzy subspace P such that

$$\mu_P([x_1, \ldots, x_n]) \geq \mu_P(x_i),$$

$$\lambda_P([x_1, \ldots, x_n]) \leq \lambda_P(x_i)$$

for all $x_1, \ldots, x_n \in L$ and $0 \leq i \leq n$.

Proposition 10.9 *A Pythagorean fuzzy subspace $P = (\mu_P, \lambda_P)$ of an n-Lie algebra L is its Pythagorean fuzzy ideal if and only if*

$$\mu_P([x_1, \ldots, x_n]) \geq \max\{\mu_P(x_1), \ldots, \mu_P(x_n)\},$$

$$\lambda_P([x_1, \ldots, x_n]) \leq \min\{\lambda_P(x_1), \ldots, \lambda_P(x_n)\}$$

for all $x_1, \ldots, x_n \in L$.

Proposition 10.10 *If $P = (\mu_P, \lambda_P)$ is a Pythagorean fuzzy ideal of an n-Lie algebra L, then*

$$L_P = \{x \in L \mid \mu_P(x) = \mu_P(0),\ \lambda_P(x) = \lambda_P(0)\}$$

is an ideal of L contained in every nonempty level subset of P.

Theorem 10.17 *Let $\phi : L_1 \longrightarrow L_2$ be an n-Lie algebra homomorphism of an n-Lie algebra L_1 onto an n-Lie algebra L_2. Then the following conditions hold:*

(1) if P is a Pythagorean fuzzy ideal of L_1, then $\phi(P)$ is a Pythagorean fuzzy ideal of L_2,
(2) if Q is a Pythagorean fuzzy ideal of L_2, then $\phi^{-1}(Q)$ is a Pythagorean fuzzy ideal of L_1.

Definition 10.9 If $P = (\mu_P, \lambda_P)$ is a Pythagorean fuzzy subset of an n-Lie algebra L, and f is a function defined on L, then the Pythagorean fuzzy subset $Q = (\mu_Q, \lambda_Q)$ of $f(L)$ defined by $\mu_Q(y) = \sup_{x \in f^{-1}(y)}\{\mu_p(x)\}$, for all $y \in f(L)$, $\lambda_Q(y) = \inf_{x \in f^{-1}(y)}\{\lambda_p(x)\}$, for all $y \in f(L)$ is called the *image* of P under f. Similarly, if Q is a Pythagorean fuzzy subset in $f(L)$, then the Pythagorean fuzzy set $P = Q \circ f$ in L is called *preimage* of Q under f.

Theorem 10.18 *An n-Lie algebra homomorphic preimage of a Pythagorean fuzzy ideal is a Pythagorean fuzzy ideal.*

Proof Let $\phi : L_1 \longrightarrow L_2$ be an n-Lie algebra homomorphism, and $Q = (\mu_Q, \lambda_Q)$ be a Pythagorean fuzzy ideal of L_2 and $P = (\mu_P, \lambda_P)$ be the preimage of Q under ϕ. Then, as it is not difficult to see, $P = (\mu_P, \lambda_P)$ is a Pythagorean fuzzy subspace of L and

$$\begin{aligned}\mu_P([x_1, \ldots, x_{i-1}, y, x_{i+1}, \ldots, x_n]) &= \mu_Q(\phi([x_1, \ldots, x_{i-1}, y, x_{i+1}, \ldots, x_n])) \\ &= \mu_Q([\phi(x_1), \ldots, \phi(x_{i-1}), \phi(y), \phi(x_{i+1}), \ldots, \phi(x_n)]) \\ &\geq \mu_Q(\phi(y)) = \mu_P(y),\end{aligned}$$

$$\begin{aligned}\lambda_P([x_1, \ldots, x_{i-1}, y, x_{i+1}, \ldots, x_n]) &= \lambda_Q(\phi([x_1, \ldots, x_{i-1}, y, x_{i+1}, \ldots, x_n])) \\ &= \lambda_Q([\phi(x_1), \ldots, \phi(x_{i-1}), \phi(y), \phi(x_{i+1}), \ldots, \phi(x_n)]) \\ &\leq \lambda_Q(\phi(y)) = \lambda_P(y),\end{aligned}$$

for all $x_1, \ldots, x_n, y \in L$ and $\alpha \in \mathbb{F}$.

Definition 10.10 A Pythagorean fuzzy set $P = (\mu_P, \lambda_P)$ of a set X is said to possess *sup property* if for every nonempty subset S of X, there exists $x_0 \in S$ such that $\mu_P(x_0) = \sup_{x\in S}\{\mu_P(x)\}$, $\lambda_P(x_0) = \inf_{x\in S}\{\lambda_P(x)\}$.

Theorem 10.19 *An n-Lie algebra homomorphism image of a Pythagorean fuzzy ideal having the sup-inf property is a Pythagorean fuzzy ideal.*

Proof Suppose that $\phi : L_1 \longrightarrow L_2$ is an n-Lie algebra homomorphism, μ_P is a Pythagorean fuzzy ideal of L_1 with the sup property, and $Q = (\mu_Q, \lambda_Q)$ is the image of $P = (\mu_P, \lambda_P)$ under ϕ. Suppose that $\phi(x), \phi(y) \in \phi(L_1)$. Let $x_0 \in \phi^{-1}(\phi((x))$ and $y_0 \in \phi^{-1}(\phi(y))$ be such that

$$\mu_P(x_0) = \sup_{t\in\phi^{-1}(\phi(x))}\{\mu_P(t)\},\ \ \lambda_P(x_0) = \inf_{t\in\phi^{-1}(\phi(x))}\{\lambda_P(t)\},$$

$$\mu_P(y_0) = \sup_{t\in\phi^{-1}(\phi(y))}\{\mu_P(t)\},\ \ \lambda_P(y_0) = \inf_{t\in\phi^{-1}(\phi(y))}\{\lambda_P(t)\}.$$

Then,

$$\begin{aligned}\mu_Q(\phi(x)+\phi(y)) &= \sup_{t\in\phi^{-1}(\phi(x)+\phi(y))}\{\mu_P(t)\} \geq \mu_P(x_0+y_0) \geq \min\{\mu_P(x_0), \mu_P(y_0)\}\\ &= \min\left\{\sup_{t\in\phi^{-1}(\phi(x))}\{\mu_P(t)\},\ \sup_{t\in\phi^{-1}(\phi(y))}\{\mu_P(t)\}\right\}\\ &= \min\{\mu_Q(\phi(x),\ \mu_Q(\phi(y))\},\end{aligned}$$

$$\begin{aligned}\lambda_Q(\phi(x)+\phi(y)) &= \inf_{t\in\phi^{-1}(\phi(x)+\phi(y))}\{\lambda_P(t)\} \leq \lambda_P(x_0+y_0) \leq \max\{\lambda_P(x_0), \lambda_P(y_0)\}\\ &= \max\left\{\inf_{t\in\phi^{-1}(\phi(x))}\{\lambda_P(t)\},\ \inf_{t\in\phi^{-1}(\phi(y))}\{\lambda_P(t)\}\right\}\\ &= \max\{\lambda_Q(\phi(x),\ \lambda_Q(\phi(y))\},\end{aligned}$$

$$\begin{aligned}\mu_Q(\phi(-x)) &= \sup_{t\in\phi^{-1}(\phi(-x))}\{\mu_P(t)\} \geq \mu_P(-x_0) = \mu_P(x_0) = \mu_Q(\phi(x)),\\ \mu_Q(\alpha\phi(x)) &= \sup_{t\in\alpha\phi^{-1}(\phi(x))}\{\mu_P(t)\} \geq \mu_P(\alpha x_0) \geq \mu_P(x_0) = \mu_Q(\phi(x)),\end{aligned}$$

$$\begin{aligned}\lambda_Q(\phi(-x)) &= \inf_{t\in\phi^{-1}(\phi(-x))}\{\lambda_P(t)\} \leq \lambda_P(-x_0) = \lambda_P(x_0) = \lambda_Q(\phi(x)),\\ \lambda_Q(\alpha\phi(x)) &= \inf_{t\in\alpha\phi^{-1}(\phi(x))}\{\lambda_P(t)\} \leq \lambda_P(\alpha x_0) = q\lambda_P(x_0) = \lambda_Q(\phi(x)).\end{aligned}$$

Finally, let $\phi(x_1), \ldots, \phi(x_n), \phi(y) \in \phi(L)$ and let $a_1 \in \phi^{-1}(\phi(x_1)), \ldots, a_n \in \phi^{-1}(\phi(x_n)), b \in \phi^{-1}(\phi(y))$ be such that

$$\mu_P(a_1) = \sup_{t\in\phi^{-1}(\phi(x_1))}\{\mu_P(t)\}, \ldots, \mu_P(a_n) = \sup_{t\in\phi^{-1}(\phi(x_n))}\{\mu_P(t)\},$$
$$\mu_P(b) = \sup_{t\in\phi^{-1}(\phi(y))}\{\mu_P(t)\},$$

$$\mu_P(a_1) = \inf_{t\in\phi^{-1}(\phi(x_1))}\{\lambda_P(t)\}, \ldots, \lambda_P(a_n) = \inf_{t\in\phi^{-1}(\phi(x_n))}\{\lambda_P(t)\},$$
$$\lambda_P(b) = \inf_{t\in\phi^{-1}(\phi(y))}\{\lambda_P(t)\}.$$

Then,

$$\begin{aligned}
&\mu_Q([\phi(x_1), \ldots, \phi(x_{i-1}), \phi(y), \phi(x_{i+1}), \ldots, \phi(x_n)]) \\
&= \mu_Q(\phi([\phi(x_1), \ldots, \phi(x_{i-1}), \phi(y), \phi(x_{i+1}), \ldots, \phi(x_n)])) \\
&= \sup_{t \in \phi^{-1}(\phi([\phi(x_1), \ldots, \phi(x_{i-1}), \phi(y), \phi(x_{i+1}), \ldots, \phi(x_n)]))}\{\mu_P(t)\} \\
&\geq \mu_P([a_1, \ldots, a_{i-1}, b, a_{i+1}, \ldots, a_n]) \geq \mu_P(b) = \mu_Q(\phi(y)),
\end{aligned}$$

$$\begin{aligned}
&\lambda_Q([\phi(x_1), \ldots, \phi(x_{i-1}), \phi(y), \phi(x_{i+1}), \ldots, \phi(x_n)]) \\
&= \lambda_Q(\phi([\phi(x_1), \ldots, \phi(x_{i-1}), \phi(y), \phi(x_{i+1}), \ldots, \phi(x_n)])) \\
&= \inf_{t \in \phi^{-1}(\phi([\phi(x_1), \ldots, \phi(x_{i-1}), \phi(y), \phi(x_{i+1}), \ldots, \phi(x_n)]))}\{\lambda_P(t)\} \\
&\leq \lambda_P([a_1, \ldots, a_{i-1}, b, a_{i+1}, \ldots, a_n]) \leq \lambda_P(b) = \lambda_Q(\phi(y)).
\end{aligned}$$

This proves that Q is a Pythagorean fuzzy ideal of $\phi(L_1)$.

References

1. S. Abdullah, M. Aslam, K. Ullah, Bipolar fuzzy soft sets and its applications in decision making problem. J. Intell. Fuzzy Syst. **27**(2), 729–742 (2014)
2. S. Abou-Zaid, On fuzzy subnear-rings and ideals. Fuzzy Sets Syst. **44**, 139–146 (1991)
3. M.I. Ali, F. Feng, X.Y. Liu, W.K. Min, M. Shabir, On some new operations in soft set theory. Comput. Math. Appl. **57**, 1547–1553 (2009)
4. M. Akram, Anti fuzzy Lie ideals of Lie algebras. Quasigroups Relat. Syst. **14**, 123–132 (2006)
5. M. Akram, Fuzzy Lie ideals of Lie algebras with interval-valued membership function. Quasigroups Relat. Syst. **16**(1), 1–12 (2008)
6. M. Akram, Intuitionistic (S, T)-fuzzy Lie ideals of Lie algebras. Quasigroups Relat. Syst. **15**, 201–218 (2007)
7. M. Akram, Redefined fuzzy Lie algebras. Quasigroups Relat. Syst. **16**(2), 133–146 (2008)
8. M. Akram, A. Farooq, m-polar fuzzy Lie ideals of Lie algebras. Quasigroups Relat. Syst. **24**, 101–110 (2016)
9. M. Akram, F. Feng, Soft intersection Lie algebras. Quasigroups Relat. Syst. **21**, 1–10 (2013)
10. M. Akram, Bipolar fuzzy soft Lie algebras. Quasigroups Relat. Syst. **21**, 11–18 (2013)
11. M. Akram, W. Chen, Y. Lin, Bipolar fuzzy Lie superalgebras. Quasigroups Relat. Syst. **20**, 139–156 (2012)
12. M. Akram, Bipolar fuzzy $\mathbb{L}$-Lie algebras. World Appl. Sci. J. **14**(12), 1908–1913 (2011)
13. M. Akram, Co-fuzzy Lie superalgebras over a co-fuzzy field. World Appl. Sci. J. **7**, 25–32 (2009)
14. M. Akram, A new structure of fuzzy Lie algebras. World Appl. Sci. J. **14**(12), 1879–1887 (2011)
15. M. Akram, Intuitionistic fuzzy Lie ideals of Lie algebras. World Appl. Sci. J. **16**(4), 991–1008 (2008)
16. M. Akram, N.O. Al-Shehrie, Vague Lie superalgebras. Ars Comb. **109**, 327–344 (2013)
17. M. Akram, W. Chen, Generalized anti fuzzy Lie algebras. Util. Math. **87**, 111–122 (2012)
18. M. Akram, B. Davvaz, K.P. Shum, Generalized fuzzy Lie ideals of Lie algebras. Fuzzy Syst. Math. **24**(4), 48–55 (2010)
19. M. Akram, B. Davvaz, F. Feng, Fuzzy soft Lie algebras. J. Multivalued Log. Soft Comput. **24**(5–6), 501–520 (2015)
20. M. Akram, W.A. Dudek, Interval-valued intuitionistic fuzzy Lie ideals of Lie algebras. World Appl. Sci. J. **7**(7), 812–819 (2009)
21. M. Akram, A. Farooq, K.P. Shum, On m-polar fuzzy Lie subalgebras. Ital. J. Pure Appl. Math. **36**, 445–454 (2016)

M. Akram, *Fuzzy Lie Algebras*, Infosys Science Foundation Series,
https://doi.org/10.1007/978-981-13-3221-0

22. M. Akram, K.P. Shum, Intuitionistic fuzzy Lie algebras. Southeast Asian Bull. Math. **31**(5), 843–855 (2007)
23. M. Akram, K.P. Shum, Fuzzy Lie ideals over fuzzy field. Ital. J. Pure Appl. Math. **27**, 281–292 (2010)
24. A. Akram, K.P. Shum, Vague Lie subalgebras over a vague field. Quasigroups Relat. Syst. **17**, 141–156 (2009)
25. N. Alshehri, M. Akram, Generalized bifuzzy Lie subalgebras. Sci. World J. (2013). https://doi.org/10.1155/2013/365065
26. P.L. Antony, P.L. Lilly, *Some properties of fuzzy Lie algebra over fuzzy field, in Proceedings, International Seminar on Recent Trends in Topology and its Application* (Irinjalakuda, St.Joseph's College, 2009), pp. 196–199
27. P.L. Antony, P.L. Lilly, Some properties of intuitionistic fuzzy Lie algebras over a fuzzy field. J. Gen. Lie Theory Appl. **5**, 5 (2011). Article ID G100802
28. P.L. Antony, P.L. Lilly, (α, β)-fuzzy Lie algebras over an (α, β)-fuzzy field. J. Gen. Lie Theory Appl. **4**, 8 (2010)
29. K.T. Atanassov, Intuitionistic fuzzy sets, in *VII ITKR's Session, Sofia (Deposed in Central Science-Technical Library of Bulgarian Academy of Science, 1697/84)* (1983) (in Bulgarian)
30. K.T. Atanassov, Intuitionistic fuzzy sets. Fuzzy Sets Syst. **20**(1), 87–96 (1986)
31. K.T. Atanassov, *Intuitionistic Fuzzy Sets: Theory and Applications, Studies in Fuzziness and Soft Computing*, vol. 35 (Physica-Verl, New York, 1999)
32. K.T. Atanassov, G. Gargov, Interval valued intuitionistic fuzzy sets. Fuzzy Sets Syst. **31**(3), 343–349 (1989)
33. A. Ayg*ü*noglu, H. Ayg*ü*n, Introduction to fuzzy soft groups. Comput. Math. Appl. **58**(6), 1279–1286 (2009)
34. T.M. Basu, N.K. Mahapatra, S.K. Mondal, On some properties of fuzzy soft sets and intuitionistic fuzzy soft sets. Int. J. Phys. Soc. Sci. **2**(5), 240–270 (2012)
35. S.K. Bhakat, P. Das, $(\in, \in \vee q)$,-fuzzy subgroup. Fuzzy Sets Syst. **80**, 359–368 (1996)
36. R. Biswas, Fuzzy subgroups and anti fuzzy subgroups. Fuzzy Sets Syst. **44**, 121–124 (1990)
37. R. Biswas, Rosenfeld's fuzzy subgroups with interval-valued membership functions. Fuzzy Sets Syst. **63**, 87–90 (1994)
38. R. Biswas, Vague groups. Int. J. Comput. Cogn. **4**(2), 20–23 (2006)
39. N. Cağman, S. Enginoglu, F. Citak, Fuzzy soft set theory and its applications. Iran. J. Fuzzy Syst. **8**(3), 137–147 (2011)
40. J. Chen, S. Li, S. Ma, X. Wang, m-polar fuzzy sets: An extension of bipolar fuzzy sets. Sci. World J. **8** (2014). https://doi.org/10.1155/2014/416530
41. W.J. Chen, Fuzzy quotient Lie superalgebras. J. Shandong Univ. Nat. Sci. Shandong Daxue Xuebao. Lixue Ban **43**, 25–27 (2008)
42. W.J. Chen, Fuzzy subcoalgebras and duality. Bull. Malays. Math. Sci. Soc. **32**, 283–294 (2009)
43. W.J. Chen, Intuitionistic fuzzy quotient Lie superalgebras. Int. J. Fuzzy Syst. **12**(4), 330–339 (2010)
44. W.J. Chen, M. Akram, Interval-valued fuzzy structures on Lie superalgebras. J. Fuzzy Math. **19**(4), 951–968 (2011)
45. W.J. Chen, S.H. Zhang, Intuitionistic fuzzy Lie sub-superalgebras and intuitionistic fuzzy ideals. Comput. Math. Appl. **58**, 1645–1661 (2009)
46. P. Coelho, U. Nunes, Lie algebra application to mobile robot control. a tutorial. Robotica **21**, 483–493 (2003)
47. P. Das, Fuzzy groups and level subgroups. J. Math. Anal. Appl. **85**, 264–269 (1981)
48. B. Davvaz, A note on fuzzy Lie algebras. JP J. Algebr., Number Theory Appl. **2**, 131–136 (2002)
49. B. Davvaz, Fuzzy Lie algebras. Int. J. Appl. Math. **6**, 449–461 (2001)
50. B. Davvaz, Fuzzy R-subgroups with thresholds of near-rings and implication operators. Soft Comput. **12**, 875–879 (2008)
51. B. Davvaz, $(\in, \in \vee q)$-fuzzy subnear-rings and ideals. Soft Comput. **10**, 206–211 (2006)

52. B. Davvaz, P. Corsini, Redefined fuzzy Hv-submodules and many valued implications. Inf. Sci. **177**(3), 865–875 (2007)
53. B. Davvaz, W.A. Dudek, Fuzzy n-Lie Algebras. J. Gen. Lie Theory Appl. **11**(2), 1–6 (2017). https://doi.org/10.4172/1736-4337.1000268
54. B. Davvaz, W.A. Dudek, Y.B. Jun, Intuitionistic fuzzy H_υ-submodules. Inf. Sci. **176**, 285–300 (2006)
55. B. Davvaz, J.M. Zhan, K.P. Shum, Generalized fuzzy H_υ-submodules endowed with interval valued membership functions. Inf. Sci. **178**, 3147–3159 (2008)
56. G. Deschrijver, Arithmetric operators in interval-valued fuzzy theory. Inf. Sci. **177**, 2906–2924 (2007)
57. D. Dubois, S. Kaci, H. Prade, Bipolarity in Reasoning and Decision, an Introduction, in *International Conference on Information Processing and Management of Uncertainty. IPMU'04*, vol. 4 (2004), pp. 959–966
58. D. Dubois, H. Prade, *Fuzzy Sets and Systems* (Academic Press, New York, 1980), pp. 959–966
59. D. Dubios, H. Prade, Rough fuzzy and fuzzy rough sets. Int. J. Gen. Syst. **17**(2–3), 191–209 (1990)
60. W.A. Dudek, Fuzzy subquasigroups. Quasigroups Relat. Syst. **5**, 81–98 (1998)
61. W.A. Dudek, Intuitionistic fuzzy approach to n-ary systems. Quasigroups Relat. Syst. **13**, 213–228 (2005)
62. W.A. Dudek, On some old and new problems in n-ary groups. Quasigroups Relat. Syst. **8**, 15–36 (2001)
63. W.A. Dudek, Y.B. Jun, Rough subalgebras of some binary algebras connected with logics. Int. J. Math. Math. Sci. **3**, 437–447 (2005)
64. W.A. Dudek, B. Davvaz, Y.B. Jun, On intuitionistic fuzzy subhyper-quasigroups of hyperquasigroups. Inf. Sci. **170**, 251–262 (2005)
65. W.A. Dudek, Fuzzification of n-ary groupoids. Quasigroups Relat. Syst. **7**, 45–66 (2000)
66. A. Farooq, G. Ali, M. Akram, On m-polar fuzzy groups. Int. J. Algebr. Stat. **5**(2), 115–127 (2016)
67. F. Feng, Y.B. Jun, X.Y. Liu, L.F. Li, An adjustable approach to fuzzy soft set based decision making. J. Comput. Appl. Math. **234**, 10–20 (2010)
68. F. Feng, Y.B. Jun, X. Zhao, Soft semirings. Comput. Math. Appl. **56**(10), 2621–2628 (2008)
69. F. Feng, X. Liu, V.L. Fotea, Y.B. Jun, Soft sets and soft rough sets. Inf. Sci. **181**(6), 1125–1137 (2011)
70. J.C.M. Ferreira, M.G.B. Marietto, Solvable and nilpotent radicals of the fuzzy Lie algebras. J. Gen. Lie Theory Appl. **6**, 7 (2012)
71. V.T. Filippov, n-Lie algebras. Sib. Math. J. **26**, 879–891 (1985). (Transl. from Sibirsk. Mat. Zh., 26 (1985) 126–140)
72. V.T. Filippov, On n-Lie algebra of Jacobians. Math. J. **39**, 573–581 (1998)
73. M.B. Gorzalczany, A method of inference in approximate reasoning based on interval-valued fuzzy sets. Fuzzy Sets Syst. **21**, 1–17 (1987)
74. M.B. Gorzalczany, An Interval-valued fuzzy inference method some basic properties. Fuzzy Sets Syst. **31**, 243–251 (1989)
75. G.N. Garrison, Quasigroups. Ann. Math. **41**, 474–487 (1940)
76. W.L. Gau, D.J. Buehrer, Vague sets. IEEE Trans. Syst., Man Cybern. **23**, 610–614 (1993)
77. W. Gu, T. Lu, Fuzzy algebras over fuzzy fields redefined. Fuzzy Sets Syst. **53**, 105–107 (1993)
78. L. Guangwen, G. Enrui, Fuzzy algebra and fuzzy quotient algebras over fuzzy field. Electron. Busefal **85**, 1–4 (2001)
79. J.W.B. Hughes, J.V.D. Jeugt, Unimodal polynomials associated with Lie algebras and superalgebras. J. Comput. Appl. Math. **37**, 81–88 (1991)
80. J.E. Humphreys, *Introduction to Lie Algebras and Representation Theory* (Springer, New York, 1972)
81. K. Hur, S.Y. Jang, H.W. Kang, Intuitionistic fuzzy ideals of a ring. J. Korean Soc. Math. Educ. Ser. B: Pure Appl. Math. **12**, 193–209 (2005)

82. Y.B. Jun, H. Chul, Intuitionistic fuzzy Lie ideals of Lie algebras. Honam Math. J. **29**, 259–268 (2007)
83. Y.B. Jun, M.A. Ozturk, C.H. Park, Intuitionistic nil radicals of intuitionistic fuzzy ideals and Euclidean intuitionistic fuzzy ideals in rings. Inf. Sci. **177**, 4662–4677 (2007)
84. Y.B. Jun, S.Z. Song, Generalized fuzzy interior ideals in semigroups. Inf. Sci. **176**, 3079–3093 (2006)
85. V.G. Kac, Lie superalgebras. Adv. Math. **26**, 8–96 (1977)
86. A.K. Katsaras, D.B. Liu, Fuzzy vector spaces and fuzzy topological vector spaces. J. Math. Anal. Appl. **58**, 135–146 (1977)
87. Q. Keyun, Q. Quanxi, C. Chaoping, Some properties of fuzzy Lie algebras. J. Fuzzy Math. **9**, 985–989 (2001)
88. A. Khan, Y.B. Jun, M. Shabir, A study of generalized fuzzy ideals in ordered semigroups. Neural Comput. Appl. **21**(1), 69–78 (2012)
89. C.G. Kim, D.S. Lee, Fuzzy Lie ideals and fuzzy Lie subalgebras. Fuzzy Sets Syst. **94**(1), 101–107 (1998)
90. M. Kondo, W.A. Dudek, On the transfer principle in fuzzy theory. Mathw. Soft Comput. **12**, 41–55 (2005)
91. V. Leoreanu-Fotea, Fuzzy hypermodules. Comput. Math. Appl. **57**(3), 466–475 (2009)
92. V. Leoreanu-Fotea, B. Davvaz, Fuzzy hyperrings. Fuzzy Sets Syst. **160**(16), 2366–2378 (2009)
93. K.M. Lee, Bipolar-valued fuzzy sets and their basic operations, in *Proceedings of the International Conference* (Bangkok, Thailand, 2000), pp. 307–317
94. K.M. Lee, Comparison of interval-valued fuzzy sets, intuitionistic fuzzy sets, and bipolar-valued fuzzy sets. J. Fuzzy Log. Intell. Sys. **14**, 125–129 (2004)
95. Z. Li, T. Xie, Roughness of fuzzy soft sets and related results. Int. J. Comput. Intell. Syst. **8**(2), 278–296 (2015)
96. W.J. Liu, Fuzzy invariant subgroups and fuzzy ideal. Fuzzy Sets Syst. **8**, 133–139 (1982)
97. P.K. Maji, R. Biswas, R. Roy, Fuzzy soft sets. J. Fuzzy Math. **9**(3), 589–602 (2001)
98. P.K. Maji, R. Biswas, R. Roy, Soft set theory. Comput. Math. Appl. **45**, 555–562 (2003)
99. D.S. Malik, J.N. Mordeson, P.S. Nair, Fuzzy normal subgroups in fuzzy groups. J. Korean Math. Soc. **29**, 1–8 (1992)
100. D.S. Malik, J.N. Mordeson, Fuzzy subfields. Fuzzy Sets Syst. **37**, 383–388 (1990)
101. D.S. Malik, J.N. Mordeson, Fuzzy vector spaces. Inf. Sci. **55**, 271–281 (1991)
102. J.M. Mendel, *Uncertain Rule-based Fuzzy Logic Systems: Introduction and New Directions* (Prentice-Hall, New Jersey, 2001)
103. D. Molodtsov, Soft set theory first results. Comput. Math. Appl. **37**, 19–31 (1999)
104. J.N. Mordeson, D.S. Malik, N. Kuroki, *Fuzzy Semigroups, Studies in fuzziness and soft computing* (Springer, New York, 2007)
105. V. Murali, Fuzzy points of equivalent fuzzy subsets. Inf. Sci. **158**, 277–288 (2004)
106. Y. Nambu, Generalized Hamiltonian mechanics. Phys. Rev. **7**, 2405–2412 (1973)
107. S. Nanda, Fuzzy algebra over a fuzzy field. Fuzzy Sets Syst. **37**, 99–103 (1990)
108. A.I. Nesterov, Some applications of quasigroups and loops in physics, in *International Conference Nonassociative Algebra and its Applications* (Mexico, 2003)
109. A.M.T. Osman, On some product of fuzzy subgroups. Fuzzy Sets Syst. **24**, 79–86 (1987)
110. W. Pan, J. Zhan, Rough fuzzy groups and rough soft groups. Ital. J. Pure Appl. Math. **36**, 617–628 (2016)
111. Z. Pawalak, Rough sets. Int. J. Comput. Inf. Sci. **11**(5), 341–356 (1982)
112. X. Peng, Y. Yang, Some results for Pythagorean fuzzy sets. Int. J. Intell. Syst. **30**(11), 1133–1160 (2015)
113. P.M. Pu, Y.M. Liu, Fuzzy topology I. Neighbourhood structures of a fuzzy point and Moore-Smith convergence. J. Math. Anal. Appl. **76**, 571–579 (1980)
114. S. Rizvi, H.J. Naqvi, D. Nadeem, Rough intuitionistic fuzzy set, in *Proceedings of the 6th Joint Conference on Information Sciences (JCIS)* (Durham, NC, 2002), pp. 101–104
115. A. Rosenfeld, Fuzzy groups. J. Math. Anal. Appl. **35**, 512–517 (1971)

116. A.R. Roy, P.K. Maji, A fuzzy soft set theoretic approach to decision making problems. J. Comput. Appl. Math. **203**, 412–418 (2007)
117. M.K. Roy, R. Biswas, l-v fuzzy relations and Sanchez's approach for medical diagnosis. Fuzzy Sets Syst. **47**, 35–38 (1992)
118. B. Schweizer, A. Sklar, Statistical metric spaces. Pac. J. Math. **10**, 313–334 (1960)
119. B. Schweizer, A. Sklar, Associative functions and abstract semigroups. Publ. Math. **10**, 69–81 (1963)
120. M. Scheunert, *The Theory of Lie Superalgebras: An Introduction LNM* (Springer, New York, 1979)
121. M. Shabir, Y.B. Jun, Y. Nawaz, Characterizations of regular semigroups by (α, β)-fuzzy ideals. Comput. Math. Appl. **59**(1), 161–175 (2010)
122. T. Som, On the theory of soft sets, soft relation and fuzzy soft relation, in *Proceedings of the National Conference on Uncertainty: A Mathematical Approach, UAMA-06* (Burdwan, 2006), pp. 1–9
123. L.A. Takhtajan, On fundation of generalized Nambu mechanics. Commun. Math. Phys. **160**, 295–315 (1994)
124. G. Takeuti, S. Titants, Intuitionistic fuzzy logic and intuitionistic fuzzy set theory. J. Symb. Log. **49**, 851–866 (1984)
125. V. Torra, Hesitant fuzzy sets. Int. J. Intell. Syst. **25**(6), 529–539 (2010)
126. V. Torra, Y. Narukawa, On hesitant fuzzy sets and decisions. IEEE Int. Conf. Fuzzy Syst. **1–3**, 1378–1382 (2009)
127. I.B. Turksen, Interval valued fuzzy sets based on normal forms. Fuzzy Sets Syst. **20**, 191–210 (1986)
128. M. Wakimoto, *Infinte-Dimensional Lie Algebras* (The American Mathematical Society, 2001)
129. M.M. Xia, Z.S. Xu, Hesitant fuzzy information aggregation in decision making. Int. J. Approx. Reason. **52**, 395–407 (2011)
130. R.R. Yager, Pythagorean fuzzy subsets, in *Proceedings of the Joint IFSA World Congress and NAFIPS Annual Meeting* (Edmonton, Canada, 2013), pp. 57–61
131. R.R. Yager, A.M. Abbasov, Pythagorean membership grades, complex numbers and decision making. Int. J. Intell. Syst. **28**(5), 436–452 (2013)
132. W. Yang, S. Li, Bipolar-value fuzzy soft sets. Comput. Eng. Appl. **48**(35), 15–18 (2012)
133. S.E. Yehia, Fuzzy ideals and fuzzy subalgebras of Lie algebras. Fuzzy Sets Syst. **80**, 237–244 (1996)
134. S.E. Yehia, The adjoint representation of fuzzy Lie algebras. Fuzzy Sets Syst. **119**, 409–417 (2001)
135. M.S. Ying, On standard models of fuzzy modal logics. Fuzzy Sets Syst. **26**, 357–363 (1988)
136. M.S. Ying, A new approach for fuzzy topology (I). Fuzzy Sets Syst. **39**, 303–321 (1991)
137. X. Yuan, C. Zhang, Y. Rena, Generalized fuzzy groups and many-valued implications. Fuzzy Sets Syst. **138**, 205–211 (2003)
138. E.M. Zadeh, R. Ameri, Some results of fuzzy Lie algebras, in *4th Iranian Joint Congress on Fuzzy and Intelligent Systems* (2015)
139. L.A. Zadeh, Fuzzy sets. Inf. Control **8**(3), 338–353 (1965)
140. L.A. Zadeh, Similarity relations and fuzzy orderings. Inf. Sci. **3**(2), 177–200 (1971)
141. L.A. Zadeh, The concept of a linguistic and application to approximate reasoning-I. Inf. Sci. **8**, 199–249 (1975)
142. J. Zhan, B. Davvaz, K.P. Shum, A new view on fuzzy hypermodule. Acta Math. Sin. (Engl. Ser.) **23**(8), 1345–1356 (2007)
143. J. Zhan, W.A. Dudek, Interval valued intuitionistic (S, T)-fuzzy Hv submodules. Acta Math. Sin. **22**, 963–970 (2006)
144. J. Zhan, Y.B. Jun, Generalized fuzzy interior ideals of semigroups. Neural Comput. Appl. **19**(4), 515–519 (2010)
145. J. Zhan, Q. Liu, T. Herawan, A novel soft rough set. soft rough hemirings and its multicriteria group decision making. Appl. Soft Comput. **54**, 393–402 (2017)
146. W.R. Zhang, Bipolar fuzzy sets, in *Proceedings of FUZZ-IEEE* (1998), pp. 835–840

147. W.R. Zhang, Bipolar fuzzy sets and relations. a computational framework forcognitive modeling and multiagent decision analysis, in *Proceedings of the IEEE Conference* (1994), pp. 305–309
148. X. Zhang, Z. Xu, Extension of TOPSIS to multiple-criteria decision making with Pythagorean fuzzy sets. Int. J. Intell. Syst. **29**(12), 1061–1078 (2014)

Glossary of Symbols

Symbol	Meaning
X	Universal set (or universe of discourse)
$[\cdot,\cdot]$	Lie bracket
V	Vector space
L	Lie algebra, n-Lie algebra
L/J	Quotient Lie algebra
$\mathscr{L}$	Lie superalgebra
$\mathbb{Q}$	The set of rational numbers
$\mathbb{R}$	The set of real numbers
$\mathbb{F}$	Field
μ	Fuzzy set
$U(\mu, t)$	t-level set
h	Hesitant fuzzy set
x_t	Fuzzy point
$A = (\mu_A, \lambda_A)$	Intuitionistic fuzzy set
$P = (\mu_P, \lambda_P)$	Pythagorean fuzzy set
$\widetilde{\mu}$	Interval-valued fuzzy set
$\tilde{A} = (\widetilde{\mu}_{\tilde{A}}, \widetilde{\lambda}_{\tilde{A}})$	Interval-valued intuitionistic fuzzy set
$x_t \in \mu$	Fuzzy point belonging to
$x_t q \mu$	Fuzzy point quasicoincident with
$x_t \prec \nu$	Anti-fuzzy point besides to
$x_t \vdash \nu$	Anti-fuzzy point non-quasicoincident with
$A = (t_A, f_A)$	Vague set
$A = (\mu_A^P, \nu_A^N)$	Bipolar fuzzy set
$C(x) = p_i \circ C(x)$	$m-$polar fuzzy set
(f, A)	Fuzzy soft set
$(\underline{R}S, \overline{R}S)$	Rough fuzzy Lie subalgebra
$(\underline{apr}_R(A), \overline{apr}_R(A))$	Fuzzy rough Lie subalgebra

M. Akram, *Fuzzy Lie Algebras*, Infosys Science Foundation Series,
https://doi.org/10.1007/978-981-13-3221-0

Index

M. Akram, *Fuzzy Lie Algebras*, Infosys Science Foundation Series,
https://doi.org/10.1007/978-981-13-3221-0

Zeitfracht Medien GmbH
Ferdinand-Jühlke-Straße 7
99095 Erfurt, Deutschland
produktsicherheit@kolibri360.de